MATH/STAT
LIBRARY
Y0-BSC-990

Recent Titles in This Series

171 **Rüdiger Göbel, Paul Hill, and Wolfgang Liebert, Editors,** Abelian group theory and related topics, 1994

170 **John K. Beem and Krishan L. Duggal, Editors,** Differential geometry and mathematical physics, 1994

169 **William Abikoff, Joan S. Birman, and Kathryn Kuiken, Editors,** The mathematical legacy of Wilhelm Magnus, 1994

168 **Gary L. Mullen and Peter Jau-Shyong Shiue, Editors,** Finite fields: Theory, applications, and algorithms, 1994

167 **Robert S. Doran, Editor,** C^*-algebras: 1943–1993, 1994

166 **George E. Andrews, David M. Bressoud, and L. Alayne Parson, Editors,** The Rademacher legacy to mathematics, 1994

165 **Barry Mazur and Glenn Stevens, Editors,** p-adic monodromy and the Birch and Swinnerton-Dyer conjecture, 1994

164 **Cameron Gordon, Yoav Moriah, and Bronislaw Wajnryb, Editors,** Geometric topology, 1994

163 **Zhong-Ci Shi and Chung-Chun Yang, Editors,** Computational mathematics in China, 1994

162 **Ciro Ciliberto, E. Laura Livorni, and Andrew J. Sommese, Editors,** Classification of algebraic varieties, 1994

161 **Paul A. Schweitzer, S. J., Steven Hurder, Nathan Moreira dos Santos, and José Luis Arraut, Editors,** Differential topology, foliations, and group actions, 1994

160 **Niky Kamran and Peter J. Olver, Editors,** Lie algebras, cohomology, and new applications to quantum mechanics, 1994

159 **William J. Heinzer, Craig L. Huneke, and Judith D. Sally, Editors,** Commutative algebra: Syzygies, multiplicities, and birational algebra, 1994

158 **Eric M. Friedlander and Mark E. Mahowald, Editors,** Topology and representation theory, 1994

157 **Alfio Quarteroni, Jacques Periaux, Yuri A. Kuznetsov, and Olof B. Widlund, Editors,** Domain decomposition methods in science and engineering, 1994

156 **Steven R. Givant,** The structure of relation algebras generated by relativizations, 1994

155 **William B. Jacob, Tsit-Yuen Lam, and Robert O. Robson, Editors,** Recent advances in real algebraic geometry and quadratic forms, 1994

154 **Michael Eastwood, Joseph Wolf, and Roger Zierau, Editors,** The Penrose transform and analytic cohomology in representation theory, 1993

153 **Richard S. Elman, Murray M. Schacher, and V. S. Varadarajan, Editors,** Linear algebraic groups and their representations, 1993

152 **Christopher K. McCord, Editor,** Nielsen theory and dynamical systems, 1993

151 **Matatyahu Rubin,** The reconstruction of trees from their automorphism groups, 1993

150 **Carl-Friedrich Bödigheimer and Richard M. Hain, Editors,** Mapping class groups and moduli spaces of Riemann surfaces, 1993

149 **Harry Cohn, Editor,** Doeblin and modern probability, 1993

148 **Jeffrey Fox and Peter Haskell, Editors,** Index theory and operator algebras, 1993

147 **Neil Robertson and Paul Seymour, Editors,** Graph structure theory, 1993

146 **Martin C. Tangora, Editor,** Algebraic topology, 1993

145 **Jeffrey Adams, Rebecca Herb, Stephen Kudla, Jian-Shu Li, Ron Lipsman, and Jonathan Rosenberg, Editors,** Representation theory of groups and algebras, 1993

144 **Bor-Luh Lin and William B. Johnson, Editors,** Banach spaces, 1993

143 **Marvin Knopp and Mark Sheingorn, Editors,** A tribute to Emil Grosswald: Number theory and related analysis, 1993

(*Continued in the back of this publication*)

Abelian Group Theory and Related Topics

CONTEMPORARY MATHEMATICS

171

Abelian Group Theory and Related Topics

Conference on Abelian Groups
August 1–7, 1993
Oberwolfach, Germany

Rüdiger Göbel
Paul Hill
Wolfgang Liebert
Editors

American Mathematical Society
Providence, Rhode Island

The Conference on Abelian Groups was held at the Mathematische Forschungsinstitut, Oberwolfach, Germany, from August 1–7, 1993.

1991 *Mathematics Subject Classification*. Primary 20KXX, 03E05, 16E50, 03C60, 16G20.

Library of Congress Cataloging-in-Publication Data

Conference on Abelian Groups (1993: Oberwolfach, Germany)
Abelian group theory and related topics/Conference on Abelian Groups, August 1–7, 1993, Oberwolfach, Germany: Rüdiger Göbel, Paul Hill, Wolfgang Liebert, editors.
p. cm. — (Contemporary mathematics; v. 171)
Includes bibliographical references.
ISBN 0-8218-5178-0 (acid-free)
1. Abelian groups—Congresses. I. Göbel, R. (Rüdiger), 1940– . II. Hill, Paul, 1933 May 8– III. Liebert, Wolfgang, 1936– . IV. Title. V. Series: Contemporary mathematics (American Mathematical Society); v. 171.
QA180.C66 1993 94-25813
512′.2—dc20 CIP

♾ The paper used in this book is acid-free and falls within the guidelines
established to ensure permanence and durability.
♻ Printed on recycled paper.
All articles in this volume were printed from copy prepared by the authors.
Some of the articles in this volume were typeset by the authors using AMS-TEX,
the American Mathematical Society's TEX macro system.

10 9 8 7 6 5 4 3 2 1 99 98 97 96 95 94

CONTENTS

Preface . . . xi

Conference Participants . . . xiii

Richard Scott Pierce
C. VINSONHALER . . . 1

On B_2-Groups
L. BICAN . . . 13

Decompositions of Almost Completely Decomposable Abelian Groups
E. A. BLAGOVESHCHENSKAYA and A. MADER . . . 21

On the Divisible Parts of Quotient Groups
A. BLASS . . . 37

Minimal Rings, Central Idempotents and the Pierce Sheaf
W. D. BURGESS . . . 51

On Endomorphisms and Automorphisms of Some Pure Subgroups of the Baer-Specker Group
A. L. S. CORNER and B. GOLDSMITH . . . 69

A Combinatorial Principle Equivalent to the Existence of Non-Free Whitehead Groups
P. C. EKLOF and S. SHELAH . . . 79

Endomorphisms of Local Warfield Groups
S. T. FILES . . . 99

Finitely Presented Modules Over the Ring of Universal Numbers
A. A. FOMIN . . . 109

A Survey of Butler Groups of Infinite Rank
L. FUCHS . . . 121

Unions of Chains of Butler Groups
L. FUCHS and K. M. RANGASWAMY . . . 141

Some Torsion-Free Groups Arising in Measure Theory
R. G. GÖBEL and R. M. SHORTT 147

Numerical Invariants for a Class of Butler Groups
H. P. GOETERS, W. ULLERY and C. VINSONHALER 159

K_0 of Regular Rings with Bounded Index of Nilpotence
K. R. GOODEARL 173

Torsion in Quotients of the Multiplicative Group of a Number Field
D. HOLLEY and R. WIEGAND 201

On p^α-Injective Abelian Groups
P. KEEF 205

Abelian Groups with Contractions I
F.-V. KUHLMANN 217

Typesets and Cotypesets of Finite-Rank Torsion-Free Abelian Groups
R. S. LAFLEUR 243

A Generalization of Butler Groups
A. MADER, O. MUTZBAUER and K. M. RANGASWAMY 257

Endomorphisms Over Incomplete Discrete Valuation Rings
W. MAY 277

$Bext^2(G,T)$ Can Be Nontrivial Even Assuming GCH
M. MAGIDOR and S. SHELAH 287

Representations and Duality
R. MINES, C. VINSONHALER and W. J. WICKLESS 295

Extending a Splitting Criterion on Mixed Modules
O. MUTZBAUER and E. TOUBASSI 305

Direct Summands of Z^κ for Large κ
J. D. O'NEILL 313

Abelian Groups as Noetherian Modules Over Their Endomorphism Rings
A. T. PARAS 325

Isomorphism of Butler Groups at a Prime
F. RICHMAN 333

The Braid Group Action on the Set of Exceptional Sequences of a Hereditary Artin Algebra
C. M. RINGEL 339

Direct Limits of Two-Dimensional Prime Spectra
C. ROTTHAUS and S. WIEGAND 353

When is an Abelian p-Group Determined by the Jacobson Radical of Its Endomorphism Ring?
P. SCHULTZ 385

Similarities and Differences Between Abelian Groups and Modules Over Non-Perfect Rings
J. TRLIFAJ 397

A Functor from Mixed Groups to Torsion-Free Groups
W. J. WICKLESS 407

A Characterization of a Class of Butler Groups II
P. D. YOM 419

PREFACE

After an Olympic styled intermission of four years, the Mathematische Forschungsinstitut Oberwolfach reconvened a conference on abelian groups August 1-7, 1993. During a week of beautiful sunshine on the Schwarzwald forest, enlightening talks and bright ideas for new research were emanating from the Institute.

The conference was organized by two local participants, Rüdiger Göbel of Essen and Wolfgang Liebert of München, along with Paul Hill of the United States. It brough together forty-seven participants from all over the world with five continents represented. For the first time in a long series of conferences on abelian groups at Oberwolfach, researchers from Russia and other Eastern bloc countries were in attendance.

The broad geographic scope of the conference yielded world-wide interaction among specialists in abelian groups. Moreover, the conference attracted researchers from other areas of mathematics as well. Indeed, as if reverting to the sixteenth-century concept of the smaller planet being the center of gravity, the conference seemed to have a variety of topics revolving around abelian groups – if only for a week. Experts from model theory, set theory, noncommutative groups, module theory, and from computer science discussed problems in their fields that relate to abelian group theory.

These international conferences on abelian groups have established a tradition of paying tribute to pioneer contributors to the subject such as Prüfer, Ulm, and Baer. This time a more contemporary figure has been selected. We have chosen someone who has been a faithful friend, dedicated advisor, and an inspiring colleague to most who are currently active including the participants of this conference and the contributors to these Proceedings. Certainly, R. S. Pierce was that and more.

Although Pierce had a wide interest in mathematics, he seemed to have a special place in his heart for abelian groups. He made significant contributions to the structure theory of both torsion and torsion-free groups. One of his favorite topics was endomorphisms of abelian groups. After Pierce had become seriously ill, he still made every effort to attend all the major conferences on abelian groups. His interest in what was going on in the subject apparently never waned even over a protracted illness.

The editors invited several people who had a special relationship to R. S. Pierce or his work to contribute papers. We express our appreciation to these and other authors who have made this volume a worthy tribute.

Rüdiger Göbel (Essen)
Paul Hill (Auburn)
Wolfgang Liebert (München)

CONFERENCE PARTICIPANTS

David M. Arnold	Baylor University
Khalid Benabdallah	University of Montreal
Ekatarina Blagoveshchenskaya	University of St. Petersburg
Andreas Blass	University of Michigan
Danny Carrol	Dublin Institute of Technology
Manfred Droste	University of Essen
Manfred Dugas	Baylor University
Theodore Faticoni	Fordham University
Temple Fay	University of Southern Mississippi
Steve Files	University of Arizona
Alexander Fomin	Pedagogical State University of Moscow
Laszlo Fuchs	Tulane University
Anthony Giovannitti	University of Southern Mississippi
Rüdiger Göbel	University of Essen
Brendan Goldsmith	Dublin Institute of Technology
Jutta Hausen	University of Houston
Paul Hill	Auburn University
Wilfrid Hodges	University of London
Kin–ya Honda	Meisei University
Patrick Keef	Whitman College
Franz–Viktor Kuhlmann	University of Heidelberg
Reiff Lafleur	University of Connecticut
Wolfgang Liebert	Technical University of Munich
Adolf Mader	University of Hawaii
Menachem Magidor	Hebrew University of Jerusalem
Warren May	University of Arizona
Claudia Metelli	University of Naples
Ray Mines	New Mexico State University
Otto Mutzbauer	University of Würzburg
Loyiso Nongxa	University of the Western Cape
Juha Oikkonen	University of Helsinki
Adalberto Orsatti	University of Padova
Mohamed Ould–Beddi	University of Montreal
Simone Pabst	University of Essen
Agnes Paras	Wesleyan University
K. M. Rangaswamy	University of Colorado
James Reid	Wesleyan University
Luigi Salce	University of Padova

Phillip Schultz	The University of Western Australia
Lutz Strüngmann	University of Essen
Jan Trlifaj	Charles University of Prague
William Ullery	Auburn University
Gert Viljoen	University of Orange Free State
Charles Vinsonhaler	University of Connecticut
William Wickless	University of Connecticut
Anatoly Yakovlev	University of St. Petersburg
Peter Yom	Fordham University

Contemporary Mathematics
Volume **171**, 1994

RICHARD SCOTT PIERCE

C. VINSONHALER

Dick Pierce was one of my mathematical heroes, and I know that I am joined in this sentiment by many of those who attended the 1993 Oberwolfach Conference honoring him. Dick was a first–rate mathematician: he knew a lot of mathematics, he was very good at doing and writing mathematics, he worked hard at mathematics and he had a real feel for elegance and quality. But the high regard in which Dick is so universally held is not a product of his mathematical prowess alone. His generosity with his time and ideas is legendary. How many of us could be the author of the following segment, which was sent by Wis Comfort to Dick's wife Marilyn after his death?

> "We submitted a paper which came back vastly improved from the editor, with such a helpful, creative, remarkable referee's report that we felt the only reasonable course of action was to suggest that the referee join us as a third co–author. The editor released the referee's name to us and authorized us to write to him with our offer: Richard S. Pierce. He declined our invitation with a nice letter, indicating that we had done the main work ... ."

A brief biography will help introduce Dick to those who were not fortunate enough to know him personally. Richard Scott Pierce was born on February 26, 1927, in Long Beach, California. He was an excellent student and finished his freshman year at Caltech in February, 1945. He was drafted into the army after his eighteenth birthday and shipped out to Japan in October, 1945, taking five mathematics and physics books with him. Dick re–entered Caltech in September of 1947, graduated in 1950 and went straight into the Ph.D. program in mathematics. He earned his doctorate, under R.P. Dilworth, in two years. His thesis was a study of translation lattices and semilattices that eventually appeared as

1991 *Mathematics Subject Classification.* Primary 01A70.
This paper is in final form and no version of it will be submitted for publication elsewhere.

[10].[1] He spent 1952–53 as an Office of Naval Research Fellow at Yale and was a Jewett Research Fellow at Harvard during 1953–1955, working mostly with Garrett Birkhoff. In June of 1953 he married Mary Ray and in the fall of 1955 they moved to Seattle, where Dick began work at the University of Washington. A son, Eric, was born in March of 1956 and by 1960 Dick had been promoted to full professor. He spent 1967–68 as part of the "Abelian Group Year" at New Mexico State University, joining Reinhold Baer, Jutta Hausen, Otto Kegel and Bob Warfield, along with the permanent faculty in Las Cruces: Wolfgang Liebert, Ray Mines, Fred Richman and Carol and Elbert Walker. In the fall of 1970, Dick moved to the University of Hawaii. Shortly thereafter, following a divorce from Mary, he married Marilyn Cayford, whom many of us know as his beloved travel agent and constant companion at meetings. In 1975, Dick left Hawaii for the University of Arizona in Tucson, where he remained until he retired in December, 1990.

Pierce's first papers on abelian groups are some of his most frequently referenced: [17, 18, 19 and 21]. These laid the foundation for several generations of work in torsion–free abelian groups of finite rank. Fields of definition are studied in "Subrings of simple algebras" [17], which also contains the important result that full subrings of simple algebras are finitely generated over a subring of the center that is an E–ring. This paper was actually an offshoot of the next one, "Torsion free rings" [18] with R.A. Beaumont, that contains a wealth of techniques and results: quotient divisible groups are introduced and characterized via the "*q.d.* invariants" obtained by tensoring the group with the p–adic integers $\hat{\mathbb{Z}}_p$. These ideas paved the way for Arnold duality and Lee Lady's extensive study of Butler modules. Algebraic number theory is used to prove the Beaumont–Pierce Principal Theorem: Any torsion–free ring R of finite rank is quasi–equal to $S \oplus N$, where N is the nil radical of R and S is a subring of R with $\mathbb{Q}S$ semisimple. The systematic use of the ring of quasi–endomorphisms of a torsion–free group is another fundamental contribution of this paper, one which inspired much of the work of J.D. Reid and W.J. Wickless, for example. A third paper, "Subrings of algebraic number fields" [21], completes a trilogy. Here the subrings of algebraic number fields are studied via localizations. One frequently quoted result is the following: In each quasi–equality class of full subrings of an algebraic number field, there is one and only one integrally closed ring. This ring is the integral closure of every ring in the class.

The Memoir [19] is an exhaustive study of torsion–free groups of rank two. It provides strong testimony on the difficulties involved in classifying torsion–free abelian groups of finite rank, even up to quasi–isomorphism. A system of quasi–isomorphism invariants for rank–two groups is developed. Some results on the sets of types that can arise from such groups are given, and the quasi–endomorphism algebras that can occur are classified. A final chapter character-

[1]Numbers refer to papers listed in Pierce's bibliography.

izes when quasi–isomorphism implies isomorphism (for rank–two groups).

The next abelian group paper, "Homomorphisms of primary abelian groups" [24], in many ways typifies Dick's mathematics. It is deep, innovative, thorough (95 pages), and uses a broad spectrum of techniques. Pierce introduces the notions of large subgroup and small homomorphism: a subgroup L of a p–group G is **large** in G if L is fully invariant and $B + L = G$ for every basic subgroup B of G. A homomorphism φ between p–groups is called **small** if the kernel of φ contains a large subgroup. This latter notion was the ancestor of the "inessential" homomorphisms that have been used so effectively in the work of A.L.S. Corner, L. Fuchs, B. Goldsmith, M. Dugas, R. Göbel and others. Pierce establishes some important properties of small homomorphisms and goes on to describe the structure of the groups $\mathrm{Hom}(G, A)$ and $\mathrm{Hom}_s(G, A)$ (the small homomorphisms) for G and A p–groups. The paper "Endomorphisms of primary abelian groups" [30] is a sequel that utilizes some topological notions to study a realization problem: When can rings in a certain class be realized as $\mathrm{End}(G)$ for G a p–group without elements of infinite height and with prescribed basic subgroup B?

The paper [25], "Centers of purity in abelian groups," is a short but elegant result inspired by work of J.D. Reid. A subgroup H of an abelian group G is called a **center of purity in** G if every subgroup of G that is maximal with respect to disjointness from H is pure. Pierce showed that H is a center of purity if and only if the following two conditions hold:

(a) the torsion subgroup of H is a center of purity in the torsion subgroup of G; and

(b) either G/H is a torsion group, or else, for all primes p, $H[p] \subseteq \bigcap_{n=0}^{\infty} p^n G$.

The centers of purity in p–groups are then classified in a separate theorem.

Next comes a sequence of four more papers with Beaumont. The first, "Isomorphic direct summands of abelian groups" [28], studies those abelian groups that contain an isomorphic proper subgroup. In particular, when is such a subgroup pure or a summand? Homological machinery is used extensively to produce results like the following: Let H be an arbitrary nonzero abelian group. Then there exists a group G containing a proper isomorphic direct summand and a monomorphism φ of G into itself such that $G/\varphi G \simeq H$ and $\varphi^{\omega} G$ is not a direct summand of G.

In "Some invariants of p–groups" [29], Beaumont and Pierce introduce a system of invariants for p–groups without elements of infinite height. The invariants are ideals in the Boolean algebra of all subsets of the set of finite ordinals. Boolean algebras, of course, are another of Pierce's specialties. The results in this paper primarily deal with existence problems. Paul Hill notes that [29] was the foundation for his work with C. Megibben and others on quasi–closed p–groups. The final two members of this quartet, [31] and [32], play a familiar theme from earlier collaboration, quasi–isomorphism, this time for p–groups. Quasi–isomorphism of two groups is characterized by certain inequalities on sums of

Ulm invariants, the second paper obtaining a stronger result in case both groups are direct sums of cyclics.

Dick returned to torsion–free abelian groups in the fall of 1978, when he came with Marilyn and her son Marty to the University of Connecticut as part of a Special Year in Algebra. Joining the Pierces as visitors were Dave Arnold, Kevin O'Meara and Roger and Sylvia Wiegand. Both Arnold and Pierce ended up publishing books based on their graduate courses that year: **Finite Rank Torsion Free Abelian Groups and Rings** by Dave and **Associative Algebras** [50] by Dick. In addition Roger and Sylvia gave a year–long course on commutative rings. It was a feast of algebra that was matched by a sequence of gourmet dinners hosted by the algebraists in turn. I still remember the Coquilles St. Jacques served by the Pierces, a meal that was followed by the famous Gurleyville Hill race: Dick on a bicycle versus Roger on foot. Details of the affair may be obtained from any of the spectators.

Another product of that special year was the paper [49], characterizing the torsion–free abelian groups of finite rank that are projective as modules over their endomorphism rings. These are precisely the groups nearly isomorphic to a direct sum of an E–ring and an E–module over that ring.

A series of ten papers, [51, 52, 54, 55, 59, 60, 61, 62, 64 and 66], was begun in the fall of 1981. I had followed Dick to the University of Hawaii on sabbatical and Lee Lady was giving a seminar on torsion–free modules over Dedekind domains that was eventually published in the Proceedings of the 1982/83 Honolulu Conference on Abelian Groups. Inspired by Lady's lectures, we looked at the following question:

(1) Let D be a finite dimensional division algebra over the rational numbers $\mathbb{Q}$. For which rational primes p is there a full p–local subgroup G of D such that $\mathbb{Q}\,\mathrm{End}(G) \simeq D$?

If p is such a prime, then D is said to be **p–realizable**. The problem has roots in Corner's Theorem, as well as earlier work by H. Zassenhaus, M.C.R. Butler and J.D. Reid. The paper [52] showed that the answer is "almost all p" when D is not a field. The case when D is a field was studied in [51], where it was shown that D is p–realizable for infinitely many primes p. More complete results involved the study of Galois groups, which led to the work in [54] and [55]. Here the groups are not abelian, and Dick used some clever arguments to isolate five "trouble–causing" finite groups. This result is then used (in [54]) to prove: Let F be a Galois extension of $\mathbb{Q}$ with Galois group G. The following are equivalent:

(a) F is p–realizable for almost all primes p.
(b) G contains no non–trivial cyclic normal subgroups and G is not isomorphic to S_4 or A_4.

The next phase of this work began at the University of Washington in the fall of 1988, when I again joined Dick on sabbatical. We decided to look at a more delicate version of (1).

(2) For which p–local subrings R of a finite dimensional $\mathbb{Q}$–division algebra D is there a full p–local R–submodule G of D with $R = \mathrm{End}(G)$?

Such a ring R is called **p–realizable**. Maximal $\mathbb{Z}_p$–orders in D are always p–realizable if D is p–realizable, as was shown in [62]. Beyond that, results are more complicated. The work in [64] even led, via representations, to a question on the structure of partially ordered sets [59]. The papers [60] and [61] concern E–rings and synthesized some of the work of the papers discussed above. For example (from [60]), an algebraic number field F is p–realizable if and only if there is a p–local E–ring that is a full subring of F. A final paper [66] investigates question (1) when D is allowed to be more general than a division algebra.

Two survey papers complete Pierce's work on abelian groups: "E–modules" [56] was presented at the Perth Conference on Abelian groups in 1987; and "Abelian groups as modules over their endomorphism rings" [58] was given at the Connecticut Workshop on Abelian Groups in 1989. The paper on E–modules contained important new results. Recall that a module M over a commutative ring R is an **E–module** over R if $\mathrm{Hom}_{\mathbb{Z}}(R, M) = \mathrm{Hom}_R(R, M)$. Denote by E–Mod(R) the class of all abelian groups that support an R–module structure making M an E–module over R. The following result from [56] illustrates once again the breadth of Pierce's mathematics. For any commutative ring R, the following are equivalent:

(a) E–Mod(R) is axiomatizable by first order sentences in the language of mod–R.
(b) Either there is a natural number m such that E–Mod(R) is the class of abelian groups A with $mA = 0$; or, R has torsion–free rank one and there are no epimorphisms from R to $Z(p^\infty)$.

Both [56] and [58] contain nice lists of unsolved problems, the second paper being cited particularly often in the work of U. Albrecht, H.P. Goeters and W. Ullery.

So far, we have discussed fewer than half the publications on Pierce's vita. Some note must be made of his work outside of abelian groups. Within algebra, he worked in rings and modules, associative algebras, semigroups, universal algebra, lattices and Boolean algebras. But he also had important papers in analysis and topology. Ralph Freese recalls that in the early 1970's, Pierce would attend two–thirds of the dozen or so ongoing seminars at the University of Hawaii. The constraints of space preclude a discussion of all of his work, so we will focus on Boolean algebras and commutative rings.

Some of Pierce's early work was on κ–complete Boolean algebras. A Boolean algebra is **κ–complete** if subsets of cardinality at most κ have joins and meets. Roman Sikorski's book on Boolean algebras includes a number of Dick's results along these lines. For example, complete (κ–complete for all κ) Boolean algebras don't necessarily satisfy the corresponding complete distributive laws for infinite joins and meets. In the paper [7], Pierce gave necessary and sufficient conditions for a Boolean algebra to be κ–distributive. He then used this characterization to give examples of the various ways in which κ–distributivity can fail.

Countable Boolean algebras were a second major focus of Pierce's work in the area. Most of these results are collected in the chapter he wrote for the **Handbook of Boolean Algebras** [57]. Countable Boolean algebras are particularly tractable because their Stone spaces are metrizable. Dick used the topological derivative on the Stone space to obtain a set of invariants for countable Boolean algebras [37]. He then used these invariants to show that every countable Boolean algebra has a nice type of basis, a lexicographic sum of well–ordered sets indexed by the rational numbers [43].

A third focus was the semiring of isomorphism types of countable Boolean algebras under the operations of direct sum and tensor product. J. Ketonen showed that the direct sum of countable Boolean algebras can be as bad as you wish: any countable commutative semigroup can be embedded into the additive part of the semiring of isomorphism types. Tensor products had been shown to behave more reasonably, but in [53] Pierce showed that for all $n \geq 1$, there exists a countable Boolean algebra A such that A^i (tensor product of i copies) is not isomorphic to A^j for $1 \leq i < j \leq n$, but A^n is isomorphic to A^{n+1}.

Another area in which Pierce made an impact was commutative ring theory. In particular, the monograph [33] is an important contribution that has spawned research in many different directions. The module structure over commutative regular rings is sufficiently complex that the techniques Pierce developed in [33] have since been applied to much more general rings. A dramatic example of this phenomenon was the solution, in 1976, of Kaplansky's famous problem: characterize those commutative rings for which every finitely generated module is a direct sum of cyclic modules. The ultimate solution combined the work of many mathematicians, but the most difficult step in the proof was to show that such a ring has only finitely many minimal prime ideals. Pierce verified this for commutative regular rings using a delicate application of some deep results in point–set topology. It turned out that the same ideas could be applied to arbitrary commutative rings and once this was done the solution of Kaplansky's problem followed swiftly. Since 1976, considerable effort has been made to find a purely algebraic solution to Kaplansky's problem, but to date none has emerged. Without Pierce's ability to pull together results from seemingly unrelated areas of mathematics, it is unlikely that the solution could have been found.

In [34], Pierce showed that the free Boolean ring on $\aleph_n$ generators has global dimension $n+1$. To prove this he used an explicit projective resolution (the Koszul resolution of a canonical simple module) and showed that the nth syzygy is not projective. This required a very difficult calculation and a beautiful refinement theorem for decompositions of projective modules over Boolean rings. Pierce continued his work on the global dimension of Boolean rings in [44], a landmark paper drawing upon deep results in set theory such as the Erdös–Rado Partition Theorem.

Every commutative ring R has an associated Boolean ring $B(R)$, namely, its set of idempotents (where the sum of idempotents e and f in R is $e + f - 2ef$). If the ring R is regular, then there are tight connections between R and $B(R)$. For example, their ideal lattices are isomorphic via extension and contraction. A natural question is whether R and $B(R)$ have the same global dimension when R is regular. If this could be answered in the affirmative, many results on the global dimensions of arbitrary commutative rings would follow from Pierce's work on Boolean rings. If either R or $B(R)$ has global dimension 0 or 1, it is not hard to see that their global dimensions are equal. In [46] Pierce showed, by a difficult and ingenious argument, that the same result holds for global dimension two. He retained an active interest in the general case until his untimely death.

Dick was diagnosed with cancer in the fall of 1985. He underwent an operation and radiation therapy, but the disease returned four years later. A second operation in January, 1990, was not a success. However, Dick faced dying with all the vitality that so characterized his approach to living. He continued to travel, socialize and do mathematics, always upbeat in spite of the problems caused by his cancer.

The last note I received from Dick is dated March 4, 1992, eleven days before his death. He had received a request for a reprint of [62], "Realizing subrings of rational division algebras," from a Dr. Tom Nordahl in the Department of Psychiatry (!) at the University of California, Davis. It is interesting to make conjectures on the reason for the request, but Dick took a different tack.

"I assume that you can take care of this chance for fame and fortune," he wrote. "Should we trade a session on the couch for the realization of a good algebra?"

His sense of humor is one of the many things about Dick that we will miss. He added much to the lives of those who knew him.

Acknowledgment. The material on Pierce's Boolean algebra work was provided by J.B. Nation, and that on commutative rings by R. Wiegand. I am also grateful to D. Arnold, W.W. Comfort, L. Fuchs, P. Hill, A. Mader, M. Pierce, J.D. Reid, and W.J. Wickless for helpful comments and suggestions.

Department of Mathematics, U–9, University of Connecticut, Storrs, CT 06269
E-mail address: vinson@uconnvm.bitnet

RICHARD SCOTT PIERCE

FEBRUARY 26, 1927 — MARCH 15, 1992

EDUCATION:

1950 – B.S., California Institute of Technology
1952 – Ph.D., California Institute of Technology

ACADEMIC POSITIONS:

1950-52 – Teaching Assistant, California Institute of Technology
1952-53 – ONR Postdoctoral Fellow, Yale University
1953-55 – Jewett Fellow, Harvard University
1955-70 – Assistant, Associate, Full Professor, University of Washington
1961-62 – NSF Senior Postdoctoral Fellow, University of California, Berkeley
1967-68 – Visiting Professor, New Mexico State University
1970-75 – Professor, University of Hawaii
1975-91 – Professor, University of Arizona
1978-79 – Visiting Professor, University of Connecticut

ADMINISTRATIVE EXPERIENCE:

1962-67 – Department Chairman, University of Washington
1984-85 – Acting Department Head, University of Arizona

PROFESSIONAL OFFICES:

1958-70 – Associate Secretary, American Mathematical Society
1967-73 – Editorial Committee, Mathematical Reviews
1969-70 – Editorial Board, Pacific Journal of Mathematics
1971-91 – Editorial Board, Algebra Universalis
1985-90 – Editorial Board, Rocky Mountain Journal of Mathematics
1985-90 – Editorial Board, Communications in Algebra

Ph.D. STUDENTS: Dorothy Christensen, Richard Mayer, Gloria Hewitt, James Reid, Robert Stringall, Joel Berman, Timothy Cramer, Frank Castagna, John Werth, Roger Wiegand, John Muth, Eliot Jacobson, Achmed Hadida

BIBLIOGRAPHY

1. *Cones and the decomposition of functionals*, Math. Mag. **24** (1951), 117–122.
2. *The Boolean algebra of regular open sets*, Can. Jour. of Math. **5** (1953), 95–100.
3. *Homomorphisms of semi–groups*, Ann. of Math **59** (1954), 287–291.
4. *Coverings of a topological space*, Trans. Amer. Math. Soc. **77** (1954), 281–298.
5. *Lattice ordered rings* (with Garrett Birkhoff), An. Acad. Brasil Ci. **28** (1956), 41–69; reprinted in Coll. Works of G. Birkhoff.
6. *Radicals in function rings*, Duke Math. J. **23** (1956), 253–261.
7. *Distributivity in Boolean algebras*, Pacific J. Math. **7** (1957), 983–992.
8. *Distributivity and the normal completion of Boolean algebras*, Pacific J. Math. **8** (1958), 133–140.
9. *A note on complete Boolean algebras*, Proc. Amer. Math. Soc. **9** (1958), 892–896.
10. *Translation lattices*, Mem. Amer. Math. Soc. **32** (1959), 66 pages.
11. *Representation theorems for certain Boolean algebras*, Proc. Amer. Math. Soc. **19** (1959), 42–50.
12. *A generalization of atomic Boolean algebras*, Pac. J. Math. **9** (1959), 175–182.
13. *Partly transitive modules and modules with proper isomorphic submodules* (with R. A. Beaumont), Trans. Amer. Math. Soc. **91** (1959), 209–219.
14. *Partly invariant submodules of a torsion module* (with R. A. Beaumont), Trans. Amer. Math. Soc. **91** (1959), 220–230.
15. *Free products of alpha–distributive Boolean algebras* (with D. J. Christensen), Math. Scand. **7** (1959), 81–105.
16. *Boolean algebras with ordered bases* (with R. D. Mayer), Pacific J. Math. **10** (1960), 925–942.
17. *Subrings of simple algebras*, Michigan Math. J. **7** (1960), 241–243.
18. *Torsion free rings* (with R. A. Beaumont), Illinois J. Math. **5** (1961), 61–98.
19. *Torsion free groups of rank two* (with R. A. Beaumont), Mem. Amer. Math. Soc. **38** (1961), 41 pages.
20. *Some questions about complete Boolean algebras*, Lattice Theory, Proc. Sympos. Pure Math. of the Amer. Math. Soc. (1961), 129–140.
21. *Subrings of algebraic number fields* (with R. A. Beaumont), Acta Sci. Math. Szeged **22** (1961), 202–216.
22. *Rings of integer-valued continuous functions*, Trans. Amer. Math. Soc. **100** (1961), 371–394.

23. *The Algebraic Foundations of Mathematics* (with R. A. Beaumont), Addison-Wesley, Inc., Reading, MA (1963), 486 pages (Undergraduate Textbook).
24. *Homomorphisms of primary abelian groups*, Topics in Abelian Groups, Scott–Foresman, Chicago (1963), 215–310.
25. *Centers of purity in abelian groups*, Pacific J. Math. **13** (1963), 215–219.
26. *A note on free products of abstract algebras*, Proc. Roy. Acad. Sci., Amsterdam **25** (1963), 401–407.
27. *A note on free algebras*, Proc. Amer. Math. Soc. **14** (1963), 845–846.
28. *Isomorphic direct summands of abelian groups* (with R. A. Beaumont), Math. Ann. **153** (1964), 21–37.
29. *Some invariants of p–groups* (with R. A. Beaumont), Mich. Math. J. **11** (1964), 137–149.
30. *Endomorphism rings of primary abelian groups*, Proc. Hungar. Colloq. on Abelian Groups, Budapest (1964), 125–137.
31. *Quasi–isomorphism of p–groups* (with R. A. Beaumont), Proc. Hungar. Colloq. on Abelian Groups, Budapest (1964), 13–27.
32. *Quasi–isomorphism of direct sums of cyclic groups* (with R. A. Beaumont), Acta Math. Acad. Sci. Hungar. **16** (1965), 33–36.
33. *Modules over Commutative Regular Rings*, Mem. Amer. Math. Soc. **70** (1967), 112 pages.
34. *The global dimension of Boolean rings*, J. Algebra **7** (1967), 91–99.
35. *Introduction to the Theory of Abstract Algebras*, Holt, Rinehart and Winston, New York (1968), 148 pages (Graduate Textbook).
36. *Classification problems*, Math. Systems Theory **4** (1970), 65–80.
37. *Existence and uniqueness theorems for extensions of zero–dimensional compact metric spaces*, Trans. Amer. Math. Soc. **148** (1970), 1–21.
38. *Topological Boolean algebras*, Queen's Papers in Pure and Appl. Math. **25** (1970), 107–130.
39. *The submodule lattice of a cyclic module*, Algebra Universalis **1** (1971), 192–199.
40. *A strategy for a class of games*, The Two Year College Math. J. **2** (1971), 55–62.
41. *Closure spaces with applications to ring theory*, Proc. of the Tulane Conf. on Ring Theory, Lecture Notes in Math. **246**, Springer–Verlag, Berlin (1972), 566–616.
42. *Compact zero–dimensional metric spaces of finite type*, Mem. Amer. Math. Soc. **130** (1972), 64 pages.
43. *Bases of countable Boolean algebras*, J. Symbolic Logic **38** (1973), 212–214.
44. *The cohomology of Boolean rings*, Adv. Math. **13** (1974), 323–381.
45. *Arithmetic properties of certain partially ordered semigroups*, Semigroup Forum **11** (1975), 115–129.

46. *The global dimension of commutative regular rings*, Houston J. Math. **2** (1976), 97–110.
47. *Symmetric groupoids*, Osaka J. Math. **15** (1978), 51–76.
48. *Symmetric groupoids*, II, Osaka J. Math. **16** (1979), 317–348.
49. *Torsion-free abelian groups of finite rank projective as modules over their endomorphism rings* (with D. M. Arnold, J. D. Reid, C. Vinsonhaler, and W. Wickless), J. Algebra **71** (1981), 1–10.
50. *Associative Algebras*, Springer–Verlag, New York (1982); Russian translation MIR, Moscow (1986), 436 pages (Graduate Textbook).
51. *Realizing algebraic number fields* (with C. Vinsonhaler), Proc. Honolulu Conf. on Abelian Group Theory, Lecture Notes in Math. **1006**, Springer–Verlag, New York (1983), 49–96.
52. *Realizing central division algebras* (with C. Vinsonhaler), Pacific J. Math. **109** (1983), 165–177; correction **130** (1987), 397–399.
53. *Tensor products of Boolean algebras*, Proc. Puebla Conf. on Universal Algebra and Lattice Theory, Lecture Notes in Math. **1004**, Springer–Verlag, New York (1983), 232–239.
54. *Realizing Galois fields*, Proc. Udine Conf. on Abelian Groups and Modules, Springer–Verlag, Vienna (1984), 291–304.
55. *Permutation representations with trivial set-stabilizers*, J. of Algebra **95** (1985), 88–95.
56. *E-modules*, Proc. Perth Conf. on Abelian Groups, Contemp. Math. **87**, Amer. Math. Soc., Providence (1989), 221–240.
57. *Countable Boolean algebras*, Chapter 21, Handbook of Boolean Algebras, North Holland, Amsterdam (1989), 777–876.
58. *Abelian groups as modules over their endomorphism rings*, Proc. Univ. of Connecticut Workshop on Abelian Groups, Univ. of Connecticut, Storrs (1990), 45–59.
59. *Realizable posets* (with C. Vinsonhaler), Order **7** (1990), 275–282.
60. *Carriers of torsion-free groups* (with C. Vinsonhaler), Rend. Sem. Mat. Univ. Padova **84** (1990), 263–281.
61. *Classifying E-rings* (with C. Vinsonhaler), Comm. in Alg. **19** (1991), 615–653.
62. *Realizing subrings of rational division algebras* (with C. Vinsonhaler), Houston J. of Math. **17** (1991), 207–226.
63. *Minimal regular rings*, Abelian Groups and Noncommutative Rings, Contemp. Math. **130**, 335–348.
64. *Realizing tiled orders* (with C. Vinsonhaler), J. Algebra (to appear).
65. *Burnside algebras of profinite groups*, Comm. in Alg. (to appear).
66. *Quasi-realizing modules* (with C. Vinsonhaler), Proc. Curaçao Conf. on Abelian Groups, Lecture Notes in Pure and Appl. Math. **146** (1993), 219–229.

Contemporary Mathematics
Volume **171**, 1994

ON B_2-GROUPS

LADISLAV BICAN

October 25, 1993

ABSTRACT. It is well known [**FMa**] that in the constructible universe (V = L) the classes of B_1-groups and B_2-groups coincide. In this note a short proof of the coincidence of these two classes under the conditions (PB) and (PBB) on the prebalanced subgroups is presented.

Introduction

All groups in this paper are abelian. If x is an element of a torsionfree group G then $|x|_G$, or simply $|x|$ is the characteristic of x in G. The letter G will usually denote a general torsionfree group, while the letter B will be used for Butler groups. For unexplained terminology and notation see [**F1**]. By a *smooth increasing union of a group G* we mean a collection of subgroups G_α indexed by an initial segment of ordinals with the property that $G_\beta \subseteq G_\alpha$ when $\beta < \alpha$ and $G_\alpha = \cup_{\beta<\alpha} G_\beta$ whenever α is a limit ordinal.

An exact sequence $E \ : \ 0 \longrightarrow H \longrightarrow G \overset{\beta}{\longrightarrow} K \longrightarrow 0$ with K torsionfree is *balanced* if the induced map β_*: $\mathrm{Hom}(J,G) \longrightarrow \mathrm{Hom}(J,K)$ is surjective for each rank one torsionfree group J. Equivalently, E is balanced if all rank one (completely decomposable) torsionfree groups are projective with respect to E. A torsionfree group B is said to be a *B_1-group* (*Butler group*) if $\mathrm{Bext}(B,T) = 0$ for all torsion groups T, where Bext is the subfunctor of Ext consisting of all balanced-exact extensions. It is known [**BS**] that this definition coincides with the familiar one if B has finite rank, i.e. if it is a pure subgroup of a completely

---

1991 *Mathematics Subject Classification.* 20K20.

Key words and phrases. Butler group, B_2-group, prebalanced subgroup, torsion extension property.

This research has partially been supported by the grant FDR-0579 of the Czech Ministry of Education.

This paper is in final form and no version of it will be submitted for publication elsewhere.

decomposable group, or, equivalently [**B**], a torsionfree homomorphic image of a completely decomposable group of finite rank.

A torsionfree group B is called a B_2-*group* if it has a B-*filtration*, i.e. B is a smooth union of pure subgroups $B_\alpha, \alpha < \lambda$, λ an ordinal, and $B_{\alpha+1} = B_\alpha + H_\alpha$ for all $\alpha < \lambda$ where H_α is a Butler group of finite rank. For the sake of brevity we shall denote by $\mathcal{B}_1, \mathcal{B}_2$ the class of all B_1-groups, B_2-groups, respectively. It is well-known [**BS**] that the class of B_2-groups is contained in the class of B_1-groups and consequently it is natural to ask whether the converse is true or not. For groups of cardinality $\leq \aleph_1$ these classes coincide (see [**DHR**], [**AH**] and [**BS**] for the countable case). However, for groups of higher cardinality, the problem is undecidable in ZFC. Assuming (CH), Dugas, Hill and Rangaswamy [**DHR**] gave the affirmative answer for groups of cardinalities up to $\aleph_\omega$, while Fuchs and Magidor recently obtained positive solution in the constructive universe L.

In [**BF; Th.4.5**] it has been proved that in the constructible universe (V = L) the class of B_1-groups is closed under prebalanced subgroups. Further, it follows from [**FMa; L.2.3**] that under (V = L) any subgroup of a torsionfree group of regular cardinality is contained in a balanced subgroup of the same cardinality. The purpose of this note is to present a short proof of the coincidence of the classes of B_1 and B_2-groups under weaker forms of these facts. More precisely, we shall work with the following two statements:

(PB) Each subgroup of a B_1-group of regular cardinality is contained in a prebalanced subgroup of the same cardinality.

(PBB) Any prebalanced subgroup of a completely decomposable group is a B_1-group.

Auxiliary results

1. LEMMA. *Let* $0 \to K \to C \xrightarrow{\pi} G \to 0$ *be a balanced-exact sequence where* C *is a completely decomposable group of uncountable cardinality* κ. *Then there are continuous well-ordered ascending chains of subgroups* $K = \cup_{\alpha<\mathrm{cof}\kappa} K_\alpha$, $C = \cup_{\alpha<\mathrm{cof}\kappa} C_\alpha$ *and* $G = \cup_{\alpha<\mathrm{cof}\kappa} G_\alpha$ *such that for each* $\alpha <$ *cof* κ *the sequence* $0 \to K_\alpha \to C_\alpha \to G_\alpha \to 0$ *is exact, the subgroups* C_α *are summands of* C *and have cardinality* $< \kappa$, *the subgroups* $K_\alpha = K \cap C_\alpha$ *are balanced in* K *(and hence in* C*) and the subgroups* $G_\alpha = \pi C_\alpha$ *are pure in* G.

PROOF. See [**F2; L.1.4**].

2. LEMMA. *Let* λ *be a limit ordinal and* $H = \cup_{\alpha<\lambda} H_\alpha$ *a smooth increasing union such that* $H_{\alpha+1}$ *is prebalanced in* G *for all* $\alpha < \lambda$. *If* cof $\lambda \neq \omega$ *then* H *is prebalanced in* G.

PROOF. Take $g \in G \setminus H$ arbitrarily. For each $0 \neq m < \omega$ there is a countable subset $S_m = \{h_0^m, h_1^m, \dots\} \subseteq H$ such that $|mg + H| = \cup_{n<\omega} |mg + h_n^m|$. Since cof $\lambda > \omega$, there is $\alpha < \lambda$ non-limit such that $\{h_n^m \mid m, n < \omega\} \subseteq H_\alpha$. Further, there is $k < \omega$ and $h_0, h_1, \dots, h_l \in H_\alpha$ with $|kg + H_\alpha| = \cup_{i=0}^{l} |kg + h_i|$, H_α being prebalanced. Now $|kg+H| = \cup_{n<\omega}|kg+h_n^k| \le |kg+H_\alpha| = \cup_{i=0}^l |kg+h_i| \le |kg+H|$ and we are through.

3. LEMMA. (i) *If A is a* TEP *subgroup of a B_1-group B, then B/A is a B_1-group;*
(ii) *if A is a prebalanced subgroup of a B_1-group B such that B/A is a B_1-group, then A is* TEP *in B.*

PROOF. (i) The exact sequence $0 \to A \xrightarrow{\alpha} B \xrightarrow{\beta} B/A \to 0$ induces the exact sequence $\mathrm{Hom}(B,T) \to \mathrm{Hom}(A,T) \to \mathrm{Ext}(B/A,T) \xrightarrow{\beta^*} \mathrm{Ext}(B,T)$ where β^* is monic, A being TEP in B. Now it is easy to see that β^* induces a monomorphism $\mathrm{Bext}(B/A,T) \to \mathrm{Bext}(B,T) = 0$.

(ii) The exact sequence given above induces $\mathrm{Hom}(B,T) \xrightarrow{\alpha^*} \mathrm{Hom}(A,T) \to \mathrm{PBext}(B/A,T)$ for each T torsion. Since $\mathrm{PBext}(B/A,T) = \mathrm{Bext}(B/A,T) = 0$ by [**FMe; Th.1.5**], α^* is an epimorphism.

The following result has been proved in [**DHR**]. The idea of the proof we borrowed from [**F2**].

4. LEMMA. *Let κ be an uncountable regular cardinal and let $B = \cup_{\alpha<\kappa} B_\alpha$ be any κ-filtration of a B_1-group B of cardinality κ. If each B_α is a B_1-group then there is a* cub *$C \subseteq \kappa$ such that B_α has* TEP *in B for each $\alpha \in C$.*

PROOF. Denote $\tilde{E} = \{\alpha < \kappa \mid B_\alpha$ is not TEP in $B\}$ and $E = \{\alpha < \kappa \mid B_\alpha$ is not TEP in B_β for some $\beta > \alpha\}$. Then $E \subseteq \tilde{E}$ and the equality holds provided E is not stationary.

Supposing E stationary, there is no loss of generality in assuming $\beta = \alpha + 1$ in the definition of E.

Given any $\alpha \in E$, select a torsion group T_α and a homomorphism $\varphi_\alpha : B_\alpha \to T_\alpha$ which has no extension to $B_{\alpha+1} \to T_\alpha$. For $\alpha < \kappa, \alpha \notin E$, we take $T_\alpha = 0, \varphi_\alpha = 0$. Setting $S_\alpha = \oplus_{\beta<\alpha} T_\beta$ and $\Pi_\alpha : S_{\alpha+1} \to T_\alpha$ the canonical projection, we can form the commutative diagram

$$\begin{array}{ccccccccc}
E_\alpha : 0 & \longrightarrow & S_\alpha & \longrightarrow & S_\alpha \oplus B_\alpha & \xrightarrow{\sigma_\alpha} & B_\alpha & \longrightarrow & 0 \\
 & & \downarrow & & \downarrow{\scriptstyle \eta_\alpha} & & \downarrow & & \\
E_{\alpha+1} : 0 & \longrightarrow & S_{\alpha+1} & \longrightarrow & S_{\alpha+1} \oplus B_{\alpha+1} & \xrightarrow{\sigma_{\alpha+1}} & B_{\alpha+1} & \longrightarrow & 0
\end{array}$$

with exact rows, where $\eta_\alpha(s_\alpha, b_\alpha) = (s_\alpha + \varphi_\alpha b_\alpha, b_\alpha)$ and the remaining maps are natural embeddings and projections. Take the direct limit $E : 0 \to S \to G \to B \to 0$ of the directed system $\{E_\alpha; \mid \alpha < \kappa\}$. By [**BF; L.1.4**] the sequence E is balanced and it is therefore splitting via $\tau : B \to G$. Now τ maps B_α into the direct sum of B_α and a set of less than κ many T_β's. Further, let $\omega_\alpha : S_\alpha \oplus B_\alpha \to G$ be the canonical direct limit mapping and denote $\vartheta_\alpha = (1_G - \tau\sigma)\omega_\alpha | B_\alpha : B_\alpha \to S$. The regularity of κ leads to the conclusion that the set $F = \{\alpha < \kappa \mid \tau B_\alpha \leq S_\alpha \oplus B_\alpha, \vartheta_\alpha B_\alpha \leq S_\alpha\}$ is a cub in κ. For $\alpha \in F \cap E$ we now have $\vartheta_\alpha b = (1_G - \tau\sigma)\omega_\alpha b = (1_G - \tau\sigma)\omega_{\alpha+1}\eta_\alpha b = \varphi_\alpha b + \vartheta_{\alpha+1} b$ for each $b \in B_\alpha$, hence $\varphi_\alpha b = -\Pi_\alpha \vartheta_{\alpha+1} b$, which contradicts the choice of φ_α and the proof is complete.

5. LEMMA. *Let H be a prebalanced subgroup of a torsionfree group G such that G/H is a B_2-group. Then $G = \cup_{\alpha<\lambda} G_\alpha$ with $G_0 = H$ and $G_{\alpha+1} = G_\alpha + B_\alpha, \alpha < \lambda$, where G_α is pure in G, B_α is a finite rank Butler group and the union is smooth.*

If, moreover, H is a B_2-group, then the B-filtration of H extends to that of G.

PROOF. Let $G/H = \cup_{\alpha<\lambda} G_\alpha/H$ be a B-filtration of G/H. Then $G_{\alpha+1}/H = G_\alpha/H + B'_\alpha/H, B'_\alpha/H$ a finite rank Butler group. Hence $B'_\alpha = H + B_\alpha$ with B_α a finite rank Butler group, H being prebalanced in G, and the assertion follows easily. The rest is obvious.

Recall [**AH**] that a pure subgroup H of a torsionfree group G is said to be *decent* if for any finite subset S of G there exists a finite number of rank one pure subgroups A_i of G such that $H + \sum A_i$ is pure in G and contains S. Further, a group G is said to have an *axiom-3 family* of decent subgroups if there is a family $\mathcal{F}(G)$ of decent subgroups such that $0, G \in \mathcal{F}(G), \sum_{i\in I} H_i \in \mathcal{F}(G)$ whenever $H_i \in \mathcal{F}(G), i \in I$, and if $H \in \mathcal{F}(G)$ and $X \subseteq G$ is countable, then there is a $K \in \mathcal{F}(G)$ containing $H \cup X$ such that K/H is countable.

6. REMARK. Looking at the proof of [**FMa; Th.7.1**] we see that to a given B-filtration of a B_2-group B it is associated an axiom-3 family of decent subgroups in the natural way – given by the closed subsets. Such a family we shall call the *axiom-3 family of decent subgroups corresponding to a given B-filtration* of B.

7. LEMMA. (i) *Let B be a B_2-group and let A be any member of a B-filtration of B. If $\mathcal{F}(A)$ and $\mathcal{F}(B)$ are the corresponding axiom-3 families of decent subgroups then $\mathcal{F}(A) \subseteq \mathcal{F}(B)$.*

(ii) *If $B = \cup_{\alpha<\lambda} B_\alpha$ smooth, is a B_2-group such that all B_α's are members of a B-filtration of B, then for the corresponding axiom-3 families of decent subgroups it is $\mathcal{F}(B_\beta) \subseteq \mathcal{F}(B_\alpha)$ whenever $\beta \leq \alpha$ and $\cup_{\beta<\alpha} \mathcal{F}(B_\beta) \subseteq \mathcal{F}(B_\alpha), \alpha$ limit. Especially, $\cup_{\alpha<\lambda} \mathcal{F}(B_\alpha) \subseteq \mathcal{F}(B)$.*

PROOF. It follows immediately from the proof of [**FMa; Th.7.1**], since any subset S closed in μ is closed in each $\lambda > \mu$.

8. LEMMA. *Let $\mathcal{F}(B)$ be an axiom-3 family of decent subgroups of a torsion-free group B. Then for each $A \in \mathcal{F}(B)$ there is a B-filtration from A to B. Especially, any B-filtration of A extends to that of B.*

PROOF. See the proof of [**AH, Th.5.3**].

<u>κ–Shelah game.</u> Let κ be a regular uncountable cardinal and let G be a torsionfree group of cardinality $|G| > \kappa^+$. We define the *κ-Shelah game* on

G in the following way: Player I picks subgroups $G_{2i}, i < \omega$, of cardinality κ and player II picks G_{2i+1} such that $G_i \subseteq G_{i+1}$ for all $i < \omega$. Player II wins if G_{2i+1} is prebalanced in G and TEP in G_{2i+3} for each $i < \omega$.

9. LEMMA. *Let κ be a regular uncountable cardinal and let B be a B_1-group of cardinality $|B| > \kappa^+$. If any subgroup of B of cardinality κ is contained in a prebalanced subgroup of the same cardinality and if any prebalanced subgroup of B of the cardinality κ^+ is a B_2-group, then player II has a winning strategy in the κ-Shelah game.*

PROOF. Lemma 1.2 in [**H**] still holds, the κ-Shelah game is determinate and so we are going to show that player I has no winning strategy. By way of contradiction let us assume that I has a winning strategy s and has picked B_0. By transfinite induction we shall construct a smooth chain $\{C_\alpha \mid \alpha < \kappa^+\}$ of subgroups of B of cardinality κ. Let C_0 be a prebalanced subgroup of B of cardinality κ containing B_0. If $0 < \alpha < \kappa^+$ and the $C_\beta, \beta < \alpha$, has been defined, then for α limit we set $C_\alpha = \cup_{\beta<\alpha} C_\beta$, while for α non-limit we select C_α as a prebalanced subgroup of B of cardinality κ containing $C_{\alpha-1}$ and all $s(C_{\alpha_1}, C_{\alpha_2}, \ldots, C_{\alpha_n}), \alpha_1 < \alpha_2 < \cdots < \alpha_n < \alpha, n < \omega$. The union $C = \cup_{\alpha<\kappa^+} C_\alpha$ is prebalanced in B by Lemma 2, hence it is a B_2-group by the hypothesis and consequently it has an axiom-3 family $\mathcal{F}(C)$ of decent subgroups ([FMa; Th.7.2]). Since each element of C is contained in a countable member of $\mathcal{F}(C)$ and $\mathcal{F}(C)$ is closed under subgroups generated by arbitrary sets of members of $\mathcal{F}(C)$, we can without loss of generality assume that all the C_α's are in $\mathcal{F}(C)$. By lemma 8, for each $\alpha < \kappa^+$ there is a B-filtration of C containing C_α and so [**DHR; L.1.3**] gives that C_α is TEP in C, while [**FV; Th.8**] implies that C_α is prebalanced in B.

Now when player I has chosen B_{2i} in the κ-Shelah game, then player II picks B_{2i+1} to be C_α, where α is the least non-limit element of U such that $B_{2i} \subseteq C_\alpha$.

Main results

PROPOSITION. *Under* (PBB), *the class of B_1-groups is closed under prebalanced subgroups.*

PROOF. Take $B \in \mathcal{B}_1$ and let $E : 0 \to K \to C \to B \to 0$ be a balanced-projective resolution of B. For a prebalanced subgroup $\tilde{B} \leq B$ consider the following commutative diagram with exact rows

$$\begin{array}{ccccccccc}
\tilde{E}: 0 & \longrightarrow & K & \longrightarrow & L & \longrightarrow & \tilde{B} & \longrightarrow & 0 \\
 & & \| & & \downarrow & & \downarrow & & \\
E: 0 & \longrightarrow & K & \longrightarrow & C & \longrightarrow & B & \longrightarrow & 0
\end{array}$$

By [**FMe; L.1.3**] the subgroup L is prebalanced in C and consequently it is a B_1-group by the hypothesis. Using lemma 3 twice, we get that E is TEP, hence $\tilde{E}$ is so, and consequently $\tilde{B}$ is a B_1-group.

THEOREM. *If the conditions* (PB) *and* (PBB) *are satisfied, then each B_1-group is a B_2-group.*

PROOF. Proving indirectly, let us assume the existence of a B_1-group B which is not a B_2-group and has the smallest possible cardinality κ.

Assuming first that κ is regular we have $\kappa > \aleph_0$ by [**BS; Th.3.4**]. If $0 \to K \to C \to B \to 0$ is a balanced-projective resolution of B, then in the notation of lemma 1 we have $K_\alpha, K \in \mathcal{B}_1$ by the hypothesis and consequently with respect to lemma 4 we can suppose that all the K_α's are TEP in K. But then K_α is TEP in C_α, hence $B_\alpha \in \mathcal{B}_1$ by lemma 3 and using lemma 4 once more we obtain that all the B_α's can be assumed TEP in B. By the choice of κ each B_α is a B_2-group and so is $B_{\alpha+1}/B_\alpha$, owing to lemma 3. A simple transfinite induction based on lemma 5 leads to a contradiction $B \in \mathcal{B}_2$.

We proceed to the singular cardinality case. There is a smooth union $\kappa = \cup_{\alpha<\mu}\kappa_\alpha$ with $\kappa_0 > \mu = \operatorname{cof}\kappa$ and κ_α regular whenever α is non-limit. Further, let $B = \cup_{\alpha<\mu}C_\alpha$ be a smooth union such that $|C_\alpha| = \kappa_\alpha$.

Set $C^0_\alpha = C_\alpha$ for each $\alpha < \mu$ and assume that C^k_α has been defined for all $0 \le k \le n$ and $\alpha < \mu$. For α limit or 0 set $B^n_\alpha = C^n_\alpha$ and for α successor take B^n_α according to the κ_α– Shelah game $C^0_\alpha, B^0_\alpha, C^1_\alpha, B^1_\alpha, \ldots$, the hypotheses of lemma 9 being obviously satisfied. For each $\alpha < \mu$ list the elements of each B^n_α as $\{b^j_\alpha \mid j < \kappa_\alpha\}$ and form $A^n_\alpha = < B^n_\alpha \cup \{b^j_\gamma, \gamma < \mu, j < \kappa_\alpha\} >$. By hypothesis, B^n_α is a B_2-group and consequently by Remark 6 it has an axiom-3 family $\mathcal{F}(B^n_\alpha)$ of decent subgroups corresponding to a given B-filtration of B^n_α. The routine set-theoretical arguments lead to the conclusion that C^{n+1}_α can be selected such that it has cardinality κ_α, contains $\cup_{\beta\le\alpha}A^n_\beta$ and $C^{n+1}_\alpha \cap B^n_{\alpha+1} \in \mathcal{F}(B^n_{\alpha+1})$.

Thus $C^n_\alpha \le B^n_\alpha \le A^n_\alpha \le C^{n+1}_\alpha$ for all $\alpha < \mu$ and $n < \omega$ and we set $B_\alpha = \cup_{n<\omega}B^n_\alpha$ for each $\alpha < \mu$. Clearly, $|B_\alpha| = \kappa_\alpha < \kappa$ and $\cup_{\alpha<\mu}B_\alpha = \cup_{\alpha<\mu}\cup_{n<\omega}C^n_\alpha \ge \cup_{\alpha<\mu}C_\alpha = B$. Further, for α non-limit $B^{n+1}_\alpha/B^n_\alpha \in \mathcal{B}_2$ by the hypothesis, hence the B-filtration of B^n_α extends to that of B^{n+1}_α by lemma 5 and so by the induction we obtain that B_α is a B_2-group for each α non-limit.

It follows now from Lemma 7 that for the corresponding axiom-3 families of decent subgroups we have $\mathcal{F}(B^n_\alpha) \subseteq \mathcal{F}(B^{n+1}_\alpha)$ and $\cup_{n<\omega}\mathcal{F}(B^n_\alpha) \subseteq \mathcal{F}(B_\alpha)$.

Let $\alpha < \mu$ be arbitrary. We have $B_\alpha = B_\alpha \cap B_{\alpha+1} = B_\alpha \cap (\cup_{n<\omega}B^n_{\alpha+1}) = \cup_{n<\omega}(B_\alpha \cap B^n_{\alpha+1}) = \cup_{n<\omega}((\cup_{k<\omega}C^k_\alpha) \cap B^n_{\alpha+1}) = \cup_{n<\omega}\cup_{k<\omega}(C^k_\alpha \cap B^n_{\alpha+1}) = \cup_{n<\omega}(C^{n+1}_\alpha \cap B^n_{\alpha+1}) \in \cup_{n<\omega}\mathcal{F}(B^n_{\alpha+1}) \subseteq \mathcal{F}(B_{\alpha+1})$. By lemma 8 the B-filtration of B_α extends to that of $B_{\alpha+1}$ and to obtain the desired contradiction $B \in \mathcal{B}_2$ it remains to show that the union $B = \cup_{\alpha<\mu}B_\alpha$ is smooth.

Let $\alpha < \mu$ be a limit ordinal and take $b \in B_\alpha$ arbitrarily. Then $b \in B^n_\alpha$ for some $n < \omega$ and consequently $b = b^j_\alpha$ for some $j < \kappa_\alpha$. Thus $j < \kappa_\beta$ for some $\beta < \alpha$, the chain $\{\kappa_\alpha \mid \alpha < \mu\}$ being assumed smooth. This yields $b \in A^n_\beta \le B_\beta$ and the proof is complete.

REMARK. If $0 \to K \to C \to B \to 0$ is a balanced projective resolution of a B_1-group B, then (PBB) yields $K \in \mathcal{B}_1$. Assuming (CH) [**FR**] gives that $K \in \mathcal{B}_2$ and [**R; Th.3**] yields $B \in \mathcal{B}_2$. But the proofs of these results are rather long, especially in the singular cardinality case. So the condition (PB), which is an easy consequence of (GCH) (see [**FMa; L.2.3**]) simplifies the proof essentially.

References

[AH] Albrecht, U., Hill, P., *Butler groups of infinite rank and axiom 3*, Czech. Math. J. **37** (1987), 293–309.

[B1] Bican, L., *Splitting in abelian groups*, Czech. Math. J. **28** (1978), 356–364.

[B2] Bican, L., *Purely finitely generated groups*, Comment. Math. Univ. Carolinae **21** (1980), 209–218.

[B3] Bican, L., *Pure subgroups of Butler groups*, Proceedings of the Udine Conference April 1984, R.Göbel, at al. (Eds), CISM Courses and Lectures Springer Verlag, Wien-New York **287** (1984), 203–213.

[BF] Bican, L., Fuchs, L., *Subgroups of Butler groups* (to appear).

[BS] Bican, L., Salce, L., *Infinite rank Butler groups,*, Proc. Abelian Group Theory Conference, Honolulu, Lecture Notes in Math., Springer-Verlag **1006** (1983), 171-189.

[B] Butler, M.C.R., *A class of torsion-free abelian groups of finite rank*, Proc. London Math. Soc. **15** (1965), 680-698.

[DHR] Dugas, M., Hill, P., Rangaswamy, K.M., *Infinite rank Butler groups II*, Trans. Amer. Math. Soc. **320** (1990), 643-664.

[F1] Fuchs, L., *Infinite Abelian groups, vol. I and II*, Academic Press, New York, 1973 and 1977.

[F2] Fuchs, L., *Infinite rank Butler groups*, preprint.

[FMa] Fuchs, L., Magidor M., *Butler groups of arbitrary cardinality*, Israel J. Math. (to appear).

[FMe] Fuchs, L., Metelli, C., *Countable Butler groups*, Contemporary Math. **130** (1992), 133-143.

[FR] Fuchs, L., Rangaswamy, K.M., *Butler groups that are unions of subgroups with countable typesets*, Arch. Math. **61** (1993), 105-110.

[FV] Fuchs, L., Viljoen, G., *Note on the extensions of Butler groups*, Bull. Austral. Math. Soc. **41** (1990), 117-122.

[H] Hodges, W., *In singular cardinality, locally free algebras are free*, Algebra Universalis **12** (1981), 205-220.

[R] Rangaswamy, K.M., *A homological characterization of abelian B_2-groups* (to appear).

Department of Algebra, Faculty of Mathematics and Physics,
Charles University,
Sokolovská 83, 186 00 Prague 8,
The Czech Republic

E-mail address: BICAN@CSPGUK11.BITNET

Contemporary Mathematics
Volume **171**, 1994

Decompositions of almost completely decomposable abelian groups

E.A. BLAGOVESHCHENSKAYA AND A. MADER

ABSTRACT. The classification of block–rigid almost completely decomposable groups with cyclic regulator quotient up to near–isomorphism is used to describe the possible direct decompositions of such groups. As an application Fuchs' Problems 67 and 68 are solved for the class of block–rigid almost completely decomposable groups with cyclic regulator quotient.

1. Introduction

Problem 67 in [**8**] states: Given an integer $r \geq 3$, find all sequences $n_1 < \cdots < n_s$ of integers, for which there is a torsion–free group of rank r having decompositions into $n_1, \ldots, n_s$ indecomposable summands, respectively. Problem 68 asks: Given positive integers $r_1, \ldots, r_k, r'_1, \ldots, r'_l$ such that $r_1 + \cdots + r_k = r'_1 + \cdots + r'_l$, under what conditions does a torsion–free group exist which has a direct decomposition into indecomposable summands of ranks $r_1, \ldots, r_k$ and $r'_1, \ldots, r'_l$, respectively. Blagoveshchenskaya [**3**], [**4**] and Blagoveshchenskaya and Yakovlev [**5**] solved Fuchs' Problems 67 and 68 concerning possible decompositions of torsion–free groups of finite rank for so–called Γ–groups. In standard terminology these groups are (special) block–rigid almost completely decomposable groups with a cyclic regulator quotient. Thanks to recent classifications of such groups up to near–isomorphism [**10**], [**7**], a complete description of their possible direct decompositions can be given (3.3). With these tools, Fuchs' Problems 67 and 68 are solved for block–rigid almost completely decomposable groups with cyclic regulator quotient (3.16, 3.19). This generalizes and supplements the results of the unpublished manuscript [**4**] of the first author and places them into the context of the current state of the theory of almost completely decomposable groups.

1991 *Mathematics Subject Classification.* Primary 20K20; Secondary 20K15.
This paper is in final form and no version of it will be submitted for publication elsewhere.

2. Preliminaries

Our notation is standard and Fuchs [**8**] and Arnold [**1**] serve as general references. We write maps on the right. The symbol $|x|$ denotes the order of the group element x. If Y is a subgroup of X, then Y_* denotes the purification of Y in X.

An **almost completely decomposable group** is a torsion–free group of finite rank which contains a completely decomposable subgroup of finite index. Let $\mathbb{T}$ denote the lattice of all types, and let X be almost completely decomposable. It is well–known ([**2**]) that for each type τ, there is a direct decomposition $X(\tau) = A_\tau \oplus X^\sharp(\tau)$ with A_τ completely decomposable and τ–homogeneous. If $A_\tau \neq 0$, then τ is a **critical type** of X and $\mathrm{T_{cr}}(X)$ denotes the set of all critical types. The group $A = \sum_{\rho\in\mathbb{T}} A_\rho$ has finite index in X, and is called a **regulating subgroup** of X. Moreover, since X is almost completely decomposable, A is completely decomposable. Regulating subgroups were first introduced by Lady [**9**] who showed that they are exactly the completely decomposable subgroups of X of least index.

Burkhardt [**6**] considered the intersection $\mathrm{R}(X)$ of all regulating subgroups of X, the **regulator** of X, and showed in particular that the regulator of an almost completely decomposable group is completely decomposable.

An almost completely decomposable group X is called **block–rigid**, if the partially ordered set $\mathrm{T_{cr}}(X)$ is an antichain, i.e. if its elements are pairwise incomparable. If in addition $\mathrm{rk}(X(\tau)/X^\sharp(\tau)) \leq 1$ for each type τ, then X is called **rigid**. In a block–rigid almost completely decomposable group, each critical type is maximal, $X^\sharp(\tau) = 0$ and $X(\tau) = A_\tau$, so that there is a unique regulating subgroup which is at the same time the regulator. If X is a block–rigid almost completely decomposable group and $U = \bigoplus_\rho U_\rho$ is a completely decomposable subgroup of finite index, then $U_\tau \subset X(\tau)$ and $\mathrm{R}(X)/U \cong \bigoplus_\rho X(\rho)/U_\rho$ is finite, hence $(U_\tau)_* = X(\tau)$ and $U = \mathrm{R}(X)$ if and only if U_τ is pure in X for each τ.

We consider block–rigid groups X with cyclic regulator quotient $X/\mathrm{R}(X)$, and call such groups **block–rigid crq–groups**. We remark that every block–rigid almost completely decomposable group X containing a completely decomposable subgroup U with finite cyclic quotient X/U is a crq–group. In fact, $X/\mathrm{R}(X)$ is an epimorphic image of X/U. Thus the class of block–rigid crq–groups coincides with the class of all groups which contain a block–rigid completely decomposable group with finite cyclic quotient.

From now on we consider block–rigid crq–groups exclusively. For such a group certain numerical invariants can be defined which describe the group up to near–isomorphism. We now define these invariants.

DEFINITION 2.1. *Let X be a block–rigid crq–group and $A = \mathrm{R}(X)$ its regulator. Let e be any positive integer such that $eX \subset A$. Then A is a completely decomposable group, and A has a unique decomposition into homogeneous com-*

ponents $A = \bigoplus_{\rho \in \mathbb{T}} A_\rho$ *since* $\mathrm{T}_{\mathrm{cr}}(A) = \mathrm{T}_{\mathrm{cr}}(X)$ *is an antichain. Let*

$$^{-} : A \to \frac{A}{eA} = \overline{A}$$

be the natural epimorphism. Choose a generator $u + A$ *of* X/A *and write*

$$eu = \sum_{\rho \in \mathbb{T}} u_\rho, \quad u_\rho \in A_\rho.$$

Set

$$m_\tau(X) = |\overline{u_\tau}| = |u_\tau + eA|.$$

We verify next that the values $m_\tau(X)$ do not depend on the choices involved in its definition.

LEMMA 2.2. *Assume the situation and notation of Definition* 2.1. *Then*
(i) $m_\tau(X)$ *is independent of the choice of* e *and* u.
(ii) $|X/A| = \mathrm{lcm}_\rho\, m_\rho(X)$.
(iii) $\overline{eX} \cap \overline{A_\tau} = 0$ *for each type* τ.

PROOF. (i) Let $e_0 = \exp(X/A)$. Then e_0 divides e, say $e = ke_0$. If $e_0 u = \sum_\rho v_\rho$, $v_\rho \in A_\rho$, then $eu = \sum_\rho u_\rho = \sum_\rho kv_\rho$ and hence $u_\tau = kv_\tau$ for all τ. Since $A_\tau/e_0A_\tau \cong kA_\tau/eA_\tau$, we conclude that $|u_\tau + eA_\tau| = |v_\tau + e_0A_\tau|$ and this establishes the independence of the choice of e. Clearly, the choice of the generator u does not matter.

(ii) This follows from (i) and the fact that $|u+A| = |eu+eA| = \mathrm{lcm}_\rho\, |u_\rho+eA|$.

(iii) Since A_τ is pure in X, we have $eX \cap A_\tau = eA_\tau$, and hence

$$\overline{eX} \cap \overline{A_\tau} = \frac{eX}{eA} \cap \frac{A_\tau + eA}{eA} = \frac{(eX \cap A_\tau) + eA}{eA} = \frac{eA_\tau + eA}{eA} = 0.$$

□

REMARK 2.3. Note that we defined $m_\tau(X)$ for all types. Of course, $A_\tau = 0$ and $m_\tau(X) = 1$ if τ is not a critical type of X. The restriction to critical types is inconvenient since we will simultaneously with X consider direct summands of X and these typically have smaller critical typesets than X. So many of the invariants $m_\tau(X) = 1$ are inconsequential and only a technical convenience. In contrast, if $m_\tau(X) = 1$ for a critical type τ, then there is a non–trivial decomposition

$$X = A_\tau \oplus \left\langle \bigoplus_{\rho \neq \tau} A_\rho \right\rangle_* .$$

The first crucial ingredient of our method is the following classification theorem.

THEOREM 2.4. (Near–Isomorphism Criterion) *Let* X *and* Y *be block–rigid crq-groups. Then* $X \cong_n Y$ *if and only if* $\mathrm{R}(X) \cong \mathrm{R}(Y)$ *and for all types* τ, $m_\tau(X) = m_\tau(Y)$.

PROOF. (1) Suppose first that $X \cong_n Y$. Then by [**10**, 4.6] we have $\mathrm{R}(X) \cong \mathrm{R}(Y)$ and $X/\mathrm{R}(X) \cong Y/\mathrm{R}(Y)$. Let $A = \mathrm{R}(X)$, let $C = X/\mathrm{R}(X) = \langle u + \mathrm{R}(X)\rangle$, and let $e = \exp X/\mathrm{R}(X)$. Then by [**10**], $X \cong X_f$ and $Y \cong X_g$ for maps $f, g \in \mathrm{ReMon}(C, \overline{A})$. By [**10**, 6.3] $m_\tau(X) = m_\tau(X_f) = |(u + A)f_\tau| = |(u + A)g_\tau| = m_\tau(X_g) = m_\tau(Y)$.

(2) Conversely, assume that $A = \mathrm{R}(X) \cong \mathrm{R}(Y)$ and $m_\tau(X) = m_\tau(Y)$. Then $|X/\mathrm{R}(X)| = \mathrm{lcm}_\rho\, m_\rho(X) = \mathrm{lcm}_\rho\, m_\rho(Y) = |Y/\mathrm{R}(Y)|$ and hence $C = X/\mathrm{R}(X) \cong Y/\mathrm{R}(Y)$. Again we have $X \cong X_f$, $Y \cong X_g$ for some $f, g \in \mathrm{ReMon}(C, \overline{A})$ and by [**10**, 6.3], $X_f \cong_t X_g$, so $X \cong_n Y$ since type–isomorphism and near–isomorphism are the same [**10**, 4.5]. □

The second essential tool is a deep theorem of Arnold [**1**, 12.9] on direct decompositions and near–isomorphism which we restate for easy reference.

THEOREM 2.5. (Arnold) *Let X, Y be torsion–free groups of finite rank and suppose that $X \cong_n Y$. If $Y = Y_1 \oplus Y_2 \oplus \cdots \oplus Y_t$, then $X = X_1 \oplus X_2 \oplus \cdots \oplus X_t$ with $X_i \cong_n Y_i$.*

3. Decompositions

The first new result and the third essential ingredient of our approach connects the invariants of a direct sum with the invariants of the direct summands.

THEOREM 3.1. (Product Formula) *Let X be a block–rigid crq–group with regulator A. Suppose $X = X_1 \oplus X_2$ is a direct decomposition of X. Then each X_i is a block–rigid crq–group with regulator $A_i = X_i \cap A$, and*

$$\frac{X}{A} \cong \frac{X_1}{A_1} \oplus \frac{X_2}{A_2}$$

is a decomposition of the cyclic group X/A. The groups X_i/A_i are cyclic and have relatively prime orders. Furthermore,

$$m_\tau(X) = m_\tau(X_1) \cdot m_\tau(X_2).$$

PROOF. Since $A = \bigoplus_{\rho \in \mathrm{T}_{\mathrm{cr}}(X)} X(\rho)$ is fully invariant in X, we obtain the decomposition

$$A = A_1 \oplus A_2, \quad \text{where} \quad A_i = X_i \cap A.$$

It follows that

$$\frac{X}{A} \cong \frac{X_1}{A_1} \oplus \frac{X_2}{A_2}$$

and since X/A is cyclic, the orders of the cyclic summands X_i/A_i are pairwise relatively prime. The rest is straightforward using 2.2. □

To show that a given crq group has certain decompositions, we will construct suitable direct sums in such a way that they are near–isomorphic to the given group. Arnold's theorem then establishes the desired decompositions. To do this, we require an existence theorem.

THEOREM 3.2. (Existence Theorem) *Given a finite antichain T of types, and a set $\{m_\rho : \rho \in T\}$ of natural numbers ≥ 1, there exists a crq–group X with $\mathrm{T}_{\mathrm{cr}}(X) \subset T$ and $m_\tau(X) = m_\tau$ if and only if the following two* ***admissibility conditions*** *are satisfied.*

A1. *If the prime p divides m_τ, then $\tau(p) < \infty$.*

A2. *For all $k \in \mathbb{N}$, if all but one of the m_σ divide k, then the remaining m_τ also divides k. Equivalently, for each prime p, if p^k divides m_σ then p^k divides m_τ for some $\tau \neq \sigma$.*

Moreover, if the admissibility conditions are satisfied, then X may be chosen to be rigid, and such that $\mathrm{T}_{\mathrm{cr}}(X) = \{\rho \in T : m_\rho > 1\}$. For any choice, $|X/\mathrm{R}(X)| = \mathrm{lcm}_\rho\, m_\rho$.

PROOF. Let X be a crq–group with invariants $m_\tau = m_\tau(X) = |\overline{u_\tau}|$ where u, u_ρ are as in 2.2.

(1) If $\tau(p) = \infty$ for a prime p then $\overline{A_\tau}$ has a trivial p–primary component and $\overline{u_\tau} \in \overline{A_\tau}$, hence p does not divide $m_\tau = |\overline{u_\tau}|$. This proves A1.

(2) We have $(\mathrm{lcm}_{\tau\neq\sigma}\, m_\tau)\overline{eu} \in \overline{eX} \cap \overline{A_\sigma} = 0$ by 2.2(iii). In particular, $(\mathrm{lcm}_{\tau\neq\sigma}\, m_\tau)\overline{u_\sigma} = 0$, and so $m_\sigma = |\overline{u_\sigma}|$ divides $\mathrm{lcm}_{\tau\neq\sigma}\, m_\tau$. This implies A2.

(3) Let T and m_τ be given satisfying the admissibility conditions. Choose a rank–one group A_τ of type τ for each $\tau \in T$. Let $A = \bigoplus_{\rho\in T} A_\rho$. Set $e = \mathrm{lcm}\{m_\rho : \rho \in T\}$ and let $\overline{} : A \to A/eA = \overline{A}$ as before. Then $\overline{A} = \sum_{\rho\in T} \overline{A_\rho}$ and $\overline{A_\tau} = \langle \overline{a_\tau} \rangle$ for some $a_\tau \in A_\tau$. Further, $|\overline{a_\tau}| = |\overline{A_\tau}| = e_\tau$ is that divisor of e which is obtained by deleting those p–power factors of e for which $\tau(p) = \infty$. This means that m_τ divides e_τ by the admissibility condition A1. Let $u_\tau = (e_\tau/m_\tau)a_\tau$, $u = \frac{1}{e}\sum_{\rho\in T} u_\rho \in \mathbb{Q}A$, and $X = A + \mathbb{Z}u$. Then X is a block–rigid crq–group with $\mathrm{T}_{\mathrm{cr}}(X) = T$, and $|\overline{u_\tau}| = m_\tau$. It remains to show that $\mathrm{R}(X) = A$. This is equivalent to showing that A_τ is pure in X for each τ. So suppose that $x = \sum_\rho x_\rho + ku$, $x_\rho \in A_\rho$, $k \in \mathbb{Z}$, is such that $nx \in A_\tau$ for some positive integer n. Then $nex = \sum_\rho (nex_\rho + nku_\rho) \in A_\tau$ and hence $ex_\sigma + ku_\sigma = 0$ for all $\sigma \neq \tau$. Thus $k\overline{u_\sigma} = -\overline{ex_\sigma} = 0$, and m_σ divides k for all $\sigma \neq \tau$. By A2, m_τ and then $e = \mathrm{lcm}_\rho\, m_\rho$ divides k. Thus $x_\sigma + (k/e)u_\sigma = 0$ for all $\sigma \neq \tau$ and

$$x = \sum_\rho x_\rho + \frac{k}{e}eu = \sum_\rho (x_\rho + \frac{k}{e}u_\rho) = x_\tau + \frac{k}{e}u_\tau \in A_\tau.$$

□

The following theorem reduces decomposing a block–rigid crq–group to a factorization problem of integers.

THEOREM 3.3. (Decomposability Criterion) *Let X be a block–rigid crq–group. If $X = X_1 \oplus X_2 \oplus \cdots \oplus X_t$ and $m_{\tau\, i} = m_\tau(X_i)$, then, for all types τ, $m_\tau(X) = \prod m_{\tau\, i}$ is a factorization such that:*

D1. *The integers $m_{\tau\, i}$ and $m_{\sigma\, j}$ are relatively prime whenever $i \neq j$.*

D2. $|\{i : m_{\tau\, i} > 1\}| < \mathrm{rk}(X(\tau))$.

Conversely, if $m_\tau(X) = \prod_{i=1}^t m_{\tau\, i}$ *is a factorization of the* $m_\tau(X)$ *such that the decomposability conditions* D1 *and* D2 *are satisfied, then there is a decomposition* $X = X_1 \oplus X_2 \oplus \cdots \oplus X_t \oplus X_{t+1}$ *such that* X_i *is rigid,* $\mathrm{T}_{\mathrm{cr}}(X_i) = \{\rho : m_{\rho\, i} > 1\}$ *and* $m_\tau(X_i) = m_{\tau\, i}$ *for* $i = 1, \dots, t$*, and* X_{t+1} *is completely decomposable.*

REMARK 3.4. Let $a_{ij} \in \mathbb{N}$ be such that the integers $e_j = \mathrm{lcm}_i\, a_{ij}$ are pairwise relatively prime. Then it is straightforward to check that $\mathrm{lcm}_i \prod_j a_{ij} = \prod_j e_j$. Hence if D1 holds then $e = |X/\mathrm{R}(X)| = \prod_i e_i$ automatically.

PROOF. (1) Suppose that $X = X_1 \oplus X_2 \oplus \cdots \oplus X_t$. Then $\mathrm{R}(X) = \bigoplus_{i=1}^t \mathrm{R}(X_i)$ and $X/R(X) = \bigoplus_{i=1}^t X_i/\mathrm{R}(X_i)$. Let $e = |X/\mathrm{R}(X)|$ and $e_i = |X_i/\mathrm{R}(X_i)|$. Since $X/R(X)$ is cyclic, so are the summands $X_i/\mathrm{R}(X_i)$, and the e_i must be pairwise relatively prime. Clearly, $e = e_1 e_2 \cdots e_t$. Let $A = \mathrm{R}(X)$ and $A_i = \mathrm{R}(X_i)$. Then $A = \bigoplus_{i=1}^t A_i$ and $A_\tau = \bigoplus_{i=1}^t (A_i)_\tau$. Furthermore, $\left(\overline{X_i}\right)_\tau \leq \overline{(A_i)_\tau}$. Hence $m_\tau(X_i) > 1$ implies that $\overline{(A_i)_\tau} \neq 0$ and $(A_i)_\tau \neq 0$. From this it follows that

$$|\{i : m_\tau(X_i) > 1\}| \leq \mathrm{rk}\, A_\tau = \mathrm{rk}\, X(\tau).$$

This establishes D2. The first condition D1 immediately follows from the facts that $e_i = \mathrm{lcm}_\rho\, m_\rho(X_i)$ and that the e_i are pairwise relatively prime.

(2) To prove the converse, we first observe that, for each i, the set $\{m_{\rho\, i} : \rho \in \mathbb{T}\}$ is admissible. In fact, $m_{\tau\, i} \mid m_\tau(X)$ and A1 follows immediately since $\{m_\tau(X)\}$ is admissible. To verify A2, suppose that some prime power p^k divides $m_{\sigma\, i}$. Then p^k divides m_σ. Applying A2 to the admissible set $\{m_\rho(X)\}$, we obtain that p^k divides $m_\tau(X)$ for some $\tau \neq \sigma$. But then p^k divides $m_{\tau\, i}$ because of D1.

We now use the Existence Theorem to obtain rigid crq–groups X_i such that $m_\tau(X_i) = m_{\tau\, i}$ and $\mathrm{T}_{\mathrm{cr}}(X_i) = \{\rho : m_{\rho\, i} > 1\}$. Here $e_i = |X_i/\mathrm{R}(X_i)| = \mathrm{lcm}_\rho\, m_{\rho\, i}$, and by D1 the e_i are relatively prime. This implies that $X' = \bigoplus_{i=1}^t X_i$ is again a crq–group. Furthermore, $\mathrm{rk}\,\mathrm{R}(X')_\tau = \bigoplus_{i=1}^t \mathrm{rk}\,\mathrm{R}(X_i)_\tau \leq |\{i : m_{\tau\, i} > 1\}| \leq \mathrm{rk}\, X(\tau)$. Hence there exists a completely decomposable group X_{t+1} such that $\mathrm{R}(X') \oplus X_{t+1} \cong \mathrm{R}(X)$. It follows that $X \cong_n X' \oplus X_{t+1}$ by the Product Formula 3.1 and the Near–Isomorphism Criterion 2.4. An application of Arnold's Theorem [**1**] concludes the proof. $\square$

A special decomposition, called **main decomposition**, always exists.

THEOREM 3.5. (Main Decomposition) *Let* X *be a block–rigid crq–group. Then there exists a decomposition* $X = Y \oplus Z$ *such that* Y *is a rigid crq–group,* Z *is completely decomposable and* $\tau \in \mathrm{T}_{\mathrm{cr}}(Y)$ *if and only if* $m_\tau(Y) = m_\tau(X) > 1$*. In the main decomposition, the group* Y *is unique up to near–isomorphism and* Z *is unique up to isomorphism.*

PROOF. The factorization $m_\tau(X) = m_{\tau\,1}$ obviously satisfies D1 and D2. By 3.3, $X = Y \oplus Z$ where Z is completely decomposable, and Y is a rigid crq–group with $\mathrm{T_{cr}}(Y) = \{\rho : m_\rho(X) > 1\}$. Therefore the regulator of Y is determined up to isomorphism by X. Also $\mathrm{R}(X) = \mathrm{R}(Y) \oplus Z$ hence Z is determined up to isomorphism by X. Further, $m_\tau(Y) = m_\tau(X)$ is determined by X and the Near–Isomorphism Criterion now says that Y is determined up to near–isomorphism. □

An easy consequence is a criterion for the existence of rank–one direct summands.

COROLLARY 3.6. *A block–rigid crq–group X has no rank–one direct summands if and only if X is rigid and $m_\tau(X) > 1$ for all $\tau \in \mathrm{T_{cr}}(X)$.*

PROOF. Suppose that X has no rank–one direct summands. Then $X = Y$ in the main decomposition 3.5, so X is rigid. By 2.3 $m_\tau(X) > 1$ for all $\tau \in \mathrm{T_{cr}}(X)$. Conversely, let X be rigid and $m_\tau(X) > 1$ for all $\tau \in \mathrm{T_{cr}}(X)$. By way of contradiction, suppose that $X = Y \oplus Z$ with $\operatorname{rk} Z = 1$. Then $\tau := \operatorname{type} Z \in \mathrm{T_{cr}}(X)$ and $m_\tau(X) = m_\tau(Y) \cdot m_\tau(Z) = m_\tau(Y) > 1$, so $\tau \in \mathrm{T_{cr}}(Y)$. Now $1 = \operatorname{rk} X(\tau) = \operatorname{rk}(Y(\tau) \oplus Z) = 2$, a contradiction. □

The next theorem tells how the indecomposability of a block–rigid crq–group can be recognized by its near–isomorphism invariants.

THEOREM 3.7. (Indecomposability Criterion) *Let X be a block–rigid crq–group. Then X is directly indecomposable if and only if X is rigid, and there is no non–trivial partition $\mathrm{T_{cr}}(X) = T_1 \cup T_2$ such that $\gcd(m_\sigma(X), m_\tau(X)) = 1$ whenever $\sigma \in T_1$ and $\tau \in T_2$.*

PROOF. (1) The existence of the main decomposition implies at once that an indecomposable block–rigid crq–group must be rigid.

(2) Suppose that X is a decomposable rigid crq–group, $X = X_1 \oplus X_2$, $X_1, X_2 \neq 0$. Then $\mathrm{R}(X) = \mathrm{R}(X_1) \oplus \mathrm{R}(X_2)$ and we obtain a non–trivial partition $\mathrm{T_{cr}}(X) = T_1 \cup T_2$ for $T_i = \mathrm{T_{cr}}(X_i)$. We now have $m_\tau(X) = m_\tau(X_1) \cdot m_\tau(X_2)$, and since T_1 and T_2 are disjoint, it follows that $m_\tau(X) = m_\tau(X_1)$ and $m_\tau(X_2) = 1$ for $\tau \in T_1$, while $m_\tau(X) = m_\tau(X_2)$ and $m_\tau(X_1) = 1$ for $\tau \in T_2$. Furthermore, $X/\mathrm{R}(X) \cong X_1/\mathrm{R}(X_1) \oplus X_2/\mathrm{R}(X_2)$, and since $X/\mathrm{R}(X)$ is cyclic, it follows that $|X_1/\mathrm{R}(X_1)|$ and $|X_2/\mathrm{R}(X_2)|$ are relatively prime. Since the $m_\tau(X_i)$ are divisors of the indices $|X_i/\mathrm{R}(X_i)|$, we conclude that $\gcd(m_\sigma(X), m_\tau(X)) = 1$ whenever $\sigma \in T_1$ and $\tau \in T_2$. Since the $m_\tau(X)$, $\tau \in T_i$, are just the invariants of X_i, they certainly are admissible.

(3) Conversely, let X be a rigid crq–group and suppose that $\mathrm{T_{cr}}(X) = T_1 \cup T_2$ is a non–trivial partition such that $\gcd(m_\sigma(X), m_\tau(X)) = 1$ whenever $\sigma \in T_1$ and $\tau \in T_2$. Set $m_{\tau\,i} = m_\tau(X)$ if $\tau \in T_i$ and set $m_{\tau\,i} = 1$ otherwise. Then the Decomposability Criterion shows that X is not directly indecomposable. □

We record an easy consequence.

COROLLARY 3.8. *Let X be a block–rigid almost completely decomposable group with primary cyclic regulating quotient. Then X is directly indecomposable if and only if it has no rank–one direct summands.*

The indecomposability of a block–rigid almost completely decomposable group is easily recognized by associating a certain graph with the group.

DEFINITION 3.9. (Blagoveshchenskaya) *Let X be a block–rigid crq–group. The **frame** of X is a graph whose vertices are the elements of $\mathrm{T}_{\mathrm{cr}}(X)$ and two vertices σ and τ are joined by an edge if and only if $\gcd(m_\sigma(X), m_\tau(X)) > 1$. More generally, the **frame** of a set of integers $\{m_1, m_2, \ldots, m_k\}$ is the graph whose vertices are $\{1, 2, \ldots, k\}$ and i and j are joined by an edge if $\gcd(m_i, m_j) > 1$.*

The following observation is clear from the Indecomposability Criterion.

LEMMA 3.10. *A rigid crq-group is indecomposable if and only if its frame is connected.*

Based on the preceding results, one can investigate the possible direct decompositions of block–rigid crq–groups. Of special interest are decompositions into indecomposable subgroups. The question boils down to the following factorization problem.

DECOMPOSITION PROBLEM:

Given

(I) r_i, $1 \le i \le k$, a set of positive integers,

(II) m_i, $1 \le i \le k$, a set of positive integers with the property that for every prime p, every power of p which divides some m_i, divides at least one more of the m_i.

find all factorizations

$$m_i = m_{1i} \cdot m_{2i} \cdots m_{ti}, \quad m_{ri} > 0,$$

such that

(i) m_{ri} and m_{sj} are relatively prime whenever $r \ne s$,

(ii) for every i, $|\{m_{ri} : m_{ri} > 1\}| \le r_i$,

(iii) for each r, the frame of $\{m_{r1}, m_{r2}, \ldots, m_{rk}\}$ is connected.

REMARK 3.11. The given data belong to an almost completely decomposable group X with critical typeset $\{\tau_1, \tau_2, \ldots, \tau_k\}$ that is an antichain and such that $\tau_i(p) < \infty$ whenever $p \mid m_i$. The m_i are the invariants $m_{\tau_i}(X)$ and the r_i are the ranks of the constituents $X(\tau_i)/X^\sharp(\tau_i) = X(\tau_i)$. The condition (II) is just the admissibility condition A2.

The factorizations correspond to direct decompositions. It is easy to see that the admissibility condition A2 for the m_{rj} is automatically satisfied because of property (i). The last condition assures that the summands are indecomposable.

REMARK 3.12. Using the unique prime factorizations of the m_j, each direct summand corresponds to a combination of prime factors of $m = \operatorname{lcm}_j m_j$.

In fact, if the set of invariants $\{m_{r1}, m_{r2}, \ldots, m_{rk}\}$ belongs to a summand, then the corresponding combination of prime factors of m is the set of prime factors of $\operatorname{lcm}_j m_{rj}$, and if $\{p_1, p_2, \ldots, p_u\}$ is a combination of prime factors of m, then we get the invariant m_{rj} as the product of the highest prime powers p_v^w which divide m_j. *In particular, the exponents of the prime powers involved in the invariants are irrelevant to decomposition up to near–isomorphism.*

We will work out an example, storing the information in tables in the following manner.

Decomposition Table

types	τ_1	τ_2	$\cdots$	τ_j	$\cdots$	τ_k
Summand 1	m_{11}	m_{12}	$\cdots$	m_{1j}	$\cdots$	m_{1k}
Summand 2	m_{21}	m_{22}	$\cdots$	m_{2j}	$\cdots$	m_{2k}
$\cdots$	$\cdots$	$\cdots$	$\cdots$	$\cdots$	$\cdots$	$\cdots$
Summand r	m_{r1}	m_{r2}	$\cdots$	m_{rj}	$\cdots$	m_{rk}
$\cdots$	$\cdots$	$\cdots$	$\cdots$	$\cdots$	$\cdots$	$\cdots$
Summand t	m_{t1}	m_{t2}	$\cdots$	m_{tj}	$\cdots$	m_{tk}
invariants	m_1	m_2	$\cdots$	m_j	$\cdots$	m_k
ranks	r_1	r_1	$\cdots$	r_j	$\cdots$	r_k

We agree to enter a value $m_{rj} = 1$ if and only if τ_j is a critical type of the summand r. Thus the critical typeset of Summand r can be seen by looking at the boxes in the Factorization Table which are filled. The rank of the indecomposable Summand r equals the number of boxes which are filled in row r of the table.

EXAMPLE 3.13. We exhibit the eight up to near–isomorphism possible direct decompositions of a block–rigid crq–group X of rank 9 with $k = 5$ critical types $\{\tau_1, \tau_2, \tau_3, \tau_4, \tau_5\}$, ranks of the homogeneous constituents $r_i = \operatorname{rk}(X(\tau_i))$ and invariants $m_i = m_{\tau_i}(X)$ as indicated in the tables.

The group is initially given in its main decomposition.

types	τ_1	τ_2	τ_3	τ_4	τ_5
Summand 1	$2^2 \cdot 3^3 \cdot 5$	$5 \cdot 7^2 \cdot 11$	$2^2 \cdot 3^3$	$3^2 \cdot 7^2$	$2 \cdot 11$
Summand 2	1				
Summand 3		1			
Summand 4			1		
Summand 5				1	
invariants	$2^2 \cdot 3^3 \cdot 5$	$5 \cdot 7^2 \cdot 11$	$2^2 \cdot 3^3$	$3^2 \cdot 7^2$	$2 \cdot 11$
ranks	2	2	2	2	1

SOLUTION. (1) The critical types are assumed to have no infinities at the primes $2, 3, 5, 7, 11$, and to be pairwise incomparable; otherwise they are arbitrary. This assures that the admissibility condition A1 is satisfied. Since each prime power of the regulator index

$$e = \operatorname{lcm}_i m_i = 2^2 \cdot 3^3 \cdot 5 \cdot 7^2 \cdot 11$$

appears in at least two of the invariants in the table, the condition A2 is also satisfied and the invariants are legitimate. The summands 2 through 5 are rank-one groups of types τ_1, τ_2, τ_3, τ_4 respectively. It is obvious that the frame of the first summand is connected, and thus the summand is indecomposable.

(2) Since the constituent of type τ_5 has rank 1, there is one summand which has the invariant $2 \cdot 11$ in type τ_5. We always take the first summand to have this property. Its invariants must contain all powers of 2 and 11. The next table contains the case in which the invariants of the first summand are the smallest possible.

types	τ_1	τ_2	τ_3	τ_4	τ_5
Summand 1	2^2	11	2^2		$2 \cdot 11$
Summand 2	$3^3 \cdot 5$	$5 \cdot 7^2$	3^3	$3^2 \cdot 7^2$	
Summand 3				1	
invariants	$2^2 \cdot 3^3 \cdot 5$	$5 \cdot 7^2 \cdot 11$	$2^2 \cdot 3^3$	$3^2 \cdot 7^2$	$2 \cdot 11$
ranks	2	2	2	2	1

(3) We can adjoin to the minimal invariants of the first summands the powers of any combination of $\{3, 5, 7\}$. There are 8 such combinations:

$$\emptyset, \quad \{3\}, \quad \{5\}, \quad \{7\}, \quad \{3,5\}, \quad \{3,7\}, \quad \{5,7\}, \quad \{3,5,7\}.$$

Each results in a decomposition into indecomposable groups. The ranks of the summands in the eight decompositions are respectively

$$[4,4,1], \ [4,4,1], \ [5,3,1], \ [5,4], \ [5,2,1,1], \ [5,2,1,1], \ [5,3,1], \ [5,1,1,1,1].$$

The number of summands range from 2 to 5.

The combination $\{3, 5, 7\}$ corresponds to the main decomposition above. The minimal invariants correspond to the combination $\emptyset$. We list in a single table the invariants of the remaining six decompositions. The rank-one summands are combined into a single completely decomposable summand

Combination	τ_1	τ_2	τ_3	τ_4	τ_5
$\{3\}$	$2^2 \cdot 3^3$	11	$2^2 \cdot 3^3$	3^2	$2 \cdot 11$
	5	$5 \cdot 7^2$		7^2	
			1		
$\{5\}$	$2^2 \cdot 5$	$5 \cdot 11$	2^2		$2 \cdot 11$
	3^3	7^2	3^3	$3^2 \cdot 7^2$	
				1	
$\{7\}$	2^2	$7^2 \cdot 11$	2^2	7^2	$2 \cdot 11$
	$3^3 \cdot 5$	5	3^3	3^2	
$\{3,5\}$	$2^2 \cdot 3^3 \cdot 5$	$5 \cdot 11$	$2^2 \cdot 3^3$	3^2	$2 \cdot 11$
		7^2		7^2	
	1		1		
$\{3,7\}$	$2^2 \cdot 3^3$	$7^2 \cdot 11$	$2^2 \cdot 3^3$	$3^2 \cdot 7^2$	$2 \cdot 11$
	5	5			$2 \cdot 11$
			1	1	
$\{5,7\}$	$2^2 \cdot 5$	$5 \cdot 7^2 \cdot 11$	2^2	7^2	$2 \cdot 11$
	3^3		3^3	3^2	
		1			

A.L.S. Corner gave the following striking illustration of pathology in decompositions of torsion–free abelian groups. *Given integers $n \geq k \geq 1$, there exists a torsion–free group X of rank n such that for any partition $n = r_1 + r_2 + \cdots + r_k$, there is a decomposition of X into a direct sum of k subgroups of ranks $r_1, r_2, \ldots, r_k$ respectively.* His examples are block–rigid crq–groups, and can be exhibited and explained easily with the machinery developed here.

EXAMPLE 3.14. (A.L.S. Corner, [**8**, Vol. II, 90.2]) Given integers $n \geq k \geq 1$, choose distinct primes

$$q_1, q_2, \ldots, q_{n-k}$$

and pairwise incomparable types

$$\tau, \tau_1, \tau_2, \ldots, \tau_{n-k}$$

with $\tau(q_j) < \infty$, and $\tau_i(q_j) < \infty$ for all i, j. Let X be the block–rigid crq–group given by the following factorization table which represents the main decomposition.

types	τ	τ_1	τ_2	$\cdots$	τ_{n-k}
Summand 1	$q_1q_2\cdots q_{n-k}$	q_1	q_2	$\cdots$	q_{n-k}
Summand 2	1			$\cdots$	
Summand 3	1			$\cdots$	
$\cdots$	$\cdots$	$\cdots$	$\cdots$	$\cdots$	$\cdots$
Summand k	1			$\cdots$	
ranks	k	1	1	$\cdots$	1

The rank of this group is $k + (n-k)\cdot 1 = n$. The main summand, Summand 1, is indecomposable by 3.7. Suppose that

$$n = r_1 + r_2 + \cdots + r_k, \quad r_i \geq 1,$$

is a partition of n. *We claim that X has a decomposition into indecomposable direct summands of ranks $r_1, r_2, \ldots, r_k$.* In fact, we have $n - k = (r_1 - 1) + (r_2 - 1) + \cdots + (r_k - 1)$. Partition the set $q_1, q_2, \ldots, q_{n-k}$ into $Q_1, Q_2, \ldots, Q_{n-k}$ subsets of sizes $r_1 - 1, r_2 - 1, \ldots, r_k - 1$ respectively. Form a factorization table with k rows as follows. The i–th row has entry $\prod\{q : q \in Q_i\}$ in the first column labeled τ, q_j in column τ_j provided $q_j \in Q_i$, and no entries otherwise. This table clearly satisfies the conditions of the Decomposition Problem and constitutes a decomposition of X into k direct summands of ranks $r_1, r_2, \ldots, r_k$. $\square$

Our main goal are two general theorems which constitute answers to Problems 67 and 68 of Fuchs [**8**, Vol.II, p. 183]. The following lemma supplies the necessity of the conditions in both theorems.

LEMMA 3.15. (Blagoveshchenskaya) *Suppose that X is a block–rigid crq–group of rank n having a direct decomposition into a sum of t rigid subgroups. Then in any other decomposition of X, the rank of an indecomposable summand cannot exceed $n - t + 1$, and the number of summands in a decomposition with indecomposable summands is greater or equal to $\frac{n}{n-t+1}$.*

PROOF. (1) Suppose that

$$X = X_1 \oplus \cdots \oplus X_k \oplus X_{k+1} \oplus \cdots \oplus X_t = Y \oplus Z$$

whereby Y is indecomposable and $X_1, \ldots, X_t$ are rigid, $\operatorname{rk} X_1 \geq \operatorname{rk} X_2 \geq \operatorname{rk} X_k \geq 2$, and $\operatorname{rk} X_{k+1} = \cdots = \operatorname{rk} X_t = 1$. By the Product Formula we have

$$m_\tau(X) = m_\tau(X_1)m_\tau(X_2)\cdots m_\tau(X_k) = m_\tau(Y)m_\tau(Z).$$

Set $T = \mathrm{T}_{\mathrm{cr}}(Y)$, and $T_i = T \cap \mathrm{T}_{\mathrm{cr}}(X_i)$. Let T_i, $i \in I$, be the non–void ones among the sets T_i, $1 \leq i \leq t$. Note that

$$\mathrm{T}_{\mathrm{cr}}(X) = \bigcup_{1\leq i\leq t} \mathrm{T}_{\mathrm{cr}}(X_i) = \mathrm{T}_{\mathrm{cr}}(Y) \cup \mathrm{T}_{\mathrm{cr}}(Z),$$

hence $T = \bigcup_{i\in I} T_i$.

(2) Suppose that

$$T = \left(\bigcup_{i\in I_1} T_i\right) \cup \left(\bigcup_{j\in I_2} T_j\right)$$

with $\bigcup_{i\in I_1} T_i \neq \emptyset$ and $\bigcup_{j\in I_2} T_j \neq \emptyset$. Assume that $\bigcup_{i\in I_1} T_i \cap \bigcup_{j\in I_2} T_j = \emptyset$. We will derive a contradiction. Let $\sigma \in \bigcup_{i\in I_1} T_i$. Then $\sigma \in T = \mathrm{T_{cr}}(Y)$ and since Y is indecomposable, $m_\sigma(Y) > 1$. On the other hand, since $\sigma \notin \bigcup_{i\in I_2} T_i$, it follows that $m_\sigma(X_j) = 1$ for all $j \in I_2$. The factorization

$$m_\sigma(X) = m_\sigma(X_1)m_\sigma(X_2)\cdots m_\sigma(X_k) = \prod_{i\in I_1} m_\sigma(X_i) = m_\sigma(Y)m_\sigma(Z)$$

now gives that $m_\sigma(Y) \mid \prod_{i\in I_1} m_\sigma(X_i)$. By symmetry, if $\tau \in \bigcup_{j\in I_2} T_j$, then $m_\tau(Y) \mid \prod_{j\in I_2} m_\tau(X_j)$. By D1, $\gcd(\prod_{i\in I_1} m_\sigma(X_i), \prod_{j\in I_2} m_\tau(X_j)) = 1$ and therefore $\gcd(m_\sigma(Y), m_\tau(Y)) = 1$. This contradicts the indecomposability of Y since $T = \left(\bigcup_{i\in I_1} T_i\right) \cup \left(\bigcup_{j\in I_2} T_j\right)$ would constitute a partition of T of the sort prohibited by 3.7.

(3) Since $T = \bigcup_{i\in I} T_i$ and $\left(\bigcup_{i\in I_1} T_i\right) \cap \left(\bigcup_{j\in I_2} T_j\right) \neq \emptyset$ whenever $T = \left(\bigcup_{i\in I_1} T_i\right) \cup \left(\bigcup_{j\in I_2} T_j\right)$, it easily follows by induction that

$$|T| \leq \sum_{i\in I} |T_i| - (|I| - 1).$$

Since the X_i are all rigid groups, $\operatorname{rk} X_i = |\,\mathrm{T_{cr}}(X_i)| \geq |T_i|$ for all i. Hence

$$\begin{aligned}\operatorname{rk} Y = |T| &\leq \textstyle\sum_{i\in I} |T_i| - |I| + 1\\ &\leq \textstyle\sum_{i\in I} |\operatorname{rk} X_i| - |I| + 1\\ &\leq \textstyle\sum_{i\in I} |\operatorname{rk} X_i| - |I| + 1 + \sum_{i\notin I} |\operatorname{rk} X_i| - (t - |I|)\\ &= \textstyle\sum_{1\leq i\leq t} |\operatorname{rk} X_i| - t + 1 = n - t + 1.\end{aligned}$$

Thus $\operatorname{rk} Y \leq n - t + 1$ as claimed. It follows that in any decomposition into a sum of indecomposables with m non–zero summands, $n \leq m(n - t + 1)$ which is the last claim. □

The first of Blagoveshchenskaya's theorems solving Fuchs' Problem 67 is a follows.

THEOREM 3.16. (Blagoveshchenskaya) *Suppose that* $1 \leq m \leq M \leq n$ *are natural numbers. If there exists a block–rigid crq–group admitting direct decompositions into sums of* m *and of* M *indecomposable subgroups, then* $m \geq n/(n - M + 1)$. *Conversely, if* $m \geq n/(n - M + 1)$, *then there is an almost completely decomposable group which has decompositions into* $m, m+1, m+2, \ldots, M$ *indecomposable subgroups.*

PROOF. The necessity of the condition is exactly the second claim of 3.15 for $t = M$. The sufficiency of the condition is proven below (Lemma 3.18). □

We first exhibit a special kind of group which will be used for the existence part of 3.16.

LEMMA 3.17. *Let a set of primes* $\{p_1, p_2, \dots, p_{k-1}\}$, $k \geq 2$, *and a set of pairwise incomparable types* $\{\tau_1, \tau_2, \dots, \tau_k\}$ *be given such that* $\tau_i(p_j) < \infty$ *for all* i, j. *Define*

$$m_1 = p_1 \cdot p_2 \cdots p_{k-1},$$

$$m_i = p_{i-1} \cdot p_i \cdots p_{k-1}, \quad 2 \leq i \leq k.$$

There exists a directly indecomposable rigid crq-group $Y[k]$ *with rank* $\operatorname{rk} Y[k] = k$, *critical typeset* $\mathrm{T}_{\mathrm{cr}}(Y[k]) = \{\tau_1, \tau_2, \dots, \tau_k\}$, *and invariants* $m_{\tau_i}(Y[k]) = m_i$.

Let X_0 *be any indecomposable rigid crq-group with* $\mathrm{T}_{\mathrm{cr}}(X_0) = \{\tau_1, \tau_2, \dots, \tau_k\}$ *such that none of the invariants of* X_0 *is divisible by any* p_i, $1 \leq i \leq k-1$. *Then*

$$X_0 \oplus Y[k] \cong_n X_1 \oplus Y[k-1] \oplus Z_1$$

where Z_1 *is a rank-one group of type* τ_k *and* X_1 *is a rigid indecomposable crq-group with* $\mathrm{T}_{\mathrm{cr}}(X_1) = \mathrm{T}_{\mathrm{cr}}(X_0)$ *and invariants* $m_{X_1}(\tau_i) = m_{X_0}(\tau_i) \cdot p_{k-1}$.

Inductively, we obtain for $1 \leq i \leq k-1$ *with* $Y[1]$ *a rank-one group of type* τ_1,

$$X_0 \oplus Y[k] \cong_n X_i \oplus Y[k-i] \oplus Z_1 \oplus Z_2 \oplus \cdots \oplus Z_i.$$

PROOF. We use 3.3 and display the inductive change from one decomposition to the next in a factorization table.

	τ_1	τ_2	$\cdots$	τ_i	$\cdots$	τ_k
$Y[k]$	$p_1 \cdots p_{k-1}$	$p_1 \cdots p_{k-1}$	$\cdots$	$p_{i-1} \cdots p_{k-1}$	$\cdots$	p_{k-1}
X_0	m_{01}	m_{02}	$\cdots$	m_{0i}	$\cdots$	m_{0k}
$Y[k-1]$	$p_1 \cdots p_{k-2}$	$p_1 \cdots p_{k-2}$	$\cdots$	$p_{i-1} \cdots p_{k-2}$	$\cdots$	
X_0	$m_{01}p_{k-1}$	$m_{02}p_{k-1}$	$\cdots$	$m_{0i}p_{k-1}$	$\cdots$	$m_{0k}p_{k-1}$
Z_1						1
ranks	2	2	$\cdots$	2	$\cdots$	2

□

LEMMA 3.18. (Blagoveshchenskaya) *Let* m, M, n *be integers such that*

$$1 \leq m \leq M \leq n, \quad m(n-M+1) \geq n.$$

Then there exists a block–rigid crq-group X *which possesses decompositions into direct sums of* $m, m+1, m+2, \dots, M$ *indecomposable groups.*

PROOF. Let s be the least integer satisfying $s(n-M+1) \geq n$. Let $Y[n-M+1]$ be the group constructed in Lemma 3.17. The group X will be a direct sum of $s-1$ groups $Y[n-M+1]$ and an additional group of the same sort such that the ranks add up to n. However, for X to have a cyclic regulator quotient, it is necessary that the summands have relatively prime regulator indices. We therefore exhibit the regulator indices and write $Y[k, P]$ for the group $Y[k]$ of 3.17 with $P = p_1 p_2 \cdots p_{k-1}$. In detail, X is constructed as follows.

Let $Y_i = Y[n - M + 1, P_i]$, $i = 1, 2, \ldots, s - 1$ such that the regulator indices P_i are pairwise relatively prime, and the critical typesets are identical. Let $Y_s = Y[n - (s - 1)(n - M + 1), P_s]$ with regulator index P_s relatively prime to all the others, and critical typeset a subset of the other common critical typeset. Let $X = Y_1 \oplus Y_2 \oplus \cdots \oplus Y_s$. Then X is a block–rigid crq–group, $X = Y_1 \oplus Y_2 \oplus \cdots \oplus Y_s$ is a decomposition into $s \leq m$ indecomposable summands, and $\operatorname{rk} X = (s-1)(n-M+1)+(n - (s - 1)(n - M + 1)) = n$. Lemma 3.17 shows how rank–one summands can be peeled off one by one to obtain decompositions into $s + 1, s + 2, \ldots$ indecomposable summands until the main decomposition is reached in which case the number of summands is $1 + (n - (n - M + 1)) = M$ as claimed. $\square$

Blagoveshchenskaya's second theorem, solving Fuchs' Problem 68, is as follows.

THEOREM 3.19. (Blagoveshchenskaya) *Suppose that* $n = r_1 + r_2 + \cdots + r_s = r'_1 + r'_2 + \cdots + r'_t$ *are two partitions of* n *into a sum of natural integers. Then there exists a block–rigid crq–group of rank* n *which is the direct sum of indecomposable subgroups of ranks* $r_1, r_2, \ldots, r_s$, *and also is the direct sum of indecomposable summands of ranks* $r'_1, r'_2, \ldots, r'_s$ *if and only if* $r_i \leq n-t+1$ *for all* $i = 1, 2, \ldots, s$ *and* $r'_j \leq n - s + 1$ *for all* $j = 1, 2, \ldots, t$.

PROOF. The necessity of the conditions is immediate from 3.15. The sufficiency of the condition is contained in [**3**]. $\square$

References

1. D. Arnold, *Finite Rank Torsion Free Abelian Groups and Rings*, Lecture Notes in Mathematics **931**, Springer Verlag, 1982.
2. M.C.R. Butler, *A class of torsion-free abelian groups of finite rank*, Proc. London Math. Soc.(3) **15** (1965), 680–698.
3. E.A. Blagoveshchenskaya, *Direct decompositions of torsion-free abelian groups of finite rank*, Zap. Nauk. Sem. Leningrad. Otdel. Mat. Inst. V. A. Steklova **132** (1983), 17–25.
4. E.A. Blagoveshchenskaya, *Graphic interpretation of some torsion-free abelian groups of finite rank*, (1992), preprint.
5. E.A. Blagoveshchenskaya and A.V. Yakovlev, *Direct decompositions of torsion-free abelian groups of finite rank*, Leningrad Math. J. **1** (1990), 117–136.
6. R. Burkhardt, *On a special class of almost completely decomposable groups I*, Abelian Groups and Modules, Proceedings of the Udine Conference 1984, CISM Courses and Lecture Notes **287**, 141–150.
7. M. Dugas and E. Oxford, *Near isomorphism invariants for a class of almost completely decomposable groups*, Abelian Groups, Proceedings of the 1991 Curaçao Conference, Lecture Notes in Pure and Applied Mathematics, **146** (1993), 129–150.
8. L. Fuchs, *Infinite Abelian Groups, Vol. I, II,* Academic Press, 1970 and 1973.
9. E.L. Lady, *Almost completely decomposable torsion-free abelian groups*, Proc. Amer. Math. Soc. **45** (1974), 41–47.
10. A. Mader and C. Vinsonhaler, *Classifying almost completely decomposable groups*, J. Algebra, to appear.

Department of Mathematics, St. Petersburg State Technical University, Polytechnicheskaya 29, St. Petersburg 195251, Russia
E-mail address: E-mail address: kat@iiii.spb.su

Department of Mathematics, 2565 The Mall, University of Hawaii, Honolulu, HI 96822, USA
E-mail address: adolf@math.hawaii.edu

Contemporary Mathematics
Volume **171**, 1994

On the Divisible Parts of Quotient Groups

ANDREAS BLASS

ABSTRACT. We study the possible cardinalities of the divisible part of G/K when the cardinality of K is known and when, for all countable subgroups C of G, the divisible part of G/C is countable.

Introduction

Let G be an abelian group that is reduced, i.e., its divisible part $\mathrm{Div}(G)$ is zero. A quotient group G/K need not be reduced. This paper is about the question how big $\mathrm{Div}(G/K)$ can be, relative to the cardinality $|K|$ of K.

John Irwin has suggested, as a natural weakening of freeness, the concept of a *fully starred* torsion-free group, i.e., a torsion-free, abelian group G such that, for all subgroups K, $|\mathrm{Div}(G/K)| \leq |K|$. He asked whether, to test whether G is fully starred, it suffices to check the definition for countable K. At first sight, this seems unlikely; how should the quotients by countable subgroups influence the quotients by larger subgroups? We shall show, however, that the answer to Irwin's question is affirmative if $|K| < \aleph_\omega$. We shall also show that it is consistent with the usual axioms of set theory (ZFC, i.e., Zermelo-Fraenkel set theory, including the axiom of choice) that the answer is negative for $|K| = \aleph_\omega$ but positive for many larger values of $|K|$. If a certain very large cardinal hypothesis is consistent, then it is also consistent that the answer to Irwin's question is affirmative for $|K| = \aleph_\omega$. It remains an open problem whether an affirmative answer, for all groups regardless of cardinality, is consistent (relative to some large cardinals).

The following table summarizes the results. The middle column gives what can be said about $|\mathrm{Div}(G/K)|$ when $|K|$ is as in the first column, when $\mathrm{Div}(G/C)$ is countable for all countable $C \leq G$, and when the set-theoretic hypothesis in the right column is satisfied. (The assumption about countable C is irrelevant in the first, very elementary line of the table.)

1991 *Mathematics Subject Classification.* 20K20, 03E05, 03E35, 03E55, 03E75.
Partially supported by NSF grant DMS-9204276 and NATO grant LG921395.
This paper is in final form, and no version of it will be submitted for publication elsewhere.

$\lvert K\rvert = \kappa$	$\lvert \mathrm{Div}(G/K)\rvert$	Hypothesis
Arbitrary	$\leq \kappa^{\aleph_0}$	None
$\kappa < \aleph_\omega$	$\leq \kappa$	None
$\mathrm{cf}(\kappa) \neq \omega$	$\leq \kappa$	No inner model with measurable cardinal
$\mathrm{cf}(\kappa) = \omega$	$\leq \kappa^+$	No inner model with measurable cardinal
$\aleph_\omega$	$\leq \aleph_\omega$	Chang's conjecture for $(\aleph_{\omega+1}, \aleph_\omega)$
$\aleph_\omega$	possibly $= \aleph_{\omega+1}$	$V = L$
$\aleph_\omega$	$< \aleph_{\aleph_4}$	None

Some of the hypotheses in the right column can be weakened. See Theorems 10 and 12 for sharper statements.

1. Background

Throughout this paper, all groups are abelian and reduced. (If we were to allow non-reduced groups, Proposition 1 would have to be amended by changing $|K|^{\aleph_0}$ to $\max(|K|^{\aleph_0}, |\mathrm{Div}(G)|)$. Our other group-theoretic results would remain valid, for their hypotheses imply that $\mathrm{Div}(G)$ is countable. Their proofs would be modified by working with the reduced part of G rather than G itself.)

In this section, we present, for motivation and orientation, some elementary results and examples. No novelty is claimed for any of this material.

As Irwin pointed out when he suggested the study of fully starred torsion-free groups, this class of groups includes all free groups. Indeed, if G is freely generated by a basis B and if K is a subgroup of G, then by expressing each element of K as a combination of (finitely many) elements of B, we obtain a subset B_0 of B, no larger than K in cardinality, such that K is included in the subgroup G_0 of G generated by B_0. Then $G/K \cong (G_0/K) \oplus F$ where F is freely generated by $B - B_0$. Therefore the divisible part of G/K coincides with that of G_0/K, whose cardinality is at most that of G_0, which equals that of K.

For non-free G, on the other hand, the divisible part of G/K may well be larger than K. For example, if G is the product of countably many infinite cyclic groups and K is their direct sum (embedded in the product in the obvious way), then the divisible part of G/K is easily seen to have the cardinality of the continuum, even though K is countable. (Notice that, if the continuum hypothesis holds, then the group G in this example is almost free in the sense that all subgroups of smaller cardinality are free [**10**].) Another example is obtained by taking G to be the additive group of p-adic integers and K the subgroup of ordinary integers; then G/K has the cardinality of the continuum and is divisible.

For countable K, the preceding examples achieve the largest possible cardinality for $\mathrm{Div}(G/K)$, as the following proposition shows.

PROPOSITION 1. *For any (reduced) group G and any subgroup K, we have* $|\mathrm{Div}(G/K)| \leq |K|^{\aleph_0}$.

PROOF. Let D be the pre-image of $\mathrm{Div}(G/K)$ in G, so $K \subseteq D \subseteq G$ and D/K is divisible. We shall prove the proposition by producing a one-to-one function from D into the set of countable sequences of elements of K. As D/K is divisible, we have, for each $d \in D$ and each positive integer n, some $a_n(d) \in D$ and some $b_n(d) \in K$ such that $d = n \cdot a_n(d) + b_n(d)$. If G is torsion-free, then the map we seek sends each $d \in D$ to the sequence $(b_n(d))_{n\in\mathbb{N}}$. To see that it is one-to-one, suppose d and d' gave rise to the same sequence. Then, for each n, we would have $d - d' = n \cdot (a_n(d) - a_n(d'))$. Thus, $d - d'$ belongs to the divisible part of G (here we use that G is torsion-free), which is zero as G is reduced.

To avoid the assumption that G is torsion-free, it suffices to include, in the sequence associated to d, not only all the $b_n(d)$ but also all $b_n(a_k(d))$, all $b_n(a_k(a_l(d)))$, etc. Now, if d and d' give rise to the same sequence, then the subgroup of G generated by $d - d'$ and all elements of the forms $a_k(d) - a_k(d')$, $a_k(a_l(d)) - a_k(a_l(d'))$, etc. is divisible and therefore zero. □

The referee has pointed out that the simpler function used in the first paragraph of the proof can also be used in the presence of torsion if one invokes the well-known structure theorem for divisible groups and considers a basis rather than the whole group.

The proposition justifies the first line of the table in the introduction. As indicated there, this result, unlike the later ones, does not depend on any assumption about $\mathrm{Div}(G/C)$ for countable C, but only on the assumption (obviously needed) that G is reduced.

We note that, by results of Cohen [**2**] and Solovay [**9**], the upper bound given by Proposition 1 is rather weak. Even when K is countable, the bound $|K|^{\aleph_0}$, which then equals the cardinality of the continuum, can be an arbitrarily large cardinal.

2. Some Combinatorial Set Theory

We follow the usual set-theoretic conventions whereby an ordinal number is the set of all smaller ordinal numbers and a cardinal number is the first ordinal of that cardinality. In particular, the cardinal $\aleph_0$ is identified with the first infinite ordinal number ω and with the set of all natural numbers.

For any set A, let $[A]^\omega$ be the set of all countably infinite subsets of A, partially ordered by the subset relation. We shall need some information about the cofinality $\mathrm{cf}([A]^\omega)$, i.e., the smallest possible size for a family of countable subsets of A such that every countable subset of A is included in one from that family. Of course this depends only on the cardinality of A, so we may assume that A is a cardinal. The following proposition summarizes the information we need about $\mathrm{cf}([\kappa]^\omega)$ for uncountable κ; it is taken from Section 4 of [**6**]. The hypothesis "the covering lemma over a model of GCH" that occurs in the second part of the proposition means that there is a class M of sets such that every member of a member of M is in M (one says M is *transitive*), all axioms of ZFC and the generalized continuum hypothesis (GCH) hold in M, and every uncountable set of ordinal numbers is included in a set of the same cardinality that is a member of M. We shall never need to use this definition of the covering lemma over a model of GCH, but it is relevant that this assumption is satisfied unless there are inner models with measurable cardinals [**3**,**4**], i.e., transitive classes M such that all axioms of ZFC and the statement "a measurable cardinal exists" hold in M. In fact, it is known that if this assumption fails then there are inner models with far stronger large cardinal properties than just a measurable cardinal; see for example [**8**].

PROPOSITION 2. *The equation* $\mathrm{cf}([\kappa]^\omega) = \kappa$ *holds for all uncountable cardinals* $\kappa < \aleph_\omega$. *If the covering lemma holds over a model of GCH, then the same equation holds for all cardinals* κ *of uncountable cofinality, while for uncountable* κ *of countable cofinality* $\mathrm{cf}([\kappa]^\omega) = \kappa^+$.

PROOF. See [**6**], Corollary 4.8, Lemma 4.10, and the proof of the latter. □

We note that the result obtained for the case of countable cofinality using the covering lemma is optimal; that is, for $\mathrm{cf}(\kappa) = \omega$, we cannot have $\mathrm{cf}([\kappa]^\omega) = \kappa$. Indeed, given a family $\mathcal{F}$ of κ countable subsets of κ and given an increasing ω-sequence of infinite cardinals θ_i with supremum κ, we can split $\mathcal{F}$ into subfamilies $\mathcal{F}_i$ of respective cardinalities θ_i, and we can find for each i an element $x_i \in \theta_{i+1}$ not in the union of $\mathcal{F}_i$ (for this union has size at most $\theta_i \cdot \aleph_0$). Then the countable set consisting of these x_i's is not included in any member of $\mathcal{F}$.

In the absence of the covering lemma, it is much more difficult to bound $\mathrm{cf}([\kappa]^\omega)$ for κ of cofinality ω. The following proposition is one of the surprising results of Shelah's recently developed theory of possible cofinalities. It is stated (with a hint about the proof) as Proposition 7.13 of [**1**], attributed to Baumgartner.

PROPOSITION 3. $\mathrm{cf}([\aleph_\omega]^\omega) < \aleph_{\aleph_4}$. □

The remainder of this section is devoted to a combinatorial principle closely related to the preceding cofinality considerations but, as we shall see, of more direct relevance to the size of the divisible parts of quotient groups.

For infinite cardinals $\kappa < \lambda$, we say that the λ *to* κ *compression principle* holds and we write $\mathrm{CP}(\lambda \to \kappa)$ if, for any λ-indexed family $(X_i)_{i\in\lambda}$ of countable subsets of κ, there is an uncountable set of indices, $Z \subseteq \lambda$, such that $\bigcup_{i\in Z} X_i$ is countable. Intuitively, this means that, if λ countable sets are packed (with overlapping) into a set of size κ (non-trivial packing, as $\kappa < \lambda$), then some uncountably many of those sets must have been packed into a countable set (so we have non-trivial packing on a smaller scale).

There is a trivial connection between the compression principle and the cofinalities considered above; we state it as a proposition for future reference.

PROPOSITION 4. *If* $\lambda > \mathrm{cf}([\kappa]^\omega)$ *then* $\mathrm{CP}(\lambda \to \kappa)$ *holds.*

PROOF. Fix a family $\mathcal{F}$ of strictly fewer than λ countable subsets of κ cofinal in $[\kappa]^\omega$, and fix λ countable subsets X_i of κ. Each X_i is included in some member of $\mathcal{F}$. As there are λ X_i's and fewer members of $\mathcal{F}$, uncountably many X_i's must be included in a single element of $\mathcal{F}$ and must therefore have a countable union. □

The converse of Proposition 4 is not provable, at least if sufficiently large cardinals are consistent. The counterexample involves Chang's conjecture. Originally, Chang's conjecture was that, for a countable first-order language, every structure of cardinality $\aleph_2$ in which a particular unary predicate symbol P denotes a set of cardinality $\aleph_1$ has an elementary submodel of cardinality $\aleph_1$ in which P denotes a set of cardinality $\aleph_0$. Generalizing this by changing $\aleph_2$ and $\aleph_1$ in the hypothesis to λ and κ, respectively, but leaving the conclusion unchanged, one has the Chang conjecture for (λ, κ). The conjecture can be restated in a form, more convenient for our purposes, that avoids model-theoretic notions. We adopt this restatement as the definition: *Chang's conjecture* for (λ, κ) is the assertion that, for any countably many functions f_n, each mapping some finite power λ^p of λ into κ (where p can depend on n), there is an uncountable subset H of λ such that each f_n restricted to H (more precisely, to H^p) has countable range. We shall need an elementary connection between Chang's conjecture and the compression principle and a (non-elementary) theorem from [**7**] giving the consistency of a particular instance of Chang's conjecture.

PROPOSITION 5. *Chang's conjecture for* (λ, κ) *implies* $\mathrm{CP}(\lambda \to \kappa)$.

PROOF. Assume Chang's conjecture for (λ, κ), and let λ countable subsets X_i, $i < \lambda$ of κ be given. Fix, for each i, an enumeration of X_i by natural numbers, and define functions $f_n : \lambda \to \kappa$ for $n \in \omega$ by letting $f_n(i)$ be the nth element in the chosen enumeration of X_i. By Chang's conjecture for (λ, κ), there is an uncountable $Z \subseteq \lambda$ whose images under all the f_n are countable. Then $\bigcup_{i \in Z} X_i$, which is also the union of these countably many countable images, is countable, as required by the compression principle. □

To state concisely the consistency theorem for Chang's conjecture for $(\lambda, \kappa) = (\aleph_{\omega+1}, \aleph_\omega)$, we introduce a short name for the large cardinal hypothesis needed. (We shall not explicitly use this hypothesis, so the reader can safely skip the definition and remember only that we are dealing with very large cardinals, much larger than measurable cardinals but not known or even widely believed to be inconsistent.) A cardinal κ will be called *huge+* if there is an elementary embedding j of the set-theoretic universe into a transitive class M such that κ is the first ordinal moved and, if μ denotes the $(\omega+1)$th cardinal after $j(\kappa)$ then every function from μ into M is an element of M. (*Huge* is defined the same way except that $\mu = j(\kappa)$.)

PROPOSITION 6. *If the existence of a huge+ cardinal is consistent with ZFC, then so is Chang's conjecture for* $(\aleph_{\omega+1}, \aleph_\omega)$.

PROOF. See Levinski, Magidor, and Shelah [**7**], Theorem 5. □

We conclude this section with an analog for the compression property of a well-known fact about $\mathrm{cf}([\kappa]^\omega)$ [**6**], Lemma 4.6.

PROPOSITION 7. *If* $\mathrm{cf}(\lambda) > \kappa^+$ *and if* $\mathrm{CP}(\lambda \to \kappa)$ *then* $\mathrm{CP}(\lambda \to \kappa^+)$. *More generally, if* $\mathrm{cf}(\lambda) > \mu$, *if* $\mathrm{cf}(\mu) > \omega$, *and if* $\mathrm{CP}(\lambda \to \kappa)$ *for all* $\kappa < \mu$, *then* $\mathrm{CP}(\lambda \to \mu)$.

PROOF. Since $\mathrm{CP}(\lambda \to \kappa)$ trivially implies $\mathrm{CP}(\lambda \to \theta)$ for all $\theta < \kappa$, it suffices to prove the second statement. Let λ countable subsets X_i of μ be given. As $\mathrm{cf}(\mu) > \omega$, each X_i has its supremum $< \mu$, and, as $\mathrm{cf}(\lambda) > \mu$, this supremum must be bounded by the same ordinal $\alpha < \mu$, for λ of the X_i's. But then $\mathrm{CP}(\lambda \to |\alpha|)$ ensures that some uncountably many of these X_i's have a countable union. □

3. Divisible Parts of Quotients Are Not Too Large

In this section, we present our positive results about Irwin's question described in the introduction. That is, we assume that $|\mathrm{Div}(G/C)|$ is countable for all countable subgroups C of G, and we deduce upper bounds for $|\mathrm{Div}(G/K)|$ in terms of $|K|$. All these results are obtained by combining the set-theoretic results in Section 2 with the following proposition which relates the divisible parts of quotient groups to the compression principle.

PROPOSITION 8. *Let G be a group such that* $\mathrm{Div}(G/C)$ *is countable for all countable subgroups C of G. Let $\lambda > \kappa$ be cardinals such that* $\mathrm{CP}(\lambda \to \kappa)$ *holds. Then for all subgroups K of G of cardinality κ,* $|\mathrm{Div}(G/K)| < \lambda$.

PROOF. Assume toward a contradiction that the hypotheses hold but the conclusion fails. Then there is a subgroup D of cardinality λ with $K \subseteq D \subseteq G$ and with D/K divisible. As in the proof of Proposition 1, associate to each $d \in D$ and each positive integer n elements $a_n(d) \in D$ and $b_n(d) \in K$ such that $d = n \cdot a_n(d) + b_n(d)$. For each $d \in D$, let X_d be the set of those elements of K obtainable by applying to d any finite composite of the functions $a_n : D \to D$ followed by any of the functions $b_n : D \to K$. (The identity function counts as a finite composite of a_n's, namely the composite of none.) Thus X_d consists of all elements of the forms $b_n(d)$, $b_n(a_k(d))$, $b_n(a_k(a_l(d)))$, etc.

As each X_d is a countable subset of the κ-element set K, the assumed λ to κ compression principle provides an uncountable subset Z of D such that all X_d for $d \in Z$ are contained in a countable subset C of K. We may assume that C is a subgroup of K, by replacing it with the subgroup it generates.

Let E be the subgroup of D generated by all the elements obtainable from elements of Z by applying any finite composite of the functions a_n. Notice that, if e is any one of these generators, then $b_n(e) \in C$ and $a_n(e) \in E$ for all positive integers n. Thus, from $e = n \cdot a_n(e) + b_n(e)$, we infer that E/C is divisible. But, as E is uncountable (containing Z) and C is countable, E/C is an uncountable subgroup of G/C, contrary to the assumption that, for countable C, $\mathrm{Div}(G/C)$ is countable. □

COROLLARY 9. *Let G be a group such that* $\mathrm{Div}(G/C)$ *is countable for all countable subgroups C of G. If K is a subgroup of G of cardinality κ, then* $|\mathrm{Div}(G/K)| \leq \mathrm{cf}([\kappa]^\omega)$.

PROOF. Combine Propositions 4 and 8. □

THEOREM 10. *Let G be a group such that $\mathrm{Div}(G/C)$ is countable for all countable subgroups C of G, and let K be a subgroup of G of cardinality κ.*

(1) *If $\kappa < \aleph_\omega$ then $|\mathrm{Div}(G/K)| \leq \kappa$.*
(2) *If the covering lemma holds over a model of GCH (in particular if there is no inner model with a measurable cardinal) and if $\mathrm{cf}(\kappa) > \omega$, then $|\mathrm{Div}(G/K)| \leq \kappa$.*
(3) *If the covering lemma holds over a model of GCH (in particular if there is no inner model with a measurable cardinal) and if $\mathrm{cf}(\kappa) = \omega$, then $|\mathrm{Div}(G/K)| \leq \kappa^+$.*
(4) *If Chang's conjecture for $(\aleph_{\omega+1}, \aleph_\omega)$ is true and if $\aleph_\omega \leq \kappa < \aleph_{\omega\cdot 2}$, then $|\mathrm{Div}(G/K)| \leq \kappa$.*
(5) *If $\kappa = \aleph_\omega$ then $|\mathrm{Div}(G/K)| < \aleph_{\aleph_4}$.*

PROOF. (1), (2), and (3) all follow immediately from Proposition 2 and Corollary 9. For (4), use Propositions 5 and 7 to deduce from Chang's conjecture for $(\aleph_{\omega+1}, \aleph_\omega)$ that $\mathrm{CP}(\kappa^+ \to \kappa)$ holds for all κ as in (4); then invoke Proposition 8. (5) follows from Proposition 3 and Corollary 9. □

It follows from (1), (4) and Proposition 6 that the statement "If G is a group such that $\mathrm{Div}(G/C)$ is countable for all countable subgroups C of G, and if K is a subgroup of G with $|K| < \aleph_{\omega\cdot 2}$, then $|\mathrm{Div}(G/K)| \leq |K|$" is consistent relative to a huge+ cardinal.

The results presented in this section complete the justification of all lines in the table in the introduction except for the next to last line. The one remaining line, asserting that the answer to Irwin's question is negative for groups of cardinality $\aleph_\omega$ if $V = L$, is the subject of the next two sections.

4. More Combinatorial Set Theory

In this section, we develop, under the assumption $V = L$ the set theory needed to produce a counterexample for Irwin's question. In fact, we do not need the full strength of $V = L$ but only the combinatorial principle $\square_{\aleph_\omega}$. For any uncountable cardinal κ, $\square_\kappa$ denotes the following assertion: There exists a sequence of sets (C_ξ) indexed by the limit ordinals $\xi < \kappa^+$ such that for each such ξ:

(1) C_ξ is a closed, cofinal subset of ξ.
(2) If $\mathrm{cf}(\xi) < \kappa$ then $|C_\xi| < \kappa$.
(3) If η is a limit point of C_ξ, then $C_\eta = \eta \cap C_\xi$.

These square principles were introduced by Jensen **[5]** who showed that they follow from $V = L$ for all κ.

In order to obtain a negative answer to Irwin's question at cardinality $\aleph_\omega$, it is necessary, according to Proposition 8, to violate $\mathrm{CP}(\aleph_{\omega+1} \to \aleph_\omega)$, i.e., to produce $\aleph_{\omega+1}$ countable subsets of $\aleph_\omega$ such that no countable set contains uncountably many of them. The following proposition shows that this and a bit more (which we shall need in the next section) can be done if $\Box_{\aleph_\omega}$ holds. I am not sure to whom to attribute this proposition. Menachem Kojman told me that $\Box_{\aleph_\omega}$ contradicts $\mathrm{CP}(\aleph_{\omega+1} \to \aleph_\omega)$. Menachem Magidor showed me the proof given below for this fact and the additional information in Proposition 11. Magidor also informed me that similar arguments were known to Saharon Shelah.

PROPOSITION 11. *Assume $\Box_{\aleph_\omega}$. There are $\aleph_{\omega+1}$ countably infinite subsets X_i of $\aleph_\omega$ such that the intersection of every two of these X_i is finite and, for each countable $Y \subseteq \aleph_\omega$, at most countably many of the X_i have infinite intersection with Y (and, a fortiori, at most countably many X_i are included in Y).*

PROOF. Fix C_ξ for limit ordinals $\xi < \aleph_{\omega+1}$ as in the definition of $\Box_{\aleph_\omega}$. We shall define functions f_ξ for $\xi < \aleph_{\omega+1}$ with the following properties.

(1) Each f_ξ is a function on ω satisfying $f_\xi(n) \in \aleph_n$ for all $n \in \omega$.
(2) If $\xi < \eta$, then $f_\xi(n) < f_\eta(n)$ for all but finitely many $n \in \omega$.
(3) If η is a limit ordinal, $|C_\eta| < \aleph_n$, and $\xi \in C_\eta$, then $f_\xi(n) < f_\eta(n)$.

(Notice that, for $\xi \in C_\eta$, (3) amplifies (2) by specifying that the finitely many exceptional n in (2) are bounded above by the q such that $|C_\eta| = \aleph_q$. Notice also that such a q exists by (2) in the definition of $\Box_{\aleph_\omega}$.) After constructing such f_ξ's, we shall show that their graphs are essentially the sets needed to establish the proposition.

The construction of the f_ξ's proceeds by induction on $\xi < \aleph_{\omega+1}$. Suppose, therefore, that f_ξ is defined for every $\xi < \eta$, and we wish to define f_η.

If η is not a limit ordinal, then (3) does not apply to η, so we need only satisfy (1) and (2), which we do as follows. Partition η (the set of ordinals smaller than η), which has cardinality at most $\aleph_\omega$, into countably many pieces $A_0, A_1, \dots$ such that each A_k has cardinality at most $\aleph_k$. Then define $f_\eta(n)$ to be any ordinal that is $< \aleph_n$ (so (1) holds) but $> f_\xi(n)$ for all $\xi \in A_0 \cup \dots \cup A_{n-1}$ (so that (2) holds, because if $\xi \in A_p$ then $f_\xi(n) < f_\eta(n)$ for all $n > p$). Such an ordinal exists, because there are only $\aleph_{n-1}$ ordinals $\xi \in A_0 \cup \dots \cup A_{n-1}$ and therefore the corresponding $f_\xi(n)$'s cannot be cofinal in $\aleph_n$.

If η is a limit ordinal, let q be the natural number such that $|C_\eta| = \aleph_q$. Then, in order to satisfy (3), we must make sure that $f_\eta(n)$ satisfies, in addition to the requirements in the preceding paragraph, $f_\xi(n) < f_\eta(n)$ if $\xi \in C_\eta$ and $n > q$. But there are only $\aleph_q$ such ξ's, by choice of q, so the corresponding $f_\xi(n)$'s cannot be cofinal in $\aleph_n$ when $n > q$. So an appropriate value for $f_\eta(n)$ can again be found. This completes the construction of the f_η's and the verification of (1), (2), and (3).

We claim that, from any uncountable $A \subseteq \aleph_{\omega+1}$, one can extract an uncountable $B \subseteq A$ and one can find $m \in \omega$ such that $f_\xi(n) < f_\eta(n)$ for all $\xi < \eta$ in B and all $n \geq m$. (That is, the finitely many exceptions in (2) are bounded by the same m, independent of ξ and η, as long as these two indices lie in B.) To prove this claim, consider an arbitrary uncountable $A \subseteq \aleph_{\omega+1}$. We may assume without loss of generality that A has order type $\aleph_1$; just replace A by the subset consisting of its first $\aleph_1$ elements. Let β be the supremum of A, so $\mathrm{cf}(\beta) = \aleph_1$. Inductively choose

$$\gamma(0) < \alpha(0) < \gamma(1) < \alpha(1) < \cdots < \gamma(\omega) < \alpha(\omega) < \gamma(\omega+1) < \ldots$$

for $\aleph_1$ steps, so that all the γ's are limit points of C_β and all the α's are in A. There is no difficulty making these choices, as both A and the set of limit points of C_β are cofinal in β. For each $\mu < \aleph_1$, (2) provides a natural number $m(\mu)$ such that

$$\forall n \geq m(\mu) \quad f_{\gamma(\mu)}(n) < f_{\alpha(\mu)}(n) < f_{\gamma(\mu+1)}(n).$$

Increasing $m(\mu)$ if necessary, we may assume that $|C_\beta| < \aleph_{m(\mu)}$. As there are uncountably many μ's and only countably many possible values for $m(\mu)$, there is an uncountable $Y \subseteq \aleph_1$ such that $m(\mu)$ has the same value m for all $\mu \in Y$. Now if $\mu < \nu$ are in Y and if $n \geq m$ then we have

$$f_{\alpha(\mu)}(n) < f_{\gamma(\mu+1)}(n) \leq f_{\gamma(\nu)}(n) < f_{\alpha(\nu)}(n),$$

where the first and third inequalities hold because $n \geq m = m(\mu) = m(\nu)$, and the middle inequality holds because of (3) and the fact that, since the γ's are limit points of C_β, either $\mu + 1 = \nu$ or $\gamma(\mu+1) \in \gamma(\nu) \cap C_\beta = C_{\gamma(\nu)}$. This means that $B = \{\alpha(\mu) \mid \mu \in Y\}$ and m are as required in the claim.

Finally, we show that the functions f_ξ (for all $\xi \in \aleph_{\omega+1}$), regarded as sets of ordered pairs $\subseteq \omega \times \aleph_\omega$, have the properties that every two have finite intersection and that no countable set C can have infinite intersections with uncountably many f_ξ's. Then, transferring these subsets of $\omega \times \aleph_\omega$ to subsets of $\aleph_\omega$ by some bijection, we have the sets required by the proposition. That every two of the graphs have finite intersection is immediate from (2). To prove the other property, suppose C were a countable set having infinite intersection with f_ξ for all ξ in some uncountable $A \subseteq \aleph_{\omega+1}$. Let $B \subseteq A$ and m be as in the claim proved in the preceding paragraph. Then by deleting the first m elements from each of the graphs f_ξ, $\xi \in B$, i.e., by forming $f_\xi \restriction (\omega - m)$, we would obtain uncountably many pairwise disjoint sets, all intersecting the countable set C. As this is absurd, the proof is complete. □

At the referee's suggestion, we point out that $\aleph_\omega$ and $\aleph_{\omega+1}$ in Proposition 11 can be replaced by κ and κ^+ where κ is any cardinal of cofinality ω. The only changes needed in the proof are to replace the cardinals $\aleph_n$ with any increasing ω-sequence of regular (e.g., successor) cardinals κ_n cofinal in κ and, in the fourth paragraph of the proof, to replace $|C_\eta|$ with the first $\kappa_q \geq |C_\eta|$.

5. A Quotient with a Large Divisible Part

In this section we construct, assuming $\square_{\aleph_\omega}$, a counterexample to Irwin's question, i.e., a group G such that $\mathrm{Div}(G/C)$ is countable for all countable C but G is not fully starred.

THEOREM 12. *Assume $\square_{\aleph_\omega}$. There exist a group G of cardinality $\aleph_{\omega+1}$ and a subgroup K of cardinality $\aleph_\omega$ such that G/K is divisible but, for each countable subgroup C of G, the divisible part of G/C is countable.*

PROOF. The required groups G and K will actually be modules over the ring R of rational numbers with odd denominators, i.e., they will be divisible by all primes except 2. We write R^* for the set of units of R, the rational numbers whose numerators and denominators (in reduced form) are both odd.

Since $\square_{\aleph_\omega}$ is assumed, let X_i for $i \in \aleph_{\omega+1}$ be as in Proposition 11. Fix, for each i a bijection between X_i and the set $\mathbb{Z}$ of integers, and write $\xi(i,n)$ for the element of X_i that corresponds to the integer n. (Notice that the same ordinal can be $\xi(i,n)$ for many different pairs (i,n) because the X_i's overlap.)

The group G will be presented as the R-module generated by certain elements subject to certain relations. The generators are of two sorts. First, there are $\aleph_\omega$ generators which, to simplify notation, we take to be the ordinals $\alpha < \aleph_\omega$. Second, there are $\aleph_{\omega+1}$ generators $g(i,n)$ indexed by all the ordinals $i < \aleph_{\omega+1}$ and all integers $n \in \mathbb{Z}$. The defining relations are

$$g(i,n) = 2 \cdot g(i,n+1) + \xi(i,n).$$

This defines G as an R-module, hence as a group. K is defined to be the submodule generated by the first sort of generators of G, the ordinals below $\aleph_\omega$.

Notice that, in any non-trivial R-linear combination of the defining relations of G, some $g(i,n)$ must occur. Indeed, of the finitely many $g(i,n)$'s that occur in the relations being combined, the ones with the largest (or the smallest) values of n cannot be canceled. So the relations impose no restrictions on the generators of K alone. This means that K is freely generated, as an R-module, by the ordinals $\alpha < \aleph_\omega$. It follows, exactly as in the argument for free groups in Section 1, that K is fully starred.

The quotient group G/K can be presented by adjoining to the defining relations of G the new relations $\alpha = 0$ for all the generators α of K. The resulting presentation is clearly equivalent to one that has only the generators $\bar{g}(i,n)$ (where the bar over a letter denotes the coset modulo K) and the relations

$$\bar{g}(i,n) = 2 \cdot \bar{g}(i,n+1).$$

Thus, G/K is divisible and is in fact the rational vector space freely generated by the $\bar{g}(i,0)$'s; here $\bar{g}(i,n)$ is identified with $2^{-n}\bar{g}(i,0)$.

To complete the proof of the theorem, it remains to show that $\mathrm{Div}(G/C)$ is countable for all countable subgroups C of G. As a preliminary step toward this, we introduce a normal form for elements of G. We claim that every element x of G can be uniquely represented in the form

$$x = \sum_i r_i g(i,n_i) + \sum_\alpha s_\alpha \alpha$$

where all $r_i \in R^*$, $s_\alpha \in R$, and the summation variables i and α range over some finite subsets (depending on x) of $\aleph_{\omega+1}$ and $\aleph_\omega$, respectively. (In particular, each i that occurs at all, as the first argument of a g, occurs in only one term of such a normal form.) To produce such a normal form for x, we first write the image of x in G/K as a rational linear combination of the $\bar{g}(i,0)$'s, then we put the rational coefficients in this combination into R^* by using $2^n\bar{g}(i,0) = \bar{g}(i,-n)$, which holds in G/K, then we remove the bars from the g's in this expression, obtaining an element $y \in G$, and we observe that the difference $d = x - y$ is in K since y has the same image as x in G/K. Now the desired normal form of x consists of y plus the expansion of d in terms of the free generators α of K. The uniqueness is proved by observing that the $\sum_i r_i g(i, n_i)$ part of a normal form of x must have the same image as x in G/K and must therefore be exactly the y constructed above.

It will be useful later to have a more computational description of how to convert an arbitrary element of G, given as an R-linear combination of generators $g(i,n)$ and α, to normal form. This means that we must convert the given expression, using the defining relations of G, to one in which, for each i, there is at most one term of the form $r \cdot g(i,n)$ and the coefficient r of any such term is in R^*. Suppose that, for a certain i, the given expression contains several terms of the form $g(i,n)$, possibly with different values of n. The defining relations of G allow us to replace $g(i,n)$ with $2 \cdot g(i,n+1) + \xi(i,n)$, so we can increase the n's involved in these terms as much as we wish. In particular, we can increase them until they are all equal to, say, the largest of the originally occurring n's. At this stage, all the $g(i,n)$'s, for the particular i under consideration, have the same n, so we can collect them into one term by adding their coefficients. If the resulting coefficient of $g(i,n)$ is divisible by 2, then we can use the defining relations "in reverse" to divide the coefficient by 2 while decreasing n by 1 and subtracting $\xi(i,n-1)$. Repeat this until the coefficient is no longer divisible by 2, i.e., until it is in R^*. By carrying out the procedure just described for each i, we clearly achieve normal form.

We point out two consequences of this procedure that will be useful later. First, notice that the element $\sum_\alpha s_\alpha \alpha \in K$ occurring in the normal form at the end consists of first the linear combination of α's in the original expression, and second some multiples of certain $\xi(i,n)$, where the i's involved in these ξ's were involved in g's in the original expression.

Second, if $x - 2z \in K$ then $g(i,n)$ occurs in the normal form of x if and only if $g(i,n+1)$ occurs in the normal form of z, and they occur with the same coefficient. To see this, start with a normal form of z, multiply all terms by 2 and add a suitable element of K to get an expression for x that is in normal form except that all the coefficients of g's are divisible once by 2; then apply the normalization algorithm to this expression.

This observation makes it clear (if it wasn't clear earlier) that K is a pure subgroup of G.

We now embark on the proof that $\operatorname{Div}(G/C)$ is countable for all countable

$C \subseteq G$. Notice first that, if C were a counterexample and if C' were any countable group such that $C \subseteq C' \subseteq G$, then C' would also be a counterexample. Indeed, there would be an uncountable D with $C \subseteq D \subseteq G$ and D/C divisible. Then, as $C \subseteq D \cap C'$, $(D + C')/C' \cong D/(D \cap C')$ is a quotient of the divisible group D/C, and is therefore divisible. As $D + C'$ is uncountable and C' is countable, we have another counterexample, as claimed.

Therefore, in proving that G/C is divisible, we may assume that C satisfies the following four conditions, since each condition amounts to being closed under countably many functions and can therefore be satisfied by a suitable countable supergroup of any given countable C.

(1) C is an R-submodule of G.
(2) If $x \in C$, then all the generators $g(i, n)$ and α that occur in the normal form of x are also in C.
(3) If $g(i, n) \in C$, then $g(i, m) \in C$ for all integers m.
(4) If $g(i, n) \in C$, then all members $\xi(i, m)$ of X_i are in C.

It follows from these closure conditions and the defining relations of G that $C/(C \cap K)$ is divisible. Therefore, if D is a subgroup of G for which D/C is divisible, then so is $D/(C \cap K)$. Thus, we may replace C with $C \cap K$ and so assume, without loss of generality, that $C \subseteq K$. (I thank Laszlo Fuchs for pointing out that this assumption can be made and that it substantially simplifies my earlier proof of this theorem.) Notice that, when we replace C with $C \cap K$, conditions (1) through (4) remain true, but the last two of them become vacuous and the first two imply that C is the free R-submodule of K generated by a certain subset of the original set $\aleph_\omega$ of generators of K.

Now suppose D is a subgroup of G with $C \subseteq D$, with C as above, and with D/C divisible. We must prove that D is countable. Call an element $i \in \aleph_{\omega+1}$ *relevant* if the normal form of some element of D contains $g(i, n)$ for some n.

LEMMA. *Only countably many i are relevant.*

PROOF. Suppose uncountably many i are relevant. Since the X_i were chosen as in Proposition 11, the countable set $C \cap \aleph_\omega$ (i.e., the set of generators of K that are in C) must have finite intersection with X_i for some relevant i. Fix such an i for the rest of the proof of the lemma.

As i is relevant, choose an element $d_0 \in D$ such that $g(i, q)$ occurs in d_0 for some q. (We think of d_0 and the other elements defined below as written in normal form, so "occurs in d_0" really means "occurs in the normal form of d_0.") Let $a \in R^*$ be the coefficient of $g(i, q)$ in d_0. Let J be the (finite) set of those $j \neq i$ such that $g(j, m)$ occurs in d_0 for some m.

Since X_i has finite intersection with C (by choice of i) and with X_j for each $j \in J$ (by the "finite pairwise intersections" part of Proposition 11), fix $p \in \mathbb{Z}$ so large that, for all $n \geq p$, we have $\xi(i, n) \notin C$ and $\xi(i, n) \notin X_j$ for all $j \in J$.

Define a sequence $(d_r)_{r \in \omega}$ of elements of D by starting with d_0 and then inductively using the divisibility of D/C to set $d_r = 2 \cdot d_{r+1} + c_r$ with $d_{r+1} \in D$

and $c_r \in C$.

We claim that (the normal form of) d_r contains $g(i, q + r)$ with coefficient a. We chose a to make this true for $r = 0$. If it is true for r, then it follows for $r + 1$ by the second consequence of the normalization algorithm, because $d_r - 2d_{r+1} \in C \subseteq K$.

The second consequence of the normalization algorithm also shows that the set J, defined above using d_0, would be unchanged if we used any d_r instead.

In view of the observations in the preceding two paragraphs, any final segment of the sequence (d_r) is again a sequence of the same sort, only with a larger value of q. Since p can obviously also be replaced by any larger number, we assume without loss of generality that $p = q$. Thus, henceforth, d_r has $g(i, p + r)$ in its normal form, with coefficient $a \in R^*$.

The definition of the d sequence gives us that

$$\begin{aligned} d_0 &= 2d_1 + c_0 \\ &= 4d_2 + 2c_1 + c_0 = \dots \\ &= 2^r d_r + 2^{r-1}c_{r-1} + \cdots + 2c_1 + c_0, \end{aligned}$$

for any r. Fixing some $r \geq 1$, let us consider the normal form of d_0 as obtained from the last line in the preceding display by first writing d_r and $c = 2^{r-1}c_{r-1} + \cdots + 2c_1 + c_0$ in normal form, then combining these normal forms to make an expression for $d_0 = 2^r d_r + c$, and finally normalizing the result. More precisely, let us consider the coefficient $s \in R$ of $\xi(i, p+r-1)$ in the resulting normal form, and in fact let us consider s modulo 2^r.

First, let us consider possible occurrences of $\xi(i, p+r-1)$ in the first expression for $2^r d_r + c$ that we got by combining normal forms of d_r and c (before normalizing the result). Any ordinal α that occurs in the normal form of $c \in C$ is itself in C (closure condition (2)), and hence is different from $\xi(i, p + r - 1)$ by our choice of p. There may be occurrences of $\xi(i, p + r - 1)$ in d_r, but these will have their coefficients multiplied by 2^r in the expression for $2^r d_r + c$, and therefore will not contribute to the coefficient s modulo 2^r.

It remains to consider occurrences of $\xi(i, p + r - 1)$ that arise during the normalization of $2^r d_r + c$. These can arise as $\xi(j, n)$ during the normalization of $g(j, m)$ terms. Consider first such terms $g(j, m)$ with $j \neq i$. Then $j \in J$, so any ξ produced by normalizing $g(j, m)$ is in X_j and hence is different from $\xi(i, p + r - 1)$ by our choice of p. Thus, such a j will not contribute to s.

We have thus seen that the coefficient s modulo 2^r must arise entirely from the normalization of the term $2^r a g(i, p+r)$ in $2^r d_r + c$. This normalization reads

$$\begin{aligned} 2^r a g(i, p + r) &= 2^{r-1} a g(i, p + r - 1) - 2^{r-1} a \xi(i, p + r - 1) \\ &= 2^{r-2} a g(i, p + r - 2) - 2^{r-2} a \xi(i, p + r - 2) - 2^{r-1} a \xi(i, p + r - 1) \\ &= \dots \\ &= a g(i, p) - a \xi(i, p) - 2a \xi(i, p + 1) - \cdots - 2^{r-1} a \xi(i, p + r - 1). \end{aligned}$$

So we see that the coefficient of $\xi(i, p+r-1)$ is $2^{r-1}a$. Since $a \in R^*$, this coefficient is not zero modulo 2^r.

This proves that the coefficient of $\xi(i, p+r-1)$ in the normal form of d_0 is not divisible by 2^r and, in particular, is not zero. But r was an arbitrary integer ≥ 1, so the normal form of d_0 contains infinitely many non-zero terms. That is absurd, and so the lemma is proved. □

It follows from the lemma just proved that the image of the projection of D to G/K is countable, for it is included in the R-module generated by $\bar{g}(i, n)$'s where i is relevant and $n \in \mathbb{Z}$. That image is $D/(D \cap K)$, so to complete the proof that D is countable and thus the proof of the theorem, it suffices to show that the kernel $D \cap K$ of that projection is also countable. We remarked earlier that K, being a free R-module, is fully starred, so we need only prove that $(D \cap K)/C$ is divisible. But this is immediate from the facts that D/C is divisible, that K is pure in G, and that $C \subseteq K$. □

Acknowledgments

I thank Laszlo Fuchs and the referee for their helpful comments on an earlier version of this paper, and I thank John Irwin for asking me the question that led to this work.

References

1. M. Burke and M. Magidor, *Shelah's pcf theory and its applications*, Ann. Pure Appl. Logic **50** (1990), 207–254.
2. P. J. Cohen, *The independence of the continuum hypothesis*, Proc. Nat. Acad. Sci. U.S.A. **50** (1963), 1143–1148.
3. A. Dodd and R. B. Jensen, *The core model*, Ann. Math. Logic **20** (1981), 43–75.
4. A. Dodd and R. B. Jensen, *The covering lemma for K*, Ann. Math. Logic **22** (1982), 1–30.
5. R. B. Jensen, *The fine structure of the constructible hierarchy*, Ann. Math. Logic **4** (1972), 229–308.
6. W. Just, A. R. D. Mathias, K. Prikry, and P. Simon, *On the existence of large p-ideals*, J. Symbolic Logic **55** (1990), 457–465.
7. J.-P. Levinski, M. Magidor, and S. Shelah, *Chang's conjecture for $\aleph_\omega$*, Israel J. Math **69** (1990), 161–172.
8. W. Mitchell, *The core model for sequences of measures, I*, Math. Proc. Cambridge Phil. Soc. **95** (1984), 229–260.
9. R. M. Solovay, $2^{\aleph_0}$ *can be anything it ought to be*, The Theory of Models (J. W. Addison, L. Henkin, and A. Tarski, eds.), North-Holland, 1964, pp. 435.
10. E. Specker, *Additive Gruppen von Folgen ganzer Zahlen*, Portugal. Math. **9** (1950), 131–140.

Mathematics Department, University of Michigan, Ann Arbor, MI 48109, U.S.A.

E-mail address: ablass@umich.edu

Contemporary Mathematics
Volume **171**, 1994

Minimal rings, central idempotents and the Pierce sheaf

W.D. BURGESS

This article is dedicated to the memory of Professor Richard S. Pierce.

Introduction. In one of his last papers, [12], R.S. Pierce returned to the study of commutative regular rings. In his original investigation of the category of rings with "conformal" homomorphisms, *RNGS*, he showed that there is a categorical equivalence between *RNGS* and the category of what have come to be called Pierce sheaves. However, the Pierce sheaves are not purely topological objects since both the stalks and the topology on the *espace étalé* depend on algebraic factors. At the other extreme, the classical Stone duality gives a contravariant equivalence between the categories of boolean algebras and of boolean spaces, the latter involving only topology. Commutative (von Neumann) regular rings are exactly those whose Pierce stalks are fields. These are rightly thought of as "generalized fields" and Pierce found it natural to consider what could be called "generalized prime fields", that is those rings whose Pierce stalks are prime fields, called *minimal regular rings.* He showed in [12] that there is a general form of Stone duality for the category, *MR*, of minimal regular rings and a category, *LBS*, of objects, called *labeled boolean spaces*, given by purely topological data. A labeled boolean space X, is a boolean space equipped with a continuous function $\beta : X \to \mathbf{P}\cup\{\infty\}$, where $\mathbf{P}\cup\{\infty\}$ is the one-point compactification of the set of primes. In effect, the subset $\beta^{-1}(\{p\})$ of X identifies the location of the stalks of the form $\mathbf{Z}/(p)$, while $\beta^{-1}(\{\infty\})$ locates the stalks $\mathbf{Q}$.

The purpose of the present article is to extend the duality to a larger category of rings, called the category of *minimal rings*, denoted *MinR*, and a category of boolean spaces with additional topological structure, *boolean spaces with decomposition*, denoted *BSD*. The minimal rings will be shown to play a fundamental role, especially when the Pierce sheaf is used. Some of the various contravariant

1991 *Mathematics Subject Classification.* 16S60, 16E50, 06E15.
This paper is in final form, and no version of it will be submitted for publication elsewhere.
This research was partially supported by grant A7539 of the NSERC.

equivalences are shown schematically as follows.

CR commutative regular rings	MR minimal regular rings	$MinR$ minimal rings
$\downarrow\uparrow$	$\downarrow\uparrow$	$\downarrow\uparrow$
the category of regular Pierce sheaves	LBS labeled boolean spaces	BSD boolean spaces with decomposition

The descriptions of $MinR$ and of BSD, as well as a presentation of the contravariant equivalence between them, are the subjects of Section 1. The starting point is the observation that given a boolean space X and a collection of rings R_x, $x \in X$, so that each R_x is either of the form $\mathbf{Z}/(p^k)$ for some prime p and $k \geq 1$ or some D, $\mathbf{Z} \subseteq D \subseteq \mathbf{Q}$, then there is at most one way of constructing a ring whose Pierce sheaf has base X and whose Pierce stalks are the R_x. The category $MinR$ is chosen from this class of rings and consists of the rings generated by their central idempotents and their characteristic rings. They have enough structure to be of use, but not so much as to make them intractable. The category of minimal regular rings is a full subcategory of $MinR$, and Pierce has already shown that there is a rich variety of such rings. As we see in Section 3, $MinR$ is complete and cocomplete, and this allows the construction of new minimal rings from given ones.

Each ring R contains a unique largest minimal ring $P(R)$, the subring generated by the characteristic subring and the central idempotents of R. It is shown that it is natural to consider R as a $P(R)$-algebra. Given a ring R, $\mathbf{B}(R)$ denotes its boolean algebra of central idempotents. Suppose the Pierce stalks of R are indecomposable. If for some boolean algebra C there is a boolean algebra homomorphism $\beta : \mathbf{B}(R) \to C$, there is a ring S and a ring homomorphism $\alpha : R \to S$ so that $\mathbf{B}(S) = C$ and α restricted to $\mathbf{B}(R)$ is β. The ring S can be realized as the ring of sections of an inverse image sheaf. However, it is more natural to use the duality of $MinR$ and BSD to find a minimal ring T with $\mathbf{B}(T) = C$ and a homomorphism $\gamma : P(R) \to T$, induced by β, so that $S \cong T \otimes_{P(R)} R$; in other words, S is constructed by a change of base ring. It will be seen that this fashion of changing the boolean algebra of central idempotents of a ring is generic, in the sense that any modification of the central idempotents of a ring (i.e., a homomorphism $\mu : R \to R'$ so that R' is generated, as a ring, by $\mu(R)$ and $\mathbf{B}(R')$) can be factored through one of the above type via the natural homomorphism $R \to S$ and a surjection $S \to R'$. Section 2 is devoted to the study of such modifications of $\mathbf{B}(R)$.

In Section 3, categorical notions are examined. The categories $MinR$ and BSD are both complete and cocomplete (have all limits and colimits, in the categorical sense). In fact $MinR$ is a coreflexive subcategory of $CRNGS$, the category of commutative rings. Products and coproducts in BSD are described in some detail, especially the finite ones. Monomorphisms and epimorpisms in these categories are studied. Monomorphisms in $MinR$ and BSD need not be injective.

Notational conventions. The symbols $\mathbf{N}$, $\mathbf{Z}$ and $\mathbf{Q}$ are used in their usual manner except that it is assumed that $0 \in \mathbf{N}$; $\mathbf{P}$ is the set of primes. The one-point compactifications of the discrete spaces $\mathbf{N}$ and $\mathbf{P}$ are denoted $\mathbf{N}\cup\{\infty\}$ and $\mathbf{P}\cup\{\infty\}$, respectively. The annihilator in $\mathbf{Z}$ of a subset S of a ring or module is denoted

$\mathbf{Z}$-ann S. All rings are unitary and ring homomorphisms preserve 1. For a ring R, $\mathbf{B}(R)$ is the subset of central idempotents of R, which is sometimes thought of as a boolean algebra. A *conformal homomorphism* of rings $\alpha : R \to S$ is a homomorphism such that $\alpha(\mathbf{B}(R)) \subseteq \mathbf{B}(S)$ (Pierce, [11, page 8]). A ring R is called *indecomposable* if $\mathbf{B}(R) = \{0, 1\}$.

The notation of Pierce sheaves is as in [11] (in capsule form, Burgess and Stephenson, [3] or Carson, [5]). In particular, if R is a ring it can be regarded as the ring of sections of its Pierce sheaf which has as base space the boolean space (i.e., compact, Hausdorff and totally disconnected) $X = \mathbf{Spec}\,\mathbf{B}(R)$ and *espace étalé* $\mathcal{R}$ with stalks $R_x = R/Rx$, for $x \in X$. The image of an element $r \in R$ or a subset $S \subseteq R$ in a stalk R_x is usually denoted by r_x or S_x, respectively. If $r \in R$, $\mathrm{supp}\,(r) = \{x \in X \mid r_x \neq 0\}$, an open set. For a boolean space X and an indecomposable ring T, the *simple sheaf* they define is the sheaf over X with *espace étalé* $X \times T$ ([11, Definition 11.2]) and its ring of sections is $C(X, T)$, the ring of continuous (i.e., locally constant) functions $X \to T$, with T given the discrete topology.

The *characteristic ring*, $\kappa(R)$ of a ring R (see Burgess and Stewart, [4] or Bousefield and Kan, [2], where it is called the *core*) is used extensively. It is the largest epimorphic image of $\mathbf{Z}$ in R ("epimorphism" in the sense of category theory, Mac Lane, [10, page 19]) and its characterization in terms of the additive structure of R is [4, Proposition 1.5]. When $\mathbf{Z}$-ann $R \neq 0$, $\kappa(R) = \mathbf{Z}/\mathbf{Z}$-ann R, and when $\mathbf{Z}$-ann $R = 0$, $\kappa(R)$ is given by a function $f : \mathbf{P} \to \mathbf{N} \cup \{\infty\}$ where, for $p \in \mathbf{P}$, when $R = t_p(R) \oplus R^{(p)}$, $t_p(R)$ the p-torsion part and $R^{(p)}$ is p-divisible, $f(p) = n$ where $\mathbf{Z}$-ann $t_p(R) = (p^n)$; otherwise, $f(p) = \infty$. Then $\kappa(R)$ is the ring whose Pierce sheaf is based on the one-point compactification of $F = \{p \in \mathbf{P} \mid 1 \leq f(p) < \infty\}$ with stalk over $p \in F$, $\mathbf{Z}/(p^{f(p)})$ and stalk $D = \mathbf{Z}[1/p \mid f(p) < \infty]$ over ∞.

Any categorical language used below is as in [10]. "Commutative π-regular ring" is a synonym for "0-dimensional ring".

1. Minimal rings.

Before minimal rings are defined an observation about a very special sort of Pierce sheaf is apropos. The rings which are indecomposable epimorphs of $\mathbf{Z}$ (in the category of rings) are of the form $\mathbf{Z}/(p^k)$ for some prime p and $k \geq 1$, or some ring D with $\mathbf{Z} \subseteq D \subseteq \mathbf{Q}$ (see [2] or Dicks and Stephenson, [6]). Note that given two such rings, A and B, there is at most one homomorphism $A \to B$.

Proposition 1.1 *Let $(X, \mathcal{R})$ be a Pierce sheaf such that for each $x \in X$ the stalk $\mathcal{R}_x$ is an epimorph of $\mathbf{Z}$. Then the topology on $\mathcal{R}$ is uniquely determined by the stalks and X.*

Proof. Since the ring of sections of $\mathcal{R}$ is commutative, each $\mathcal{R}_x$ is indecomposable (e.g., [3, Proposition 1]). Let $x \in X$ and pick integers a and b such that (the image of) ab^{-1} makes sense in $\mathcal{R}_x$. Then there is a neighbourhood U of x such that for all $y \in U$, ab^{-1} is defined in $\mathcal{R}_y$. Any section σ with $\sigma(x) = ab^{-1}$ will be "constantly" ab^{-1} on some neighbourhood of x in U (where ab^{-1} is thought of as a rational number or as in some $\mathbf{Z}/(p^k)$ for $p \in \mathbf{P}$ and $1 \leq k$, depending on the stalk), i.e., $b\sigma(y) = a$ and b is invertible for all y in some neighbourhood of x . On the other hand, all values of a section can be put into this form because of the nature of the stalks. Hence, the locally "constant" sections give all sections and determine the topology. □

Definition 1.2 *For any ring R, let $P(R)$ denote the subring generated by the characteristic ring, $\kappa(R)$ and $\mathbf{B}(R)$. The subring $P(R)$ is called the* **minimal subring of** R. *Any ring R such that $P(R) = R$ is called a* **minimal ring**.

Notice that when $\kappa(R)$ is finite or when $\kappa(R)$ is given by f with $f(p) = \infty$ for all $p \in \mathbf{P}$, then $P(R)$ is just the subring generated by $\mathbf{B}(R)$. Other cases yield rings strictly larger than that generated by $\mathbf{B}(R)$. At the level of Pierce stalks, $\kappa(R_x) \supseteq P(R)_x$, and the inclusion may be strict, but not when $P(R)$ is π-regular.

In order to justify the definition, it will have to be shown that the minimal subring of a ring is a minimal ring. It will then be observed that the minimal rings form a full subcategory of the category of rings with conformal homomorphisms.

Proposition 1.3 *For any ring R, the minimal subring $P(R)$ is a minimal ring. If $\gamma : R \to S$ is a conformal homomorphism of rings then $\gamma_{|P(R)}$ is a homomorphism $P(R) \to P(S)$.*

Proof. Clearly $\mathbf{B}(P(R)) = \mathbf{B}(R)$ since both $\mathbf{B}(R)$ and $\kappa(R)$ are central in R. To show that $P(P(R)) = P(R)$, it must be shown that $\kappa(P(R)) = \kappa(R)$. If $\kappa(R)$ is infinite, it is given by $f : \mathbf{P} \to \mathbf{N} \cup \{\infty\}$. If for $p \in \mathbf{P}$, $f(p) < \infty$ then $R = t_p(R) \oplus R^{(p)}$, where $t_p(R)$ is the p-torsion part of R and $R^{(p)}$ is p-divisible ([4, Proposition 1.5]). Further, $\mathbf{Z}$-ann $t_p(R) = (p^{f(p)})$. The subgroups $t_p(R)$ and $R^{(p)}$ are seen to be ideals, so there is $e \in \mathbf{B}(R)$ with $eR = R^{(p)}$ and $(1-e)R = t_p(R)$. Hence $e \in P(R)$ with $(1-e)P(R) = t_p(P(R))$. Clearly $\mathbf{Z}$-ann $(1-e)P(R) = \mathbf{Z}$-ann $(1-e) = \mathbf{Z}$-ann $(1-e)R$.

If $f(p) = \infty$ then R has no such splitting. Suppose $P(R)$ had a splitting, $P(R) = t_p(P(R)) \oplus P(R)^{(p)}$. As before there is $e \in \mathbf{B}(P(R)) = \mathbf{B}(R)$ such that $eP(R)$ is p-divisible. Hence $px = e$ has a solution in $eP(R)$, and for some $k \geq 0$, $p^k(1-e) = 0$. This means that eR is p-divisible and there is a splitting of R. This is a contradiction.

When $\kappa(R)$ is finite (i.e., when $\mathbf{Z}$-ann $R \neq 0$), it is the image of $\mathbf{Z}$ in R, and so is also in $P(R)$.

From this it follows that $\kappa(P(R)) = \kappa(R)$ and that $P(P(R)) = P(R)$.

By the definition of a conformal homomorphism, $\gamma(\mathbf{B}(R)) \subseteq \mathbf{B}(S)$. By [4, Lemma 1.2], $\gamma(\kappa(R)) \subseteq \kappa(S)$. □

It will be seen that, from the point of view of Pierce sheaves, it is natural to regard a ring R as a $P(R)$-algebra. The above proposition shows that the class of minimal rings with ring homomorphisms forms a category, call it *MinR*. (The trivial ring, that is, the one with $1 = 0$, is included in *MinR*, for reasons made clear in Section 3.) It also says that P : *RNGS* $\to$ *MinR* is a subfunctor of the identity functor on *RNGS*. When P is restricted to the category of regular rings, it takes values in the category of minimal regular rings, *MR*. The minimal regular rings are exactly those whose Pierce sheaves have stalks which are prime fields; they are the object of study in Pierce's [12]. In fact, by [4, Lemmas 4.1 and 1.6], $\Gamma(R)$ is regular if all the ideals of R are idempotent or if R is a right V-ring. If R is π-regular, then so is $P(R)$. In any of these cases, if $P(R)$ has any infinite stalks, they are all $\mathbf{Q}$.

At this stage it can be observed that a subring of a minimal ring need not be minimal; for example $\mathbf{Z} \times \mathbf{Z}[1/2]$ is not in *MinR* but it is a subring of the minimal ring $\mathbf{Z}[1/2] \times \mathbf{Z}[1/2]$. On the other hand, a homomorphic image of a minimal ring is minimal.

The next result is a Pierce sheaf characterization of minimal rings.

Theorem 1.4 *Let R be a ring. Then R is a minimal ring if and only if (i) each Pierce stalk of R is an epimorph of $\mathbf{Z}$, and (ii) if there are any infinite stalks, they are all the same. Moreover, when R is a minimal ring, each stalk of $\kappa(R)$ appears as a stalk of R.*

Proof. For any ring R, set $X = \mathbf{Spec}\,\mathbf{B}(R)$ and $Y = \mathbf{Spec}\,\mathbf{B}(\kappa(R))$. The inclusion $\kappa(R) \to R$ induces a surjection $\alpha : X \to Y$ (in the category of boolean spaces, an epimorphism is surjective – cf. Proposition 3.4). If $\alpha(x) = y$, then, as sets, $y \subseteq x$ and so there is a commutative diagram

$$\begin{array}{ccc} \kappa(R) & \longrightarrow & R \\ \downarrow & & \downarrow \\ \kappa(R)_y & \longrightarrow & R_x \end{array}$$

where the lower arrow is induced by the upper one. Now suppose R is a minimal ring. If $r \in R$ then there is $e \in \mathbf{B}(R)$ and $t \in \kappa(R)$ so that $(et)_x = r_x$. Hence the lower arrow is always surjective. If $\kappa(R)_y$ is finite, say of the form $\mathbf{Z}/(p^k)$, then, of necessity, R_x has the form $\mathbf{Z}/(p^m)$, $m \leq k$. At most one stalk of $\kappa(R)$ can be infinite. If $\kappa(R)_y$ is infinite, it has the form D, with $\mathbf{Z} \subseteq D \subseteq \mathbf{Q}$. Here R_x may be finite, but, if it is infinite, it must be D.

For the converse, there are two cases to be considered. First, suppose R has an infinite stalk D. It will follow that D is a stalk of $\kappa(R)$. Consider $p \in \mathbf{P}$ and assume that p is invertible in D. The set $U = \{u \in X \,|\, p \textit{ is invertible in } R_u\}$ is open in X. For y in the closed set $X \setminus U$, p is not invertible in R_y, which implies that R_y is finite. Hence for each $y \in X \setminus U$, there is some $k \geq 1$ so that $p^k = 0$ in R_y. By the compactness of $X \setminus U$, there is $m \geq 1$, chosen minimal with this property, so that $p^m \in Ry$ for all $y \in X \setminus U$. It follows that $X \setminus U$ is clopen (closed and open). In the function $f : \mathbf{P} \to \mathrm{N}\cup\{\infty\}$ defining $\kappa(R)$, $f(p) = m$. On the other hand, if p is not invertible in D and there were a splitting $R = R_p \oplus R^{(p)}$ into p-torsion and p-divisible parts, then the infinite stalks must be attached to the p-divisible part, which is impossible. Hence, $f(p) < \infty$ if and only if p is invertible in D, and thus $\kappa(R)$ has the infinite stalk D.

The next task is to show that $P(R) = R$. By the nature of the stalks of R, each $r \in R$ can be expressed in the form $r = r_1e_1 + \cdots r_ke_k$, where $e_1, \ldots, e_k$ is a complete orthogonal set of idempotents and each r_i is either in $\mathbf{Z}$ or in D. The terms with $r_i \in \mathbf{Z}$ are in $P(R)$ since, of course, $\mathbf{Z} \subseteq \kappa(R)$. Consider a term, say de, $e \in \mathbf{B}(R), d \in D$, where at least one of the stalks over the support of e is D. Let $f \in \mathbf{B}(\kappa(R))$ be any idempotent whose support contains $\alpha^{-1}(\mathrm{supp}\,(e))$ and over which d makes sense. Then $df \in \kappa(R)$ and $de = dfe$.

When all the stalks of R are finite, R is clearly generated by $\mathbf{B}(R)$ and $\mathbf{Z}$.

To prove the last statement, the finite and infinite cases are treated separately. If $y \in Y$ suppose first that $\kappa(R)_y = \mathbf{Z}/(p^k)$. Assume that for all $x \in X$, $x \supseteq y$, and all $e \in \mathbf{B}(R) \setminus x$ that $p^{k-1}e \in Rx$. Then $p^{k-1}e \in \bigcap_{x \supseteq y} Rx = Ry$ (the equality is by [11, Proposition 1.7] using $M = R/Ry$). But there is some $f \in \mathbf{B}(\kappa(R))$ with $p^{k-1}f \notin \kappa(R)y = Ry \cap \kappa(R)$. This is impossible since f cannot be in all $x \in X$ with $x \supseteq y$. Thus, for some $x \supseteq y$, $R_x = \mathbf{Z}/(p^k)$. Now suppose $\kappa(R)$ has an infinite stalk $\kappa(R)_y = D$. The set $F = \{x \in X \mid x \supseteq y\}$ is closed in X. For each prime p and $k \geq 1$, $\{x \in X \mid p^k \in Rx\}$ is open in F. If for all $x \in F$, R_x is finite, then for some $n \geq 1$, $n \in \bigcap_{x \in F} Rx = Ry$, by the compactness of F, and so $n \in \kappa(R)y$, which is impossible. □

Theorem 1.4 and Proposition 1.1 suggest that minimal rings can be determined by purely topological data.

If R is in $MinR$ then R is π-regular if and only if R has no infinite stalk or if its infinite stalks are $\mathbf{Q}$. In fact, the Krull dimension of R is either 0 (in the π-regular case) or 1.

Let us look at some simple examples.

Examples 1.5 (a) Let $R_1 = \mathbf{Z}/(2) \times \mathbf{Z}[1/2]$, $R_2 = \mathbf{Z}/(2) \times \mathbf{Z}/(3) \times \mathbf{Z}[1/2, 1/3]$ and $R_3 = \mathbf{Z}/(2) \times \mathbf{Z}/(3) \times \mathbf{Z}[1/2]$. These are all examples of minimal rings. The characteristic rings are $\kappa(R_1) = R_1$, $\kappa(R_2) = R_2$ and $\kappa(R_3) = R_1$, respectively.

(b) A slightly more complicated example is the minimal ring $R = C(X, T)$, the ring of continuous functions on a boolean space X with values in an indecomposable epimorph T of $\mathbf{Z}$ (or, equivalently, the ring corresponding to the simple Pierce sheaf on X with stalks all T). Here $\kappa(R) = T$.

(c) Given $q \in \mathbf{P}$, let S be the subring of $\prod_{k \geq 1} \mathbf{Z}/(q^k)$ consisting of those sequences $(a_1, a_2, \ldots)$ which are eventually "constant" in $\mathbf{Z}$. That is, there is $z \in \mathbf{Z}$ and $n \geq 1$ so that for all $m \geq n$, $a_m = z + (q^m)$. This is a minimal ring with $\kappa(S) = \mathbf{Z}$.

The ring R_3 in Examples 1.5(a) shows that a minimal ring R may have stalks which are proper homomorphic images of those of $\kappa(R)$. In order to emulate the construction of [12] in this more general setting, it is necessary to split up **Spec** (R) into subspaces over which all the stalks of R are the same.

It is now possible to look at the topological side of minimal rings.

Definition 1.6 *The following data define a* **boolean space with decomposition**

(i) a boolean space X and a function $f : \mathbf{P} \to \mathbf{N} \cup \{\infty\}$,

(ii) a closed subset X_0 of X, and for $p \in \mathbf{P}$, if $1 \leq f(p) < \infty$ there are open subsets of X, $X_{p,1} \subseteq \cdots \subseteq X_{p,f(p)}$ so that $X_{p,f(p)}$ is clopen and non-empty,

(iii) if for some $p \in \mathbf{P}$, $f(p) = \infty$ then $X_0 \neq \emptyset$ and, moreover, there are open subsets $X_{p,1} \subseteq X_{p,2} \subseteq \cdots$,

(iv) if $p, q \in \mathbf{P}$, $p \neq q$, and $i, j \geq 1$ then $X_{p,i} \cap X_{q,j} = \emptyset$, and $X_{p,i} \cap X_0 = \emptyset$, and

(v) $\bigcup_{1 \leq f(p) < \infty} X_{p,f(p)} \cup \bigcup_{f(p)=\infty} \bigcup_{k \geq 1} X_{p,k} \cup X_0 = X$.

Such a boolean space with decomposition is denoted $\mathcal{X} = (X, f, \{X_{p,k}\}, X_0)$. The following additional notation will be handy: for $p \in \mathbf{P}$, $X_p = \bigcup_{k\geq 1} X_{p,k}$, $X_{p,0} = \emptyset$ and for $k \geq 1$, $\dot{X}_{p,k} = X_{p,k} \setminus X_{p,k-1}$. When $f(p) < \infty$, we take $X_{p,k} = X_{p,f(p)}$ for $k \geq f(p)$. Notice that for $p \in \mathbf{P}$, $\overline{X_p} \setminus X_p \subseteq X_0$. There is always a continuous function $\phi_{\mathcal{X}} : X \to \mathbf{P} \cup \{\infty\}$, given by

$$\phi_{\mathcal{X}}(x) = \begin{cases} \infty \textit{ if } x \in X_p \textit{ with } f(p) = \infty \\ \infty \textit{ if } x \in X_0 \\ p \textit{ if } x \in X_p \textit{ with } f(p) < \infty \end{cases} .$$

It does not carry a great deal of information about $\mathcal{X}$, but has enough to be of use.

Definition 1.7 *Let $\mathcal{X} = (X, f, \{X_{p,k}\}, X_0)$ and $\mathcal{Y} = (Y, g, \{Y_{p,k}\}, Y_0)$ be boolean spaces with decompostion, a morphism $A : \mathcal{X} \to \mathcal{Y}$ is defined only if for $p \in \mathbf{P}$, $f(p) \leq g(p)$. In such a case, a* **morphism** *A is a continuous function $\alpha : X \to Y$ such that*

1) $\alpha(X_0) \subseteq Y_0$;

2) for $p \in \mathbf{P}$ and $k \geq 1$, $\alpha(\dot{X}_{p,k}) \subseteq Y_p \setminus Y_{p,k-1} \cup Y_0$; and

3) for $p \in \mathbf{P}$ and $k \geq 1$, if $g(p) < \infty$ then $\alpha(\dot{X}_{p,k}) \subseteq Y_p \setminus Y_{p,k-1}$.

The category of boolean spaces with decomposition is denoted BSD.

A contravariant equivalence between *MinR* and *BSD* will be established. Before this is done, it can be noticed that if a minimal ring R has no infinite stalks or its infinite stalks are all $\mathbf{Q}$, then the structure is simpler since the corresponding function $f : \mathbf{P} \to \mathbf{N} \cup \{\infty\}$ has $f(p) < \infty$ for all $p \in \mathbf{P}$. In this case R is π-regular (and conversely). Here the structure is much like in the minimal reguar case studied by Pierce, except that the clopen subset attached to a prime p has a decomposition as a finite union of open subsets as in Definition 1.6(ii).

Theorem 1.8 *The categories* MinR *and* BSD *are contravariantly equivalent. The equivalence is given by a pair of functors: Ω : MinR $\to$ BSD with $\Omega(R) = (X, f, \{X_{p,k}\}, X_0)$, where $X = \mathbf{Spec}\,\mathbf{B}(R)$, f is a function attached to $\kappa(R)$, $\dot{X}_{p,k} = \{x \in X \mid R_x = \mathbf{Z}/(p^k)\}$ and $X_0 = \{x \in X \mid R_x \text{ is infinite}\}$; and Λ : BSD $\to$ MinR where $\Lambda(\mathcal{X})$, $\mathcal{X} = (X, f, \{X_{p,k}\}, X_0)$, is the ring of sections of the Pierce sheaf over X whose stalks are $\mathbf{Z}/(p^k)$ for $x \in \dot{X}_{p,k}$ and $D = \mathbf{Z}[1/p \mid f(p) < \infty]$ for $x \in X_0$.*

Proof. Let R be in *MinR*. Put $X = \mathbf{Spec}\,\mathbf{B}(R)$, a boolean space. If R has an infinite stalk, then f is the function defining $\kappa(R)$ as in [4, Proposition 1.5]. If $\mathbf{Z}$-ann $R = (n) \neq 0$, where the prime decompostion of n is $p_1^{m_1} \cdots p_u^{m_u}$, then $f(p) = \begin{cases} m_i \textit{ if } p = p_i \\ 0 \textit{ otherwise} \end{cases}$. For $p \in \mathbf{P}$ and $k \geq 1$, set $\dot{X}_{p,k} = \{x \in X \mid R_x = \mathbf{Z}/(p^k)\}$ and $X_0 = \{x \in X \mid R_x \textit{ is infinite}\}$. Most of the properties of Definition 1.6 are clear. For example, $X_0 = \bigcap_{p\in\mathbf{P}} \bigcap_{k\geq 1} \{x \in X \mid (p^k)_x \neq 0\}$ is an intersection of closed sets. Each $X_{p,k}$ is open since it is the set $\{x \in X \mid (p^k)_x = 0\}$, and if $1 \leq f(p) < \infty$ then $X_{p,f(p)} = \{x \in X \mid p_x \textit{ is not invertible}\} \cap \{x \in X \mid (p^{f(p)-1})_x \neq 0\}$ is closed as well as open.

Now suppose $\gamma : R \to S$ is a homomorphism in *MinR* which induces $\alpha : X = \mathbf{Spec}\,\mathbf{B}(S) \to Y = \mathbf{Spec}\,\mathbf{B}(R)$. Put $\Omega(S) = \mathcal{X} = (X, f, \{X_{p,k}\}, X_0)$ and $\Omega(R) =$

$\mathcal{Y} = (Y, g, \{Y_{p,k}\}, Y_0)$. For $x \in X$ there is an induced homomorphism $\gamma_x : R_{\alpha(x)} \to S_x$. If $x \in \dot{X}_{p,k}$ for some $p \in \mathbf{P}$ and $k \geq 1$, then $\alpha(x) \in \dot{Y}_{p,m}$, for some $m \geq k$ or $\alpha(x) \in X_0$; hence, in particular, $f(p) \leq g(p)$. If $X_0 \neq \emptyset$ then for $x \in X_0$, $\gamma_x : R_{\alpha(x)} \to S_x$ is injective. Hence, $\alpha(X_0) \subseteq Y_0$. The remaining properties of the definition of a morphism are readily verified.

In order to construct the functor Λ, the Pierce sheaf of the ring $\Lambda(\mathcal{X})$, for $\mathcal{X} = (X, f, \{X_{p,k}\}, X_0)$, is built directly. In case $X_0 \neq \emptyset$, we will need the ring $D = \mathbf{Z}[1/p \mid f(p) < \infty]$. For each $x \in \dot{X}_{p,k}$ let $\mathcal{R}_x = \mathbf{Z}/(p^k)$ and for $x \in X_0$, $\mathcal{R}_x = D$. The *espace étalé* of the sheaf will be $\mathcal{R} = \dot{\bigcup}_{x \in X} \mathcal{R}_x$ over X. We know from Proposition 1.1 that there is at most one way of turning $(X, \mathcal{R})$ into a Pierce sheaf. Given $a, b \in \mathbf{Z}$ with $b \neq 0$, N a clopen set in X so that b^{-1} is defined in $\mathcal{R}_x$ for all $x \in N$, then a basic open set of $\mathcal{R}$ is of the form $\sigma_{N,a,b} = \{ab^{-1} \in \mathcal{R}_x \mid x \in N\}$. The verification of the conditions of [11, Definitions 3.1(a) and 4.1] is easy. The essential point is that given $0 \neq b \in \mathbf{Z}$ and $x \in X$ so that b^{-1} makes sense in $\mathcal{R}_x$, then there is a clopen neighbourhood N of x on which b^{-1} is defined. To see this, write $b = p_1^{m_1} \cdots p_u^{m_u}$, its prime factorization. If $x \in X_{p,k}$, for some $p \in \mathbf{P}$ and $k \geq 1$, then $p \neq p_1, \ldots, p_u$ and we can take a clopen neighbourhood N in $X_{p,k}$. If $x \in X_0$, it follows that $f(p_i) < \infty$ for $i = 1, \ldots, u$. Then b is invertible over the clopen set $X \setminus \big(X_{p_1, f(p_1)} \cup \cdots \cup X_{p_u, f(p_u)}\big)$.

Let $R = \Lambda(\mathcal{X})$ be the ring of sections of the sheaf just constructed. Before looking at morphisms in *BSD*, let us observe that R is indeed in *MinR*. This follows since the Pierce sheaf of R satisfies the conditions of Theorem 1.4.

We next look at a morphism $A : \mathcal{X} \to \mathcal{Y}$ with associated continuous function $\alpha : X \to Y$. Put $S = \Lambda(\mathcal{X})$ and $R = \Lambda(\mathcal{Y})$. Then $f(p) \leq g(p)$ for all $p \in \mathbf{P}$, so, in particular, the infinite stalk of R (if $X_0 \neq \emptyset$) is included in a unique way in the infinite stalk E of S. By definition of a morphism, $\alpha(\dot{X}_{p,k}) \subseteq \bigcup_{m \geq k} \dot{Y}_{p,m} \cup Y_0$ so that there is always a unique surjection $R_{\alpha(x)} \to S_x$ of $\mathbf{Z}/(p^m) \to \mathbf{Z}/(p^k)$ or $D \to \mathbf{Z}/(p^k)$. Given $r \in R$ (a locally constant function on Y), r_y may be thought of as some $a(y)b(y)^{-1}$, for integers $a(y)$ and $b(y)$, interpreted in the stalk R_y. A function on X may be defined by $\zeta(r)(x) = a(\alpha(x))b(\alpha(x))^{-1}$, but now interpreted as an element of S_x, via the unique homomorphism $R_{\alpha(x)} \to S_x$. If r is constantly ab^{-1} on some clopen $N \subseteq Y$, then $\zeta(r)$ is constantly ab^{-1} on the clopen set $\alpha^{-1}(N) \subseteq X$. Hence ζ is a ring homomorphism and we define $\Lambda(A) = \zeta$. The conditions for a functor are easy to verify.

If we start with some R in *MinR* then for $x \in X = \mathbf{Spec}\,\mathbf{B}(R)$, $\Lambda(\Omega(R))_x = R_x$ by construction. From this, Proposition 1.1 shows that $\Lambda(\Omega(R)) \cong R$, since R is naturally isomorphic to the ring of sections of its Pierce sheaf. On the other hand, for $\mathcal{X} = (X, f, \{X_{p,k}\}, X_0)$, $\Omega(\Lambda(\mathcal{X}))$ has underlying space $\mathbf{Spec}\,\mathbf{B}(\Omega(\Lambda(\mathcal{X}))) = X$, by construction. We have already seen that f and whether or not $X_0 = \emptyset$ determine $\kappa(\Omega(\Lambda(\mathcal{X})))$, which again gives rise to f. The other components of $\Omega(\Lambda(\mathcal{X}))$ are clear from the manner in which the sheaf for $\Lambda(\mathcal{X})$ was built. $\square$

In the examples 1.5, we have the following. In (a), $\Omega(R_1) = (\{x, y\}, f_1, \{X_{p,k}\}, \{y\})$, where $f_1(2) = 1$ and $f_1(p) = \infty$ for $p \neq 2$, and all $X_{p,k} = \emptyset$ except $X_{2,1} = \{x\}$; $\Omega(R_2) = (\{x, y, z\}, f_2, \{X_{p,k}\}, \{z\})$, where $f_2(2) = f_2(3) = 1$ and $f_2(p) = \infty$ for $p \neq 2, 3$ and $X_{p,k} = \emptyset$ except for $X_{2,1} = \{x\}$ and $X_{3,1} = \{y\}$; $\Omega(R_3)$

is similar except that $f_3(3) = \infty$. In the example (b), $\Omega(R)$ depends on T. If, for example, $T = \mathbf{Q}$ then all $f(p) = 0$ and $X_0 = X$. If $T = \mathbf{Z}[1/2]$, then $f(p) = \infty$ for all $p \neq 2$ and $f(2) = 0$. In example (c), $\Omega(S) = (X, g, \{X_{p,k}\}, X_0)$, where $X = \mathbf{N} \cup \{\infty\}$, $X_{p,k} = \emptyset$ for $p \neq q$ and $X_{q,k} = \{1, \ldots, k\}$, while $X_0 = \{\infty\}$ and $g(p) = \infty$ for all $p \in \mathbf{P}$.

One special case of the construction of Theorem 1.8 will be particularly useful but a preliminary lemma is worth noting.

Lemma 1.9 *Suppose that $(X, \mathcal{R})$ is any Pierce sheaf with ring of sections R and V a closed subspace of X; let $(V, \mathcal{R}')$ be the restriction to V and S its ring of sections.*

(i) The restriction map $R \to S$ is a surjection (that is, a Pierce sheaf is a soft *sheaf).*

(ii) The ring S is a localization, $T^{-1}R$, for a multiplicatively closed set of idempotents T.

Proof. The first part is yet another application of the usual Pierce technique for constructing sections. Suppose $s \in S$ is a section over V. For each $v \in V$ there is $r^{(v)} \in R$ so that $s_v = (r^{(v)})_v$. Let G_v be any open set in X such that $G_v \cap V = \{y \in V \mid (r^{(v)})_y = s_y\}$. Then $\{X \setminus V, \{G_v\}_{v \in V}\}$ is an open cover of X. Let $\{X \setminus V = G_0, G_{v_1}, \ldots, G_{v_m}\}$ be a finite subcover. Then there exists a cover consisting of disjoint clopen sets $N_1, \ldots, N_l$ so that each N_i is in some G_{v_j} or is in G_0. For each $i = 1, \ldots, l$, if N_i is in some G_{v_j}, pick one so that $N_i \subseteq G_{v_{j(i)}}$, or, otherwise, put $j(i) = 0$. Then define $t \in R$ by $t_x = \begin{cases} 0 \; \textit{if}\, x \in N_i \; \textit{with}\, j(i) = 0 \\ (r^{(v_{j(i)})})_x \; \textit{if}\, x \in N_i \; \textit{and}\, j(i) \neq 0 \,. \end{cases}$ Then $t_{|V} = s$.

Take $T = \{e \in \mathbf{B}(R) \mid e \notin v, \forall v \in V\}$. Then $\eta : R \to T^{-1}R$ defined by $r \mapsto r1^{-1}$ has kernel $\{r \in R \mid \exists e \in T \; \textit{such that}\, re = 0\} = \{r \in R \mid r_{|V} = 0\}$. $\square$

The restriction of a Pierce sheaf $(X, \mathcal{R})$ to a closed set V is not necessarily a Pierce sheaf, but it will be one if the stalks of $\mathcal{R}$ over V are indecomposable.

Proposition 1.10 *Given a ring R in* MinR *with* $\mathbf{Spec}\,\mathbf{B}(R) = X$ *and a boolean space Y equipped with a continuous function $\alpha : Y \to X$, the object $\Omega(R)$ of* BSD *induces a decomposition of Y as follows. Set $\Omega(R) = (X, f, \{X_{p,k}\}, X_0)$ and then define $\mathcal{Y} = (Y, \tilde{f}, \{\alpha^{-1}(X_{p,k})\}, \alpha^{-1}(X_0))$ where $\tilde{f}(p) = f(p)$ when $\alpha^{-1}(X_0) \neq \emptyset$, and when $\alpha^{-1}(X_0) = \emptyset$, $\tilde{f}(p) = f(p)$ if $f(p) < \infty$ and $\tilde{f}(p) = 0$ if $f(p) = \infty$. Then $\mathcal{Y}$ is in* BSD *and α induces a morphism $A : \mathcal{Y} \to \mathcal{X}$. Moreover, the ring $\Lambda(\mathcal{Y}) = S$ is flat as an R-module and the image of R in S is a pure R-submodule of S. Finally, if α is surjective, S is faithfully flat as an R-module.*

Proof. It is easy to check that the data for $\mathcal{Y}$ satisfy the definition of an objcct in *BSD* and that α is a morphism in *BSD*. Flatness is a local property by the construction of the tensor product of R-modules when viewed as sheaves of modules ([11, page 41]). This reduces the question of the flatness of S over R to that of the stalk S_x of S as an R_x-module, for $x \in X$. But by the lemma, $S_x = C(\alpha^{-1}(x), R_x)$, the restriction of S to $\alpha^{-1}(x)$, as an R_x-module. Then the usual criterion is used (e.g., Stenström, [13, 10.7]). Purity is also a local property and it is readily verified that $R_x \to C(\alpha^{-1}(x), R_x)$ is pure. Now suppose that α is surjective. If for some R-module M, $M \otimes_R S = 0$, then for all $x \in X$, $M_x \otimes S_x = 0$. By purity, $M_x = 0$ for all $x \in X$. Then by [11, Proposition 1.7], $M = 0$. $\square$

Given $\mathcal{X} = (X, f, \{X_{p,k}\}, X_0)$ in BSD, and a continuous surjection $\alpha : X \to Y$, Y a boolean space, there is no natural "direct image" construction unless stringent conditions are imposed on α. However, it is always possible to define $\mathcal{Y} = (Y, g, \{\emptyset\}, Y)$ with $g(p) = \infty$ for all $p \in \mathbf{P}$. Then α defines a morphism $\mathcal{X} \to \mathcal{Y}$.

2. Changing the central idempotents of a ring.

In this section it will be seen that for a ring R, with indecomposable Pierce stalks, it is fruitful to think of R as a $P(R)$-algebra. This point of view will turn a sheaf construction into a simple change of scalars procedure and allows easy modifications of the set of central idempotents.

Goodearl and Warfield have shown in [8] that it is useful to know when a ring R can be regarded as an algebra over a commutative π-regular ring. For that reason the following remark is noted. It says, in particular, that the existence of such an algebra structure is a property of the additive structure of the ring.

Remark 2.1 *Let R be a ring. Then R has the structure of an S-algebra for some commutative π-regular ring S if and only if $P(R)$ is π-regular.*

Proof. If $P(R)$ is π-regular, it can play the role of S. Suppose there is a homomorphism $\phi : S \to R$ with S commutative π-regular and $\phi(S)$ in the centre of R. If $\mathbf{Z}$-ann $R \neq 0$, then $P(R)$ is π-regular. Otherwise, it needs to be shown that any infinite stalk of $P(R)$ is $\mathbf{Q}$. Put $T = \phi(\kappa(S))$ and let the inclusion of T in $P(R)$ induce $\alpha : X = \mathbf{Spec}\,\mathbf{B}(R) \to Y = \mathbf{Spec}\,\mathbf{B}(T)$. Suppose for $x \in X$ that $P(R)_x$ is infinite, then so is $T_{\alpha(x)} = \mathbf{Q}$ and $T_{\alpha(x)}$ injects into $P(R)_x$. $\square$

If R is a ring with indecomposable Pierce stalks (see [3, Proposition 1] for classes of rings where this occurs), put $X = \mathbf{Spec}\,\mathbf{B}(R)$ and denote by $(X, \mathcal{R})$ its Pierce sheaf. Suppose that there is continuous function $\alpha : Y \to X$, for some boolean space Y. This corresponds to a homomorphism of boolean algebras $\beta : \mathbf{B}(R) \to C$. The function α gives rise to an inverse image sheaf, $\alpha^{-1}(X, \mathcal{R})$ (e.g., Godement, [7, § 2.11]). It is readily seen, using the hypothesis on the stalks of $(X, \mathcal{R})$, that the new sheaf is again a Pierce sheaf, say $(Y, \mathcal{S})$ with ring of sections S.

Proposition 2.2 *Let R be a ring whose Pierce stalks are indecomposable and whose Pierce sheaf is $(X, \mathcal{R})$. Suppose that $\alpha : Y \to X$ is a continuous function, where Y is a boolean space. Denote by $(Y, \mathcal{S})$ the inverse image sheaf and by S its ring of sections. Set $\Omega(R) = \mathcal{X}$ and let the object of* BSD *induced by α (as in 1.10) be $\mathcal{Y}$. Put $T = \Lambda(\mathcal{Y})$. Then*

$$S \cong T \otimes_{P(R)} R\,.$$

Moreover, the ring S is (right) R-flat. If α is surjective, S is (right) R-faithfully flat. Finally, if C is the boolean algebra dual to Y then $\mathbf{B}(S) \cong C$ and the natural homomorphism $\mathbf{B}(R) \to \mathbf{B}(S)$ is dual to α.

Proof. All of this is straightforward from the inverse image construction, except the isomorphism $S \cong T \otimes_{P(R)} R$. If $P(R)$ has an infinite stalk, denote it by D. Let S be the ring constructed from the inverse image sheaf. There is a $P(R)$-bilinear function $\eta : T \times R \to S$ defined by $\eta(t, r) = s$, where $s_y = t_y r_{\alpha(y)}$, for each $y \in Y$, (recall from 1.10 that T_y is a subring of $R_{\alpha(y)}$). This is a section of $(Y, \mathcal{S})$ since t is locally constant, and every section is a sum of such. Then η induces a surjection (for convenience, with the same name), $\eta : T \otimes_{P(r)} R \to S$. If $\eta(\sum_{i=1}^m t_i \otimes r_i) = 0$ then $\sum_{i=1}^m (t_i)_y (r_i)_{\alpha(y)} = 0$ for each $y \in Y$. Then there is a clopen neighbourhood N of y on which all the t_i are "constant", say $t_i = a_i b_i^{-1}$ in D or in $\mathbf{Z}$. Let $e \in \mathbf{B}(R)$ be such that $ea_i b_i^{-1}$ is an element of $P(R)$ and the support of e covers $\alpha(N)$. Then $\sum_{i=1}^m (t_i)_z (r_i)_{\alpha(z)} = 0$ for all $z \in N$. Let $f \in \mathbf{B}(T)$ be the idempotent with support N. Denote by $\zeta : P(R) \to T$, the homomorphism defined by α. It follows that $\zeta(e) f = f$. Then $\sum_{i=1}^m t_i f \otimes r_i = \sum_{i=1}^m a_i b_i^{-1} \zeta(e) f \otimes r_i = \sum_{i=1}^m f \otimes e a_i b_i^{-1} r_i = 0$. The compactness of Y now shows that $\sum_{i=1}^m t_i \otimes r_i = 0$. The faithful flatness statement follows from 1.10 and a change of rings theorem (e.g., Atiyah and MacDonald, [1, 2.20]).

There is a natural homomorphism $\nu : R \to S$, $r \mapsto 1 \otimes r \overset{\eta}{\mapsto} s$, where for $y \in Y$, $s_y = r_{\alpha(y)}$. If $e \in \mathbf{B}(R)$, then $\operatorname{supp}(\nu(e)) = \alpha^{(-1)}(\operatorname{supp}(e))$. □

This shows that any boolean algebra homomorphism $\beta : \mathbf{B}(R) \to C$ can be realized as the restriction of a ring homomorphism $\nu : R \to S$, in a canonical way; indeed, the Pierce stalks of S are from among those of R. There are, however, many ways of changing the central idempotents of a ring other than what we have just seen. If $\mu : R \to R'$ is a conformal homomorphism, then the subring of R' generated by $\mu(R)$ and $\mathbf{B}(R')$, denoted by $\langle \mu(R), \mathbf{B}(R') \rangle$, is a ring generated by (an image of) R and some central idempotents. However, $\beta = \mu_{|\mathbf{B}(R)} : \mathbf{B}(R) \to \mathbf{B}(R')$ also gives rise to some $\mathcal{Y}$ in *BSD* and to the ring $S = \Lambda(\mathcal{Y}) \otimes_{P(R)} R$, as constructed above. We shall see that S and the homomorphism $\nu : R \to S$, as in 2.2, are *generic* as a realization of β by means of a ring homomorphism.

Proposition 2.3 *Let R be a ring whose Pierce stalks are indecomposable, $\mu : R \to R'$ a conformal homomorphism and $\beta = \mu_{|\mathbf{B}(R)} : \mathbf{B}(R) \to \mathbf{B}(R')$. Put $X = \mathbf{Spec}\,\mathbf{B}(R)$ and $Y = \mathbf{Spec}\,\mathbf{B}(R')$. Then β induces $\alpha : Y \to X$ and a morphism $A : \mathcal{Y} \to \mathcal{X}$ in* BSD*, where $\mathcal{X} = \Omega(R)$ and $\mathcal{Y}$ is the induced object in* BSD*. Define U to be the subring of R' generated by $\mu(R)$ and $\mathbf{B}(R')$, and put $S = \Lambda(\mathcal{Y}) \otimes_{P(R)} R$. Then μ factors as $\mu = \eta\nu$, where $\nu : R \to S$ is as in Proposition 2.2 and η is a surjection.*

Proof. The situation is summarized in the following diagram:

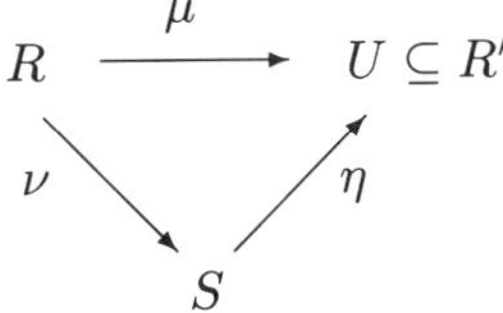

The homomorphism $\beta = \mu_{|P(R)} : P(R) \to P(R')$ gives rise to a morphism $B : \mathcal{Y}' \to \mathcal{X}$ in *BSD*, where $\mathcal{X} = \Omega(P(R))$ and $\mathcal{Y}' = \Omega(P(R'))$. Write $\mathcal{X} =$

$(X, f, \{X_{p,k}\}, X_0)$ and $\mathcal{Y}' = (Y, g, \{Y_{p,k}\}, Y_0)$. Now $\alpha : Y \to X$ produces a morphism $A : \mathcal{Y} \to \mathcal{X}$, where $\mathcal{Y} = (Y, \tilde{f}, \{\alpha^{-1}(X_{p,k})\}, \alpha^{-1}(X_0))$, as in Proposition 1.10. It is readily seen that the identity $\iota : Y \to Y$ gives a morphism $I : \mathcal{Y}' \to \mathcal{Y}$, since f and g already satisfy the required conditions for a morphism. Put $\delta = \Lambda(I)$. The homomorphism $\eta : S \to U$ is defined as follows: $\eta(v \otimes r) = \delta(v)\mu(r)$. For $r \in R$, $\eta\nu(r) = \eta(1 \otimes r) = \mu(r)$. Hence $\mu = \eta\nu$ and $\mu(R) \subseteq \eta(S)$. Finally, δ restricted to $\mathbf{B}(\Lambda(\mathcal{Y}))$ is an isomorphism onto $\mathbf{B}(R')$ and so $\mathbf{B}(R') \subseteq \eta(S)$, and thus η is a surjection. □

Certainly, the ideal $\Lambda(\mathcal{Y}) \otimes_{P(R)} \ker(\mu) \subseteq \ker(\eta)$, but as we see in Examples 2.5, $\ker(\eta)$ may be larger; however, when R is commutative regular, the case is simpler, reflecting the fact that the stalks of R are fields.

Remark 2.4 *Consider the situation of Proposition 2.3 where, now, R is a commutative regular ring. Then $\ker(\eta) = 0$ and $S \cong U$.*

Proof. Suppose $s = \sum_i t_i \otimes r_i \in \ker(\eta)$. Just as in the proof of Proposition 2.2, s can be written in the form $\sum_j f_j \otimes s_j$, where the f_j form a complete set of orthogonal idempotents and $s_j \in R$. Then $\eta(s) = \sum_j f_j\mu(s_j) = 0$. Since η restricted to $\mathbf{B}(S)$ is an isomorphism, it follows that each term $f_j\mu(s_j) = 0$. Now if for some $f \in \mathbf{B}(R')$ and $r \in R$, $f\mu(r) = 0$, let $e \in \mathbf{B}(R)$ have the same support as r, then $\mu(e)f = 0$ and so $f = (1 - \mu(e))f$. Hence, $f \otimes r = f(1 - \mu(e)) \otimes r = f \otimes (1 - e)r = 0$. □

Examples 2.5 (a) Let $R = \mathbf{Z}/(4)$ and $R' = \mathbf{Z}/(4) \times \mathbf{Z}/(2)$ and $\mu : R \to R'$ be given by $\mu(z) = (z, \bar{z})$. Then $U = R'$, $S \cong \mathbf{Z}/(4) \times \mathbf{Z}/(4)$ and $\eta(z, w) = (z, \bar{w})$.

(b) Let R be the ring of all sequences of integers which are either all even or all odd. Here $P(R) = \kappa(R) = \mathbf{Z}$. Set $R' = \prod_{\mathbf{N}} \mathbf{Q}$, the complete ring of quotients of R (see [13]). Here, using the notation of the proposition, $U = \prod_{\mathbf{N}} \mathbf{Z}$ and $S = C(\beta\mathbf{N}, \mathbf{Z}) \otimes_{\mathbf{Z}} R \cong C(\beta\mathbf{N}, R)$. Further, for $f \in C(\beta\mathbf{N}, \mathbf{Z})$ and $r \in R$, f is written $f = \sum_{i=1}^m e_i z^{(i)}$, where $\{e_i\}$ forms a complete set of orthogonal idempotents and $z^{(i)} \in \mathbf{Z}$; then for $j \in \operatorname{supp}(e_i) \cap \mathbf{N} \subseteq \mathbf{N} \subset \beta\mathbf{N}$, the j-component of $\eta(f \otimes r)$ is $z^{(i)} r_j$, where r_j is the j-component of r. In other words, if $\tilde{e_i}$ is the restriction of e_i to $\mathbf{N}$, then $\eta(f \otimes r) = \sum_{i=1}^m \tilde{e_i} z^{(i)} r$. Note that η is not one-to-one and S is not a ring of quotients of R. □

3. The categories *MinR* and *BSD*.

The two categories, *MinR* and *BSD*, are contravariantly equivalent and so the study of one is the essentially the same as the study of the other. However, as usually happens in such cases, one gets more information by switching from one category to the other, depending on the topic. The first observation is obvious but worth noting (cf.,[12, Corollary 2.5]). The terminology is that of [10].

Proposition 3.1 *The category* MinR *is a coreflexive subcategory of* RNGS *and of* CRNGS*. Moreover, the functor P :* CRNGS $\to$ MinR *is a subfunctor of the identity functor on* CRNGS*. The categories* MinR *and* BSD *are complete and cocomplete.*

Proof. The functor P on *RNGS* (or on *CRNGS*) is clearly a right adjoint of the inclusion of *MinR* into *RNGS* (or into *CRNGS*). Since *CRNGS* is a variety in the sense of universal algebra, it is complete and cocomplete. The fact that P is a subfunctor of the identity on *CRNGS* gives that *MinR* is complete and cocomplete. The statement about *BSD* follows by the equivalence of Theorem 1.8. In fact, the limit of a diagram in *MinR* is obtained by taking its limit in *CRNGS* and then applying the functor P; the colimit of a diagram in *CRNGS* all of whose objects are in *MinR* is already in *MinR*, and is its colimit there. □

The category *MinR* is not reflexive since there are rings, such as $\mathbf{Z}[X]$, with no canonical homomorphic image in *MinR*. One consequence of Proposition 3.1 is that the $\mathbf{Z}$-tensor product of two rings in *MinR* is again in *MinR*, and, similarly, for other coproducts.

Even though duality implies that *BSD* is cocomplete, it is instructive to see what coproducts in it are like. Suppose that $\mathcal{X} = (X, f, \{X_{p,k}\}, X_0)$ and $\mathcal{Y} = (Y, g, \{Y_{p,k}\}, Y_0)$ are objects in *BSD*. Set $Z = X\dot{\cup}Y$, the disjoint union, and $h : \mathbf{P} \to \mathbf{N}\cup\{\infty\}$ by $h(p) = \max(f(p), g(p))$. Moreover, for $p \in \mathbf{P}$ and $k \geq 1$, we define $Z_{p,k} = X_{p,k} \cup Y_{p,k}$ and $Z_0 = X_0 \cup Y_0$. The claim is that $\mathcal{Z} = (Z, h, \{Z_{p,k}\}, Z_0)$ is in *BSD*, and, with the natural inclusions of $\mathcal{X}$ and $\mathcal{Y}$ in $\mathcal{Z}$, is the coproduct of $\mathcal{X}$ and $\mathcal{Y}$. The details of the verification are omitted.

The case of a coproduct of infinitely many terms is similar, but the boolean compactification ([12, page 342]) of the disjoint union of the underlying spaces must be used. Suppose $\mathcal{X}^a = (X^a, f^a, \{X^a_{p,k}\}, X^a_0), a \in A$, is a family of objects from *BSD*. Put $Z = \delta(\dot{\bigcup}_{a\in A} X^a)$, the boolean compactification, and define $h : \mathbf{P} \to \mathbf{N}\cup\{\infty\}$ by $h(p) = \sup_{a\in A}\{f^a(p)\}$. An object $\mathcal{Z} = (Z, h, \{Z_{p,k}\}, Z_0)$ is defined as follows. If $h(p) = \infty$ and $k \geq 1$, $Z_{p,k} = \bigcup_{a\in A} X^a_{p,k}$, an open set in Z since $\dot{\bigcup}_{a\in A} X^a$ is open in its compactification. If $h(p) < \infty$ and $1 \leq k < h(p)$ then $Z_{p,k} = \bigcup_{a\in A} X^a_{p,k}$,while $Z_{p,h(p)} = \overline{\bigcup_{a\in A} X^a_{p,h(p)}}$. Finally, $Z_0 = \overline{\bigcup_{a\in A} X^a_0}$.

Lemma 3.2 *The quadruple $\mathcal{Z} = (Z, h, \{Z_{p,k}\}, Z_0)$ is an object in* BSD.

Proof. It remains to verify that when $h(p) < \infty$, $Z_{p,h(p)}$ is clopen, and that the components $\{Z_p, p \in \mathbf{P}, Z_0\}$ are disjoint. A function like $\phi_{\mathcal{X}}$ mentioned after Definition 1.6 is of use. Define $\phi : \bigcup_{a\in A} X^a \to \mathbf{P}\cup\{\infty\}$ by

$$\phi(z) = \begin{cases} \infty \textit{ if } z \in X^a_p, \textit{for some } a, \textit{ and } h(p) = \infty \\ \infty \textit{ if } z \in X^a_0, \textit{for some } a \\ p \textit{ if } z \in X^a_p, \textit{for some } a, \textit{ and } h(p) < \infty \end{cases} .$$

Note that if $h(p) < \infty$ then $\bigcup_{a\in A} X^a_{p,f(p)} = \bigcup_{a\in A} X^a_p$ is closed (as well as open) in $\bigcup_{a\in A} X^a$. Hence, ϕ is continuous. By the universal property of the boolean compactification, ϕ extends uniquely to a continuous function $\psi : Z \to \mathbf{P}\cup\{\infty\}$. When $h(p) < \infty$, $\psi^{-1}(p)$ is a clopen set containing $\bigcup_{a\in A} X^a_p$ and excluding $\bigcup_{a\in A} X^a \setminus \bigcup_{a\in A} X^a_p$. Hence $\psi^{-1}(p) = Z_{p,h(p)}$. Now that we know that each Z_p is open, the disjointness of the components is clear. □

It is easy to check that the inclusions $X^a \to Z$ give morphisms in BSD. The universal property of the boolean compactification comes up again in the verification that $\mathcal{Z}$ is the coproduct.

The product of rings in $MinR$ can be strictly smaller than the cartesian product; for example $P(\mathbf{Z} \times \mathbf{Z}[1/2]) = \mathbf{Z} \times \mathbf{Z}$, or $P(\prod_{\mathbf{N}} \mathbf{Z}) = C(\beta\mathbf{N}, \mathbf{Z})$, where $\beta\mathbf{N}$ is the Stone-Čech compactification of $\mathbf{N}$.

Let us now look at the product of two objects in BSD. Suppose that $\mathcal{X} = (X, f, \{X_{p,k}\}, X_0)$ and $\mathcal{Y} = (Y, g, \{Y_{p,k}\}, Y_0)$ are objects in BSD. The product, $\mathcal{Z} = (Z, h, \{Z_{p,k}\}, Z_0)$, will be based on the following subspace Z of $X \times Y$:

$$\begin{aligned} Z &= \bigcup_{p\in\mathbf{P}} \{(x,y) \,|\, x \in X_p, y \in Y_p\} \cup \\ &\bigcup_{p\in\mathbf{P}, g(p)=\infty} \{(x,y) \,|\, x \in X_p, y \in Y_0\} \cup \\ &\bigcup_{p\in\mathbf{P}, f(p)=\infty} \{(x,y) \,|\, x \in X_0, y \in Y_p\} \cup X_0 \times Y_0 \ . \end{aligned}$$

Notice that Z can be empty (and, to accommodate this, we have allowed the trivial ring to be in $MinR$). The appropriate subspaces are as follows: $Z_{p,k} = \{(x,y) \in Z \,|\, x \in X_{p,k} \ or \ y \in Y_{p,k}\}$ and $Z_0 = X_0 \times Y_0$. The function h is $h(p) = \min\{f(p), g(p)\}$ for all $p \in \mathbf{P}$.

We need to show that $\mathcal{Z}$ is in BSD. The first step is to verify that Z is closed in $X \times Y$. Indeed, we get the union of open sets

$$(X \times Y) \setminus Z = \bigcup_{p,q\in\mathbf{P},\, p\neq q} X_p \times Y_q \cup \bigcup_{g(p)<\infty} X_p \times (Y \setminus Y_p) \cup \bigcup_{f(p)<\infty} (X \setminus X_p) \times Y_p \ .$$

It must now be shown that the subsets $Z_{p,k}$ are open. When $f(p)$ and $g(p)$ are both finite, $Z_{p,k} = X_{p,k} \times Y_p \cup X_p \times Y_{p,k}$, which is open, even in $X \times Y$. When $f(p) < \infty$ and $g(p) = \infty$, $Z_{p,k} = X_p \times Y_{p,k} \cup X_{p,k} \times Y_p \cup X_{p,k} \times Y_0$ and the first term is open in $X \times Y$ while the union of the last two is $Z \cap (X_{p,k} \times Y)$. The remaining two cases are similar. Clearly, also, $Z_{p,1} \subseteq Z_{p,2} \subseteq \cdots$.

If $h(p) < \infty$, suppose for convenience that $h(p) = f(p)$. Then we have $Z_{p,f(p)} = X_{p,f(p)} \times Y_{p,g(p)}$ if $g(p) < \infty$ and when $g(p) = \infty$,

$$Z_{p,f(p)} = X_{p,f(p)} \times Y_p \cup X_{p,f(p)} \times Y_0 = Z \cap \big(X_{p,f(p)} \times (\bigcup_{q\in\mathbf{P}, g(q)=\infty} Y_q \cup Y_0)\big) \ ,$$

which is closed because $\bigcup_{q\in\mathbf{P}, g(q)=\infty} Y_q \cup Y_0$ is closed in Y by Definition 1.6.

It is easy to verify that the projections $\pi_1 : Z \to X$ and $\pi_2 : Z \to Y$ give rise to morphisms $P_1 : \mathcal{Z} \to \mathcal{X}$ and $P_2 : \mathcal{Z} \to \mathcal{Y}$. It remains to show that $\mathcal{Z}$, along with its projections, satifies the universal property of a product. Suppose that $\mathcal{W} = (W, l, \{W_{p,k}\}, W_0)$ is in BSD and that $T_1 : \mathcal{W} \to \mathcal{X}$ and $T_2 : \mathcal{W} \to \mathcal{Y}$ are morphisms based on functions τ_1 and τ_2, respectively. Define $\sigma : W \to Z$ by $\sigma(w) = (\tau_1(w), \tau_2(w))$. This yields a morphism in BSD. First, σ is well-defined since if $w \in W_0$ then $\sigma(w) \in X_0 \times Y_0 \subseteq Z_0$ and if $w \in W_{p,k}$, then Definition 1.7 gives the following possibilities: for $1 \leq l(p) < \infty$ and $f(p), g(p) < \infty$ then $\sigma(w) \in X_{p,m} \times Y_{p,n}$

for some $m, n \geq k$ so that $\sigma(w) \in Z_{p,\min(m,n)}$, if $f(p) < \infty$ and $g(p) = \infty$ then $\tau_1(w) \in X_{p,m}$ for some $m \geq k$ and $\tau_2(w) \in Y_p \cup Y_0$, again giving the result; the remaining cases are similar. For $l(p) = \infty$, if $w \in W_{p,k}$ then $\tau_1(w) \in X_p \cup X_0$ and $\tau_2(w) \in Y_p \cup Y_0$ giving $\sigma(w) \in Z_{p,m}$, for some $m \geq k$, or $\sigma(w) \in Z_0$.

That σ gives rise to a morphism $S : \mathcal{W} \to \mathcal{Z}$ is easily verified, as is its uniqueness.

Now let us turn briefly to the general case. The following notation will be convenient: For $\mathcal{X} = (X, f, \{X_{p,k}\}, X_0)$, let $X_{p,\infty} = \emptyset$ if $f(p) < \infty$ and $X_{p,\infty} = X_0$ if $f(p) = \infty$. Suppose that $\mathcal{X}^a = (X^a, f^a, \{X^a_{p,k}\}, X^a_0)$ is an object in *BSD* for all $a \in A$. A new object $\mathcal{Z} = (Z, h, \{Z_{p,k}\}, Z_0)$ is defined as follows. Put $Z = \{(x_a) \in \prod_{a\in A} X^a \mid \textit{for some } a \in A, x_a \in X^a_p \textit{ and for all } b \in A, x_b \in X^b_p \cup X^b_{p,\infty}\} \cup \prod_{a\in A} X^a_0$. The subset $Z_{p,k} = \{(x_a) \in \prod_{a\in A} X^a \mid \textit{for some } a \in A, x_a \in X^a_{p,k} \textit{ and for all } b \in A, x_b \in X^b_p \cup X^b_{p,\infty}\}$ and $Z_0 = \prod_{a\in A} X^a_0$. Define $h : \mathbf{P} \to \mathbf{N} \cup \{\infty\}$ by $h(p) = \inf\{f^a(p) : a \in A\}$.

The verification that $\mathcal{Z}$ is the product proceeds much as in the special case. For example, Z is closed in $\prod_{a\in A} X^a$ since

$$\prod_{a\in A} X^a \setminus Z = \bigcup_{a,b\in A, a\neq b} \bigcup_{p\in\mathbf{P}} X^a_p \times (X^b \setminus (X^b_p \cup X^b_{p,\infty})) \times \prod_{c\neq a,b} X^c .$$

The final remarks about our categories concern their monomorphisms and epimorphisms ([10, page 19]). Both categories are unusual in having monomorphisms which are not injective (cf. the classical Stone duality, e.g., Koppelberg, [9, Theorem 8.2]).

Proposition 3.3 *Let $\mathcal{X} = (X, f, \{X_{p,k}\}, X_0)$ and $\mathcal{Y} = (Y, g, \{Y_{p,k}\}, Y_0)$ be objects in* BSD *and $B : \mathcal{X} \to \mathcal{Y}$ a morphism given by $\beta : X \to Y$. Then B is a monomorphism if and only if for $x, x' \in X$, with $x \neq x'$, $\beta(x) = \beta(x')$ implies that either (i) $x \in X_{p,k}, x' \in X_{q,m}$ for primes $p \neq q$ and some $k, m \geq 1$, or (ii) $x \in X_{p,k}$ for some prime p and $k \geq 1$, $x' \in X_0$ and $f(p) < \infty$.*

Proof. Suppose the condition on β. Let $\mathcal{Z} = (Z, h, \{Z_{p,k}\}, Z_0) \in BSD$ and $G, D : \mathcal{Z} \to \mathcal{X}$ be morphisms given by $\gamma, \delta : Z \to X$ so that $\beta\gamma = \beta\delta$. Assume that for some $z \in Z$, $\gamma(z) = x \neq \delta(z) = x'$. The condition applies to this pair. We may assume $x \in X_{p,k}$ for some prime p and $k \geq 1$. Hence $z \in Z_{p,l}$ for some $l \leq k$. Then $x' \notin X_q$ for any $q \neq p$ and so we must have $x' \in X_0$ and $f(p) < \infty$. But this possiblity is excluded by Definition 1.7(3).

For the converse, suppose $B : \mathcal{X} \to \mathcal{Y}$ is a monomorphism. If $x, x' \in X$, $x \neq x'$, $\beta(x) = \beta(x')$ and neither of the possibilities of the condition prevails then in each of the resultant situations, the definition of a monomorphism will be contradicted. If $x, x' \in X_0$ then let $Z = \{z\}$ and define $\mathcal{Z} = (Z, h, \{\emptyset\}, Z)$, where $h(p) = 0$ for all $p \in \mathbf{P}$. Then γ and δ sending z to x and z to x', respectively will work. If $x \in X_{p,k}$ for some prime p and some $k \geq 1$ and $x' \in X_0$ but $f(p) = \infty$, then take $\mathcal{Z}$ with $Z = \{z\}$ but now with $h(p) = k$ and for $q \neq p$, $h(q) = 0$. Put $Z = Z_{p,k}$. Again, sending z to x and sending z to x' both define morphisms in *BSD*. The final case is where $x \in X_{p,k}$ and $x' \in X_{p,m}$ for some $p \in \mathbf{P}$ and $k, m \geq 1$. Here $\mathcal{Z}$ is modified to have $Z = Z_{p,\min\{k,m\}}$ and $h(p) = \min\{k, m\}$ and for $q \neq p$, $h(q) = 0$. Then the same trick works. □

The two possibilities in the condition on β are illustrated in the duals of the following examples: $\mathbf{Z} \to \mathbf{Z}/(2) \times \mathbf{Z}/(3)$ and $\mathbf{Z} \to \mathbf{Z}[1/2] \times \mathbf{Z}/(2)$, given by $z \mapsto (\bar{z}, \bar{z})$ and $z \mapsto (z, \bar{z})$, respectively, both of which are epimorphisms in *MinR*.

Proposition 3.4 *Let $\mathcal{X} = (X, f, \{X_{p,k}\}, X_0)$ and $\mathcal{Y} = (Y, g, \{Y_{p,k}\}, Y_0)$ be objects in* BSD *and $B : \mathcal{X} \to \mathcal{Y}$ a morphism given by $\beta : X \to Y$. Then B is an epimorphism if and only if β is a surjection.*

Proof. Suppose β is not a surjection and let N be a non-empty clopen set of Y in $Y \setminus \beta(X)$. Define $\mathcal{Z} = (Z, h, \{\emptyset\}, Z)$ where $h(p) = \infty$ for all $p \in \mathbf{P}$ and $Z = \{z_1, z_2\}$ has the discrete topology. Then is follows readily that the following continuous functions $Y \to Z$ yield morphisms in *BSD*: $\gamma(y) = z_1$ for all $y \in Y$, and $\delta(y) = z_1$ for all $y \in Y \setminus N$ and $\delta(y) = z_2$ for all $y \in N$. □

Dually, this says that a homomorphism $\eta : R \to S$ in *MinR* is a monomorphism if and only if the induced function $\mathbf{Spec}\,\mathbf{B}(S) \to \mathbf{Spec}\,\mathbf{B}(R)$ is onto. A simple example of a non-injective monomorphism is $\mathbf{Z} \to \mathbf{Z}/(2)$, another is $\mathbf{Z} \to \mathbf{Z}/(2) \times \mathbf{Z}/(3)$. The category **MR** of minimal regular rings is quite different with respect to monomorphisms and epimorphisms: The former can be seen to be injective and the latter surjective. More exactly, we have the following, which is easily proved.

Proposition 3.5 *Let $\alpha : R \to S$ be a homomorphism in* MinR*. If R is π-regular then α is a monomorphism if and only if α is an injection. If S is π-regular then α is an epimorphism if and only if α is a surjection.*

References

[1] M.F. Atiyah and I.G. MacDonald, *Introduction to Commutative Algebra.* Addison-Wesley, Reading, Mass., Menlo Park, Cal., London, Don Mills, Ont., 1969.

[2] A.K. Bousfield and D.M. Kan, The core of a ring, *J. Pure Appl. Algebra* **2** (1972), 73-81.

[3] W.D. Burgess and W. Stephenson, Pierce sheaves of non-commutative rings, *Comm. Algebra* **4** (1976), 51-75.

[4] W.D. Burgess and P.N. Stewart, The characteristic ring and the "best" way to adjoin a one, *J. Austral.Math. Soc. (Series A)* **47** (1989), 483-496.

[5] A.B. Carson, *Model Completions, Ring Representation and the Topology of the Pierce Sheaf.* Pitman Research Notes in Mathematics, **209**, Longman Scientific and Technical, Harlow, U.K., 1989.

[6] W. Dicks and W. Stephenson, Epimorphs and dominions of a Dedekind domain, *J. London Math. Soc.* **24** (1984), 224-228.

[7] R. Godement, *Théorie des faisceaux.* Hermann, Paris, 1964.

[8] K.R. Goodearl and R.B. Warfield, Jr., Algebras over zero-dimensional rings, *Math. Ann.* **223** (1976), 157-168.

[9] S. Koppelberg, *A Handbook of Boolean Algebras, Vol. 1* (J.D. Monk, ed.). North-Holland, Amsterdam, New York, Oxford, Tokyo, 1989.

[10] S. Mac Lane, *Categories for the Working Mathematician.* Graduate Texts in Mathematics, vol. 5, Springer-Verlag, New York, Heidelberg, Berlin, 1971.

[11] R.S. Pierce, Modules over commutative regular rings, *Memoires Amer. Math. Soc.* **70** (1967).

[12] R.S. Pierce, Minimal regular rings, *Contemporary Math.* **130** (1992), 335-348.

[13] B. Stenström, *Rings of Quotients.* Die Grundlagen der mathematischen Wissenschaft, Band 217, Springer-Verlag, New York, Heidelberg, Berlin, 1975.

DEPARTMENT OF MATHEMATICS, UNIVERSITY OF OTTAWA,
OTTAWA, CANADA K1N 6N5 *wdbsg@acadvm1.uottawa.ca*

Contemporary Mathematics
Volume **171**, 1994

On Endomorphisms and Automorphisms of Some Pure Subgroups of the Baer-Specker Group

A.L.S.CORNER and B. GOLDSMITH

ABSTRACT. The endomorphism algebras of certain pure subgroups of the Baer-Specker group are investigated to answer a question of Irwin. Automorphism groups of these Abelian groups are investigated and a realization theorem is established.

§1. Introduction.

The endomorphism rings of homogeneous separable torsion-free Abelian groups have been extensively studied by a number of authors: Corner [3] initiated the study, while Dugas and Göbel obtained "large" realization results in [8]. More recently, pure subgroups of the Baer-Specker group P have received renewed attention ([9], [11] and [12]). The first part of the present work is in this vein and was motivated by a question of John Irwin at the recent Caribbean Conference in Curaçao(1991).

Despite the attention given to endomorphism rings of such groups, very little attention has been paid to the study of their automorphism groups. In the second part of this work, we investigate these automorphism groups. Our viewpoint is very much that of Corner's approach to the so-called Realization Problem for Abelian p-groups ([4]). A vital ingredient in such an approach is the identification, within the automorphism group, of a normal subgroup which will play a role analogous to that played by small endomorphisms (or more generally inessentials) in the theory of p-groups. Recent work of Liebert [15] , which is based on earlier work of Dieudonné [7], assigns a special role to transvections.

1991 *Mathematics Subject Classification.* Primary 20K30, 20K20.

This work was supported by a joint British Council - Eolas Science and Technology Grant.

This paper is in final form and no version of it will be submitted for publication elsewhere.

With this in mind, we define a similar concept for pure subgroups of the Baer-Specker group and this turns out to be precisely what we require.

We show, *inter alia*, that if the finite (not necessarily commutative) group H occurs as the group of automorphisms of a torsion-free Abelian group, then there is a pure subgroup G of P such that Aut$G$, the automorphism group of G, is a split extension of H and the subgroup of transvections.

We conclude this introduction by establishing some notation and conventions; throughout P shall denote the Baer-Specker group, $P = \prod_{n=1}^{\infty} < e_n >$ and S the corresponding direct sum, $S = \sum_{n=1}^{\infty} < e_n >$. We reserve the letter D for the **Z**-adic closure of S in P i.e. D is the subgroup of P such that D/S is the maximal divisible subgroup of P/S. All groups, with the exception of automorphism groups, are additively written Abelian groups. Our terminology and notation are standard and may be found in [10]; an exception is that maps are written on the right.

§2. A Family of Pure Subgroups of P.

In this section we show the existence of a family $\mathcal{G}'$ of $2^{2^{\aleph_0}}$ non-isomorphic pure subgroups of P having some interesting properties: each group is slender, essentially indecomposable and the family is essentially-rigid in the sense that the only homomorphisms between distinct members, are homomorphisms of finite rank. Our construction of such a family was motivated by a question of J.Irwin at the Curaçao conference on Abelian Groups(1991),where he asked:

if $D' \subseteq D$ is such that $D/S = D'/S \oplus \mathbf{Q}$, are D' and D necessarily isomorphic? The existence of the family $\mathcal{G}'$ gives a strong negative answer to this question. We remark that our construction is similar to that given in [13] which was, in effect, a local version of the present result. (It is perhaps worth remarking that the use of GCH in [13] has been eliminated in this global result.)

We begin with a simple result on vector spaces.

Lemma 2.0. Suppose U is a subspace of codimension 1 in the infinite dimensional vector space V (of dimension α), $V = U \oplus < e >$ and $\delta : \alpha \to 2$ is any function. If $\{u_i, v_i : i \in \alpha\}$ is a basis for U and δU is defined by $\delta U = < u_i - i\delta e, v_i - i\delta e : i \in \alpha >$, then

(i)$V = \delta U \oplus < e >$

(ii) if $\delta \neq \epsilon$ is another function : $\alpha \to 2$ then $\delta U \neq \epsilon U$.

Proof. To establish (i) , it suffices to show $e \notin \delta U$ but that $u_i, v_i \in \delta U \oplus < e >$ for all i. However if $e \in \delta U$, then $e = \Sigma r_i(u_i - i\delta e) + \Sigma s_j(v_j - j\delta e)$, which forces $r_i, s_j = 0$ for all i, j and so $e = 0$ -contradiction. Moreover, for each i, $i\delta = 0$ or 1; if $i\delta = 0$ then $u_i = u_i - i\delta e \in \delta U$, while if $i\delta = 1$, then $u_i = (u_i - i\delta e) + e \in \delta U \oplus < e >$. Similarly for v_i. Finally to establish (ii), suppose $\delta U = \epsilon$U. Then, for all $i, u_i - i\delta e \in \epsilon U$ and so

$$u_i - i\delta e = \Sigma r_j(u_j - j\epsilon e) + \Sigma s_k(v_k - k\epsilon e).$$

However, a simple argument using linear independence gives that $s_k = 0$ for all k and $r_j = 1$, if $j = i$ and 0 otherwise. But this forces $i\delta = i\epsilon$ for all i, whence $\delta = \epsilon$.

Our next result is an extension of Lemma 5.2. in [1].

Lemma 2.1. Let V be a vector space over a field F with $\dim V = \alpha$, an infinite cardinal. Let $\{W_i : i < \alpha\}$ be a family of subspaces of V such that $\dim W_i = \alpha$ for all i. Then there exists a family $\{U_j : j \in 2^\alpha\}$ of 2^α subspaces of V such that each subspace U_j is of codimension 1 in V and no subspace W_i is contained in any subspace U_j.

Proof. Let $0 \neq e \in V$. We prove by induction the existence of a sequence of pairs $\{(x_\xi, y_\xi) : \xi < \alpha\}$ of linearly independent elements of W_ξ such that the space $U_\alpha =< e - x_\xi, y_\xi : \xi < \alpha >$ does not contain e. Choose $x_0, y_0 \in W_0$ such that $\{x_0, y_0, e\}$ is a linearly independent set. Then certainly $U_1 =< e - x_0, y_0 >$ does not contain e. (For $e = r(e - x_0) + sy_0$ would imply that $0 = (r-1)e + sy_0 - rx_0$, whence r is simultaneously both 0 and 1.) Assume that the pairs (x_η, y_η) have been specified for $\eta < \xi (\xi < \alpha)$ in such a way that the spaces spanned by $\{e - x_\eta, y_\eta : \eta \leq \zeta\} (\zeta < \xi)$ do not contain e. Then the space $U_\xi = < e - x_\eta, y_\eta : \eta < \xi >$ does not contain e. Moreover for any y the space $U_\xi + Fe + Fy$ has dimension at most

$$| \xi | + | \xi | + 2 = | \xi | + 2. \text{ Since } \xi < \alpha, \text{ we have that}$$

$$| \xi | + 2 < \alpha + 2 = \alpha = \dim W_\xi.$$

Choose linearly independent elements x_ξ, y_ξ in W_ξ such that $x_\xi \notin U_\xi + Fe + Fy_\xi$; this is possible by the dimension argument above. Then $e \notin U_\xi + F(e - x_\xi) + Fy_\xi$. Thus the inductive assumption on pairs has been preserved. Set $U_\alpha =< e - x_\xi, y_\xi : \xi < \alpha >$, so that $e \notin U_\alpha$ but $e - x_\xi \in U_\alpha$ for all $\xi < \alpha$. Now choose U maximal subject to $U \supseteq U_\alpha$ and $e \notin U$. Then V/U is one dimensional and $x_\xi \notin U$ for all ξ. (Otherwise $e = x_\xi + (e - x_\xi) \in U$.) Hence $W_\xi \not\leq U$ for all ξ. Note however, that for all ξ, y_ξ belongs to both W_ξ and U.

Now consider the family $\{\delta U : \delta \in^\alpha 2\}$ of 2^α subspaces of codimension 1 in V, where we have chosen a basis of U in such a way that $e - x_\xi = u_\xi$ and $y_\xi = v_\xi$ are basis elements used in the construction of δU. We claim that for each $\delta \in^\alpha 2, W_\xi \not\leq \delta U$. For suppose not, then $x_\xi = e - (e - x_\xi) = e - u_\xi \in \delta\text{U}$. Hence we have

$$e - u_\xi = \Sigma r_j(u_j - j\delta e) + \Sigma s_j(v_j - j\delta e)$$

which gives, by a simple linear independence argument, that $s_j = 0$ for all j, while $r_j = 0$ except when $j = \xi$, when it is -1. But then $e - u_\xi = \xi\delta e - u_\xi$ which implies that $\xi\delta = 1$. However $y_\xi = v_\xi \in W_\xi$ and so, by assumption, $v_\xi \in \delta\text{U}$. Thus we can write

$$v_\xi = \Sigma r_j(u_j - j\delta e) + \Sigma s_j(v_j - j\delta e)$$

which forces $r_j = 0$ for all j and s_j is then equal to 1 if $j = \xi$, and 0 otherwise. Hence $v_\xi = v_\xi - \xi\delta e$ and so $\xi\delta = 0$ - contradiction. The result now follows from Lemma 2.0.

Proposition 2.2. There exists a family $\mathcal{G}$ of $2^{2^{\aleph_0}}$ subgroups G of D such that each $G \in \mathcal{G}$ is slender and $D/G \cong \mathbf{Q}$.

Proof. Since our subgroups are constructed in D, it suffices , by Nunke's characterization of slenderness, to show that G does not contain a copy of P. Let

$\{W_k : k \in K\}$ denote the set of subgroups of D which are isomorphic to P. Then each W_k is determined by an endomorphism of P. However a standard slenderness argument (e.g. [10, $p.162$]) shows that the endomorphism ring, End P, of P has cardinality $2^{\aleph_0}$; hence $| K |\leq 2^{\aleph_0}$.

Let $\bar{W}_k =< W_k + S >_* /S \subseteq D/S$ so that $\{\bar{W}_k : k \in K\}$ is a family of at most $2^{\aleph_0}$ subspaces of the $\mathbf{Q}$-vector space D/S. Moreover since $| W_k |=| P |= 2^{\aleph_0}$, it follows that each $\bar{W}_k$ is a subspace of D/S of dimension $2^{\aleph_0}$. Now it follows from Lemma 2. 1. that there is a family of $2^{2^{\aleph_0}}$ subspaces U of D/S , such that no $\bar{W}_k$ is contained in any U. Moreover each U has codimension 1 in D/S. Choose $G \subseteq D$ so that $G/S = U$; the family $\mathcal{G}$ of all such subgroups is a family of $2^{2^{\aleph_0}}$ pure subgroups of D(and hence of P) with $D/G \cong D/S/G/S \cong \mathbf{Q}$. Moreover no W_k is contained in any G since no $\bar{W}_k$ is contained in U and so we conclude that each $G \in \mathcal{G}$ is slender.

Remark. This proposition suffices to give a negative answer to Irwin's question: if $G \in \mathcal{G}$, then $D/S = G/S \oplus \mathbf{Q}$ but D and G are not isomorphic since G is slender while D is clearly not. (The subgroup $P' = \Pi < e_n n! >$ is an isomorphic copy of P contained in D.)

The family $\mathcal{G}$ has some additional interesting properties which we now investigate.

Definition. If G is a pure subgroup of D containing S then a homomorphism

$\phi : G \to G$ is said to be *inessential* if its unique extension $\hat{\phi} : D \to D$ is such that $D\hat{\phi} \leq G$.

If there is no danger of ambiguity, we shall refer to the extended map as ϕ. It is easily checked that the set of inessential endomorphisms of G form a 2-sided ideal in End G; denote this ideal by Iness G.

Theorem 2.3. If $G \in \mathcal{G}$, then the endomorphism ring of G is a ring split extension of $\mathbf{Z}$ by Iness G, End $G = \mathbf{Z} \oplus$ Iness G.

Proof. Since $\mathbf{Z}$ + Iness $G \leq$ End G, it suffices to show the reverse inequality and that $\mathbf{Z} \cap$ Iness $G = \{0\}$. As $G \in \mathcal{G}$, there exists $d \in D$ such that $< G, d >_* = D$. If $\phi \in$ End G, then there are $r, s \in \mathbf{Z}$ such that $d(r\phi) = g + sd$ for some $g \in$ G. We distinguish two cases.

Case 1.$r \mid s$ Writing $s = r\alpha$, we see that $\phi - \alpha : G \to G$ and also $r(d(\phi - \alpha)) = g + sd - r\alpha d = g \in$ G. Since G is pure in D, we deduce that $d(\phi - \alpha) \in G$, so that $\phi - \alpha$ is inessential. Hence End $G \leq \mathbf{Z}$ + Iness G in this case.

Case 2.$r \not| s$ In this case $r\phi - s$ maps D into G since $d(r\phi - s) = g \in$ G. Now consider Ker$(r\phi - s)$; if $0 \neq x$ is in Ker$(r\phi - s)$, then $r(x\phi) = sx$. Since we are in a torsion-free group, we may assume $(r, s) = 1$. Hence there is a prime p with $p \mid r$ but $p \not| s$. However a simple calculation shows that for each $n \in \omega, s^n x = r^n x\phi^n$ and so $ht_p(x) = ht_p(s^n x) = ht_p(r^n x\phi^n) \geq n$. Thus $ht_p(x)$ is infinite, which is impossible since G is homogeneous of type $\mathbf{Z}$. So we conclude that Ker$(r\phi - s) = 0$. However it now follows that $D \cong D/$Ker$(r\phi - s)$ is isomorphic to a subgroup of G - contradiction, since G is slender while D is not. Thus Case 2 does not arise.

It remains only to show that $\mathbf{Z}\cap$ Iness $G = \{0\}$, but this is immediate since $G \subset D$ and G is pure in D. Hence we have the desired splitting which is clearly a ring split extension.

We now show that the inessential endomorphisms are, in this case, rather tightly prescribed.

Recall that a homomorphism $\phi : G \to G$ is said to be of *finite rank* if $G\phi$ is of finite rank. When $G \leq P$ it follows that if ϕ is of finite rank, then $G\phi$ is free.

Proposition 2.4. If H is a slender group and $\phi : D \to H$ is a homomorphism, then $D\phi$ is free of finite rank.

Proof. This is well known, see e.g. [12, Corollary 2.5(1)].

Corollary 2.5. If $G \in \mathcal{G}$, then

(i) End $G = \mathbf{Z} \oplus E_0(G)$, where $E_0(G)$ is the ideal of finite rank endomorphisms.

(ii) G is essentially indecomposable i.e. if $G = A \oplus B$, then one of A, B is free of finite rank.

Proof. Since $G \in \mathcal{G}$ implies G is slender, it follows from Proposition 2.4. that every inessential homomorphism has finite rank and then End $G = \mathbf{Z}$ + Iness $G \subseteq \mathbf{Z} + E_0(G) \subseteq$ End G. Thus End $G = \mathbf{Z} + E_0(G)$; in particular the modular law yields Iness $G = E_0(G)$.

Suppose that $G = A \oplus B$ with projections π_1, π_2 respectively. Then, letting bars denote images in End $G/E_0(G)$, we have that $\bar{\pi}_1\bar{\pi}_2 = \bar{0}$. However the quotient End $G/E_0(G) \cong \mathbf{Z}$ so that one of $\bar{\pi}_1, \bar{\pi}_2$ is zero, say $\bar{\pi}_1$. Thus $\pi_1 \in E_0(G)$ and so $A = G\pi_1$ has finite rank. Since $G \leq P, A$ is then free.

We conclude this section by investigating homomorphisms between different members of the family $\mathcal{G}$.

Lemma 2.6. If $G \in \mathcal{G}$, then there are at most $2^{\aleph_0}$ subgroups $G_i \in \mathcal{G}$ with the property that there is a homomorphism from G_i to G which is not of finite rank.

Proof. Suppose $\{G_i : i < \lambda\}$ is a subfamily of $\mathcal{G}$ with $\lambda > 2^{\aleph_0}$ such that there is a homomorphism $\phi_i : G_i \to G$ which is not of finite rank. Consider the extension $\hat{\phi}_i : D \to D$. Since S is dense in D, every endomorphism of D is determined by its action on S, and so $|\text{End } D| = 2^{\aleph_0}$. It follows from a pigeon-hole argument that $\hat{\phi}_i = \hat{\phi}_j$ for some $i \neq j < \lambda$. This is impossible since the maximality of the G_i in D and the fact that $G_i + G_j$ is automatically pure in D gives $D = G_i + G_j$ for $i \neq j$, whence

$$D\hat{\phi}_i = (G_i + G_j)\hat{\phi}_i \leq G_i\hat{\phi} + G_j\hat{\phi} \leq G$$

and so ϕ_i has finite rank. This establishes the lemma.

We note that the proof of Lemma 2.6. is almost identical to that of Lemma 3.5. in [13]. Our next result corresponds very closely to [13, Lemma 3.4.] and its proof follows an exactly similar argument (replacing the reference to Lemma 3.2. in [13] by our Lemma 2.1.). Consequently we omit its proof.

Lemma 2.7. If $\{G_\alpha\}$ is a subfamily of $\mathcal{G}$ of size at most $2^{\aleph_0}$, then there are $2^{2^{\aleph_0}}$ subgroups G in $\mathcal{G}$ such that for all α, every homomorphism $:G_\alpha \to G$ has finite rank.

We can easily combine these results to obtain the essentially-rigid family $\mathcal{G}'$ mentioned above.

Theorem 2.8. There exists a family $\mathcal{G}' \subseteq \mathcal{G}$ of $2^{2^{\aleph_0}}$ subgroups $\{G_i\}$ of D such that

(i) End $G_i = \mathbf{Z} \oplus E_0(G_i)$ for each i.

(ii) if $i \neq j$, then every homomorphism : $G_i \to G_j$ is of finite rank.

Proof. Observe that (i) follows directly from Corollary 2.5. if we show $\mathcal{G}' \subseteq \mathcal{G}$. We construct the family by transfinite induction. Choose G_0 to be any member of $\mathcal{G}$. Suppose the subgroups $\{G_\alpha : \alpha < \beta\}$ have been constructed for $\beta < 2^{2^{\aleph_0}}$. By Lemma 2.7. there are $2^{2^{\aleph_0}}$subgroups G in $\mathcal{G}$ such that every homomorphism : $G_\alpha \to G$ has finite rank. However, for each $\alpha < \beta$ there are, by Lemma 2.6., at most $2^{\aleph_0}$ of these G having a homomorphism : $G \to G_\alpha$ which is not of finite rank. Deleting all such G removes at most $\beta..2^{\aleph_0} < 2^{2^{\aleph_0}}$ subgroups. Hence there exists $G \in \mathcal{G}$ such that $\mathrm{Hom}(G, G_\alpha)$ and $\mathrm{Hom}(G_\alpha, G)$ consist of finite rank mappings for all $\alpha < \beta$. Set $G_\beta = G$. The proof follows by transfinite induction.

§3. Automorphism Groups.

In this section we investigate which groups (non-commutative in general) can occur as automorphism groups of pure subgroups G of the Baer-Specker group P. Since such groups are necessarily separable, it is clear that they will have many automorphisms. Hence any characterization of their automorphism groups is bound to be quite complex (and so unlikely to be of great use in applications). Consequently we approach the problem along similar lines to that developed by Corner [4] for endomorphism rings of Abelian p-groups: we try to realize a given group A as a component of a split extension which is the full automorphism group. Our approach is much influenced by recent work on the so-called Realization Problem for endomorphism algebras (see [6], [11] and [12]).

Throughout G shall be a pure subgroup of the Baer-Specker group P. We begin with an *ad-hoc* definition.

Definition. An automorphism ϕ of G is said to be a *transvection* if $\phi - 1$ is of finite rank; ϕ is said to be a *quasi-transvection* if either $\phi - 1$ or $\phi + 1$ has finite rank.

Let T_G(respectively S_G) denote the set of transvections (respectively quasi-transvections) of the group G.

Remark. The motivation for the terminology transvection in this context is clear : classically a transvection is an automorphism which is the identity on a hyperplane (= summand of corank 1)- see [7] and [15]. In our context we are using the term to denote an automorphism which is the identity on a summand of finite corank.

Proposition 3.1. T_G and S_G are both normal subgroups of Aut$G$.

Proof. Suppose ϕ_1 and ϕ_2 are transvections, so $\phi_1 - 1$ and $\phi_2 - 1$ both have finite rank. Now consider the automorphism $\phi_1\phi_2$ and observe the identity

$$\phi_1\phi_2 = \phi_1 + \phi_2 + \alpha - 1, \text{ where } \alpha = (\phi_1 - 1)(\phi_2 - 1).$$

Thus $\phi_1\phi_2 - 1 = (\phi_1 - 1) + (\phi_2 - 1) + \alpha$, which has finite rank since each of the components has finite rank. So $\phi_1\phi_2$ is a transvection.

Suppose ψ is the inverse of the transvection ϕ. Then $\beta = (\psi - 1)(\phi - 1)$ is of finite rank and we note the identity

$$\psi - 1 = (1 - \phi) - \beta$$

Thus $\psi - 1$ has finite rank and so inverses of transvections are again transvections. Hence $T_G \leq \text{Aut}G$.

Suppose now that ϕ_1 and ϕ_2 are quasi-transvections. If both $\phi_1 + 1$ and $\phi_2 + 1$ have finite rank or both $\phi_1 - 1$ and $\phi_2 - 1$ have finite rank, it follows easily that $\phi_1\phi_2$ is again a quasi-transvection. Without loss we may take the remaining case to be where $\phi_1 + 1$ and $\phi_2 - 1$ have finite rank. Then if we set $\beta = (\phi_1 + 1)(\phi_2 - 1)$, wehave$\phi_1\phi_2 + 1 = \beta + (\phi_1 + 1) - (\phi_2 - 1)$ and so $\phi_1\phi_2 + 1$ is of finite rank.Thus in all cases the product of two quasi-transvections is a quasi-transvection. Finally the inverse of a quasi-transvection is a quasi-transvection: it suffices to consider the case where $\phi + 1$ is of finite rank and ψ is the inverse of ϕ. But then if $\chi = (\psi + 1)(\phi + 1)$, χ has finite rank and so $\psi + 1 = \chi - (\phi + 1)$ also has finite rank. Thus ψ is a quasi-transvection.

We complete the proof of 3.1. by establishing normality of the subgroups T_G, S_G in Aut$G$. Suppose $\phi \in T_G, \psi \in S_G$ and $\theta \in \text{Aut}G$. Then $\theta^{-1}\phi\theta - 1 = \theta^{-1}(\phi - 1)\theta$ has finite rank since $\phi - 1$ has finite rank. Moreover $\theta^{-1}\psi\theta \pm 1 = \theta^{-1}(\psi \pm 1)\theta$ has finite rank when the appropriate choice ± 1 is made. This completes the proof.

Observe that since $-1 \notin T_G, T_G$ is always a proper subgroup of Aut$G$. It is , however, possible that $S_G = \text{Aut}G$; if this happens we say that G is a QT-group. Later we shall give examples of QT and non$-QT$ subgroups of the Baer-Specker group P.

Definition. If A is an unital ring, then a subgroup G of P is said to be A-*realizable* if End G is the split ring extension of A by the ideal $E_0(G)$ of all finite rank endomorphisms of G.

We say G is *realizable* if it is A-realizable for some unital ring A.

Thus if $\mathcal{G}$ is the collection of groups described in §2, then each $G \in \mathcal{G}$ is **Z**-realizable. For realizable groups we can improve on Proposition 3.1.

Proposition 3.2. If G is an A-realizable group, then AutG/T_G is isomorphic to $U(A)$, the group of units of A and AutG/S_G is isomorphic to $U(A)/\{\pm 1\}$.

Proof. Since G is A-realizable, End $G = A \oplus E_0(G)$ and so any automorphism ϕ of G can be written uniquely in the form $a + \theta$, where $a \in A, \theta \in E_0(G)$. Since $E_0(G)$ is a 2-sided ideal, it is easy to see that a must be a unit in A. Define a map :Aut$G \to U(A)$ by $\phi \mapsto a$, where ϕ was uniquely of the form $\phi = a + \theta$. Since $(a + \theta)(a_1 + \theta_1) = aa_1 + \theta'$, some $\theta' \in E_0(G)$, this is easily seen to be a homomorphism. Moreover, as every unit in A is an automorphism of G, this mapping is onto $U(A)$. The kernel is precisely the set of automorphisms of the form $1 + \theta$ with $\theta \in E_0(G)$ i.e. the set T_G. Hence $\mathrm{Aut}G/T_G \cong U(A)$. Finally consider the composite homomorphism

$$\mathrm{Aut}G \to U(A) \to U(A)/\{\pm 1\};$$

this composite has kernel consisting of all automorphisms ϕ of the form $\pm 1 + \theta$ with $\theta \in E_0(G)$ i.e. the set S_G. So $\mathrm{Aut}G \cong U(A)/\{\pm 1\}$ as required.

Corollary 3.3. If $\mathcal{G}$ is as in §2, then every $G \in \mathcal{G}$ is a QT-group.

Proof. If $G \in \mathcal{G}$, then End $G = \mathbf{Z} \oplus E_0(G)$ and so

$$\mathrm{Aut}G/S_G \cong U(\mathbf{Z})/\{\pm 1\} \cong \{1\}.$$

Thus $\mathrm{Aut}G = S_G$ and G is a QT-group.

With the help of recent work on the Realization Problem for endomorphism rings, we can realize a wide variety of groups as the "essential automorphism group" of a subgroup of P.

Theorem 3.4. Let H be a countable (not necessarily commutative) group which admits a total order, then there exists a pure subgroup G of the Baer-Specker group P such that $\mathrm{Aut}G$ is a split extension of H and the group of quasi-transvections, S_G, of G

$$\mathrm{Aut}G = S_G.H.$$

Proof. The integral group ring $\mathbf{Z}H$ is additively a free Abelian group of countable rank and so it follows from [9], [11] or [12] that there is a $\mathbf{Z}H$-realizable subgroup G of P. Moreover $U(\mathbf{Z}H) = \{\pm 1\} \times H$ since H admits a total order. The result now follows directly from Proposition 3.2.

We remark that groups which admit a total order are abundant : torsion-free Abelian and nilpotent groups are examples.

Corollary 3.5. There exist subgroups G of P which are not QT-groups.

Proof. Choose any non-trivial countable group which admits a total order and apply Theorem 3.4.

By restricting to finite (not necessarily commutative) groups H we can say a little more. First we establish a well-known result relating finite groups of automorphisms to groups of units in certain rings.

Lemma 3.6. For a finite group H the following are equivalent:

(i) H is the automorphism group of a torsion-free Abelian group

(ii) H is the group of units of a ring A with $(A,+)$ free Abelian of finite rank.

Proof. To show (i) implies (ii), let $H = \mathrm{Aut}B$ where B is torsion-free Abelian and set A to be the subring of End G generated by H. Certainly H is the group of units of A. However since H is finite and the additive group of End B is torsion-free, $(A,+)$is finitely generated torsion-free and so of finite rank. The reverse implication follows immediately from Corner's Finite Rank Theorem [2, Theorem B].

Proposition 3.7. If H is a finite group which is the automorphism group of a torsion-free Abelian group, then there exists a pure subgroup G of the Baer-Specker group P with Aut$G$ a split extension, $T_G.H$, of H and the group of transvections of G.

Proof. It follows from Lemma 3.6. and the hypothesis on H that $H = U(A)$ for some ring A with $(A,+)$ free of finite rank. But then we can deduce from [9], [11] or [12] that there is an A-realizable subgroup G of P. An appeal to Proposition 3.2. yields the result.

In fact the above proposition can be extended to countable groups.

Proposition 3.8. Let H be a countable torsion group which is the automorphism group of a torsion-free Abelian group, then there is a pure subgroup G of P with Aut$G = T_G.H$.

Proof. If a torsion group H is the automorphism group of a torsion-free Abelian group, then it follows from [5, Theorem C6] that H is the group of units of a ring A with $(A,+)$ free Abelian of rank $=\mid H \mid$. As in the proof of 3.7. we can find an A-realizable subgroup G of P and the result follows from Proposition 3.2.

We remark that all possible finite groups which occur as automorphism groups of torsion-free Abelian groups, have been classified by Corner [5], correcting earlier work of Hallett and Hirsch [14].

Finally we draw attention to some connections between essentially indecomposable subgroups of P and QT-groups: if $G \leq_* P$ is a QT-group then a simple calculation shows that G must be essentially indecomposable. The converse is not true: let G be a pure subgroup of P such that End $G = J[i] \oplus E_0(G)$, where $J[i]$ is the ring of Gaussian integers - the existence of such a G follows from [9],[11] or [12]. Since $J[i]$ is an integral domain, G is necessarily essentially indecomposable. However it follows from Proposition 3.2. that Aut$G/T_G \cong U(J[i]) = \{\pm 1, \pm i\}$ and so G is not a QT-group.

References

[1] R.A. Beaumont and R.S. Pierce, "Some invariants of p-groups", Michigan Math. J. 11(1964) 137-149.

[2] A.L.S. Corner, "Every countable reduced torsion-free ring is an endomorphism ring", Proc.London Math. Soc. 13(1963) 687-710.

[3] A.L.S. Corner, "A class of pure subgroups of the Baer-Specker group", unpublished talk given at the Montpellier Conference on Abelian Groups 1967.

[4] A.L.S. Corner, "On endomorphism rings of primary abelian groups", Quart. J. Math. Oxford 20(1969) 277-296.

[5] A.L.S. Corner, "Finite automorphism groups of torsion-free abelian groups", to appear.

[6] A.L.S. Corner and R. Göbel, "Prescribing endomorphism algebras, a unified treatment", Proc. London Math. Soc. 50(1985) 447- 479.

[7] J. Dieudonné, *La Géométrie des Groupes Classiques,* Ergebn. Mathematik, Springer, Berlin 1963.

[8] M. Dugas and R. Göbel, "Endomorphism rings of separable torsion-free abelian groups", Houston J. Math. 11(1985) 471 - 483.

[9] M. Dugas, J. Irwin and S. Khabbaz, "Countable rings as endomorphism rings", Quart. J. Math. Oxford 39(1988) 201 - 211.

[10] L. Fuchs, *Infinite Abelian Groups, Vol I and II,* Academic Press, New York, 1970 and 1973.

[11] R. Göbel and B. Goldsmith, "On separable torsion-free modules of countable density character", J. Algebra 144(1991) 79 - 87.

[12] R. Göbel and B. Wald, "Separable torsion-free modules of small type", Houston J. Math. 16(1990) 271 - 287.

[13] B. Goldsmith, "Essentially indecomposable modules over a complete discrete valuation ring", Rend. Sem. Mat. Univ. Padova 70(1983) 21 - 29.

[14] J.T. Hallett and K.A. Hirsch, "Finite groups of exponent 12 as automorphism groups", Math Z. 155(1977) 43 - 53.

[15] W. Liebert, "Isomorphic automorphism groups of primary abelian groups", in *Abelian Group Theory (Oberwolfach),* Gordon and Breach, 1987, 9 - 31.

WORCESTER COLLEGE, OXFORD, ENGLAND

E-mail address: corner@vax.oxford.ac.uk

DUBLIN INSTITUTE OF TECHNOLOGY, KEVIN STREET, DUBLIN 8

and

DUBLIN INSTITUTE FOR ADVANCED STUDIES, DUBLIN, IRELAND

E-mail address: bgoldsmith@dit.ie

Contemporary Mathematics
Volume **171**, 1994

A Combinatorial Principle Equivalent to the Existence of Non-free Whitehead Groups

PAUL C. EKLOF AND SAHARON SHELAH

ABSTRACT. As a consequence of identifying the principle described in the title, we prove that for any uncountable cardinal λ, if there is a λ-free Whitehead group of cardinality λ which is not free, then there are many "nice" Whitehead groups of cardinality λ which are not free.

1. Introduction

Throughout, "group" will mean abelian group; in particular, "free group" will mean free abelian group.

Two problems which have been shown to be undecidable in ZFC (ordinary set theory) for some uncountable λ are the following:

- Is there a group of cardinality λ which is λ-free (that is, every subgroup of cardinality $< \lambda$ is free), but is not free?
- Is there a Whitehead group G (that is, $\mathrm{Ext}(G, \mathbb{Z}) = 0$) of cardinality λ which is not free?

(See [**6**] for the first, [**7**] for the second; also [**2**] is a general reference for unexplained terminology and further information.)

The second author has shown that the first problem is equivalent to a problem in pure combinatorial set theory (involving the important notion of a λ-system; see Theorem 3.) This not only makes it easier to prove independence results (as in [**6**]), but also allows one to prove (in ZFC!) group-theoretic results such as:

> (∇) if there is a λ-free group of cardinality λ which is not free, then there is a strongly λ-free group of cardinality λ which is not free.

1991 *Mathematics Subject Classification*. Primary 20K20, 20K35; Secondary 03E05.

Thanks to Rutgers University for funding the authors' visits to Rutgers.

The second author was partially supported by the *Basic Research Foundation* administered by The Israel Academy of Sciences and Humanities. Pub. No. 505. Final version.

(See [**10**] or [**2**, Chap. VII].) A group G is said to be *strongly λ-free* if every subset of G of cardinality $< \lambda$ is contained in a free subgroup H of cardinality $< \lambda$ such that G/H is λ-free. One reason for interest in this class of groups is that they are precisely the groups which are equivalent to a free group with respect to the infinitary language $L_{\infty\lambda}$ (see [**1**]).

There is no known way to prove ($\bigtriangledown$) except to go through the combinatorial equivalent.

As for the second problem, the second author has shown that for $\lambda = \aleph_1$, there is a combinatorial characterization of the problem:

> there is a non-free Whitehead group of cardinality $\aleph_1$ if and only if there is a ladder system on a stationary subset of $\aleph_1$ which satisfies 2-uniformization.

(See the Appendix to this paper; a knowledge of the undefined terminology in this characterization is not needed for the body of this paper.) Again, there are group-theoretic consequences which are provable in ZFC:

> ($\bigtriangledown\bigtriangledown$) if there is a non-free Whitehead group of cardinality $\aleph_1$, then there is a strongly $\aleph_1$-free, non-free Whitehead group of cardinality $\aleph_1$.

(See [**2**, §XII.3]. It is consistent with ZFC that there are Whitehead groups of cardinality $\aleph_1$ which are not strongly $\aleph_1$-free.)

Our aim in this paper is to generalize ($\bigtriangledown\bigtriangledown$) to cardinals $\lambda > \aleph_1$ by combining the two methods used to prove ($\bigtriangledown$) and ($\bigtriangledown\bigtriangledown$). Since the existence of a non-free Whitehead group G of cardinality $\aleph_1$ implies that for every uncountable cardinal λ there exist non-free Whitehead groups of cardinality λ (e.g. the direct sum of λ copies of G — which is not λ-free), the appropriate hypothesis to consider is:

- there is a λ-free Whitehead group of cardinality λ which is not free.

By the Singular Compactness Theorem (see [**8**]), λ must be regular. It can be proved consistent that there are uncountable $\lambda > \aleph_1$ such that the hypothesis holds. (See [**4**] or [**11**].) In particular, for many λ (for example, $\lambda = \aleph_{n+1}$, $n \in \omega$) it can be proved consistent with ZFC that λ is the smallest cardinality of a non-free Whitehead group; hence there is a λ-free Whitehead group of cardinality λ which is not free. (See [**4**].)

Our main theorem is then:

THEOREM 1. *If there is a λ-free Whitehead group of cardinality λ which is not free, then there are 2^λ different strongly λ-free Whitehead groups of cardinality λ.*

Our proof will proceed in three steps. First, assuming the hypothesis — call it (A) — of the Theorem, we will prove

> (B) there is a combinatorial object, consisting of a λ-system with a type of uniformization property.

Second, we will show that the combinatorial property (B) can be improved to a stronger combinatorial property (B+), which includes the "reshuffling property". Finally, we prove that (B+) implies

(A+): the existence of many strongly λ-free Whitehead groups.

Note that we have found, in (B) or (B+), a combinatorial property which is equivalent to the existence of a non-free λ-free Whitehead group of cardinality λ, answering an open problem in [**2**, p. 453]. Certainly this combinatorial characterization is more complicated than the one for groups of cardinality $\aleph_1$ cited above. This is not unexpected; indeed the criterion for the existence of λ-free groups in Theorem 3 implies that the solution to the open problem is inevitably going to involve the notion of a λ-system. A good reason for asserting the interest of the solution is that it makes possible the proof of Theorem 1.

In an Appendix we provide a simpler proof than the previously published one of the fact that the existence of a non-free Whitehead group of cardinality $\aleph_1$ implies the existence of a ladder system on a stationary subset of $\aleph_1$ which satisfies 2-uniformization.

We thank Michael O'Leary for his careful reading of and comments on this paper.

2. Preliminaries

The following notion, of a λ-set, may be regarded as a generalization of the notion of a stationary set.

DEFINITION 2. *(1) The set of all functions*

$$\eta\colon n = \{0, \dots, n-1\} \to \lambda$$

($n \in \omega$) is denoted ${}^{<\omega}\lambda$; the domain of η is denoted $\ell(\eta)$ and called the length *of η; we identify η with the sequence*

$$\langle \eta(0), \eta(1), \dots, \eta(n-1) \rangle.$$

Define a partial ordering on ${}^{<\omega}\lambda$ by: $\eta_1 \le \eta_2$ if and only if η_1 is a restriction of η_2. This makes ${}^{<\omega}\lambda$ into a tree. *For any $\eta = \langle \alpha_0, \dots, \alpha_{n-1} \rangle \in {}^{<\omega}\lambda$, $\eta \frown \langle \beta \rangle$ denotes the sequence $\langle \alpha_0, \dots, \alpha_{n-1}, \beta \rangle$. If S is a subtree of ${}^{<\omega}\lambda$, an element η of S is called a* final node *of S if no $\eta \frown \langle \beta \rangle$ belongs to S. Denote the set of final nodes of S by S_f.*

(2) A λ-set *is a subtree S of ${}^{<\omega}\lambda$ together with a cardinal λ_η for every $\eta \in S$ such that $\lambda_\emptyset = \lambda$, and:*

(a) for all $\eta \in S$, η is a final node of S if and only if $\lambda_\eta = \aleph_0$;

(b) if $\eta \in S \setminus S_f$, then $\eta \frown \langle \beta \rangle \in S$ implies $\beta \in \lambda_\eta$, $\lambda_{\eta \frown \langle \beta \rangle} < \lambda_\eta$ and $E_\eta = \{\beta < \lambda_\eta \colon \eta \frown \langle \beta \rangle \in S\}$ is stationary in λ_η.

(3) A λ-system is a λ-set together with a set B_η for each $\eta \in S$ such that $B_\emptyset = \emptyset$, and for all $\eta \in S \setminus S_f$:

(a) for all $\beta \in E_\eta$, $\lambda_{\eta\frown\langle\beta\rangle} \leq |B_{\eta\frown\langle\beta\rangle}| < \lambda_\eta$;

(b) $\{B_{\eta\frown\langle\beta\rangle} : \beta \in E_\eta\}$ is a continuous chain of sets, i.e. if $\beta \leq \beta'$ are in E_η, then $B_{\eta\frown\langle\beta\rangle} \subseteq B_{\eta\frown\langle\beta'\rangle}$, and if σ is a limit point of E_η, then $B_{\eta\frown\langle\sigma\rangle} = \bigcup\{B_{\eta\frown\langle\beta\rangle} : \beta < \sigma,\, \beta \in E_\eta\}$;

(4) For any λ-system $\Lambda = (S,\, \lambda_\eta,\, B_\eta : \eta \in S)$, and any $\eta \in S$, let $\bar{B}_\eta = \bigcup\{B_{\eta\restriction m} : m \leq \ell(\eta)\}$. Say that a family $\mathcal{S} = \{s_\zeta : \zeta \in S_f\}$ of countable sets is based on Λ *if $\mathcal{S}$ is indexed by S_f and for every $\zeta \in S_f$, $s_\zeta \subseteq \bar{B}_\zeta$.*

(5) A family $\mathcal{S} = \{s_i : i \in I\}$ is said to be free *if it has a transversal, that is, a one-one function $T : I \to \bigcup\mathcal{S}$ such that for all $i \in I$, $T(i) \in s_i$. We say $\mathcal{S}$ is* λ-free *if every subset of $\mathcal{S}$ of cardinality $< \lambda$ has a transversal.*

It can be proved that a family $\mathcal{S}$ which is based on a λ-system is not free. (See [**2**, Lemma VII.3.6].)

The following theorem now gives combinatorial equivalents to the existence of a λ-free group of cardinality λ which is not free. (See [**10**] or [**2**, §VII.3].)

THEOREM 3. *For any uncountable cardinal λ, the following are equivalent:*

(i) *there is a λ-free group of cardinality λ which is not free;*
(ii) *there is a family $\mathcal{S}$ of countable sets such that $\mathcal{S}$ has cardinality λ and is λ-free but not free;*
(iii) *there is a family $\mathcal{S}$ of countable sets such that $\mathcal{S}$ has cardinality λ, is λ-free, and is based on a λ-system.*

DEFINITION 4. *A subtree S of ${}^{<\omega}\lambda$ is said to have* height n *if all the final nodes of S have length n. A λ-set or λ-system is said to have height n if its associated subtree S has height n.*

A λ-set of height 1 is essentially just a stationary subset of λ. Not every λ-set has a height, but the following lemma implies that every λ-set contains one which has a height. It is a generalization, and a consequence, of the fact that if a stationary set E is the union of subsets E_n $(n \in \omega)$, then for some n, E_n is stationary (cf. [**2**, Exer. 2, p. 238]).

LEMMA 5. *If $(S, \lambda_\eta : \eta \in S)$ is a λ-set, and $S_f = \bigcup_{n\in\omega} S_f^n$, let $S^n = \{\eta \in S : \eta \leq \tau$ for some $\tau \in S_f^n\}$. Then for some n, $(S^n, \lambda_\eta : \eta \in S^n)$ contains a λ-set.*

If $\Lambda = (S, \lambda_\eta, B_\eta : \eta \in S)$ is a λ-system of height n, and $\mathcal{S} = \{s_\zeta : \zeta \in S_f\}$ is a family of countable sets based on Λ, let $s_\zeta^k = s_\zeta \cap B_{\zeta\restriction k}$ for $1 \leq k \leq n$.

The following is useful in carrying out an induction on λ-systems.

DEFINITION 6. *Given a λ-system $\Lambda = (S, \lambda_\eta, B_\eta : \eta \in S)$ and a node η of S, let $S^\eta = \{\nu \in S : \eta \le \nu\}$. We will denote by Λ^η the λ_η-system which is naturally isomorphic to $(S^\eta, \lambda_\nu, B'_\nu : \nu \in S^\eta)$ where $B'_\eta = \emptyset$ and $B'_\nu = B_\nu$ if $\nu \neq \eta$.*

(That is, we replace the initial node, η, of S^η by $\emptyset$, and translate the other nodes accordingly.)

If $\mathcal{S} = \{s_\zeta : \zeta \in S_f\}$ is a family of countable sets based on Λ *and $\zeta \in S^\eta_f$, let $s^\eta_\zeta = \cup\{s^k_\zeta : k > \ell(\eta)\}$. Let $\mathcal{S}^\eta = \{s^\eta_\zeta : \zeta \in S^\eta_f\}$; it is a family of countable sets based on Λ^η.*

In order to construct a (strongly) λ-free group from a family of countable sets based on a λ-system, we need that the family have an additional property:

DEFINITION 7. *A family $\mathcal{S}$ of countable sets based on a λ-system Λ is said to have the* reshuffling property *if for every $\alpha < \lambda$ and every subset I of S_f such that $|I| < \lambda$, there is a well-ordering $<_I$ of I such that for every τ, $\zeta \in I$, $s_\zeta \setminus \bigcup_{\nu <_I \zeta} s_\nu$ is infinite, and $\tau(0) \le \alpha < \zeta(0)$ implies that $\tau <_I \zeta$.*

It can be shown (in fact it is part of the proof of the theorem) that the three equivalent conditions in Theorem 3 are equivalent to:

> (iv) *there is a family $\mathcal{S}$ of countable sets such that $\mathcal{S}$ has cardinality λ, is λ-free, is based on a λ-system, and has the reshuffling property.*

Finally, for future reference, we observe the following simple fact:

LEMMA 8. *Suppose that for some integers $r \ge 0$, and d^ℓ_m, and some primes q_m ($\ell < r$, $m \in \omega$), H is the abelian group on the generators $\{z_j : j \in \omega\}$ modulo the relations*

$$q_m z_{m+r+1} = z_{m+r} + \sum_{\ell < r} d^\ell_m z_\ell$$

for all $m \in \omega$. Then H is not free.

Conversely, if C is a torsion-free abelian group of rank $r+1$ which is not free but is such that every subgroup of rank $\le r$ is free, then C contains a subgroup H which is given by generators and relations as above.

PROOF. Let H be as described in the first part. If H is free, then H is finitely generated, since it clearly has rank $\le r+1$. Let L be the subgroup of H generated by (the images of) $z_0, \dots, z_{r-1}$. By comparing coefficients of linear combinations in the free group on $\{z_j : j \in \omega\}$, one can easily verify that L is a pure subgroup of H, and that H/L is a rank one group which is not free (because $z_r + L$ is non-zero and divisible by $q_0 q_1 \cdot \dots \cdot q_m$ for all $m \in \omega$) and hence not finitely-generated. But this is impossible if H is free.

Conversely, let C be as stated, and let L be a pure subgroup of rank r. Then L is free (say with basis $z_0, \dots, z_{r-1}$) and C/L is a non-free torsion-free group of rank 1. Thus C/L contains a subgroup with a non-zero element $z_r + L$ such that either: $z_r + L$ is divisible by all powers of p for some prime p (in which case we let $q_m = p$ for all m); or $z_r + L$ is divisible by infinitely many primes (in which case we let $\{q_m : m \in \omega\}$ be an infinite set of primes dividing z_r). It is then easy to see that H exists as desired. □

3. (A) implies (B)

THEOREM 9. *For any regular uncountable cardinal λ, if*
(A) there is a Whitehead group of cardinality λ which is λ-free but not free, then
(B) there exist integers $n > 0$ and $r \geq 0$, and:

(i) *a λ-system $\Lambda = (S, \lambda_\eta, B_\eta : \eta \in S)$ of height n;*
(ii) *one-one functions φ^k_ζ ($\zeta \in S_f$, $1 \leq k \leq n$) with dom$(\varphi^k_\zeta) = \omega$;*
(iii) *primes $q_{\zeta,m}$ ($\zeta \in S_f$, $m \in \omega$); and*
(iv) *integers $d^\ell_{\zeta,m}$ ($\zeta \in S_f$, $m \in \omega$, $\ell < r$)*

such that

(a) if we define $s_\zeta = \bigcup_{k=1}^n rge(\varphi^k_\zeta)$, then $\mathcal{S} = \{s_\zeta : \zeta \in S_f\}$ is a λ-free family of countable sets based on Λ; in particular, $rge(\varphi^k_\zeta) \subseteq B_{\zeta\restriction k}$;

and

(b) for any functions $c_\zeta : \omega \to \mathbb{Z}$ ($\zeta \in S_f$), there is a function $f : \bigcup \mathcal{S} \to \mathbb{Z}$ such that for all $\zeta \in S_f$ there are integers $a_{\zeta,j}$ ($j \in \omega$) such that for all $m \in \omega$,

$$c_\zeta(m) = q_{\zeta,m} a_{\zeta,m+r+1} - a_{\zeta,m+r} - \sum_{\ell<r} d^\ell_{\zeta,m} a_{\zeta,\ell} - \sum_{k=1}^{n} f(\varphi^k_\zeta(m)).$$

PROOF. We shall refer to the data in (B), which satisfies (a) and (b), as a *λ-system with data for the Whitehead problem* or more briefly a *Whitehead λ-system.* Given a Whitehead group G of cardinality λ which is λ-free but not free, we begin by defining a λ-system and a family of countable sets based on the λ-system following the procedure given in [**2**, VII.3.4]; we review that procedure here.

Choose a *λ-filtration* of G, that is, write G as the union of a continuous chain

$$G = \bigcup_{\alpha<\lambda} B_\alpha$$

of pure subgroups of cardinality $< \lambda$ such that if G/B_α is not λ-free, then $B_{\alpha+1}/B_\alpha$ is not free. Since G is not free,

$$E_\emptyset = \{\alpha < \lambda \colon B_\alpha \text{ is not } \lambda\text{-pure in } G\}$$

is stationary in λ. For each $\alpha \in E_\emptyset$, let λ_α ($< \lambda$) be minimal such that $B_{\alpha+1}/B_\alpha$ has a subgroup of cardinality λ_α which is not free; λ_α is regular by the Singular Compactness Theorem (see [**8**] or [**2**, IV.3.5]). If λ_α is countable, then let $\langle\alpha\rangle$

be a final node of the tree; otherwise choose $G_\alpha \subseteq B_{\alpha+1}$ of cardinality λ_α such that

$$H_\alpha = (G_\alpha + B_\alpha)/B_\alpha$$

is not free. Then H_α is λ_α-free, and we can choose a λ_α-filtration of G_α,

$$G_\alpha = \bigcup_{\beta<\lambda_\alpha} B_{\alpha,\beta}$$

such that for all β, $(B_{\alpha,\beta} + B_\alpha)/B_\alpha$ is pure in H_α and if $(B_{\alpha,\beta} + B_\alpha)/B_\alpha$ is not λ_α-pure in H_α, then

$$(B_{\alpha,\beta+1} + B_\alpha/B_\alpha)/(B_{\alpha,\beta} + B_\alpha/B_\alpha) \cong (B_{\alpha,\beta+1} + B_\alpha)/(B_{\alpha,\beta} + B_\alpha)$$

is not free. Since H_α is not free,

$$E_\alpha = \{\beta < \lambda_\alpha \colon B_{\alpha,\beta} + B_\alpha/B_\alpha \text{ is not } \lambda_\alpha\text{-pure in } H_\alpha\}$$

is stationary in λ_α. For each $\beta \in E_\alpha$, choose $\lambda_{\alpha,\beta}$ ($< \lambda_\alpha$) minimal such that there is a subgroup $G_{\alpha,\beta}$ of $B_{\alpha,\beta+1}$ of cardinality $\lambda_{\alpha,\beta}$ so that

$$H_{\alpha,\beta} = (G_{\alpha,\beta} + B_{\alpha,\beta} + B_\alpha)/(B_{\alpha,\beta} + B_\alpha)$$

is not free. If $\lambda_{\alpha,\beta}$ is countable, let $\langle\alpha,\beta\rangle$ be a final node; otherwise choose a $\lambda_{\alpha,\beta}$-filtration of $G_{\alpha,\beta}$. Continue in this way along each branch until a final node is reached.

As we have just done, we will use, when convenient, the notation $G_{\alpha,\beta}$ instead of $G_{\langle\alpha,\beta\rangle}$, etc.; thus for example we will write $G_{\eta,\delta}$ instead of $G_{\eta\frown\langle\delta\rangle}$.

In this way we obtain a λ-system $\Lambda = (S, \lambda_\eta, B_\eta : \eta \in S)$ where for each $\zeta \in S_f$, there is a countable subgroup G_ζ of G such that

$$G_\zeta + \langle\bar{B}_\zeta\rangle/\langle\bar{B}_\zeta\rangle$$

is not free. We can assume that for each $\eta \in S \setminus S_f$ and each $\delta \in E_\eta$,

$$B_{\eta,\delta+1} + \langle\bar{B}_\eta\rangle = G_{\eta,\delta} + B_{\eta,\delta} + \langle\bar{B}_\eta\rangle.$$

We can also assume that for all $\zeta \in S_f$, G_ζ has been chosen so that $G_\zeta + \langle\bar{B}_\zeta\rangle/\langle\bar{B}_\zeta\rangle$ has finite rank $r_\zeta + 1$ for some r_ζ such that every subgroup of rank $\leq r_\zeta$ is free. By restricting to a sub-λ-set, we can assume that there is an r such that $r_\zeta + 1 = r + 1$ for all $\zeta \in S_f$ and that there is an n such that Λ has height n (cf. Lemma 5). Moreover, we can assume (easing the purity condition, if necessary) that $G_\zeta + \langle\bar{B}_\zeta\rangle/\langle\bar{B}_\zeta\rangle$ is as described in Lemma 8, that is, it is generated modulo $\langle\bar{B}_\zeta\rangle$ by the cosets of elements $z_{\zeta,j}$ which satisfy precisely the relations which are consequences of relations

$$q_{\zeta,m} z_{\zeta,m+r+1} = z_{\zeta,m+r} + \sum_{\ell<r} d^\ell_{\zeta,m} z_{\zeta,\ell}$$

(modulo $\langle \bar{B}_\zeta \rangle$) for some primes $q_{\zeta,m}$ and integers $d^\ell_{\zeta,m}$. Fix $g_{\zeta,m} \in \langle \bar{B}_\zeta \rangle$ such that in G

$$q_{\zeta,m} z_{\zeta,m+r+1} = z_{\zeta,m+r} + \sum_{\ell<r} d^\ell_{\zeta,m} z_{\zeta,\ell} + g_{\zeta,m}.$$

There is a countable subset t_ζ of $\bar{B}_\zeta$ such that $G_\zeta \cap \langle \bar{B}_\zeta \rangle$ is contained in the subgroup generated by t_ζ. Let $s_\zeta = t_\zeta \times \omega$. Then it is proved in [**2**, VII.3.7] that $\{s_\zeta \colon \zeta \in S_f\}$ is λ-free and based on the λ-system $(S, \lambda_\eta, B'_\eta : \eta \in S)$ where $B'_\eta = B_\eta \times \omega$.

Let $s^k_\zeta = s_\zeta \cap B'_{\zeta\restriction k}$ and let $\nu^k_\zeta : \omega \to s^k_\zeta$ enumerate s^k_ζ without repetition. We can write each $g_{\zeta,m}$ as a sum $\sum^n_{k=1} g^k_{\zeta,m}$ where $g^k_{\zeta,m} \in B_{\zeta\restriction k}$. Now define

$$\varphi^k_\zeta(m) = \langle \nu^k_\zeta(m), g^k_{\zeta,m} \rangle \in B'_{\zeta\restriction k} \times B_{\zeta\restriction k}.$$

Then $s''_\zeta = \bigcup^n_{k=1} \mathrm{rge}(\varphi^k_\zeta)$ is based on the λ-system $(S, \lambda_\eta, B''_\eta : \eta \in S)$ where $B''_\eta = B'_\eta \times B_\eta$. Moreover $\{s''_\zeta : \zeta \in S_f\}$ is a λ-free family because of the choice of the first coordinate of $\varphi^k_\zeta(m)$. Thus we have defined the data in (B) such that (a) holds. It remains to verify (b). So let $c_\zeta : \omega \to \mathbb{Z}$ $(\zeta \in S_f)$ be given. We are going to define a short exact sequence

$$0 \longrightarrow \mathbb{Z} \longrightarrow M \xrightarrow{\pi} G \longrightarrow 0$$

and then use a splitting of π to define the function $f : \bigcup \mathcal{S} \to \mathbb{Z}$.

We will use the lexicographical ordering, $<_\ell$, on S defined as follows: $\eta_1 <_\ell \eta_2$ if and only if either η_1 is a restriction of η_2 or $\eta_1(i) < \eta_2(i)$ for the least i such that $\eta_1(i) \neq \eta_2(i)$. Note that if $\eta_1 <_\ell \eta_2$, then $\langle \bar{B}_{\eta_1} \rangle \subseteq \langle \bar{B}_{\eta_2} \rangle$. The lexicographical ordering is a well-ordering of S, so there is an order-preserving bijection $\theta : \tau \to \langle S, <_\ell \rangle$ for some ordinal τ. If for each $\sigma < \tau$ we let $A_\sigma = \langle \bar{B}_{\theta(\sigma)} \rangle$, then $G = \cup_{\sigma<\tau} A_\sigma$ represents G as the union of a chain of subgroups. However, we must exercise caution since, as we will see, this chain is not necessarily continuous.

The kernel of π will be generated by an element $e \in M$. We will define π to be the union of a chain of homomorphisms $\pi_\sigma : M_\sigma \to A_\sigma \to 0$ with kernel $\mathbb{Z}e$. The π_σ will be defined by induction on σ. At the same time, we will also define, as we go along, a chain of set functions $\psi_\sigma : A_\sigma \to M_\sigma$ such that $\pi_\sigma \circ \psi_\sigma = \mathrm{id}_{A_\sigma}$. Let π_0 be the zero homomorphism $: \mathbb{Z}e \to A_0 = \{0\}$.

Suppose that π_ρ and ψ_ρ have been defined for all $\rho < \sigma$ for some $\sigma < \tau$; say $\theta(\sigma) = \eta$ where $\eta = \langle \nu, \delta \rangle$ for some $\nu \in S$, $\delta \in E_\nu$. Suppose first that σ is a limit ordinal. Let $\pi'_\sigma : M'_\sigma \to \cup_{\rho<\sigma} A_\rho$ be the direct limit of the π_ρ $(\rho < \sigma)$ and let ψ'_σ be the direct limit of the ψ_ρ. In particular, $M'_\sigma = \varinjlim\{M_\rho : \rho < \sigma\}$. If $\cup_{\rho<\sigma} A_\rho = A_\sigma$, then we can let $\pi_\sigma = \pi'_\sigma$ and $\psi_\sigma = \psi'_\sigma$; this will happen, for example, if δ is a limit point of E_ν.

But it may be that δ has an immediate predecessor $\delta_1 \in E_\nu$. (Since σ is a limit ordinal, it follows that η is not a final node of S.) Then

$$\cup_{\rho<\sigma} A_\rho = \cup_{\gamma<\lambda_{\nu,\delta_1}} B_{\nu,\delta_1,\gamma} + B_{\nu,\delta_1} + \langle \bar{B}_\nu \rangle = B_{\nu,\delta_1+1} + \langle \bar{B}_\nu \rangle.$$

Notice that $\cup_{\rho<\sigma} A_\rho$ will be a proper subgroup of A_σ if $\delta_1 + 1 < \delta$ (i.e. if $\delta_1 + 1 \notin E_\nu$). We can extend π'_σ to $\pi_\sigma : M_\sigma \to A_\sigma$ because the inclusion of $\cup_{\rho<\sigma} A_\rho$ into A_σ induces a surjection of $\mathrm{Ext}(A_\sigma, \mathbb{Z})$ onto $\mathrm{Ext}(\cup_{\rho<\sigma} A_\rho, \mathbb{Z})$. Finally, extend ψ'_σ to ψ_σ in any way such that $\pi_\sigma \circ \psi_\sigma$ is the identity on A_σ.

Now let us consider the case when $\sigma = \rho + 1$ is a successor ordinal. Recall that $\theta(\sigma) = \langle \nu, \delta \rangle$. There are two subcases. In the first, δ is the least element of E_ν, so $\theta(\rho) = \nu$ and $A_\rho = \langle \bar{B}_\nu \rangle$, $A_\sigma = B_{\nu,\delta} + \langle \bar{B}_\nu \rangle$. In this subcase, we extend π_ρ to π_σ using the surjectivity of $\mathrm{Ext}(A_\rho, \mathbb{Z}) \to \mathrm{Ext}(A_\sigma, \mathbb{Z})$.

In the second and last subcase, δ has an immediate predecessor δ_1 in E_ν; then $\theta(\rho) = \langle v, \delta_1 \rangle$, a final node of S. Let ζ denote $\langle v, \delta_1 \rangle$; then $B_{\nu,\delta_1+1} + \langle \bar{B}_\zeta \rangle / \langle \bar{B}_\zeta \rangle$ is as described in Lemma 8, that is, it is generated modulo $\langle \bar{B}_\zeta \rangle$ by the cosets of elements $z_{\zeta,j}$ which satisfy the relations

$$(1) \qquad q_{\zeta,m} z_{\zeta,m+r+1} = z_{\zeta,m+r} + \sum_{\ell<r} d^\ell_{\zeta,m} z_{\zeta,\ell} + \sum_{k=1}^{n} g^k_{\zeta,m}$$

in G for some primes $q_{\zeta,m}$, integers $d^\ell_{\zeta,m}$ and elements $g^k_{\zeta,m} \in B_{\zeta\restriction k}$. It is at this point that we use the function c_ζ. Define M'_σ to be generated over M_ρ by elements $z'_{\zeta,j}$ modulo the relations

$$(2) \qquad q_{\zeta,m} z'_{\zeta,m+r+1} = z'_{\zeta,m+r} + \sum_{\ell<r} d^\ell_{\zeta,m} z'_{\zeta,\ell} + \sum_{k=1}^{n} \psi_\rho(g^k_{\zeta,m}) + c_\zeta(m)e$$

and define

$$\pi'_\sigma : M'_\sigma \to B_{\nu,\delta_1+1} + \langle \bar{B}_\zeta \rangle$$

to be the homomorphism extending π_ρ which takes $z'_{\zeta,j}$ to $z_{\zeta,j}$. One can verify that π'_σ is well-defined and has kernel $\mathbb{Z}e$. Extend ψ_ρ to ψ'_σ in any way such that $\pi'_\sigma \circ \psi'_\sigma$ is the identity. We extend π'_σ to $\pi_\sigma : M_\sigma \to A_\sigma = \langle \bar{B}_{\langle \nu,\delta \rangle} \rangle$ by using the surjectivity of $\mathrm{Ext}(A_\sigma, \mathbb{Z}) \to \mathrm{Ext}(B_{\nu,\delta_1+1} + \langle \bar{B}_\zeta \rangle, \mathbb{Z})$; finally we extend ψ'_σ.

This completes the definition of $\pi : M \to G$ and of the set map $\psi : G \to M$ (= the direct limit of the ψ_σ). Since G is a Whitehead group, there is a homomorphism ρ: $G \to M$ such that $\pi \circ \rho$ is the identity on G. In order to define f, consider an element x of $\cup S$; x is an ordered pair equal to $\varphi^k_\zeta(m)$ (possibly for many different (ζ, k, m)). If g is the second coordinate of x, let $f(x)$ be the unique integer such that

$$\psi(g) - \rho(g) = f(x)e.$$

Also for any $\zeta \in S_f$ and $j \in \omega$ define $a_{\zeta,j}$ such that

$$z'_{\zeta,j} - \rho(z_{\zeta,j}) = a_{\zeta,j}e.$$

Then applying ρ to the equation (1) and subtracting the result from equation (2), we obtain

$$q_{\zeta,m}(z'_{\zeta,m+r+1} - \rho(z_{\zeta,m+r+1})) = (z'_{\zeta,m+r} - \rho(z_{\zeta,m+r})) + \sum_{\ell<r} d^\ell_{\zeta,m}(z'_{\zeta,\ell} - \rho(z_{\zeta,\ell})) + \sum_{k=1}^{n} (\psi(g^k_{\zeta,m}) - \rho(g^k_{\zeta,m})) + c_\zeta(m)e$$

from which, comparing coefficients in $\mathbb{Z}e$, we get

$$q_{\zeta,m}a_{\zeta,m+r+1} = a_{\zeta,m+r} + \sum_{\ell<r} d^{\ell}_{\zeta,m}a_{\zeta,\ell} + \sum_{k=1}^{n} f(\varphi^k_\zeta(m)) + c_\zeta(m).$$

□

4. (B) implies (B+)

Now we are going to move from one combinatorial property, (B), to a stronger one, (B+), which will allow us to construct Whitehead groups that are strongly λ-free. Recall that in section 2 we defined the reshuffling property (Definition 7).

THEOREM 10. *Suppose that for some regular uncountable cardinal λ, there is a Whitehead λ-system. Then the following also holds:*
(B+) there exist integers $n > 0$ and $r \geq 0$, and:

(i) *a λ-system $\Lambda = (S, \lambda_\eta, B_\eta : \eta \in S)$ of height n ;*
(ii) *one-one functions φ^k_ζ $(\zeta \in S_f,\ 1 \leq k \leq n)$ with $dom(\varphi^k_\zeta) = \omega$;*
(iii) *primes $q_{\zeta,m}$ $(\zeta \in S_f,\ m \in \omega)$;*
(iv) *integers $d^{\ell}_{\zeta,m}$ $(\zeta \in S_f,\ m \in \omega,\ \ell < r)$*

satisfying (a) and (b) as in (B), with the additional properties:

- *$\mathcal{S} = \{s_\eta : \eta \in S_f\}$ has the reshuffling property; and*
- *for all $\zeta \in S_f$ and $k, i \in \omega$, $rge(\varphi^i_\zeta) \cap rge(\varphi^k_\zeta) = \emptyset$ if $i \neq k$.*

PROOF. We shall refer to the data in (B+), with the given properties, as a *strong Whitehead λ-system.*

Suppose that $\Lambda' = (S', \lambda'_\eta, B'_\eta : \eta \in S')$, φ'^k_ζ, $q'_{\zeta,m}$, and $d'^{\ell}_{\zeta,m}$ is a Whitehead λ-system (as in (B)); in particular, $\mathcal{S}' = \{s'_\zeta : \zeta \in S'_f\}$ is a family of countable sets based on Λ' , where $s'_\zeta = \bigcup_{k=1}^{n} \mathrm{rge}(\varphi'^k_\zeta)$. In [**2**, §VII.3A] is contained a proof that if there exists a family $\mathcal{S}'$ of countable sets based on a λ-system Λ' which is λ-free, then there is a family $\mathcal{S}$ of countable sets based on a λ-system Λ which has the reshuffling property. Our task is to examine the proof and show how the transformations carried out in the proof can be done in such a way that the additional data and properties given in (B) — namely the existence of the functions φ^k_ζ, primes $q_{\zeta,m}$, and integers $d^{\ell}_{\zeta,m}$ satisfying (b) — continue to hold. The transformations in question change the given $\mathcal{S}'$ and Λ' into $\mathcal{S}$ and Λ which are *beautiful,* that is, they satisfy the following six properties:

(i) for η, $\nu \in S$, if $B_\eta \cap B_\nu \neq \emptyset$, then there are $\tau \in S$ and α, β so that $\eta = \tau \frown \langle\alpha\rangle$ and $\nu = \tau \frown \langle\beta\rangle$;
(ii) for ζ, $\nu \in S_f$ and k, $i \in \omega$, if $s^k_\zeta \cap s^i_\nu \neq \emptyset$ then $k = i$, $\ell(\zeta) = n = \ell(\nu)$ for some n and for all $j \neq k - 1$,[1] $\zeta(j) = \nu(j)$;

[1]Note that this corrects an error in [**2**]. A list of errata for [**2**] is available from the first author.

(iii) for each k and ζ, s^k_ζ is infinite and has a tree structure; that is, for each ζ there is an enumeration $t^{k\zeta}_0, t^{k\zeta}_1, \ldots$ of s^k_ζ so that for all ν, $\zeta \in S_f$ and $n \in \omega$, if $t^{k\zeta}_{n+1} \in s^k_\nu$, then $t^{k\zeta}_n \in s^k_\nu$;

(iv) $\mathcal{S}$ is λ-free;

(v) for all $\alpha \in E_\emptyset$, $\Lambda^{\langle\alpha\rangle}$ and $\mathcal{S}^{\langle\alpha\rangle}$ are beautiful; and

(vi) one of the following three possibilities holds:

(a) every $\gamma \in E_\emptyset$ has cofinality ω and there is an increasing sequence of ordinals $\{\gamma_n : n \in \omega\}$ approaching γ such that for all $\zeta \in S_f$ if $\zeta(0) = \gamma$ then $s^1_\zeta = \{\langle \gamma_n, t_n\rangle : n \in \omega\}$ for some t_n's; moreover, these enumerations of the s^1_ζ satisfy the tree property of (iii);

(b) there is an uncountable cardinal κ and an integer $m > 0$ so that for all $\gamma \in E_\emptyset$ the cofinality of γ is κ and for all $\zeta \in S_f$, $\lambda_{\zeta\restriction m} = \kappa$; moreover, for each $\gamma \in E_\emptyset$ there is a strictly increasing continuous sequence $\{\gamma_\rho : \rho < \kappa\}$ cofinal in γ such that for all $\zeta \in S_f$ if $\zeta(0) = \gamma$ then $s^1_\zeta = \{\gamma_{\zeta(m)}\} \times X_\zeta$ for some X_ζ;

(c) each $\gamma \in E_\emptyset$ is a regular cardinal and $\lambda_{\langle\gamma\rangle} = \gamma$; moreover, for every $\zeta \in S_f$, $s^1_\zeta = \{\zeta(1)\} \times X_\zeta$ for some X_ζ.

By [**2**, Thm. VII.3A.6], if $\mathcal{S}$ and Λ are beautiful, then $\mathcal{S}$ has the reshuffling property. Thus it is enough to show that we can transform $\mathcal{S}'$ and Λ' into a beautiful $\mathcal{S}$ and Λ and at the same time preserve the additional structure of (B).

Let us begin with property (i). We do not change S' (the tree), but for every $\tau \in S' \setminus S_f$ and $\alpha \in E'_\tau$, we replace $B'_{\tau,\alpha}$ with $B'_{\tau,\alpha} \times \{\tau\}$. Define

$$\varphi^k_\zeta(m) = \langle \varphi'^k_\zeta(m), \zeta \restriction k-1\rangle \in B'_{\zeta\restriction k} \times \{\zeta \restriction k-1\}.$$

The definitions of the rest of the data are unchanged. Then (B)(b) continues to hold since given the c_ζ, define f by $f(\varphi^k_\zeta(m)) = f'(\varphi'^k_\zeta(m))$, where f' is the function associated with the original data (and the same c_ζ). The function f is well-defined because if $\varphi^k_{\zeta_1}(m_1) = \varphi^k_{\zeta_2}(m_2)$, then $\varphi'^k_{\zeta_1}(m_1) = \varphi'^k_{\zeta_2}(m_2)$. Note that property (i) implies that $\mathrm{rge}(\varphi^i_\zeta) \cap \mathrm{rge}(\varphi^k_\zeta) = \emptyset$ if $i \neq k$.

Property (ii) of the definition of beautiful is handled similarly.

To obtain property (iii), we do not change S', but for all $\tau \in S'$ we replace B'_τ with ${}^{<\omega}B'_\tau$, the set of all finite sequences of elements of B'_τ. Enumerate s'^k_ζ as $\{x^k_{\zeta,j} : j \in \omega\}$. If $\varphi'^k_\zeta(m) = x^k_{\zeta,j_m}$, define

$$\varphi^k_\zeta(m) = \langle x^k_{\zeta,i} : i \leq j_m\rangle \in {}^{<\omega}B'_{\zeta\restriction k}.$$

Given c_ζ ($\zeta \in S_f$), define $f(\varphi^k_\zeta(m)) = f'(\varphi'^k_\zeta(m))$, where f' is the function associated with the original data (and the same c_ζ). Again, f is well-defined.

So we can suppose that $\Lambda' = (S', \lambda'_\eta, B'_\eta : \eta \in S')$ and $\mathcal{S}' = \{s'_\zeta : \zeta \in S'_f\}$ satisfy also properties (i), (ii) and (iii). The proof of [**2**, Thm. VII.3A.5] shows that one can define Λ and $\mathcal{S}$ which are beautiful and such that there is a one-one order-preserving map ψ of S into S' such that for all $\eta \in S$, $\lambda_\eta = \lambda'_{\psi(\eta)}$; and for each $\zeta \in S_f$, there is a level-preserving bijection $\theta_\zeta : s_\zeta \to s'_{\psi(\zeta)}$ such that for all $\zeta, \nu \in S_f$, if $x \in s'_{\psi(\zeta)}$, $y \in s'_{\psi(\nu)}$ and $x \neq y$, then $\theta_\zeta^{-1}(x) \neq \theta_\nu^{-1}(y)$.[2]

Observe that $\eta \in S_f$ if and only if $\psi(\eta) \in S'_f$ since $\lambda_\eta = \lambda'_{\psi(\eta)}$. We use the functions ψ and θ_ζ to define the additional data in (B): let $q_{\zeta,m} = q'_{\psi(\zeta),m}$ and $d^\ell_{\zeta,m} = d'^\ell_{\psi(\zeta),m}$; moreover, define $\varphi^k_\zeta(m) = \theta_\zeta^{-1}(\varphi'^k_{\psi(\zeta)}(m))$. Given c_ζ for $\zeta \in S_f$, define $c'_{\psi(\zeta)} = c_\zeta$ and let c_ν be arbitrary for $\nu \in S'_f \setminus \psi[S]$. Then since the original data satisfy (B), $f' : \bigcup S' \to \mathbb{Z}$ and $a'_{\nu,j}$ ($\nu \in S'_f$, $j \in \omega$) exist. Let $f(\varphi^k_\zeta(m)) = f'(\varphi'^k_{\psi(\zeta)}(m))$; the (contrapositive of the) final hypothesis on θ_ζ implies that f is well-defined. Let $a_{\zeta,m} = a'_{\psi(\zeta),m}$. Then for each $\zeta \in S_f$, the equation

$$q'_{\psi(\zeta),m}a'_{\psi(\zeta),m+r+1} = a'_{\psi(\zeta),m+r} + \sum_{\ell<r} d'^\ell_{\psi(\zeta),m}a'_{\psi(\zeta),\ell} + \sum_{k=1}^{n} f'(\varphi'^k_{\psi(\zeta)}(m)) + c'_{\psi(\zeta)}(m)$$

is the desired equation

$$q_{\zeta,m}a_{\zeta,m+r+1} = a_{\zeta,m+r} + \sum_{\ell<r} d^\ell_{\zeta,m}a_{\zeta,\ell} + \sum_{k=1}^{n} f(\varphi^k_\zeta(m)) + c_\zeta(m).$$

□

5. (B+) IMPLIES (A+)

THEOREM 11. *Let λ be a regular uncountable cardinal such that (B+) holds, i.e., there is a strong Whitehead λ-system. Then*

(A+) there are 2^λ strongly λ-free Whitehead groups of cardinality λ.

PROOF. Given a strong Whitehead λ-system $(S, \lambda_\eta, B_\eta : \eta \in S)$ together with $\varphi^k_\zeta, q_{\zeta,m}, d^\ell_{\zeta,m}$, we use them to define a group G in terms of generators and relations. Our group G will be the group F/K where F is the free abelian group with basis

$$\bigcup S \cup \{z_{\zeta,j} : \zeta \in S_f,\ j \in \omega\} \tag{3}$$

and K is the subgroup of F generated by the elements $w_{\zeta,m} =$

$$q_{\zeta,m}z_{\zeta,m+r+1} - z_{\zeta,m+r} - \sum_{\ell<r} d^\ell_{\zeta,m}z_{\zeta,\ell} - \sum_{k=1}^{n} \varphi^k_\zeta(m) \tag{4}$$

for all $m \in \omega$, and $\zeta \in S_f$. Let us see first that G is a Whitehead group. (For this we need only (B).) It suffices to show that every group homomorphism $\psi : K \longrightarrow \mathbb{Z}$ extends to a homomorphism from F to $\mathbb{Z}$. (See, for example, [**2**, p.8].) Given ψ, define $c_\zeta(m) = \psi(w_{\zeta,m})$ for all $m \in \omega$, and $\zeta \in S_f$. Then by

[2]Note that this is a clarification and correction of the first paragraph of the proof of [2, Thm. VII.3A.5, p. 219]. Also, in the third paragraph of that proof, ψ should be ψ^{-1}.

(B)(b), there are integers $a_{\zeta,j}$ ($\zeta \in S_f$, $j \in \omega$) and a function $f : \bigcup \mathcal{S} \to \mathbb{Z}$ such that for all $\zeta \in S_f$ and $m \in \omega$,

$$(5) \qquad c_\zeta(m) = q_{\zeta,m} a_{\zeta,m+r+1} - a_{\zeta,m+r} - \sum_{\ell < r} d^\ell_{\zeta,m} a_{\zeta,\ell} - \sum_{k=1}^{n} f(\varphi^k_\zeta(m)).$$

Define $\theta : F \longrightarrow \mathbb{Z}$ by setting $\theta \restriction \bigcup \mathcal{S} = f$ and $\theta(z_{\zeta,j}) = a_{\zeta,j}$. We just need to check that θ extends ψ. But for all $\zeta \in S_f$ and $m \in \omega$, we have

$$\theta(w_{\zeta,m}) = q_{\zeta,m} a_{\zeta,m+r+1} - a_{\zeta,m+r} - \sum_{\ell<r} d^\ell_{\zeta,m} a_{\zeta,\ell} - \sum_{k=1}^{n} f(\varphi^k_\zeta(m))$$

by the definitions of θ and of $w_{\zeta,m}$. Thus $\theta(w_{\zeta,m}) = c_\zeta(m) = \psi(w_{\zeta,m})$, by (5).

Next let us show that G is not free. (Here again, we need only (B).) The proof is essentially the same as that of Lemma VII.3.9 of [**2**, pp. 205f], but we will give a somewhat different version of the proof here. The proof proceeds by induction on n where n is the height of our λ-system. For each $\alpha < \lambda$, let G_α be the subgroup of G generated by

$$\{z_{\zeta,j} : \zeta \in S_f, \zeta(0) < \alpha,\ j \in \omega\} \cup \bigcup \{s_\zeta : \zeta \in S_f, \zeta(0) < \alpha\}.$$

It suffices to prove that for all α in a stationary subset of λ, $G_{\alpha+1}/G_\alpha$ is not free (cf. [**2**, IV.1.7]). In fact, we will show that $G_{\alpha+1}/G_\alpha$ is not free when α is a limit point of $E_\emptyset$ and belongs to $C \cap E_\emptyset$, where C is the cub

$$\{\alpha < \lambda : \text{whenever } \varphi^1_\zeta(m) \in \bigcup\{B_{\langle\beta\rangle} : \beta < \alpha\} \text{ then } \exists \sigma \in S_f \text{ with } \sigma(0) < \alpha \text{ and } \varphi^1_\zeta(m) \in \operatorname{rge}(\varphi^1_\sigma)\}.$$

We begin with the case $n = 1$. Then for all $\alpha \in C \cap E_\emptyset$ such that α is a limit point of $E_\emptyset$, $G_{\alpha+1}/G_a$ is non-free because it is as described in the first part of Lemma 8 (with generators $\{z_{\langle\alpha\rangle,j} : j \in \omega\}$), since for all $m \in \omega$, $\varphi^1_{\langle\alpha\rangle}(m) \in B_{\langle\alpha\rangle} = \bigcup\{B_{\langle\beta\rangle} : \beta < \alpha\}$ by the definition of a λ-system (because α is a limit point of $E_\emptyset$) and hence $\varphi^1_{\langle\alpha\rangle}(m) \in G_\alpha$ since $\alpha \in C$.

Now suppose $n > 1$ and the result is proved for $n - 1$. Again, let $\alpha \in C \cap E_\emptyset$ such that α is a limit point of $E_\emptyset$. Again we have that $\varphi^1_\zeta(m) \in G_\alpha$ for all $m \in \omega$ when $\zeta(0) = \alpha$. We will consider the $\lambda_{\langle\alpha\rangle}$-system $\Lambda^{\langle\alpha\rangle}$. (See Definition 6.) Note that $\Lambda^{\langle\alpha\rangle}$ has height $n - 1$, and the group $G_{\alpha+1}/G_\alpha$ is defined as in (3) and (4) relative to this $\lambda_{\langle\alpha\rangle}$-system. Hence by induction $G_{\alpha+1}/G_\alpha$ is not free.

Finally, we will use the reshuffling property given by (B+) to prove that G is strongly λ-free. As in the proof of [**2**, VII.3.11], we will prove that for all $\alpha \in \lambda \cup \{-1\}$ and all $\beta > \alpha$, $G_\beta/G_{\alpha+1}$ is free. Let $I = \{\zeta \in S_f : \zeta(0) < \beta\}$, and let $<_I$ be the well-ordering given by the reshuffling property for I and α. Let s^k_ζ denote $\operatorname{rge}(\varphi^k_\zeta)$. We claim that there is a basis $\mathcal{Z}_{\beta,\alpha}$ of $G_\beta/G_{\alpha+1}$ consisting of the cosets of the members of the following two sets:

$$\{z_{\zeta,j} : \alpha < \zeta(0) < \beta, \text{ and either } j < r \text{ or } \exists k \text{ s.t. } \varphi^k_\zeta(j - r) \notin \bigcup\{s^k_\nu : \nu <_I \zeta\}\}$$

and

$$\{\varphi^k_\zeta(m) : \varphi^k_\zeta(m) \notin \bigcup\{s^k_\nu : \nu <_I \zeta\} \text{ and } \\ \exists i < k[\varphi^i_\zeta(m) \notin \bigcup\{s^i_\nu : \nu <_I \zeta\}]\}.$$

To see that the elements of $\mathcal{Z}_{\beta,\alpha}$ generate $G_\beta/G_{\alpha+1}$, we proceed by induction with respect to $<_I$ to show that the coset of every $z_{\zeta,j}$ ($\zeta(0) < \beta, j \in \omega$) and the coset of every element of s_ζ ($\zeta(0) < \beta$) is a linear combination of the elements of $\mathcal{Z}_{\beta,\alpha}$. Since $s_\zeta \setminus \bigcup\{s_\nu : \nu <_I \zeta\}$ is infinite, for each $j \in \omega$ such that $z_{\zeta,j} + G_{a+1} \notin \mathcal{Z}_{\beta,\alpha}$, there is $t > j$ such that $z_{\zeta,t} + G_{\alpha+1}$ belongs to $\mathcal{Z}_{\beta,\alpha}$. Without loss of generality, $t = j + 1$. Then

$$z_{\zeta,j} = q_{\zeta,j-r} z_{\zeta,t} - \sum_{\ell<r} d^\ell_{\zeta,j-r} z_{\zeta,\ell} - \sum_{k=1}^{n} \varphi^k_\zeta(j-r)$$

by (4). By induction each $\varphi^k_\zeta(j-r) + G_{a+1}$ is a linear combination of members of $\mathcal{Z}_{\beta,\alpha}$ (because $\varphi^k_\zeta(j-r) \in \bigcup\{s^k_\nu : \nu <_I \zeta\}$ since $z_{\zeta,j} + G_{a+1} \notin \mathcal{Z}_{\beta,\alpha}$); hence $z_{\zeta,j} + G_{a+1}$ belongs to the subgroup generated by the members of $\mathcal{Z}_{\beta,\alpha}$.

For each $m, i \in \omega$, if $\varphi^i_\zeta(m) \in \bigcup\{s^k_\nu : \nu <_I \zeta\}$, then by induction $\varphi^i_\zeta(m) + G_{\alpha+1}$ is a linear combination of elements of $\mathcal{Z}_{\beta,\alpha}$. Otherwise, $\varphi^i_\zeta(m) + G_{\alpha+1}$ belongs to $\mathcal{Z}_{\beta,\alpha}$ unless i is minimal such that $\varphi^i_\zeta(m) \notin \bigcup\{s^k_\nu : \nu <_I \zeta\}$. But in the latter case,

$$\sum_{k=1}^{n} \varphi^k_\zeta(m) = q_{\zeta,m} z_{\zeta,m+r+1} - z_{\zeta,m+r} - \sum_{\ell<r} d^\ell_{\zeta,m} z_{\zeta,\ell}$$

so its coset is a linear combination of elements of $\mathcal{Z}_{\beta,\alpha}$; thus since $\varphi^k_\zeta(m) + G_{\alpha+1} \in \mathcal{Z}_{\beta,\alpha}$ for all $k \neq i$, $\varphi^i_\zeta(m) + G_{\alpha+1}$ belongs to the subgroup generated by the elements of $\mathcal{Z}_{\beta,\alpha}$. This completes the proof that $\mathcal{Z}_{\beta,\alpha}$ is a generating set. To see that the elements of $\mathcal{Z}_{\beta,\alpha}$ are independent, compare coefficients in F.

To construct not just one but 2^λ different strongly λ-free Whitehead groups, we use a standard trick: write $E_\emptyset$ as the disjoint union $\coprod_{\sigma<\lambda} X_\sigma$ of λ stationary sets; then for every non-empty subset W of λ, do the construction above for the generalized λ-system $\Lambda = (S_W, \lambda_\zeta, B_\zeta : \zeta \in S_W)$ with $E_\emptyset = \coprod_{\sigma\in W} X_\sigma$, i.e., where $S_W = \{\zeta \in S : \zeta(0) \in \coprod_{\sigma \in W} X_\sigma\}$. □

6. Appendix: Non-free Whitehead implies 2-uniformization

A *ladder system* on a stationary subset E of ω_1 is an indexed family of functions $\{\eta_\alpha : \alpha \in E\}$ such that each $\eta_\alpha : \omega \to \alpha$ is strictly increasing and $\sup(\mathrm{rge}(\eta_\alpha)) = \alpha$. If $\{\varphi_\alpha : \alpha \in I\}$ is an indexed family of functions each with domain ω, we say that it has the *2-uniformization property* provided that for every family of functions $c_\alpha : \omega \to 2 = \{0, 1\}$ ($\alpha \in I$), there exists a function H such that for all $\alpha \in I$, $H(\varphi_\alpha(n))$ is defined and equals $c_\alpha(n)$ for all but finitely many n. It is not hard, given a ladder system on E which has the 2-uniformization property, to construct, explicitly (by generators and relations), a non-free Whitehead group. (See [**2**, Prop. XII.3.6].) It is more difficult to go the other way: starting with an arbitrary non-free Whitehead group of cardinality

$\aleph_1$ to show that there exists a ladder system on a stationary subset of ω_1 which has the 2-uniformization property. This was left to the reader in the original paper by the second author [**9**, Thm. 3.9, p. 277]. The only published proof is a rather complicated one in [**2**, §XII.3]; so considering the importance of this result, it seems to us worthwhile to give another proof which is conceptually and technically simpler than that one. The proof given here resembles the original proof found by the second author, which was also the basis of the proofs in [**3**] and in this paper.

Our goal is to prove the following.

THEOREM 12. *If there is a non-free Whitehead group A of cardinality $\aleph_1$, then there is a ladder system $\{\eta_\alpha : \alpha \in E\}$ on a stationary set E which has the 2-uniformization property.*

We begin with an observation. It is sufficient to show that the hypothesis of Theorem 12 implies that there is a family $\{\varphi_\alpha : \alpha \in E\}$ of functions which has the 2-uniformization property and is *based on an ω_1-filtration*, that is, indexed by a stationary subset E of ω_1 and such that there is a continuous ascending chain $\{B_\nu : \nu \in \omega\}$ of countable sets such that for all $\alpha \in E$, $\varphi_\alpha : \omega \longrightarrow B_\alpha$. (Note that what we are talking about, in the language of the preceding sections, is a family of countable sets based on an $\aleph_1$-system.) Indeed, by a suitable coding we can assume that $B_\alpha = \alpha$ and if the range of φ_α is not cofinal in α, we can choose a ladder η'_α on α, replace $\varphi_\alpha(n)$ by $\langle\varphi_\alpha(n), \eta'_\alpha(n)\rangle$, and re-code, to obtain a ladder system on $E \cap C$, (for some cub C) which has the 2-uniformization property.

From now on, let A denote a non-free Whitehead group of cardinality $\aleph_1$. Then we can write A as the union, $A = \cup_{\nu<\omega_1} A_\nu$, of a continuous chain of countable free subgroups; since A is not free, we can assume that there is a stationary subset E of ω_1(consisting of limit ordinals) such that for all $\alpha \in E$ $A_{\alpha+1}/A_\alpha$ is not free. By Pontryagin's Criterion we can assume without loss of generality that $A_{\alpha+1}/A_\alpha$ is of finite rank and, in fact, that every subgroup of $A_{\alpha+1}/A_\alpha$ of smaller rank is free. Since

($\star$) *whenever $E = \cup_{n\in\omega} E_n$, at least one of the E_n is stationary*

(cf. [**2**, Cor. II.4.5]) we can also assume that all of the $A_{\alpha+1}/A_\alpha$ (for $\alpha \in E$) have the same rank $r+1$ ($r \geq 0$). In order to make clear the ideas involved in the proof of the Theorem, we will give the proof first in the special case when $r = 0$, i.e., $A_{\alpha+1}/A_\alpha$ is a rank one non-free group when $\alpha \in E$, and then describe how to handle the extension to the general case. In fact this special case divides into two subcases: using ($\star$) and replacing $A_{\alpha+1}$ by a subgroup if necessary, we can assume that either

(i) for all $\alpha \in E$, $A_{\alpha+1}/A_\alpha$ has a type all of whose entries are 0's or 1's [and there are infinitely many 1's]; or

(ii) there is a prime p such that for all $\alpha \in E$, the type of $A_{\alpha+1}/A_\alpha$ is $(0, 0, ...0, \infty, 0, ...)$ where the ∞ occurs in the pth place.

(See [**5**, pp. 107ff].) We next give the easy combinatorial lemmas needed for the first, and simplest, subcase.

LEMMA 13. *Suppose Y and Y' are finite subsets of an abelian group G such that $|Y|^2 < |Y'|$. Then there exists $b \in Y'$ such that Y and $b + Y$ are disjoint. [Here $b + Y = \{b + y : y \in Y\}$.]*

PROOF. Choose $b \in Y' \setminus \{x - y : x, y \in Y\}$. □

LEMMA 14. *For any positive integer $p > 1$ there are integers a_0 and a_1 and a function $F_p : \mathbb{Z}/p\mathbb{Z} \to 2 = \{0, 1\}$ such that for all $m \in \mathbb{Z}$ with $(2|m| + 1)^2 < p$, $F_p(m + a_\ell + p\mathbb{Z}) = \ell$, for $\ell = 0, 1$.*

PROOF. Let $a_0 = 0$ and let $a_1 = b$ as in Lemma 13, where $G = \mathbb{Z}/p\mathbb{Z} = Y'$ and $Y = \{m + p\mathbb{Z} : (2|m| + 1)^2 < p\}$. Then since $Y = a_0 + Y$ and $a_1 + Y$ are disjoint, we can define F_p. (Note that F_p is a set function, not a homomorphism.) □

PROOF OF THEOREM 12 (*in special subcase* (i)): For all $\alpha \in E$ there is an infinite set P_α of primes such that

$$A_{\alpha+1}/A_\alpha \cong \{\frac{m}{n} \in \mathbb{Q} : n \text{ is a product of } \textit{distinct} \text{ primes from } P_\alpha\}.$$

Then if $P_\alpha = \{p_{\alpha,n} : n \in \omega\}$, $A_{\alpha+1}$ is generated over A_α by a subset $\{y_{\alpha,n} : n \in \omega\}$ satisfying the relations (and only the relations)

$$(\dagger) \quad p_{\alpha,n} y_{\alpha,n+1} = y_{\alpha,0} - g_{\alpha,n}$$

for some $g_{\alpha,n} \in A_\alpha$. We define $\varphi_\alpha(n) = \langle p_{\alpha,n}, g_{\alpha,n}\rangle$. Then $\{\varphi_\alpha : \alpha \in E\}$ is based on an ω_1-filtration, in fact on the chain $\{\mathbb{Z} \times A_\alpha : \alpha < \omega_1\}$.

Given functions $c_\alpha : \omega \to 2$, we are going to define a homomorphism $\pi : A' \to A$ with kernel $\mathbb{Z}e$ and then use the splitting $\rho : A \to A'$ to define the uniformizing function H.

We define $\pi_\nu : A'_\nu \to A_\nu$ inductively along with a set function ψ_ν: $A_\nu \to A'_\nu$ such that $\pi_\nu \circ \psi_\nu = 1_{A_\nu}$. The crucial case is when π_α and ψ_α have been defined and $\alpha \in E$. (When $\alpha \notin E$ we can use the fact that $\mathrm{Ext}(A_{\alpha+1}, \mathbb{Z}) \to \mathrm{Ext}(A_\alpha, \mathbb{Z})$ is onto.) We define $A'_{\alpha+1}$ by generators $\{y'_{\alpha,n} : n \in \omega\}$ over A'_α satisfying relations

$$(\dagger\dagger) \quad p_{\alpha,n} y'_{\alpha,n+1} = y'_{\alpha,0} - \psi_\alpha(g_{\alpha,n}) + a_\ell e$$

where a_ℓ is as in Lemma 14 for $p = p_{\alpha,n}$ and $\ell = c_\alpha(n)$.

In the end we let $\pi = \cup_\nu \pi_\nu : A' = \cup_\nu A'_\nu \to A$ and $\psi = \cup_\nu \psi_\nu$. Then since A is a Whitehead group, there exists a homomorphism ρ such that $\pi \circ \rho = 1_A$. For any $g \in A$, $\psi(g) - \rho(g) \in \ker(\pi) = \mathbb{Z}e$; we will abuse notation and identify $\psi(g) - \rho(g)$ with the unique integer k such that $\psi(g) - \rho(g) = ke$. For any $w \in \cup_{\alpha \in E} \mathrm{rge}(\varphi_\alpha)$, if $w = \langle p, g\rangle$, let $H(w) = F_p(\psi(g) - \rho(g) + p\mathbb{Z})$.

Note that w may equal $\varphi_\alpha(n)$ $(= \langle p_{\alpha,n}, g_{\alpha,n}\rangle)$ for many pairs (α, n). To see that this definition of H works, fix $\alpha \in E$. For any $n \in \omega$, applying ρ to equation (†) and subtracting from equation (††), we have

$$p_{\alpha,n}(y'_{\alpha,n+1} - \rho(y_{\alpha,n+1})) = y'_{\alpha,0} - \rho(y_{\alpha,0}) - (\psi_\alpha(g_{\alpha,n}) - \rho(g_{\alpha,n})) + a_\ell$$

so that $\psi_\alpha(g_{\alpha,n}) - \rho(g_{\alpha,n})$ is congruent to $y'_{\alpha,0} - \rho(y_{\alpha,0}) + a_\ell \bmod p_{\alpha,n}$. Then if n is large enough, $(2|\, y'_{\alpha,0} - \rho(y_{\alpha,0})| + 1)^2 < p_{\alpha,n}$ so by choice of $F_{p_{\alpha,n}}$ and a_ℓ,

$$H(\varphi_\alpha(n)) = F_{p_{\alpha,n}}(\psi_\alpha(g_{\alpha,n}) - \rho(g_{\alpha,n}) + p_{\alpha,n}\mathbb{Z}) = \\ F_{p_{\alpha,n}}(y'_{\alpha,0} - \rho(y_{\alpha,0}) + a_\ell + p_{\alpha,n}\mathbb{Z}) = \ell = c_\alpha(n).$$

This completes the proof in the first special subcase.

For the purposes of the second special subcase we need another combinatorial lemma.

LEMMA 15. *Fix a positive integer $p > 1$. Define a strictly increasing sequence of positive integers t_i inductively, as follows. Let $t_0 = 0$. If t_{i-1} has been defined for some $i \geq 1$, let $t_i = t_{i-1} + d_i$ where d_i is the least positive integer such that $(2p^{t_{i-1}} + 1)^2 p^{2t_{i-1}} < p^{d_i}$. Then for every $i \geq 1$ there exists a function*

$$F_i : \mathbb{Z}/p^{t_i}\mathbb{Z} \to 2$$

and integers $a_n^\ell \in \{0, ..., p-1\}$ $(t_{i-1} \leq n < t_i, \ell = 0, 1)$ such that whenever $|m_0| \leq p^{t_{i-1}}$ and $a_j \in \{0, ..., p-1\}$ for $j < t_{i-1}$, then for $\ell = 0, 1$

$$F_i(m_0 + \sum_{j<t_{i-1}} p^j a_j + \sum_{n=t_{i-1}}^{t_i - 1} p^n a_n^\ell + p^{t_i}\mathbb{Z}) = \ell.$$

PROOF. We apply Lemma 13 to the sets $Y = \{m_0 + \sum_{j<t_{i-1}} p^j a_j + p^{t_i} Z : |m_0| \leq p^{t_{i-1}}, a_j \in \{0, ..., p-1\}\}$ (which has cardinality $\leq (2p^{t_{i-1}} + 1)p^{t_{i-1}}$) and $Y' = \{\sum_{n=t_{i-1}}^{t_i-1} p^n x_n + p^{t_i} Z : x_n \in \{0, ..., p-1\}\}$ (which has cardinality p^{d_i}), to get $b \in Y'$. Then choose $a_n^0 = 0$ for all n, and a_n^1 so that $\sum_{n=t_{i-1}}^{t_i-1} p^n a_n^1 = b$ and define F_i as in Lemma 14. □

PROOF OF THEOREM 12 (*in special subcase* (ii)): For all $\alpha \in E$

$$A_{\alpha+1}/A_\alpha \cong \{\frac{m}{n} \in \mathbb{Q} : n \text{ is a power of } p\}$$

Then $A_{\alpha+1}$ is generated over A_α by a subset $\{y_{\alpha,n} : n \in \omega\}$ satisfying the relations (and only the relations)

$$(\dagger) \quad p y_{\alpha,n+1} = y_{\alpha,n} - g_{\alpha,n}$$

for some $g_{\alpha,n} \in A_\alpha$. Let $\varphi_\alpha(m) = \langle g_{\alpha,j} : j < t_{m+1}\rangle$ for all $m \in \omega$. Given functions $c_\alpha : \omega \to 2$, we define $\pi_\nu : A'_\nu \to A_\nu$ with kernel $\mathbb{Z}e$ inductively along with a set function ψ_ν: $A_\nu \to A'_\nu$ such that $\pi_\nu \circ \psi_\nu = 1_{A_\nu}$. The crucial case is when π_α and ψ_α have been defined and $\alpha \in E$. Then we define $A'_{\alpha+1}$ by generators $\{y'_{\alpha,n} : n \in \omega\}$ over A'_α satisfying relations

$$(\dagger\dagger)\ \ py'_{\alpha,n+1} = y'_{\alpha,n} - \psi_\alpha(g_{\alpha,n}) + a_n^{\ell(n)}e$$

where $\ell(n)$ is taken to be $c_\alpha(i-1)$ when $t_{i-1} \leq n < t_i$. In the end we let $\pi = \cup_\nu \pi_\nu : A' = \cup_\nu A'_\nu \to A$ and $\psi = \cup_\nu \psi_\nu$. Then since A is a Whitehead group, there exists a homomorphism ρ such that $\pi \circ \rho = 1_A$. For any $w \in \cup_{\alpha\in E}\mathrm{rge}(\varphi_\alpha)$, if $w = \langle g_j : j < t_i\rangle$, let $H(w) = F_i(\sum_{n<t_i} p^n(\psi(g_n) - \rho(g_n)) + p^{t_i}\mathbb{Z})$. To see that this works, fix $\alpha \in E$ and for $i \geq 1$, consider $w_i = \varphi_\alpha(i-1) = \langle g_{\alpha,j} : j < t_i\rangle$. From the equations $(\dagger)$, for $n \leq t_i$ we obtain that

$$p^{t_i} y_{\alpha,t_i} = y_{\alpha,0} - \sum_{n<t_i} p^n g_{\alpha,n}$$

If we apply ρ to this and subtract from the corresponding equation derived from $(\dagger\dagger)$ we obtain that $\sum_{n<t_i} p^n(\psi(g_{\alpha,n}) - \rho(g_{\alpha,n}))$ is congruent to

$$(y'_{\alpha,0} - \rho(y_{\alpha,0})) + \sum_{n<t_i} p^n a_n^{\ell(n)}$$

mod p^{t_i}. So if $|y'_{\alpha,0} - \rho(y_{\alpha,0})| \leq p^{t_{i-1}}$, then by our choice of F_i and the $a_n^{\ell(n)}$ for $t_{i-1} \leq n < t_i$, $H(w_i)$ equals $c_\alpha(i-1)$.

This completes the proof of Theorem 12 when $r = 0$.

PROOF OF THEOREM 12 (*in the general case*): In the general case without loss of generality we have either

(i) for all $\alpha \in E$, $A_{\alpha+1}/A_\alpha$ has a free subgroup L_α/A_α of rank r such that $A_{\alpha+1}/L_\alpha$ has a type all of whose entries are 0's or 1's [and there are infinitely many 1's]; or

(ii) there is a prime p such that for all $\alpha \in E$, $A_{\alpha+1}/A_\alpha$ has a free subgroup L_α/A_α of rank r such that the type of $A_{\alpha+1}/L_\alpha$ is $(0, 0, ...0, \infty, 0, ...)$ where the ∞ occurs in the pth place.

In other words, $A_{\alpha+1}$ is generated by A_α and a subset $\{z_{\alpha,k} : k = 1, ..., r\} \cup \{y_{\alpha,n} : n \in \omega\}$ modulo (only) the relations in A_α plus relations:

(i) $(\dagger)$ $p_{\alpha,n} y_{\alpha,n+1} = y_{\alpha,0} + \sum_{k=1}^r \mu_{\alpha,k}(n) z_{\alpha,k} - g_{\alpha,n}$ for some family of distinct primes $p_{\alpha,n}$ and $\mu_{\alpha,k}(n) \in \mathbb{Z}$, $g_{\alpha,n} \in A_\alpha$; or

(ii) $(\dagger)$ $py_{\alpha,n+1} = y_{\alpha,n} + \sum_{k=1}^r \mu_{\alpha,k}(n) z_{\alpha,k} - g_{\alpha,n}$ for some $\mu_{\alpha,k}(n) \in \mathbb{Z}$ and $g_{\alpha,n} \in A_\alpha$ for each $n \in \omega$.

For use in (the harder) subcase (ii), define a strictly increasing sequence of positive integers t_i inductively, as follows. Let $t_0 = 0$. If t_{i-1} has been defined for some $i \geq 1$, let $t_i = t_{i-1} + d_i$ where d_i is the least positive integer such that

$$(2p^{t_{i-1}} + 1)^{2r+2} p^{2t_{i-1}} < p^{d_i}.$$

Then we have the following generalization of Lemma 15. (Note that when $r = 0$ the sequence μ is empty.)

LEMMA 16. *Fix $p > 1$ and $r \geq 0$. For every sequence of functions $\mu = \langle \mu_1, ..., \mu_r \rangle$, where $\mu_k : \omega \to \mathbb{Z}$ and every $i \geq 1$ there exists a function*

$$F_{i,\mu} : \mathbb{Z}/p^{t_i}\mathbb{Z} \to 2$$

and integers $a^{\ell}_{n,\mu} \in \{0, ..., p-1\}$ $(t_{i-1} \leq n < t_i, \ell = 0, 1)$ such that $F_{i,\mu}$ and $a^{\ell}_{n,\mu}$ depend only on $\mu \restriction t_i$ $(= \langle \mu_1 \restriction t_i, ..., \mu_r \restriction t_i \rangle)$ and are such that whenever $m_0, ..., m_r$ are integers with $|m_k| \leq p^{t_{i-1}}$ for all $k \leq r$ and $a_j \in \{0, ..., p-1\}$ for $j < t_{i-1}$, then

$$F_{i,\mu}(m_0 + \sum_{k=1}^{r} (\sum_{j<t_i} p^j \mu_k(j)) m_k + \sum_{j<t_{i-1}} p^j a_j + \sum_{n=t_{i-1}}^{t_i - 1} p^n a^{\ell}_{n,\mu} + p^{t_i}\mathbb{Z}) = \ell.$$

PROOF. We apply Lemma 13 with $G = \mathbb{Z}/p^{t_i}\mathbb{Z}$,

$$Y = \{m_0 + \textstyle\sum_{k=1}^{r} (\sum_{j<t_i} p^j \mu_k(j)) m_k + \sum_{j<t_{i-1}} p^j a_j + p^{t_i}\mathbb{Z} : \\ |m_k| \leq p^{t_i}\text{ , for all } k \leq r,\ a_j \in \{0, ..., p-1\}\}$$

and

$$Y' = \{\sum_{n=t_{i-1}}^{t_i - 1} p^n x_n + p^{t_i}\mathbb{Z} : x_n \in \{0, ..., p-1\}\}.$$

and proceed as in the proof of Lemma 15. □

Similarly we have the following generalization of Lemma 14 for use in subcase (i).

LEMMA 17. *Given $p > 1$ and $r \geq 0$, and a sequence of integers $\mu = \langle \mu_1, ..., \mu_r \rangle$, let t_p be maximal such that $(2t_p + 1)^{2r+2} < p$. Then there exists a function*

$$F_{p,\mu} : \mathbb{Z}/p\mathbb{Z} \to 2$$

and integers $a^{\ell}_{p,\mu} \in \{0, ..., p-1\}$ $(\ell = 0, 1)$ such that whenever $m_0, ..., m_r$ are integers such that $|m_k| \leq t_p$ for all $k \leq r$, then

$$F_{p,\mu}(m_0 + \sum_{k=1}^{r} \mu_k m_k + a^{\ell}_{p,\mu} + p\mathbb{Z}) = \ell.$$

□

Now define the function φ_α with domain ω by letting

(i) $\varphi_\alpha(m) = \langle \langle \mu_{\alpha,k}(m) : k = 1, ..., r \rangle, p_{\alpha,m}, g_{\alpha,m} \rangle$; or
(ii) $\varphi_\alpha(m) = \langle \langle \mu_{\alpha,k}(n) : k = 1, ..., r \rangle, g_{\alpha,n} : n < t_{m+1} \rangle$.

Given functions $c_\alpha : \omega \to 2$, we define $\pi_\nu : A'_\nu \to A_\nu$ inductively along with a set function ψ_ν: $A_\nu \to A'_\nu$ such that $\pi_\nu \circ \psi_\nu = 1_{A_\nu}$. The crucial case is when π_α and ψ_α have been defined and $\alpha \in E$. Then we define $A'_{\alpha+1}$ by generators $\{z'_{\alpha,k} : k = 1, ..., r\} \cup \{y'_{\alpha,n} : n \in \omega\}$ over A'_α satisfying relations

(i) (††) $p_{\alpha,n} y'_{\alpha,n+1} = y'_{\alpha,0} + \sum_{k=1}^r \mu_{\alpha,k}(n) z'_{\alpha,k} - \psi_\alpha(g_{\alpha,n}) + a^\ell_{p_{\alpha,n},\mu} e$; or
(ii) (††) $p y'_{\alpha,n+1} = y'_{\alpha,n} + \sum_{k=1}^r \mu_{\alpha,k}(n) z'_{\alpha,k} - \psi_\alpha(g_{\alpha,n}) + a^\ell_{\alpha,n,\mu} e$

where $a^\ell_{p_{\alpha,n},\mu}$ (respectively, $a^\ell_{\alpha,n,\mu}$) is as in Lemma 17 (respectively, Lemma 16) for $\ell = c_\alpha(n)$ (respectively, $\ell = c_\alpha(i-1)$ if $t_{i-1} \leq n < t_i$) (and the appropriate prime or primes are used).

In the end we use a splitting ρ of $\pi = \cup_\nu \pi_\nu : A' = \cup_\nu A'_\nu \to A$ to define $H(w)$ as follows:

(i) if $w = \langle\langle \mu_k : k = 1, ..., r\rangle, p, g\rangle$, let $H(w) = F_{p,\mu}(\psi(g) - \rho(g) + p\mathbb{Z})$; or
(ii) if $w = \langle\langle \mu_k(n) : k = 1, ..., r\rangle, g_n : n < t_i\rangle$, let $H(w) =$
$F_{i,\mu}(\sum_{n<t_i} p^n(\psi(g_n) - \rho(g_n)) + p^{t_i}\mathbb{Z})$.

Then we check as before that this definition works. □

References

1. P. C. Eklof, *Infinitary equivalence of abelian groups*, Fund. Math. **81** (1974), 305–314.
2. P. C. Eklof and A. H. Mekler, **Almost Free Modules**, North-Holland (1990).
3. P. C. Eklof, A. H. Mekler and S. Shelah, *Uniformization and the diversity of Whitehead groups*, Israel J. Math **80** (1992), 301-321.
4. P. C. Eklof and S. Shelah, *On Whitehead Modules*, J. Algebra **142** (1991), 492–510.
5. L. Fuchs, **Infinite Abelian Groups**, vol II, Academic Press (1973).
6. M. Magidor and S. Shelah, *When does almost free imply free? (For groups, transversal etc.)*, to appear in Journal Amer. Math. Soc.
7. S. Shelah, *Infinite abelian groups, Whitehead problem and some constructions*, Israel J. Math **18** (1974), 243–25.
8. S. Shelah, *A compactness theorem for singular cardinals, free algebras, Whitehead problem and transversals*, Israel J. Math., **21**, 319–349.
9. S. Shelah, *Whitehead groups may not be free even assuming CH, II*, Israel J. Math. **35** (1980), 257–285.
10. S. Shelah, *Incompactness in regular cardinals*, Notre Dame J. Formal Logic **26** (1985), 195–228.
11. J. Trlifaj, *Non-perfect rings and a theorem of Eklof and Shelah,* Comment. Math. Univ. Carolinae **32** (1991), 27–32.

Department of Mathematics, University of California, Irvine, Irvine, CA 92717
E-mail address: pceklof@uci.edu

Institute of Mathematics, Hebrew University, Jerusalem 91904, Israel
E-mail address: shelah@math.huji.ac.il

Contemporary Mathematics
Volume **171**, 1994

Endomorphisms of Local Warfield Groups

Steve Todd Files

1. Introduction

In this paper, we prove that local Warfield groups are determined up to isomorphism by their endomorphism rings. Throughout, p is a fixed prime and all groups are p-local abelian groups (modules over the p-adic rationals $\mathbf{Z}_p$). If G is a group, $End(G)$ denotes its endomorphism ring and tG its torsion subgroup. A ring isomorphism $\Phi : End(G) \rightarrow End(H)$ is *induced* by an isomorphism $\phi : G \rightarrow H$ if $\Phi(\psi) = \phi\psi\phi^{-1}$ for all ψ in $End(G)$. We state two precedents to the theorem we shall prove here.

Theorem [3]. *If G and H are torsion groups, then every isomorphism of $End(G)$ with $End(H)$ is induced by an isomorphism of G with H.*

Theorem [4]. *If G and H are groups of torsion-free rank 1 and tG is totally projective, then every isomorphism of $End(G)$ with $End(H)$ is induced by an isomorphism of G with H.*

These are examples of *isomorphism theorems* for the endomorphism rings of groups from given classes. Few such theorems are known in the case of mixed groups, and no generalizations of the result in [4] have come forth. In [6], May proves that for any reduced, mixed group G of countable rank there are $2^{2^{\aleph_0}}$ nonisomorphic groups H with $End(H) \cong End(G)$. Such behavior owes largely to the incompleteness of $\mathbf{Z}_p$ in its p-adic topology.

Warfield groups are defined in Section 2. Within this special class, many problems normally caused by the incompleteness of $\mathbf{Z}_p$ disappear and an isomorphism theorem results:

1991 *Mathematics subject classifications.* Primary 20K30.

This paper is in final form, and no version of it will be submitted for publication elsewhere.

THEOREM. *Assume G and H are Warfield groups, G is reduced, and Φ is an isomorphism of End(G) with End(H). Then Φ is topological and G is isomorphic to H.*

Here *topological* means that Φ and Φ^{-1} are continuous when the endomorphism rings are equipped with their finite topologies. In general, if Φ is induced by an isomorphism of the underlying groups then Φ is topological, but the converse is false. In the theorem, if G has torsion-free rank greater than 1 then Φ may not be induced.

The proof we present in Section 4 is an adaption of methods first employed in May [5], where an isomorphism theorem is established for the endomorphism algebras of a broad class of modules over $\hat{\mathbf{Z}}_p$ (the p-adic completion of $\mathbf{Z}_p$). In our case, certain $\hat{\mathbf{Z}}_p$-modules will serve as an aid in determining the structure of groups.

2. Local Warfield Groups

If G is a p-local group and $x \in G$, we let $h_G(x)$ or simply $h(x)$ denote the p-height of x in G. A torsion-free basis $X = \{x_i\}$ of G is a *decomposition basis* if $h(\Sigma r_i x_i) = min\{h(r_i x_i)\}$ for all finite subsets $\{r_i\} \subseteq \mathbf{Z}_p$, and a *nice* decomposition basis if the subgroup $\langle X\rangle$ generated by X is also nice in G, i.e., $p^\mu(G/\langle X\rangle) = (p^\mu G + \langle X\rangle)/\langle X\rangle$ for all ordinals μ. We call G a *Warfield group* if it possesses a nice decomposition basis X with $G/\langle X\rangle$ totally projective. As shown in [2], G is a Warfield group if and only if there is a torsion group T such that $G \oplus T$ is simply presented.

A *subordinate* to a subset $\{x_i\}$ of G is any set of the form $\{p^{n_i} x_i\}$ for numbers $n_i \geq 0$. Results in [2] show that every subordinate to a nice decomposition basis X for G with $G/\langle X\rangle$ totally projective retains those properties.

In the proof of our isomorphism theorem in Section 4, we will need to choose a specific kind of decomposition basis. Let $\ell(G)$ denote the p-length of a group G.

LEMMA 2.1. *If G is a reduced Warfield group with torsion subgroup T and $\lambda = \ell(T)$, then G has a nice decomposition basis X such that $G/\langle X\rangle$ is totally projective and $p^\lambda G \subseteq \langle X\rangle$.*

PROOF: Let Y be a nice decomposition basis for G with $G/\langle Y\rangle$ totally projective, and put $Y_0 = \{y \in Y : h(p^k y) \geq \lambda$ for some $k \geq 0\}$. By replacing Y_0 by a subordinate, we may assume $Y = Y_0 \cup Y_1$, where

$h(y) \geq \lambda$ for all $y \in Y_0$ and $\langle Y_1 \rangle \cap p^\lambda G = 0$. For each $y \in Y_0$, there is a unique $x \in G$ such that $h(x) = \lambda$ and $y = p^k x$ for some k. Let X_0 be the collection of these elements x and set $X = X_0 \cup Y_1$.

We first show X is a decomposition basis for G. If $x_1, \cdots, x_n$ are distinct elements of X and $0 \neq r_i \in \mathbf{Z}_p$ for $1 \leq i \leq n$, then clearly $h(\Sigma r_i x_i) = min\{h(r_i x_i)\}$ if some x_i is in Y_1. Thus, assume all x_i are in X_0 and choose $m \geq 0$ so that $p^m x_i \in Y_0$ for all i. Since there can be no jumps in height above λ, we have $h(\Sigma r_i x_i) + m = h(\Sigma r_i p^m x_i) = min\{h(r_i p^m x_i)\} = min\{h(r_i x_i) + m\}$, so that $h(\Sigma r_i x_i) = min\{h(r_i x_i)\}$ as desired.

To see that $p^\lambda G \subseteq \langle X \rangle$, let $z \in p^\lambda G$ and choose $n \geq 0$ so that $p^n z \in \langle X \rangle$. Then $p^n z \in \langle X_0 \rangle$, say $p^n z = \Sigma r_i x_i$ for distinct $x_i \in X_0$. Thus $\lambda + n \leq h(p^n z) = min\{h(r_i x_i)\} = min\{\lambda + h_{\mathbf{Z}_p}(r_i)\}$, so that p^n divides r_i for all i. It follows that $z = \Sigma(p^{-n} r_i) x_i \in \langle X_0 \rangle$, and we have $\langle X \rangle = p^\lambda G \oplus \langle Y_1 \rangle$. Since $\langle Y \rangle = \langle Y_0 \rangle \oplus \langle Y_1 \rangle$ is nice in G, it is easy to check that $\langle Y_1 \rangle$ is also nice. Thus $\langle X \rangle / \langle Y_1 \rangle = p^\lambda (G/\langle Y_1 \rangle)$ is nice in $G/\langle Y_1 \rangle$, and it follows that $\langle X \rangle$ is nice in G. Finally, note $\langle X \rangle = p^\lambda G + \langle Y \rangle$ so that $G/\langle X \rangle \cong (G/\langle Y \rangle)/p^\lambda(G/\langle Y \rangle)$ is totally projective because $G/\langle Y \rangle$ is totally projective. □

A structure theorem for Warfield groups is given in [2] and [7]. An easy consequence of the theorem will be useful to us in Section 3: two Warfield groups G and H with decomposition bases X and Y are isomorphic if $tG \cong tH$ and there is an isomorphism of $\langle X \rangle$ onto $\langle Y \rangle$ which preserves heights in G and H.

3. Modules Over the Completion

If $\hat{\mathbf{Z}}_p$ denotes the p-adic completion of $\mathbf{Z}_p$, then there is a homological way to embed reduced groups G in modules over $\hat{\mathbf{Z}}_p$. The cotorsion hull $G^\bullet = Ext(\mathbf{Z}(p^\infty), G)$ of any group G is reduced, and has a unique $\hat{\mathbf{Z}}_p$-module structure extending its normal group structure. When G is reduced, G embeds in $G^\bullet$ and we let $\hat{\mathbf{Z}}_p G$ denote the $\hat{\mathbf{Z}}_p$-submodule of $G^\bullet$ generated by G. The relation between G and $\hat{\mathbf{Z}}_p G$ is discussed in depth in [1]; here, it will suffice to note that $\hat{\mathbf{Z}}_p G/G$ is torsion-free and divisible since $G^\bullet/G$ is torsion-free and $\hat{\mathbf{Z}}_p/\mathbf{Z}_p$ is divisible. Thus $t(\hat{\mathbf{Z}}_p G) = tG$, and G is an isotype subgroup of $\hat{\mathbf{Z}}_p G$.

LEMMA 3.1. *If G and H are reduced Warfield groups, then $G \cong H$ if and only if $\hat{\mathbf{Z}}_p G \cong \hat{\mathbf{Z}}_p H$.*

PROOF: Obviously $\hat{\mathbf{Z}}_pG = G$ if G is a torsion group, hence we may assume G and H are nontorsion. Let ϕ be an isomorphism of $\hat{\mathbf{Z}}_pG$ with $\hat{\mathbf{Z}}_pH$, and choose a torsion group T so that $G \oplus T = \oplus_{i\in I} G_i$ and $H \oplus T = \oplus_{j\in J} H_j$ are sums of rank 1 groups G_i and H_j. For each $i \in I$, choose a torsion-free element $x_i \in G_i \cap G$. Then $X = \{x_i : i \in I\}$ is a decomposition basis for G, and also a decomposition basis for $\hat{\mathbf{Z}}_pG$ as a $\hat{\mathbf{Z}}_p$-module because $\hat{\mathbf{Z}}_pG \oplus T = \oplus_{i\in I}\hat{\mathbf{Z}}_pG_i$ and each $\hat{\mathbf{Z}}_pG_i$ is a module of torsion-free rank 1. Similarly, by choosing $y_j \in H_j \cap H$ torsion-free for each $j \in J$ we obtain a decomposition basis $Y = \{y_j : j \in J\}$ for both H and $\hat{\mathbf{Z}}_pH$. Since $\phi(X)$ and Y are decomposition bases for $\hat{\mathbf{Z}}_pH$, Lemma 4.6 in [7] guarantees the existence of subordinates $X' = \{x_i' : i \in I\}$ to X and $Y' = \{y_j' : j \in J\}$ to Y together with a bijection $\tau : I \to J$ such that $h_{\hat{\mathbf{Z}}_pH}\big(p^k\phi(x_i')\big) = h_{\hat{\mathbf{Z}}_pH}\big(p^k y'_{\tau(i)}\big)$ for all $i \in I$ and $k \geq 0$. Since G and H are isotype in $\hat{\mathbf{Z}}_pG$ and $\hat{\mathbf{Z}}_pH$, we have $h_G(p^k x_i') = h_H\big(p^k\phi(x_i')\big) = h_H\big(p^k y'_{\tau(i)}\big)$ for all $i \in I$ and $k \geq 0$. Thus, by setting $\tilde{\phi}(x_i') = y'_{\tau(i)}$ for each $i \in I$, we induce a surjective homomorphism $\tilde{\phi} : \langle X' \rangle \to \langle Y' \rangle$ that preserves heights in G and H. It follows from the remark in Section 2 that $G \cong H$. □

In the next section, we will show that $End(G) \cong End(H)$ implies $\hat{\mathbf{Z}}_pG \cong \hat{\mathbf{Z}}_pH$ for reduced Warfield groups G and H, and thus conclude $G \cong H$ by Lemma 3.1. Another $\hat{\mathbf{Z}}_p$-module from [6] will also play a role in our proof: if G is a reduced group, then the *$\hat{\mathbf{Z}}_p$-core* $C(G)$ of G is the maximal $\hat{\mathbf{Z}}_p$-module contained in G.

The next lemma shows that $C(G)$ cannot properly contain tG if G is a Warfield group. We will write $\hat{\mathbf{Z}}_p\langle x \rangle$ for the cyclic submodule of $\hat{\mathbf{Z}}_pG$ generated by an element $x \in G$.

LEMMA 3.2. *If G is a reduced group which contains a nice torsion-free basis, then $C(G) = tG$.*

PROOF: Let X be a nice torsion-free basis for G. Clearly, $tG \subseteq C(G)$. If $C(G) \neq tG$, we may choose a nonzero element $x \in C(G) \cap \langle X \rangle$. Then $\hat{\mathbf{Z}}_p\langle x \rangle \subseteq G$, so that $\hat{\mathbf{Z}}_p/\mathbf{Z}_p \cong \hat{\mathbf{Z}}_p\langle x \rangle/\langle x \rangle \subseteq G/\langle x \rangle$ and $G/\langle x \rangle$ contains a copy of $\mathbf{Q}$. But $G/\langle X \rangle \cong (G/\langle x \rangle)/(\langle X \rangle/\langle x \rangle)$ and $\langle X \rangle/\langle x \rangle$ is reduced, hence $G/\langle X \rangle$ cannot be reduced. This contradicts the niceness of $\langle X \rangle$ in G. □

4. Isomorphism of Endomorphism Rings

For much of this section, our focus will be on reduced groups G with G/tG divisible. Setting $T = tG$ in this case, we have $T^\bullet = G^\bullet$ and every endomorphism ϕ of G has a unique extension to an endomorphism $\phi^\bullet$ of $T^\bullet$. Thus $End(G)$ embeds in $End(T^\bullet)$, as does $End(H)$ for any pure subgroup H of $T^\bullet$ containing T. We have $End(G) = End(H)$ if and only if $\phi^\bullet(H) \subseteq H$ for all $\phi \in End(G)$ and $\psi^\bullet(G) \subseteq G$ for all $\psi \in End(H)$.

LEMMA 4.1. *Let T be a reduced torsion group and suppose G and H are pure subgroups of $T^\bullet$ containing T which possess nice torsion-free bases. If $End(G) = End(H)$, then $Hom(G,T) = Hom(H,T)$.*

PROOF: We claim $Hom\big(G, C(G)\big) = Hom\big(H, C(H)\big)$. If $\phi(G) \subseteq C(G)$, then $\alpha\phi(G) \subseteq G$ for all $\alpha \in \hat{\mathbf{Z}}_p$, hence $\alpha\phi^\bullet(H) \subseteq H$ for all α. This shows $\phi^\bullet(H) \subseteq C(H)$, and by symmetry we obtain $Hom\big(G, C(G)\big) = Hom\big(H, C(H)\big)$. Lemma 3.2 implies $C(G) = C(H) = T$, hence $Hom(G,T) = Hom(H,T)$. □

We will soon require several facts relating heights of torsion-free elements of $T^\bullet$ to $\ell(T)$. Let T be a reduced, unbounded torsion group and assume $T \subseteq \tilde{T}$ for a totally projective group $\tilde{T}$ with $\ell(\tilde{T}) < \ell(T) + \omega$. Then Lemma 4 in [5] implies $p^\mu(T^\bullet)$ is nontorsion for any ordinal μ with $\mu + \omega \leq \ell(T^\bullet)$, and $\ell(T) = \ell(T^\bullet)$ if and only if $\ell(T) \in [\sigma, \sigma + \omega)$ for an ordinal σ of cofinality greater than ω. Setting $\tilde{T} = G/\langle X\rangle$ in Lemma 2.1 shows the validity of these facts for the torsion subgroup T of any reduced Warfield group G with G/T divisible.

We handle the brunt of our theorem's proof in the next lemma. If $x \in T^\bullet$ is torsion-free, then $\langle x\rangle^\bullet = \hat{\mathbf{Z}}_p\langle x\rangle$ because the cotorsion hull of a torsion-free reduced group is its completion. Let E_* denote the purification in $T^\bullet$ of a subgroup $E \subseteq T^\bullet$, i.e., $E_*/E = t(T^\bullet/E)$.

LEMMA 4.2. *Let T be a reduced torsion group and suppose G and H are pure Warfield subgroups of $T^\bullet$ containing T such that $End(G) = End(H)$. Then $\hat{\mathbf{Z}}_pG = \hat{\mathbf{Z}}_pH$.*

PROOF: If G is torsion, then $\hat{\mathbf{Z}}_p \subseteq End(T) = End(H)$ and $H = C(H) = T$ by Lemma 3.2. Assume G is nontorsion, and note $Hom(G,T) = Hom(H,T)$ by Lemma 4.1. Let $\lambda = \ell(T)$, and choose a nice decomposition basis $X \subseteq G$ with $G/\langle X\rangle$ totally projective and $p^\lambda G \subseteq \langle X\rangle$ by Lemma 2.1. Denote $B = \langle X\rangle$, and observe $B^\bullet$ embeds in $T^\bullet$ with $(G/B)^\bullet = T^\bullet/B^\bullet$ because G/B is reduced.

We first show $H \subseteq (B^\bullet)_* \subseteq T^\bullet$. Let $x \in T^\bullet \setminus (B^\bullet)_*$, and denote $\overline{G} = \langle x+B^\bullet\rangle_* \subseteq T^\bullet/B^\bullet$. Since $\overline{G}$ is reduced and $t\overline{G} = G/B$ has length λ, we have $h_{\overline{G}}(p^k x+B^\bullet) < \lambda+\omega$ for all $k \geq 0$ because $x+B^\bullet$ is torsion-free in $\overline{G}$. We claim there exists an ordinal μ with $\mu+\omega \leq \ell(T^\bullet)$ such that $h_{\overline{G}}(p^k x + B^\bullet) \leq \mu + k$ for all $k \geq 0$. First suppose $\ell(T) \neq \ell(T^\bullet)$, so that $\ell(T^\bullet) = \ell(T) + \omega$. If $h_{\overline{G}}(p^k x + B^\bullet) = \lambda + m \geq \lambda$ for some $k \geq 0$, put $\mu = \lambda + m$ and note $h_{\overline{G}}(p^k x + B^\bullet) \leq \mu + k$ for all $k \geq 0$ since $\ell(t\overline{G}) = \lambda$. Clearly, $\mu + \omega \leq \ell(T^\bullet)$. For the other case, assume $\ell(T) = \ell(T^\bullet)$ so that $\sigma \leq \ell(T) < \sigma + \omega$ for an ordinal σ of cofinality greater than ω. Then $\ell(t\overline{G}) = \ell\big((t\overline{G})^\bullet\big) = \lambda$, hence $h_{\overline{G}}(p^k x + B^\bullet) < \lambda$ for all $k \geq 0$. Put $\mu = sup\{h_{\overline{G}}(p^k x + B^\bullet) : k \geq 0\}$ and note $\mu + \omega < \sigma \leq \ell(T^\bullet)$ by the cofinality of σ. This establishes the existence of μ.

By the remarks preceding the lemma, we may choose a torsion-free element $z \in p^\mu(T^\bullet)$. Since $\overline{G}$ has rank 1 and $t\overline{G}$ is totally projective, $\langle x + B^\bullet\rangle$ is nice in $\overline{G}$ and $\overline{G}/\langle x + B^\bullet\rangle$ is totally projective. Since the map $rx + B^\bullet \longmapsto rz$ from $\langle x + B^\bullet\rangle$ to $T^\bullet$ does not decrease heights relative to $\overline{G}$ and $T^\bullet$, it extends to a homomorphism $\overline{G} \to T^\bullet$. Composing this with the natural map $G \to G/B \subseteq \overline{G}$ and extending the result to an endomorphism of $T^\bullet$, we have $\phi(x) = z$ and $\phi(B) = 0$. Thus $\phi \in Hom(G,T) = Hom(H,T)$ and $\phi(x) \notin T$, so that $x \notin H$. This establishes $H \subseteq (B^\bullet)_*$.

Next, we show $G \subseteq \hat{\mathbf{Z}}_p H$. It suffices to show $X \subseteq \hat{\mathbf{Z}}_p H$. Let $y \in X$ and write $B = \langle y\rangle \oplus C$, so that $B^\bullet = \hat{\mathbf{Z}}_p\langle y\rangle \oplus C^\bullet$. Since X is a nice decomposition basis for G and G/B is totally projective, the projection $B \to \langle y\rangle$ extends to an endomorphism π of G. Since $\pi(G) \nsubseteq T$ we have $\pi^\bullet(H) \nsubseteq T$, say $\pi^\bullet(w) \notin T$ for an element $w \in H$. We may assume $w \in B^\bullet$ because $H \subseteq (B^\bullet)_*$. Now $\pi^\bullet$ projects $B^\bullet$ onto $\hat{\mathbf{Z}}_p\langle y\rangle$, hence we may write $w = \alpha r y + c$, where α is a unit of $\hat{\mathbf{Z}}_p$, $0 \neq r \in \mathbf{Z}_p$, and $c \in C^\bullet$. We have $\pi^\bullet(H) \subseteq H$ since $\pi \in End(G)$, hence $\alpha r y = \pi^\bullet(\alpha r y+c) = \pi^\bullet(w) \in H$. Thus $ry = \alpha^{-1}\pi^\bullet(w) \in \hat{\mathbf{Z}}_p H$, and it follows that $X \subseteq \hat{\mathbf{Z}}_p H$ by purity.

By symmetry we obtain $H \subseteq \hat{\mathbf{Z}}_p G$, so that $\hat{\mathbf{Z}}_p G = \hat{\mathbf{Z}}_p H$ as desired.
□

Before proving our theorem, we prepare for the case where G is a Warfield group with torsion T and G/T is not divisible. Choose a torsion group S so that $G \oplus S = \oplus G_i$ is a sum of rank 1 groups, and note G/T has a summand isomorphic to $\mathbf{Z}_p$ because some G_i/tG_i must be reduced. It follows that G has a torsion-free, cyclic direct summand.

THEOREM 4.3. *Assume G and H are Warfield groups and G is reduced. If Φ is an isomorphism of $End(G)$ with $End(H)$, then Φ is topological and $G \cong H$.*

PROOF: If H is not reduced, then by mapping projections back to $End(G)$ we see that G has a summand isomorphic to $\mathbf{Z}(p^\infty)$, $\mathbf{Q}$ or $\hat{\mathbf{Z}}_p$. The first two are impossible since G is reduced, and the last violates $C(G) = tG$ (Lemma 3.2). Thus, H is reduced.

If G/tG is not divisible, let π be projection from G onto a torsion-free, cyclic summand. Then H has a summand $\Phi(\pi)(H)$ of the same type, and the technique used in the proof of Kaplansky's theorem [3] can be used to construct an isomorphism from G to H which induces Φ. In particular, Φ is topological.

Assume G/tG is divisible. If G is torsion, then $\hat{\mathbf{Z}}_p \subseteq End(H)$ and Lemma 3.2 implies H is torsion since $H = C(H)$. Kaplansky's theorem then implies Φ is induced by an isomorphism of G with H. Therefore assume G is nontorsion, and note H/tH is divisible because H cannot have a cyclic, torsion-free summand. Denote $T = tH$. Kaplansky's method can be used to construct an isomorphism $\phi : tG \to T$ such that $\Phi(\gamma)$ and $\phi\gamma\phi^{-1}$ agree on T for all $\gamma \in End(G)$ (for details see [5]). Let $G' = \phi^\bullet(G) \subseteq T^\bullet$. We claim $End(G') = End(H)$. If $\psi \in End(G')$, then $\tilde{\psi} = (\phi^\bullet)^{-1}\psi\phi^\bullet \in End(G)$, hence $\Phi(\tilde{\psi}) \in End(H)$. But $\Phi(\tilde{\psi})$ and $\phi^\bullet\tilde{\psi}(\phi^\bullet)^{-1} = \psi$ agree on T, hence $\psi^\bullet = \Phi(\tilde{\psi})^\bullet$ and we have $\psi^\bullet \in End(H)$. Thus $End(G') \subseteq End(H)$, and a similar argument shows the reverse inclusion. Lemma 4.3 now implies $\hat{\mathbf{Z}}_p G' = \hat{\mathbf{Z}}_p H$, so that $G \cong G' \cong H$ by Lemma 3.1.

To see that Φ is topological, let $y \in H$ and consider the subbasic open neighborhood $\mathcal{O} = \{\theta \in End(H) : \theta(y) = 0\}$ of 0 in $End(H)$. Since $\hat{\mathbf{Z}}_p G' = \hat{\mathbf{Z}}_p H$, we have $y = \alpha_1 x_1 + \cdots + \alpha_n x_n$ for elements

$\alpha_i \in \hat{\mathbf{Z}}_p$ and $x_i \in G'$. Thus

$$\mathcal{O} = \bigcup_{\theta \in \mathcal{O}} \{\eta \in End(G') : \eta(x_i) = \theta^\bullet(x_i),\ 1 \leq i \leq n\}$$

is an open set in $End(G')$, and it becomes evident that the identification $End(G') = End(H)$ is topological. Since Φ is the composition of this identification with the isomorphism $End(G) \to End(G')$ induced by $\phi^\bullet$, Φ is topological. □

As a special case, the theorem applies to reduced, simply presented groups G and H. There is another obvious corollary as well.

Corollary 4.4. *If G is a reduced Warfield group, then every ring automorphism of $End(G)$ is topological.*

Lemma 3.2 implies that no Warfield group G can have a summand isomorphic to $\hat{\mathbf{Z}}_p$. If C is a summand of G with $End(C) \cong \hat{\mathbf{Z}}_p$, it follows that $C \cong \mathbf{Z}(p^\infty)$, and it is now straightforward to show that the divisible subgroups of two Warfield groups G and H are isomorphic if $End(G) \cong End(H)$. In this case Theorem 4.4 shows the reduced parts of G and H are isomorphic as well, and we conclude $G \cong H$.

Before stating a final corollary, we remark that all of our arguments carry over virtually unchanged to modules over any *incomplete* discrete valuation ring R. By pooling our results with others from [5] (the case where R is complete), we obtain a broader result:

Corollary 4.5. *If M and N are Warfield modules over a discrete valuation ring R and $End_R(M) \cong End_R(N)$, then $M \cong N$.*

When R is incomplete, it is vital that we assume both modules in the Corollary are Warfield modules. In contrast, only M need be assumed of that type when R is complete.

5. Example

A brief construction shows that the isomorphism Φ in Theorem 4.4 may not be induced by an isomorphism of the groups.

Let $\{\lambda_i^{(k)} : i < \omega\}$ $(k = 1, 2)$ be two increasing sequences of finite ordinals such that the set $\{\lambda_{i+m}^{(1)} - \lambda_{i+n}^{(2)} : i < \omega\}$ contains a positive and a negative number for every m, $n \geq 0$. If $T = \oplus\{\mathbf{Z}(p^i) :$

$1 \le i < \omega\}$, then there exist rank 1 groups G_1 and G_2 with torsion T such that $\{\lambda_i^{(k)} : i < \omega\}$ is equivalent to the height sequence of an element of G_k for $k = 1, 2$. We have $Hom(G_i, G_j) = Hom(G_i, T)$ if $i \neq j$. Choose a unit $\alpha \in \hat{\mathbf{Z}}_p$ transcendental over $\mathbf{Z}_p$, and take $G = G_1 \oplus G_2$, $H = \alpha G_1 \oplus G_2 \subseteq T^\bullet \oplus T^\bullet$. Then G and H are reduced Warfield groups of torsion-free rank 2, and $End(G) = End(H)$ because $End(\alpha G_1) = End(G_1)$, $Hom(G_2, \alpha G_1) = Hom(G_2, T)$ and $Hom(\alpha G_1, G_2) = Hom(G_1, T)$. If $\phi : G \to H$ is an isomorphism inducing the identity map $\Phi : End(G) \to End(H)$, then $\phi|_{T \oplus T}$ commutes with the bounded idempotents in $End(T \oplus T)$, hence acts as multiplication by a unit $\beta \in \hat{\mathbf{Z}}_p$. Thus $\phi = \beta$, so that $H = \beta G$. This impossibility shows that no isomorphism ϕ can induce the identity Φ, and the example stands complete.

REFERENCES

1. S. T. Files, Mixed modules over incomplete discrete valuation rings, Comm. Alg. (to appear).

2. R. Hunter, F. Richman and E. Walker, Warfield modules, in *Abelian Group Theory*, Springer Lecture Notes **616**, Berlin, 1977, 87-123.

3. I. Kaplansky, *Infinite Abelian Groups*, University of Michigan Press, Ann Arbor, 1969.

4. W. May and E. Toubassi, Endomorphisms of rank one mixed modules over discrete valuation rings, Pac. J. Math. **108** (1983), 155-163.

5. W. May, Isomorphism of endomorphism algebras over complete discrete valuation rings, Math Z. **204** (1990), 485-499.

6. W. May, Endomorphism algebras of not necessarily cotorsion-free modules, Contemp. Math. **130**, (1992), 257-264.

7. R. B. Warfield, Jr., Classification of abelian groups II: local theory, in *Abelian Group Theory*, Springer Lecture Notes **874**, Heidelberg, 1981, 322-349.

University of Connecticut, Storrs, Connecticut, 06268

Contemporary Mathematics
Volume **171**, 1994

Finitely Presented Modules Over the Ring of Universal Numbers

ALEXANDER A. FOMIN

ABSTRACT. A description of finitely presented modules over the ring of universal numbers is given. This yields a new category, which is dual to the category of torsion-free abelian groups of finite rank with quasi-homomorphisms as morphisms.

Introduction and Preliminaries

Consider a direct product of one copy each of the fields of p-adic numbers over the prime numbers p. An element $\alpha = (\alpha_p)$ of this direct product is called a *universal number*, if all but a finite number of its p-components α_p are p-adic integers. All the universal numbers form a ring which is called the *ring of universal numbers* and denoted by $\mathbf{K}$. The purpose of this paper is a description of finitely presented $\mathbf{K}$-modules and an application of this description to the theory of torsion-free abelian groups of finite rank. We introduce a category connected with finitely presented $\mathbf{K}$-modules, which is dual to the category of torsion-free abelian groups of finite rank with quasi-homomorphisms as morphisms. Let $\mathbf{Z}_{p^\infty}$ denote the ring of p-adic integers. Then $\mathbf{K} = \mathbf{Q} \otimes \prod\limits_p \mathbf{Z}_{p^\infty}$, where $\mathbf{Q}$ is the field of rational numbers. The ring $\prod\limits_p \mathbf{Z}_{p^\infty}$ is a subring of the ring $\mathbf{K}$, and it is called the ring of *universal integers*. For a universal integer $\alpha = (\alpha_p) \in \prod\limits_p \mathbf{Z}_{p^\infty}$, let h_p denote the greatest exponent of the prime p for which p^{h_p} is a divisor of α_p in the ring $\mathbf{Z}_{p^\infty}$. If $\alpha_p = 0$, then $h_p = \infty$. We obtain a characteristic (h_p), i. e. a sequence of non-negative integers and symbols infinity indexed by the set of all prime numbers. As for torsion-free groups of rank one, the characteristic (h_p) determines a type $[(h_p)]$ which is called the *type of the universal integer* α, $[(h_p)] = \text{type}(\alpha)$. For an arbitrary universal number α, there exists a non-zero

---

1991 *Mathematics Subject Classification.* 20K15.
This paper is in final form and no version of it will be submitted for publication elsewhere.

integer m such that $m\alpha$ is a universal integer. Let type(α) = type$(m\alpha)$. It is easy to see that the type of α does not depend on the choice of m. The definitions imply the following properties of universal numbers.

U1 A universal number α is a divisor of a universal number β if and only if type$(\alpha) \leq$ type(β).

U2 For every finite set of universal numbers, there exists a greatest common divisor of these numbers.

U3 Each finitely generated ideal of $\mathbf{K}$ is principal.

U4 For universal numbers $\alpha_1, ..., \alpha_n$, there exist idempotents $\varepsilon_1, ..., \varepsilon_n \in \mathbf{K}$ such that:

1. $\alpha_1\varepsilon_1 + ... + \alpha_n\varepsilon_n = \gcd(\alpha_1, ..., \alpha_n)$,
2. $\varepsilon_i\varepsilon_j = 0$ for all $i \neq j$,
3. $\varepsilon_1 + ... + \varepsilon_n = 1$.

U5 Let a finite system of linear equations with universal coefficients be given

$$\begin{cases} \alpha_{11}x_1 + ... + \alpha_{1n}x_n = 0 \\ \quad \cdot \quad \cdot \quad \cdot \\ \alpha_{m1}x_1 + ... + \alpha_{mn}x_n = 0 \end{cases}$$

Then the module of solutions of this system is a finitely generated $\mathbf{K}$-submodule of the $\mathbf{K}$-module $\mathbf{K}^n$.

1. Finitely Presented Modules Over the Ring of Universal Numbers

Let $\mathbf{M}$ be a $\mathbf{K}$-module with a system of generators $v_1, ..., v_n$. Then the $\mathbf{K}$-module $\{(\alpha_1, ..., \alpha_n) \mid \alpha_1 v_1 + ... + \alpha_n v_n = 0\} \subset \mathbf{K}^n$ is called the *module of relations* of $\mathbf{M}$ with respect to this system of generators. A finitely generated $\mathbf{K}$-module is said to be *finitely presented* (or *finitely related*), if the module of relations with respect to some finite system of generators is also finitely generated over $\mathbf{K}$.

EXAMPLE 1.1. For each type τ the set of universal numbers

$$I(\tau) = \{\alpha \mid \text{type}(\alpha) \geq \tau\}$$

is an ideal of the ring $\mathbf{K}$. The ring $\mathbf{Q}(\tau) = \mathbf{K}/I(\tau)$ is called the *ring of τ-adic numbers*. As a $\mathbf{K}$-module the ring $\mathbf{Q}(\tau)$ is a finitely presented module for each type τ.

EXAMPLE 1.2. The direct sum of fields of p-adic numbers over all the prime numbers p is an ideal I of the ring $\mathbf{K}$, which is evidently not finitely generated. Therefore the $\mathbf{K}$-module $\mathbf{K}/I$ is finitely generated, even cyclic, but not finitely presented.

LEMMA 1.1. *Let $\mathbf{M}_1$ be a finitely generated submodule of a finitely presented $\mathbf{K}$-module $\mathbf{M}$. Then the two $\mathbf{K}$-modules $\mathbf{M}_1$ and $\mathbf{M}/\mathbf{M}_1$ are finitely presented.*

PROOF. Let $v_1, ..., v_n$ be a system of generators of the module $\mathbf{M}$, such that the following collection of elements of $\mathbf{K}^n$

$$(1) \qquad \begin{cases} (\alpha_{11}, ..., \alpha_{1n}) \\ \quad . \quad . \quad . \\ (\alpha_{m1}, ..., \alpha_{mn}) \end{cases}$$

is a system of generators of the module of relations with respect to $v_1, ..., v_n$. Let the submodule $\mathbf{M}_1$ be generated by the following elements

$$w_1 = \beta_{11}v_1 + ... + \beta_{1n}v_n,$$

$$. \quad . \quad .$$

$$w_k = \beta_{k1}v_1 + ... + \beta_{kn}v_n.$$

Then the rows (1) together with the rows $(\beta_{i1}, ..., \beta_{in})$, $i = 1, ..., k$, generate the module of relations of the module $\mathbf{M} / \mathbf{M}_1$ with respect to the system of generators $v_1 + \mathbf{M}_1, ..., v_n + \mathbf{M}_1$. Hence the $\mathbf{K}$-module $\mathbf{M} / \mathbf{M}_1$ is finitely presented.

On the other hand, $\gamma_1 w_1 + ... + \gamma_k w_k = 0$ in the module $\mathbf{M}_1$ if and only if the equality

$$\sum_{i=1}^{k} \gamma_i (\sum_{j=1}^{n} \beta_{ij} v_j) = \sum_{i=1}^{m} \delta_i (\sum_{j=1}^{n} \alpha_{ij} v_j)$$

holds in the free $\mathbf{K}$-module $v_1 \mathbf{K} \oplus ... \oplus v_n \mathbf{K}$ for suitable universal numbers $\delta_1, ..., \delta_m$. We obtain the following finite system of linear equations with universal coefficients

$$\sum_{i=1}^{k} x_i \beta_{ij} = \sum_{i=1}^{m} y_i \alpha_{ij}, \quad j = 1, ..., n.$$

According to **U5**, the module of solutions of this system is generated by the finite set of solutions

$$(\gamma_{i1}, ..., \gamma_{ik}, \delta_{i1}, ..., \delta_{im}), \quad i = 1, ..., s,$$

in the $\mathbf{K}$-module $\mathbf{K}^{k+m}$. Then the system of relations $(\gamma_{i1}, ..., \gamma_{ik})$, $i = 1, ..., s$, generates the module of relations of the module $\mathbf{M}_1$ with respect to $w_1, ..., w_k$. □

COROLLARY 1.2. *The module of relations of a finitely presented module is finitely generated with respect to every finite system of generators.*

2. The Structure of Finitely Presented Modules

The structure theorems of finitely presented modules over the ring of universal numbers resemble the structure theorems of finitely generated abelian groups in a surprising way.

LEMMA 2.1. *Every finitely presented cyclic* $\mathbf{K}$*-module is isomorphic to a* $\mathbf{K}$*-module* $\mathbf{Q}(\tau)$ *for some type* τ.

PROOF. Let $\mathbf{M}$ be a finitely presented $\mathbf{K}$-module generated by one element v. Then the module of relations $I = \{\alpha \in \mathbf{K} \mid \alpha v = 0\}$ is a finitely generated ideal of the ring $\mathbf{K}$. Property **U3** implies that $I = \gamma\mathbf{K}$ for $\gamma \in \mathbf{K}$ and $\mathbf{M} \cong \mathbf{K}/I \cong \mathbf{Q}(\tau)$, where $\tau = \text{type}(\gamma)$. □

THEOREM 2.2. *Every finitely presented* $\mathbf{K}$*-module decomposes into a finite direct sum of finitely presented cyclic modules.*

PROOF. We prove the theorem by induction on the number of generators. The base of the induction is obvious. Let the finitely presented $\mathbf{K}$-module $\mathbf{M}$ have the system of generators $v_1, ..., v_n$, $n > 1$, and the corresponding system of generating relations

$$(2)\qquad \begin{cases} \alpha_{11}v_1 + ... + \alpha_{1n}v_n = 0 \\ \quad \cdot \quad \cdot \quad \cdot \\ \alpha_{m1}v_1 + ... + \alpha_{mn}v_n = 0. \end{cases}$$

Due to **U3**, a greatest common divisor β_1 of the universal numbers $\alpha_{11}, ..., \alpha_{m1}$ may be presented in the form

$$\beta_1 = \xi_1\alpha_{11} + ... + \xi_m\alpha_{m1}$$

for some universal numbers $\xi_1, ..., \xi_m$. Then the linear combination of relations (2) with the coefficients $\xi_1, ..., \xi_m$ is a relation of $\mathbf{M}$ with respect to the system $v_1, ..., v_n$

$$\beta_1 v_1 + ... = 0.$$

Dealing with analogously the set of coefficients $\alpha_{1i}, ..., \alpha_{mi}$ for each $i = 2, ..., n$, we obtain a system of relations

$$(3)\qquad \begin{cases} \beta_1 v_1 + & \cdots & \cdots & = 0 \\ \cdots & +\beta_2 v_2 + & \cdots & = 0 \\ \cdots & \cdots & \cdots & \\ \cdots & \cdots & +\beta_n v_n & = 0 \end{cases}$$

having the following property: For every relation $\alpha_1 v_1 + ... + \alpha_n v_n = 0$, each coefficient α_i is divisible by β_i, $i = 1, ..., n$. Now we construct one more relation in the following way. According to **U4**, there exists a system of orthogonal idempotents $\varepsilon_1, ..., \varepsilon_n \in \mathbf{K}$ such that

$$(4)\qquad \beta = \varepsilon_1\beta_1 + ... + \varepsilon_n\beta_n$$

is a greatest common divisor of the universal numbers $\beta_1, ..., \beta_n$. Consider the following linear combination of the relations (3)

$$(\gamma_1, ..., \gamma_n) = \varepsilon_1(\beta_1, ...) + ... + \varepsilon_n(..., \beta_n).$$

It is easy to see that $\varepsilon_1\gamma_1 = \varepsilon_1\beta_1, ..., \varepsilon_n\gamma_n = \varepsilon_n\beta_n$ and hence $\varepsilon_1\gamma_1 + ... + \varepsilon_n\gamma_n = \beta$. Therefore we obtain a relation of the **K**-module **M**

$$(5) \qquad \gamma_1 v_1 + ... + \gamma_n v_n = 0,$$

for which $\varepsilon_1\gamma_1 + ... + \varepsilon_n\gamma_n = \beta$ is the greatest common divisor of all the coefficients α_{ij} of the system of relations (2). Moreover β is evidently a divisor of each coefficient of every relation of **M** with respect to $v_1, ..., v_n$. Further we change the system of generators. Consider the matrix **T** over the ring **K**,

$$\mathbf{T} = \begin{pmatrix} \varepsilon_1 & \varepsilon_2 & \varepsilon_3 & \cdots & \varepsilon_n \\ \varepsilon_2 & \varepsilon_3 & \varepsilon_4 & \cdots & \varepsilon_1 \\ \varepsilon_3 & \varepsilon_4 & \varepsilon_5 & \cdots & \varepsilon_2 \\ \cdot & \cdot & \cdot & \cdots & \cdot \\ \varepsilon_n & \varepsilon_1 & \varepsilon_2 & \cdots & \varepsilon_{n-1} \end{pmatrix}.$$

A new system of elements $v'_1, ..., v'_n$ of the module **M** defined with help of the matrix equality

$$\begin{pmatrix} v'_1 \\ \cdots \\ v'_n \end{pmatrix} = \mathbf{T} \begin{pmatrix} v_1 \\ \cdots \\ v_n \end{pmatrix}$$

is also a system of generators of **M**, because the matrix **T** is invertible, $\mathbf{T}^2 = E$. Moreover, a vector $(\alpha_1, ..., \alpha_n) \in \mathbf{K}^n$ is a relation of **M** with respect to the system $v_1, ..., v_n$ if and only if the vector

$$(6) \qquad (\alpha'_1, ..., \alpha'_n) = (\alpha_1, ..., \alpha_n)\,\mathbf{T}$$

is a relation of **M** with respect to the system $v'_1, ..., v'_n$.

In particular by (5), $(\gamma'_1, ..., \gamma'_n) = (\gamma_1, ..., \gamma_n)\,\mathbf{T}$ is a relation of **M** with respect to the system $v'_1, ..., v'_n$. The first coefficient $\gamma'_1 = \gamma_1\varepsilon_1 + ... + \gamma_n\varepsilon_n$ is equal to the universal number β defined in (4). The equality (6) shows that β is a divisor of each coefficient of every relation of **M** with respect to $v'_1, ..., v'_n$ (and generally with respect to every system of generators). In particular, $\gamma'_i = \beta\delta_i$, $i = 2, ..., n$. Eventually we obtain the following relation $\beta(v'_1 + \delta_2 v'_2 + ... + \delta_n v'_n) = 0$.

Let $w = v'_1 + \delta_2 v'_2 + ... + \delta_n v'_n$. Obviously, the system $w, v'_2, ..., v'_n$ is also a system of generators of the module **M**. Let $\langle w\rangle_{\mathbf{K}}$ and $\langle v'_2, ..., v'_n\rangle_{\mathbf{K}}$ be **K**-modules generated by w and $v'_2, ..., v'_n$, respectively. Then $\mathbf{M} = \langle w\rangle_{\mathbf{K}} + \langle v'_2, ..., v'_n\rangle_{\mathbf{K}}$ and $\langle w\rangle_{\mathbf{K}} \cap \langle v'_2, ..., v'_n\rangle_{\mathbf{K}} = 0$, because an equality $\xi w = \xi_2 v'_2 + ... + \xi_n v'_n$ holds, only if ξ is divisible by β, but $\beta w = 0$. Hence $\mathbf{M} = \langle w\rangle_{\mathbf{K}} \oplus \langle v'_2, ..., v'_n\rangle_{\mathbf{K}}$ and, according to Lemma 1.3, both modules $\langle w\rangle_{\mathbf{K}}$ and $\langle v'_2, ..., v'_n\rangle_{\mathbf{K}}$ are finitely presented. The first of them is cyclic and the second one is decomposed into a finite direct sum of finitely presented cyclic modules by the induction hypothesis. □

THEOREM 2.3. *For every finitely presented* $\mathbf{K}$*-module* $\mathbf{M}$ *there exists a unique increasing sequence* $\tau_1 \leq \dots \leq \tau_m$ *of types with* $\tau_1 \neq 0$ *such that*

$$\mathbf{M} \cong \mathbf{Q}(\tau_1) \oplus \dots \oplus \mathbf{Q}(\tau_m).$$

PROOF. According to Lemma 2.1 and Theorem 2.2, $\mathbf{M} \cong \mathbf{Q}(\sigma_1) \oplus \dots \oplus \mathbf{Q}(\sigma_n)$ for types $\sigma_i = [(k_{ip})]$, $i = 1, \dots, n$. Note that $\mathbf{Q}(\sigma_i) \cong \mathbf{Q} \otimes \prod \mathbf{Z}(p^{k_{ip}})$, where $\mathbf{Z}(p^k)$ is either the ring of integers modulo p^k or the ring of p-adic integers if $k = \infty$. Hence $\mathbf{M} \cong \mathbf{Q} \otimes \prod\limits_p \left(\mathbf{Z}(p^{k_{1p}}) \oplus \dots \oplus \mathbf{Z}(p^{k_{np}})\right)$. It is possible to order the exponents $k_{1p}, \dots, k_{np}$ for each prime number p

$$l_{1p} \leq l_{2p} \leq \dots \leq l_{np}.$$

Then we obtain an increasing sequence of types $[(l_{1p})] \leq \dots \leq [(l_{np})]$. Some of the first types may be equal to zero. Therefore, beginning with the first non-zero type, we obtain eventually the sequence $\tau_1 \leq \dots \leq \tau_m$, for which

$$\mathbf{M} \cong \mathbf{Q}(\tau_1) \oplus \dots \oplus \mathbf{Q}(\tau_m).$$

Moreover, the type τ_m is uniquely determined, because an ideal

$$I(\tau_m) = \{\alpha \in \mathbf{K} \mid \text{type}(\alpha) \geq \tau_m\}$$

of the ring $\mathbf{K}$ of universal numbers coincides with the annihilator $\operatorname{ann} \mathbf{M}$ of the $\mathbf{K}$-module $\mathbf{M}$. By induction, the number m and the other types $\tau_1 \leq \dots \leq \tau_{m-1}$ are uniquely determined, too. □

3. The Category of Linear Mappings to Finitely Presented K-Modules

We construct a new category $\mathcal{L}$. The objects of this category are linear mappings $g : U \to \mathbf{M}$ from finite dimensional vector spaces U over the field of rational numbers to finitely presented modules $\mathbf{M}$ over the ring of universal numbers, which are considered as vector spaces over $\mathbf{Q}$, such that the image of g generates $\mathbf{M}$ as a $\mathbf{K}$-module.

The morphisms from an object $g : U \to \mathbf{M}$ to an object $g_1 : U_1 \to \mathbf{M}_1$ are pairs (f, φ), where $f : U \to U_1$ is a linear mapping and $\varphi : \mathbf{M} \to \mathbf{M}_1$ is a $\mathbf{K}$-module homomorphism, such that the following diagram is commutative

$$\begin{array}{ccc} U & \xrightarrow{f} & U_1 \\ g\downarrow & & \downarrow g_1 \\ \mathbf{M} & \xrightarrow[\varphi]{} & \mathbf{M}_1 \end{array}$$

i. e. $\varphi g = g_1 f$

Let (f_1, φ_1) be a morphism from the object $g_1 : U_1 \to \mathbf{M}_1$ to an object $g_2 : U_2 \to \mathbf{M}_2$. Then the product of the morphisms is defined in the following way

$$(f_1, \varphi_1)(f, \varphi) = (f_1 f, \varphi_1 \varphi).$$

It is easy to see that $\mathcal{L}$ is a category. We call $\mathcal{L}$ the *category of linear mappings to finitely presented modules over the ring of universal numbers.*

Two categories $\mathcal{A}$ and $\mathcal{B}$ are called *dual*, if there exist two contravariant functors $S : \mathcal{A} \to \mathcal{B}$ and $T : \mathcal{B} \to \mathcal{A}$ such that $TS \sim 1_{\mathcal{A}}$ and $ST \sim 1_{\mathcal{B}}$. The equivalence $TS \sim 1_{\mathcal{A}}$ means that for every object A of the category $\mathcal{A}$ there exists an isomorphism $h(A) : A \to TS(A)$ of the category $\mathcal{A}$ such that for each morphism $f : A \to B$ of the category $\mathcal{A}$ the following diagram is commutative

$$\begin{array}{ccc} A & \xrightarrow{f} & B \\ h(A)\downarrow & & \downarrow h(B) \\ TS(A) & \xrightarrow[TS(f)]{} & TS(B) \end{array}$$

We use the following well-known proposition.

PROPOSITION 3.1. (Criterion of duality) *Two categories $\mathcal{A}$ and $\mathcal{B}$ are dual, if and only if there exists a contravariant functor $S : \mathcal{A} \to \mathcal{B}$ such that*

1. *The mapping* $\mathrm{Hom}_{\mathcal{A}}(A, B) \to \mathrm{Hom}_{\mathcal{B}}(SB, SA) : f \mapsto S(f)$ *is a one-to-one correspondence for every two objects A and B of the category $\mathcal{A}$.*
2. *For every object B of the category $\mathcal{B}$, there exists an object A of the category $\mathcal{A}$ such that B is isomorphic to $S(A)$.*

The next theorem is the main result of this paper.

THEOREM 3.2. *The category of torsion-free abelian groups of finite rank with quasi-homomorphisms as morphisms is dual to the category $\mathcal{L}$.*

PROOF. In fact, this theorem was proved in [1] for the special case, where only coreduced groups having a fixed Richman type were considered. That proof needs only a few modifications. Keeping the terminology and notations of [1], we define a contravariant functor Φ from the category of torsion-free abelian groups of finite rank with quasi-homomorphisms as morphisms to the category of linear mappings to finitely presented modules over $\mathbf{K}$.

Let A be a torsion-free abelian group with a maximal linearly independent system $x_1, \ldots, x_n$ and let $F = \langle x_1, \ldots, x_n \rangle$. According to [1], a decomposition

$$A/F \cong \bigoplus_p (Z(p^{i_{1p}}) \oplus \ldots \oplus Z(p^{i_{np}}))$$

provides the following system of τ-adic relations

$$(7)\qquad \begin{cases} \alpha_{11}x_1 + ... + \alpha_{1n}x_n = 0 & ; \quad \alpha_{11}, ..., \alpha_{1n} \in \mathbf{Q}(\tau_1) \\ \alpha_{21}x_1 + ... + \alpha_{2n}x_n = 0 & ; \quad \alpha_{21}, ..., \alpha_{2n} \in \mathbf{Q}(\tau_2) \\ \quad . \quad . \quad . & ; \quad \quad . \quad . \quad . \\ \alpha_{n1}x_1 + ... + \alpha_{nn}x_n = 0 & ; \quad \alpha_{n1}, ..., \alpha_{nn} \in \mathbf{Q}(\tau_n), \end{cases}$$

where $\tau_1 = [(i_{1p})], ..., \tau_n = [(i_{np})]$.

Unlike in [1], we don't put in order the types $\tau_1, ..., \tau_n$ and don't eliminate zero types. Nevertheless, the defect (see [1]) of the system (7) is equal to zero. This means that

(i) the columns

$$(8)\qquad u_1 = \begin{pmatrix} \alpha_{11} \\ \cdots \\ \alpha_{n1} \end{pmatrix}, \ldots, u_n = \begin{pmatrix} \alpha_{1n} \\ \cdots \\ \alpha_{nn} \end{pmatrix}$$

generate the whole module $\mathbf{Q}(\tau_1) \oplus ... \oplus \mathbf{Q}(\tau_n)$ over $\mathbf{K}$, and

(ii) every τ-adic relation $\alpha_1 x_1 + ... + \alpha_n x_n = 0$, where $\alpha_1, ..., \alpha_n \in \mathbf{Q}(\tau)$, of the group A is a τ-combination [1] of relations (7) with uniquely determined coefficients $\gamma_1, ..., \gamma_n \in \mathbf{Q}(\tau)$, i. e.

$$(\alpha_1, ..., \alpha_n) = \gamma_1 \circ (\alpha_{11}, ..., \alpha_{1n}) + ... + \gamma_n \circ (\alpha_{n1}, ..., \alpha_{nn}),$$

where

$$\text{type}(\gamma_1) \geq \tau - (\tau \wedge \tau_1), ..., \text{type}(\gamma_n) \geq \tau - (\tau \wedge \tau_n).$$

Let U be a divisible hull of the group A, i. e. U is a vector space over $\mathbf{Q}$ with a basis $x_1, ..., x_n$. Consider the dual vector space U^* with the dual basis $x_1^*, ..., x_n^*$. The following mapping of this basis to the columns (8) $x_1^* \mapsto u_1, ..., x_n^* \mapsto u_n$ determines a linear mapping $g : U^* \to \mathbf{Q}(\tau_1) \oplus ... \oplus \mathbf{Q}(\tau_n)$. We define $\Phi(A) = g$.

Let B be another torsion-free abelian group with a maximal linearly independent system $y_1, ..., y_m$ and with the following system of relations

$$(9)\qquad \begin{cases} \beta_{11}y_1 + ... + \beta_{1m}y_m = 0 & ; \quad \beta_{11}, ..., \beta_{1m} \in \mathbf{Q}(\sigma_1) \\ \quad . \quad . \quad . & ; \quad \quad . \quad . \quad . \\ \beta_{m1}y_1 + ... + \beta_{mm}y_m = 0 & ; \quad \beta_{m1}, ..., \beta_{mm} \in \mathbf{Q}(\sigma_m) \end{cases}$$

having the following columns

$$v_1 = \begin{pmatrix} \beta_{11} \\ \cdots \\ \beta_{m1} \end{pmatrix}, \ldots, v_m = \begin{pmatrix} \beta_{1m} \\ \cdots \\ \beta_{mm} \end{pmatrix} \in \mathbf{Q}(\sigma_1) \oplus \cdots \oplus \mathbf{Q}(\sigma_m),$$

$\Phi(B) = h : V^* \to \mathbf{Q}(\sigma_1) \oplus ... \oplus \mathbf{Q}(\sigma_m)$, where $h(y_1^*) = v_1, ..., h(y_m^*) = v_m$.

Let $f : A \to B$ be a quasi-homomorphism of groups. It determines uniquely a matrix

$$T = \begin{pmatrix} t_{11} \dots & t_{1m} \\ \cdot \quad \cdot & \cdot \\ t_{n1} \dots & t_{nm} \end{pmatrix}$$

with rational entries,

$$\begin{pmatrix} fx_1 \\ \vdots \\ fx_n \end{pmatrix} = T \begin{pmatrix} y_1 \\ \vdots \\ y_m \end{pmatrix}.$$

According to the Criterion of quasi-homomorphism [1], the following matrix equality holds

$$\begin{pmatrix} \alpha_{11} \dots & \alpha_{1n} \\ \cdot \quad \cdot & \cdot \\ \alpha_{n1} \dots & \alpha_{nn} \end{pmatrix} \begin{pmatrix} fx_1 \\ \vdots \\ fx_n \end{pmatrix} = \begin{pmatrix} 0 \\ \vdots \\ 0 \end{pmatrix}.$$

Hence

$$\begin{pmatrix} \alpha_{11} \dots & \alpha_{1n} \\ \cdot \quad \cdot & \cdot \\ \alpha_{n1} \dots & \alpha_{nn} \end{pmatrix} T \begin{pmatrix} y_1 \\ \vdots \\ y_m \end{pmatrix} = \begin{pmatrix} 0 \\ \vdots \\ 0 \end{pmatrix}.$$

The last equality is a collection of n relations of the group B with coefficients from $\mathbf{Q}(\tau_1), ..., \mathbf{Q}(\tau_n)$, respectively. Each of them is a uniquely determined τ_i-combination of the relations (9), $i = 1, ..., n$. The coefficients of the τ_i-combination form the i-th row of the matrix

$$\Delta = \begin{pmatrix} \gamma_{11} \cdots & \gamma_{1m} \\ \cdot \quad \cdot & \cdot \\ \gamma_{n1} \cdots & \gamma_{nm} \end{pmatrix},$$

where $\gamma_{ij} \in \mathbf{Q}(\tau_i)$ and $\text{type}(\gamma_{ij}) \geq \tau_i - (\tau_i \wedge \sigma_j)$.

As in [1] we have

$$\begin{pmatrix} \alpha_{11} \dots & \alpha_{1n} \\ \cdot \quad \cdot & \cdot \\ \alpha_{n1} \dots & \alpha_{nn} \end{pmatrix} T = \Delta \begin{pmatrix} \beta_{11} \dots & \beta_{1m} \\ \cdot \quad \cdot & \cdot \\ \beta_{m1} \dots & \beta_{mm} \end{pmatrix}. \tag{10}$$

The matrix Δ determines a $\mathbf{K}$-module homomorphism

$$\varphi : \mathbf{Q}(\sigma_1) \oplus ... \oplus \mathbf{Q}(\sigma_m) \to \mathbf{Q}(\tau_1) \oplus ... \oplus \mathbf{Q}(\tau_n),$$

which maps an element $v \in \mathbf{Q}(\sigma_1) \oplus ... \oplus \mathbf{Q}(\sigma_m)$ written as a column

$$v = \begin{pmatrix} \beta_1 \\ \dots \\ \beta_m \end{pmatrix}, \quad \beta_1 \in \mathbf{Q}(\sigma_1), ..., \beta_m \in \mathbf{Q}(\sigma_m),$$

to an element

$$\varphi(v) = \Delta \begin{pmatrix} \beta_1 \\ \dots \\ \beta_m \end{pmatrix} = \begin{pmatrix} \gamma_{11} \circ \beta_1 + \dots + \gamma_{1m} \circ \beta_m \\ \dots \\ \gamma_{n1} \circ \beta_1 + \dots + \gamma_{nm} \circ \beta_m \end{pmatrix} \in \mathbf{Q}(\tau_1) \oplus ... \oplus \mathbf{Q}(\tau_n).$$

Let $f^* : V^* \to U^*$ be the dual linear mapping of dual vector spaces, i. e. f^* is defined by a matrix equality

$$\begin{pmatrix} f^*(y_1^*) \\ \dots \\ f^*(y_m^*) \end{pmatrix} = T^t \begin{pmatrix} x_1^* \\ \dots \\ x_n^* \end{pmatrix},$$

where T^t is the transposed matrix. We define $\Phi(f) = (f^*, \varphi)$. Consider the diagram

$$\begin{array}{ccc} U^* & \xleftarrow{f^*} & V^* \\ g\downarrow & & \downarrow h \\ \langle u_1, \dots, u_n \rangle_{\mathbf{K}} & \xleftarrow{\varphi} & \langle v_1, \dots, v_m \rangle_{\mathbf{K}} \end{array}$$

The equality (10) implies the commutativity of the diagram. Indeed

$$gf^*(y_i) = g(t_{1i}x_1^* + ... + t_{ni}x_n^*) = t_{1i}u_1 + ... + t_{ni}u_n = \sum_{k=1}^{n} u_k t_{ki}$$

is the i-th column of the matrix on the left side of (10), $i = 1, ..., m$. Due to (10), this column is equal to $\Delta \begin{pmatrix} \beta_{1i} \\ \dots \\ \beta_{mi} \end{pmatrix} = \varphi(v_i) = \varphi h(y_i^*)$. So the desired functor Φ is defined. We need only to verify that the functor Φ satisfies Proposition 3.1. This is routine and may be done analogously to [1]. □

COROLLARY 3.3. *Let $g : U \to \mathbf{M}$ and $h : V \to \mathbf{M}$ be objects of the category of linear mappings to finitely presented modules over $\mathbf{K}$. A and B are corresponding groups, i. e. $\Phi(A) = g$, $\Phi(B) = h$. Then the following statements are true:*

1. *If $\dim U = \dim V$ and $\operatorname{Im}(g) = \operatorname{Im}(h)$, then $A \doteq B$.*
2. *If $\dim U - \dim \operatorname{Im}(g) = k > 0$, then $A = F \oplus C$, where F is free, $\operatorname{rank} F = k$, and C is coreduced, i. e. C has no free direct summands.*

EXAMPLES 3.4.

1. The group F is a free group of rank n if and only if $\Phi(F) : U \to 0$, where $\dim U = n$. The property, that every mapping from a system of free generators of the group F to a torsion-free group A of finite rank may be extended to a homomorphism $F \to A$, means that for every object $h : V \to \mathbf{M}$ and for every linear mapping $f : V \to U$ the following diagram is commutative:

$$\begin{array}{ccc} U & \xleftarrow{f} & V \\ \downarrow & & \downarrow h \\ 0 & \longleftarrow & \mathbf{M} \end{array} .$$

2. The group D is a divisible torsion-free group of rank n if and only if $\Phi(D) = g : U \to \mathbf{K}^n$, where g maps every basis of U to a system of free generators of the free $\mathbf{K}$-module $\mathbf{K}^n$. The property, that every mapping from a maximal linearly independent system of a torsion-free group A of finite rank to the group D may be extended to a homomorphism $A \to D$, means that for every object $h : V \to \mathbf{M}$ and for every linear mapping $f : U \to V$ there exists a $\mathbf{K}$-module homomorphism $\varphi : \mathbf{K}^n \to \mathbf{M}$ such that the following diagram is commutative:

$$\begin{array}{ccc} U & \xrightarrow{f} & V \\ g\downarrow & & \downarrow h \\ \mathbf{K}^n & \xrightarrow[\varphi]{} & \mathbf{M} \end{array} .$$

3. Let $\mathbf{K}_p$ be the field of p-adic numbers. It is, so to say, a minimal $\mathbf{K}$-module. Let $g : U \to \mathbf{K}_p$ be an injective linear mapping and $y_1, ..., y_n$ a basis of U, $n > 0$. Now we construct a group A such that $\Phi(A) = g$. Let $D = \mathbf{Q}\,x_1 \oplus ... \oplus \mathbf{Q}\,x_n$ be a divisible group of rank n. We can assume that $\alpha_1 = g(y_1), ..., \alpha_n = g(y_n)$ are p-adic integers. Then A is generated by all elements

$$\frac{a_1^{(k)}x_1 + ... + a_n^{(k)}x_n}{p^k},$$

where $a_1^{(k)}, ..., a_n^{(k)} \in \mathbf{Z}$ are integer approximations of the α_i, i. e.

$$a_1^{(k)} \equiv \alpha_1 \;(mod\, p^k)\,, \;\; ... \;\; , \;\; a_n^{(k)} \equiv \alpha_n \;(mod\, p^k)\,, \;\; k = 0, 1, 2... \,.$$

These groups were considered in [2]. The object $g : U \to \mathbf{K}_p$ has the following obvious property: For every object $h : V \to \mathbf{M}$ of the category $\mathcal{L}$ and for every epimorphisms $f : U \to V$ and $\varphi : \mathbf{K}_p \to \mathbf{M}$, such that the diagram

$$\begin{array}{ccc} U & \xrightarrow{f} & V \\ g\big\downarrow & & \big\downarrow h \\ \mathbf{K}_p & \xrightarrow[\varphi]{} & \mathbf{M} \end{array}$$

is commutative, the homomorphism φ is either an isomorphism or zero. The corresponding property for the group A is that every subgroup of A with infinite index is free.

References

1. A. A. Fomin, *The category of quasi-homomorphisms of torsion free abelian groups of finite rank*, Contemp. Math., v.131, 1992 (Part 1), 91–111.
2. A. A. Fomin, *Abelian groups with free subgroups of infinite index and their endomorphism rings*, Math. Zam., **36**, N2 (1984), 179–187.

Moscow Pedagogical State University, Math. Department, Krasnoprudnaya ul., 14, 107140 Moscow, Russia

E-mail address: fomin.algebra@mpgu.msk.su

Contemporary Mathematics
Volume **171**, 1994

A Survey of Butler Groups of Infinite Rank

L. FUCHS

Introduction

The study of Butler groups, both in the finite and in the infinite rank cases, is a most active area of Abelian Group Theory. There are numerous challenging problems many of which can be dealt with successfully by using available methods or by developing new machinery. While the finite rank case is intimately related to the representation of finite posets, the special flavor of the infinite rank case is its connection with infinite combinatorics. The present survey is devoted to selected questions for Butler groups of arbitrary cardinalities and its aim is to report on the current state of affairs in this area.

All groups in the following are abelian. They are torsion-free unless stated otherwise. For unexplained notation and terminology we refer to Fuchs [F1].

A torsion-free group B of finite rank is called a *Butler group* if it satisfies either of the following equivalent conditions:

(a) B is a pure subgroup of a completely decomposable group (of finite rank);

(b) B is an epimorphic image of a completely decomposable group of finite rank.

The equivalence of these conditions has been proved by Butler [Bu]. For more on finite rank Butler groups, we refer to [AV]. See also [B2].

Bican-Salce [BS] noticed that (a) and (b) were equivalent to the condition that $\mathrm{Bext}^1(B,T)=0$ for all torsion groups T (where Bext^1 denotes the group of equivalence classes of all balanced extensions of T by B). This led them to a generalization to torsion-free groups of infinite rank: a torsion-free group B (of any rank) is called a

1. a B_1-*group* if $\mathrm{Bext}^1(B,T)=0$ for all torsion groups T;

2. a B_2-*group* if there is a continuous well-ordered ascending chain of pure subgroups,

$$0=B_0<B_1<\ldots<B_\alpha<\ldots<B_\tau=B=\cup B_\alpha$$

1991 *Mathematics Subject Classification*: 20K20.

This paper is in final form, and no version of it will be submitted for publication elsewhere.

with rank 1 factors such that, for each $\alpha < \tau$, $B_{\alpha+1} = B_\alpha + G_\alpha$ for some finite rank Butler group G_α. (There is no general agreement as to which of these should be called Butler group, so if we do not specify then we might understand either or both.)

Bican-Salce [BS] proved that these definitions are equivalent for countable groups B and, without any cardinality restriction, every B_2-group is a B_1-group. The major problems in the theory of infinite rank Butler groups are concerned with the following questions:

1. Is every B_1-group a B_2-group?
2. Is $\operatorname{Bext}^2(G,T) = 0$ for every torsion-free group G and torsion group T?
3. When is a pure subgroup of a Butler group again a Butler group?

We will discuss these questions with various other related problems. In several cases one has to impose conditions on the groups which can be either group theoretical or set theoretical. For detailed proofs we refer to the original sources, but often we are able to indicate the approach to the problem.

We do not touch upon generalizations to modules over domains, see e.g. Fuchs and Monari-Martinez [FMM].

1. Balanced and prebalanced subgroups

A pure subgroup A of the torsion-free group G is said to be a *balanced* subgroup if every coset $g + A$ $(g \in G)$ contains an element $g + a$ $(a \in A)$ (called *proper with respect to* A) such that $\chi(g + a) \geq \chi(g + x)$ for each $x \in A$. Here $\chi(a)$ denotes the characteristic of an element a in a given group G.

An exact sequence $0 \to A \to G \to C \to 0$ is *balanced-exact* if the image of A in G is a balanced subgroup of G. It is routine to check that balanced-exactness of the last exact sequence is equivalent to the following property: for every rank 1 torsion-free group J, every homomorphism $J \to C$ can be lifted to a map $J \to G$.

We collect here a few useful facts on balanced subgroups which are relevant in the proofs of the main theorems.

Unions of countable ascending chains of balanced subgroups need not be balanced again, however, we have:

LEMMA 1.1. *Let* $0 = A_0 < A_1 < \ldots < A_\nu < \ldots (\nu < \lambda)$ *be a (not necessarily continuous) well-ordered ascending chain of balanced subgroups of the torsion-free group* G. *If* $\operatorname{cf}\lambda \geq \omega_1$, *then* $A = \cup_\nu A_\nu$ *is again balanced in* G. □

Another property of balanced subgroups related to uncountable chains is the following.

LEMMA 1.2. [F2] *Let* $H = \cup_{\alpha<\omega_1} H_\alpha$ *be a continuous well-ordered ascending chain of pure subgroups of a torsion-free group* H. *If* B *is a subgroup of* H *such that* $B \cap H_\alpha = B_\alpha$ *is balanced in* H_α *for each* $\alpha < \omega_1$, *then* B *is balanced in* H. □

The next lemma provides information about the existence of balanced subgroups in torsion-free groups. It does not sound interesting in the stated generality, but this stronger form is needed in several proofs.

LEMMA 1.3. [DHR] *Let* A *and* $H_m (m < \omega)$ *be pure subgroups of the torsion-free group* G $(H_0 = 0)$ *such that the subgroups* H_m, $A + H_m$ *are balanced in* G *for all* $m < \omega$. *Given a subgroup* C *of* A, *there exists a subgroup* B *of* A *such that*

(i) $C \leq B \leq A$;

(ii) $|B| \leq |C|^{\aleph_0}$;
(iii) $B + H_m$ *is balanced in* G *for each* $m < \omega$. □

An immediate consequence is of particular interest: a torsion-free group which contains no balanced subgroups other than the trivial ones must have cardinality at most the continuum.

Balanced subgroups play an important role in the theory of torsion-free groups, but the homological machinery they provide do not fit well the theory of Butler groups, since not even finite rank Butler groups contain many balanced subgroups. However, there is a natural generalization of balancedness which can furnish us with a sought-after homological machinery. The main properties of this notion (introduced by Richman [Ri] and rediscovered by Fuchs-Viljoen [FV]) were developed by Fuchs-Metelli [FM2].

If the group B is torsion-free, then the *prebalancedness* of a pure subgroup A in B means that for each rank 1 pure subgroup H/A of B/A there is a finite rank Butler subgroup B' of B such that $A + B' = H$. The same property required for all finite rank pure subgroups H/A of B/A yields the definition of *decent* subgroup; this concept was introduced by Albrecht-Hill [AH].

A lemma sheds light on a noteworthy difference between balancedness and prebalancedness.

LEMMA 1.4. [FMa, F2] *A pure (prebalanced) subgroup* A *of a torsion-free group* G *is balanced in* G *if and only if, for every* $g \in G$ *and for every countable subset* $\{a_n\}_{n<\omega}$ *(for every finite subset* $\{a_n\}_{n<k}$*) of* A *there exists an* $a \in A$ *such that*

$$\chi(a + a_n) \geq \chi(g + a_n) \quad \text{for every} \quad n < \omega \quad (\text{for every} \quad n < k). \ \square$$

The homological properties of prebalancedness were discussed by Fuchs-Metelli [FM2]. It is easily seen that, just as the balanced-exact sequences, the prebalanced-exact sequences form a proper class. Hence we derive that the equivalence classes of extensions of A by a torsion-free C represented by prebalanced-exact sequences form a subgroup $\mathrm{PBext}^1(C, A)$ in $\mathrm{Ext}^1(C, A)$, which evidently contains the subgroup $\mathrm{Bext}^1(C, A)$ of balanced extensions of A by C. Hence we are led to:

THEOREM 1.5. [FM2] *Let* $0 \to A \to B \to C \to 0$ *be a prebalanced-exact sequence of torsion-free groups. For a torsion-free* G *and arbitrary* H *we have the following exact sequences:*

$$\begin{aligned} 0 \to &\mathrm{Hom}(G, A) \to \mathrm{Hom}(G, B) \to \mathrm{Hom}(G, C) \to \\ &\mathrm{PBext}^1(G, A) \to \mathrm{PBext}^1(G, B) \to \mathrm{PBext}^1(G, C) \to \dots, \\ 0 \to &\mathrm{Hom}(C, H) \to \mathrm{Hom}(B, H) \to \mathrm{Hom}(A, H) \to \\ &\mathrm{PBext}^1(C, H) \to \mathrm{PBext}^1(B, H) \to \mathrm{PBext}^1(A, H) \to \dots . \ \square \end{aligned}$$

A drawback of prebalancedness is that it does not admit enough prebalanced-projectives (only the free groups), therefore prebalanced-projective resolutions do not exist in general.

Though $\mathrm{PBext}^1(C, A)$ is in general much larger than $\mathrm{Bext}^1(C, A)$, for torsion groups A these two groups coincide. This is a crucial fact for applications.

THEOREM 1.6. [FM2] *A prebalanced-exact sequence* $0 \to T \to G \xrightarrow{\alpha} C \to 0$ *with* C *torsion-free is necessarily balanced-exact whenever* T *is torsion.* □

Prebalancedness is particularly suited to deal with B_1-groups. The following (which is readily verified for B_2-groups) can be proved easily by using (1.5) above.

LEMMA 1.7. [FM2] *Prebalanced extensions of* B_1*-groups by* B_1*-groups are* B_1*-groups. Moreover, if* $0 = B_0 < B_1 < \ldots < B_\alpha < \ldots < B_\tau = B$ *is a continuous well-ordered ascending chain of subgroups in* B *such that each subgroup is pebalanced in its successor and all the factors* $B_{\alpha+1}/B_\alpha$ *are* B_1*-groups, then* B *is likewise a* B_1*-group.* □

Hence the following conclusion is immediate:

THEOREM 1.8. [BS] B_2*-groups of any rank are* B_1*-groups.* □

The concept of prebalancedness was extended to infinite cardinals by Bican-Fuchs [BF2]. Let B be any group and A a subgroup such that the factor group B/A is torsion-free of rank 1. Consider the types $\mathbf{t}(J)$ of those rank 1 (torsion-free) subgroups J of B which are not contained in A. In the lattice of all types, we form the ideal $\mathcal{I}_{B|A}$ generated by all these types $\mathbf{t}(J)$. Let κ be a cardinal such that $\aleph_{-1} \leq \kappa \leq 2^{\aleph_0}$. A is said to be *κ-prebalanced in B*, if $\mathcal{I}_{B|A}$ is at most κ-generated (i.e. it has a generating set of cardinality $\leq \kappa$); here $\aleph_{-1}$-generated means finitely generated. More explicitly, A κ-prebalanced in B means that we can write

$$B = A + \Sigma_{\alpha \in I} J_\alpha$$

where I is an index set of cardinality $\leq \kappa$ and the groups J_α are rank 1 pure subgroups of B not in A such that every rank 1 pure subgroup J of B not in A satisfies $\mathbf{t}(J) \leq \mathbf{t}(J_{\alpha_1}) \cup \ldots \cup \mathbf{t}(J_{\alpha_m})$ for some integer m, $\alpha_j \in I$.

Let us point out rightaway that, moreover, we may assume that the J_α are selected such that for every rank 1 pure subgroup J of B not in A, there are a finite number of J_α satisfying $J \leq J_{\alpha_1} + \ldots + J_{\alpha_j}$. In the special case $\kappa = \aleph_{-1}$, we have $B = A + J_1 + \ldots + J_m$ for some integer m, and A is *prebalanced* in B.

If B/A is torsion-free of arbitrary rank, then we define A to be *κ-prebalanced* in B whenever A is κ-prebalanced in every subgroup C of B that contains A as a corank 1 subgroup. We shall say that an exact sequence $0 \to A \to B \to C \to 0$ is *κ-prebalanced-exact* if A maps onto a κ-prebalanced subgroup of B.

It is worthwhile comparing this notion with Hill's notion of separability. A pure subgroup A of G is called *separative* - or separable in the sense of Hill - if for each $g \in G$ there is a countable subset $\{a_n \mid n < \omega\} \subset A$ such that $\{\chi(g + a_n) \mid n < \omega\}$ is cofinal in the partially ordered set $\{\chi(g + a) \mid a \in A\}$. From the definitions it is evident that separative subgroups are necessarily $\aleph_0$-prebalanced. Easy examples show that the converse is not true.

Since $\aleph_0$-prebalancedness will be most relevant for our survey, we list some of its main properties in the following lemma.

LEMMA 1.9. [BF2] *Let* $C \leq B$ *be pure subgroups of the torsion-free group* A. *Then the following holds:*

(i) *If* C *is* $\aleph_0$*-prebalanced in* B *and* B *is* $\aleph_0$*-prebalanced in* A*, then* C *is* $\aleph_0$*-prebalanced in* A.

(ii) *Let C be prebalanced in A. B is $\aleph_0$-prebalanced in A if and only if B/C is $\aleph_0$-prebalanced in A/C.*
(iii) *If C is $\aleph_0$-prebalanced in A and if B/C is countable, then B is $\aleph_0$-prebalanced in A.*
(iv) *The union of a countable ascending chain of $\aleph_0$-prebalanced subgroups of A is again $\aleph_0$-prebalanced in A.* □

2. Relative balanced-projective resolutions

One of the most useful devices in the discussion of infinite rank Butler groups is what we call relative balanced-projective resolution; this was introduced by Bican-Fuchs [BF2].

Let A be a subgroup of the group B such that B/A is torsion-free, and let J_α run over all rank one pure subgroups of B not contained in A. If we write $X = \bigoplus_{\alpha \in I} X_\alpha$ where $X_\alpha \cong J_\alpha$, then the map $A \oplus X \longrightarrow B$ induced by the chosen isomorphisms between X_α and J_α, along with the inclusion map of A in B, gives rise to a balanced-exact sequence $0 \longrightarrow K \longrightarrow A \oplus X \xrightarrow{\phi} B \longrightarrow 0$. It is readily seen that if A has corank 1 in B, then the kernel K is isomorphic to a corank 1 pure subgroup of X; see [BF2]. (If A has arbitrary corank in B, then nothing similar can be said of K.) Such a balanced-exact sequence is referred to as a *balanced-projective resolution of B relative to A.*

THEOREM 2.1. [BF2] *Let $0 \longrightarrow A \longrightarrow B \longrightarrow C \longrightarrow 0$ be an exact sequence where C is a torsion-free group. Then there is a balanced-exact sequence*

$$0 \longrightarrow K \longrightarrow A \oplus X \xrightarrow{\phi} B \longrightarrow 0 \tag{1}$$

where $X = \oplus X_i$ is completely decomposable, $\phi|A : A \longrightarrow B$ is the inclusion map and K is isomorphic to a pure subgroup of X. □

Relative balanced-projective resolutions are especially useful in the study of $\aleph_0$-prebalancedness, as we shall see in (2.3).

One of the motivations for studying $\aleph_0$-prebalancedness stems from the following result which was proved by Fuchs-Metelli-Rangaswamy [FMR].

THEOREM 2.2. *A corank 1 pure subgroup A of the completely decomposable group $G = \oplus_{\alpha<\tau} G_\alpha$ (each G_α is torsion-free of rank 1) that does not contain any G_α is a B_2-group if and only if A is $\aleph_0$-prebalanced in G.*

If CH *is assumed, then the same criterion applies to A being a B_1-group.* □

This result leads to the following basic fact.

THEOREM 2.3. [BF2], [F2] *Let A be a subgroup of a group B such that B/A is torsion-free of rank 1. A is $\aleph_0$-prebalanced in B exactly if there is a prebalanced-exact sequence (1) where X, K are B_2-groups, and $\phi|A$ is the inclusion map.* □

3. κ-prebalanced chains

κ-prebalanced chains in torsion-free groups provide us with a powerful tool to study Butler groups, especially when $\kappa = \aleph_0$.

Let A be a pure subgroup of the torsion-free group G. By a *κ-prebalanced chain* from A to G is meant a continuous well-ordered ascending chain

$$A = G_0 < G_1 < \ldots < G_\alpha < \ldots < G_\tau = G \quad \text{(for some ordinal } \tau)$$

of κ-prebalanced subgroups of G, where all the factors $G_{\alpha+1}/G_\alpha$ are of rank one. We say that G *admits a κ-prebalanced chain* if there is a κ-prebalanced chain from 0 to G. As countable extensions of κ-prebalanced subgroups are again κ-prebalanced, it suffices to require that the factors $G_{\alpha+1}/G_\alpha$ be countable (or of cardinality $\leq \aleph_1$).

Paul Hill introduced the 3rd Axiom of Countability in order to characterize totally projective p-groups. This concept turned out most relevant in various other applications as well [H2]. Recall that by an *Axiom-3 family* of subgroups in G is meant a collection $\mathcal{C}$ of subgroups of G such that (i) $0, G \in \mathcal{C}$; (ii) $\mathcal{C}$ is closed under taking arbitrary unions of its members; (iii) if $A \in \mathcal{C}$ and X is any countable subset of G, then there is an $A' \in \mathcal{C}$ that contains both A and X such that A'/A is countable. We say that a family $\mathcal{F}$ of subgroups of G is an Axiom-3 family over a subgroup A of G if every member F of $\mathcal{F}$ contains A and $\{F/A \mid F \in \mathcal{F}\}$ is an Axiom-3 family in G/A.

Let us formulate here a similar concept. A *$G(\lambda)$-family* (λ is an infinite cardinal) in the group G means a collection $\mathcal{G}$ of subgroups of G such that (i) $0, G \in \mathcal{G}$; (ii) $\mathcal{G}$ is closed under unions of chains; (iii) if $B \in \mathcal{G}$ and X is any subset of G of cardinality $\leq \lambda$, then there is a $B' \in \mathcal{G}$ that contains both B and X, and satisfies $|B'/B| \leq \lambda$. Manifestly, every Axiom-3 family is a $G(\aleph_0)$-family, but the converse is false.

Using ideas due to Hill [H2], one can show:

THEOREM 3.1. [F2] *A torsion-free group G that has a single $\aleph_0$-prebalanced chain from a pure subgroup A to G admits an Axiom-3 family over A of $\aleph_0$-prebalanced subgroups.* □

By making use of the observation that every pure subgroup of a completely decomposable group, whose typeset has cardinality κ^+, admits a κ-prebalanced chain, one can easily show that, in the presence of CH, pure subgroups of completely decomposable groups admit $\aleph_0$-prebalanced chains.

One can also derive the following result.

THEOREM 3.2. [F2] *If there is a κ^+-prebalanced chain from a pure subgroup A of a torsion-free group G to the group G itself, then in a relative balanced-projective resolution $0 \to B \to A \oplus C \to G \to 0$ (where C is completely decomposable) the group B admits a κ-prebalanced chain.* □

Our primary concern is when $\kappa = \aleph_0$, in which case (3.2) can be improved, and we get the following most useful result.

THEOREM 3.3. [F2] *There is an $\aleph_0$-prebalanced chain from a pure subgroup A of a torsion-free group G to G itself if and only if in a relative balanced-projective resolution $0 \to B \to A \oplus C \to G \to 0$ (where C is completely decomposable), B is a B_2-group.* □

It is routine to verify that the condition that B be a B_2-group is independent of the particular choice of the relative prebalanced resolution in the theorem.

The last theorem does not hold (in ZFC) for κ-prebalancedness for uncountable κ. In fact, if $2^{\aleph_0} \geq \aleph_3$, then (9.3) infra yields a corank one pure subgroup G of a completely decomposable group that has no $\aleph_1$-prebalanced chain, but the kernel of its balanced-projective resolution admits an $\aleph_0$-prebalanced chain.

The special case $A = 0$ in the preceding theorem leads at once to an important criterion we have been looking for:

COROLLARY 3.4. *A torsion-free group G admits an $\aleph_0$-prebalanced chain if and only if in a balanced-projective resolution $0 \to B \to C \to G \to 0$ of G (where C is completely decomposable) the subgroup B is a B_2-group.* □

The crucial lemma is the following whose proof involves relative balanced-projective resolutions. It is used in the proof of (11.1).

LEMMA 3.5. [BF2] *Suppose $0 \to T \to G \to A \to 0$ is a balanced-exact sequence where T is a torsion group. If A is a subgroup in a torsion-free group B such that there is an $\aleph_0$-prebalanced chain from A to B, then there exists a commutative diagram with balanced-exact rows*

$$\begin{array}{ccccccccc} 0 & \longrightarrow & T & \longrightarrow & G & \longrightarrow & A & \longrightarrow & 0 \\ & & \| & & \downarrow & & \downarrow & & \\ 0 & \longrightarrow & T & \longrightarrow & H & \longrightarrow & B & \longrightarrow & 0. \end{array}$$ □

4. B_1-groups with $\aleph_0$-prebalanced chains

From the very definition of B_2-groups it is clear that every B_2-group admits a (prebalanced, and hence an) $\aleph_0$-prebalanced chain. It turns out that this property alone characterizes B_2-groups within the class of B_1-groups - no matter what model of ZFC is chosen.

THEOREM 4.1. [F2] *A B_1-group is a B_2-group if and only if it admits an $\aleph_0$-prebalanced chain.* □

It is worth while mentioning the following result which is, in a certain sense, a generalization of (4.1).

COROLLARY 4.2. [BF2] *If a B_2-group A is a pure subgroup of a B_1-group G such that there is an $\aleph_0$-prebalanced chain from A to G, then G is a B_2-group, too.* □

Let λ be an infinite cardinal. A continuous well-ordered ascending chain of subgroups of G, $0 = G_0 < \ldots < G_\alpha < G_{\alpha+1} < \ldots$ with union G, will be called a *λ-filtration* of G if the cardinalities of the quotients $G_{\alpha+1}/G_\alpha$ do not exceed λ for any α. From (4.1) it is readily seen that a B_1-group that admits an $\aleph_1$-filtration with (pre)balanced subgroups is actually a B_2-group.

Hence and from (3.2) it is straightforward to derive the main result in Fuchs-Rangaswamy [FR1], which uses an algebraic, rather than a set-theoretic, condition to ensure that B_1-groups are B_2-groups:

THEOREM 4.3. [FR1] *Let G be a B_1-group which is the union of a continuous well-ordered ascending chain of pure subgroups G_α. If each G_α has a countable typeset, then G is a B_2-group.* □

In the next three corollaries we assume that CH holds. This hypothesis is needed to guarantee that the set of all types has cardinality $\aleph_1$, which fact is essential in the proofs.

COROLLARY 4.4. [FR1] (CH) *A pure subgroup of a completely decomposable group is a B_2-group if and only if it is a B_1-group.* □

The terminologies for B_1- and B_2-groups have been chosen in the right way as is shown by the next corollary which follows from (3.4) and (4.4) at once:

COROLLARY 4.5. [R1] (CH) *A torsion-free group B is a B_2-group if and only if both* $\mathrm{Bext}^1(B,T) = 0$ *and* $\mathrm{Bext}^2(B,T) = 0$ *for all torsion groups T.* □

From (3.4) we can easily derive:

COROLLARY 4.6. (CH) *A torsion-free group G admits an $\aleph_0$-prebalanced chain exactly if* $\mathrm{Bext}^2(G,T) = 0$ *for all torsion groups T.* □

The next simple result is most useful in proofs in which fine distinction has to be made between the B_1- and B_2-properties.

LEMMA 4.7. [R1], [F2] *Suppose that $0 \to H \to C \to G \to 0$ is a balanced-exact sequence where C is a B_2-group and H, G are B_1-groups. If one of H and G is a B_2-group, then so is the other.* □

Another corollary which is worth while recording is as follows:

COROLLARY 4.8. [F2] *If B is a B_2-group, then* $\mathrm{Bext}^i(B,T) = 0$ *for all $i \geq 1$ and for all torsion groups T.* □

By a *long balanced-projective resolution of* the torsion-free group G we mean a long exact sequence

$$0 \to K_n \to C_n \to \ldots \to C_1 \to C_0 \to G \to 0$$

in which the groups C_i are completely decomposable and in each C_i the image of the preceding map is a balanced subgroup; in other words, the long sequence is balanced-exact. From (3.2) it is easy to conclude by straightforward induction that if G admits an $\aleph_n$-prebalanced chain for some integer $n \geq 0$, then the kernel K_n is a B_2-group.

LEMMA 4.9. [F2] *If a torsion-free group B admits an $\aleph_k$-prebalanced chain for some $k \geq 0$, and if* $\mathrm{Bext}^i(B,T) = 0$ *for all torsion groups T and for $1 \leq i \leq k+1$, then B is a B_2-group.* □

5. A homological characterization of B_2-groups

A characterization of B_2-groups in terms of the functor Bext seems to be very doubtful, unless an extra set-theoretical hypothesis is imposed on the underlying sets. We shall work under the hypothesis that the continuum is equal to $\aleph_n$ for some integer $n \geq 1$.

The next lemma leads to a kind of characterization we are looking for.

LEMMA 5.1. [F2] *Suppose that $2^{\aleph_0} = \aleph_n$ for an integer $n \geq 1$. Then a torsion-free group G has an $\aleph_0$-prebalanced chain if and only if*

$$\mathrm{Bext}^2(G,T) = \ldots = \mathrm{Bext}^{n+1}(G,T) = 0 \quad \textit{for all torsion groups } T.$$

Equivalently, if and only if $\mathrm{Bext}^i(G,T) = 0$ *holds for all $i \geq 2$ and all torsion groups T.* □

This result is essential in deriving the next theorem which generalizes Rangaswamy's theorem [R1] proved under CH.

THEOREM 5.2. [F2] *Let* $2^{\aleph_0} = \aleph_n$ *for an integer* $n \geq 1$. *Then a torsion-free group* B *is a* B_2*-group if and only if*

$$\mathrm{Bext}^1(B,T) = \mathrm{Bext}^2(B,T) = \ldots = \mathrm{Bext}^{n+1}(B,T) = 0$$

for all torsion groups T. *Equivalently, if and only if* $\mathrm{Bext}^i(B,T) = 0$ *for all* $i \geq 1$ *and all torsion groups* T. □

An immediate consequence is that if a torsion-free group G satisfies $\mathrm{Bext}^i(G,T) = 0$ for all $i \geq 1$ and all torsion groups T, then either G and all the kernels in its long balanced-projective resolution are B_2-groups, or none of them is a B_2-group. It is still an open problem whether or not the second alternative can occur. Needless to say that then $2^{\aleph_0} \geq \aleph_\omega$ must hold and none of the kernels can be a B_2-group.

A result similar to the one above is as follows:

THEOREM 5.3. [F2] *A torsion-free group* B *of cardinality* $\aleph_n$ *(for an integer* $n \geq 1$*) is a* B_2*-group if and only if*

$$\mathrm{Bext}^1(B,T) = \mathrm{Bext}^2(B,T) = \ldots = \mathrm{Bext}^{n+1}(B,T) = 0$$

for all torsion groups T. □

6. TEP-subgroups

A property which is extremely relevant in the study of Butler groups is the Torsion Extension Property (TEP) which was, in an equivalent form, introduced by Procházka [P] and Bican [B1] for groups of finite rank. In the infinite rank case, the properties of TEP subgroups were developed by Dugas-Rangaswamy [DR] and Dugas-Hill-Rangaswamy [DHR].

Here is the definition: a subgroup A of a torsion-free group G is called a *TEP-subgroup* of G, or to *have TEP in* G, if every homomorphism $A \to T$ (T any torsion group) extends to a homomorphism $G \to T$; that is, if the induced map $\mathrm{Hom}(G,T) \to \mathrm{Hom}(A,T)$ is surjective for every torsion group T. This property is easily seen to be equivalent to the following one: for every subgroup C of A such that A/C is torsion, A/C is a summand of G/C.

Observe that a TEP subgroup is necessarily pure. In fact, if A is not pure in G, say $pA < A \cap pG$ for some prime p, then it is easy to find a homomorphism of A into $\mathbb{Z}(p)$ which does not extend to G.

From the exact sequences in (1.5) it follows readily:

LEMMA 6.1. *Let* $0 \to A \to B \to C \to 0$ *be a prebalanced-exact sequence with* B *a* B_1*-group.* A *is a TEP-subgroup of* B *if and only if* C *is a* B_1*-group.* □

Here the "if" part holds for any torsion-free group B. For the "only if" part prebalancedness is not really needed; in fact, the exactness of $0 \to \mathrm{Ext}^1(C,T) \to \mathrm{Ext}^1(B,T)$ implies by (1.5) that $0 \to \mathrm{PBext}^1(C,T) \to \mathrm{PBext}^1(B,T) = 0$ is exact (see [DHR]).

It has been observed by Dugas-Rangaswamy [DR] that TEP-subgroups of countable B_1-groups are exactly the decent subgroups. The main step in the proof of this important fact is the content of our next theorem which we phrase in a more general setting needed for large cardinalities.

THEOREM 6.2. [DR], [FM2] *Let* B *be a* B_1*-group, and* A *an* $\aleph_0$*-prebalanced subgroup of finite corank in* B. A *is a TEP-subgroup of* B *if and only if* $B = A + B'$ *holds for a finite rank Butler subgroup* B' *of* B *(i.e.* A *is decent in* B*).* □

7. Countable Butler groups

The general results which were proved without any cardinality restrictions can be used to derive some relevant facts on countable Butler groups (see [FM2]); these are otherwise more cumbersome to prove. In the proofs, the only result which is taken for granted is the equivalence of (a) and (b) for finite rank Butler groups mentioned in the introduction.

It is convenient to start with the finite rank case. The equivalence of (i) with (ii)-(iii) is due to Bican-Salce [BS]; a simpler proof for (iii) $\Rightarrow$ (ii) can be found in Mines-Vinsonhaler [MV].

THEOREM 7.1. *For a torsion-free group B of finite rank, the following are equivalent:*

(i) *B is a Butler group;*
(ii) *B is a B_2-group;*
(iii) *B is a B_1-group;*
(iv) *every pure subgroup of B is prebalanced.* □

Let us observe that the condition "all pure subgroups are TEP-subgroups" does not imply that a finite rank group has to be a Butler group. In fact, in Pontryagin's rank two indecomposable group (which is evidently not a Butler group), all pure subgroups have TEP.

Turning to the countable case, note that pure subgroups of completely decomposable groups are *finitely Butler* in the sense that every finite rank pure subgroup is a Butler group. The following theorem, due to Bican-Salce [BS], was the starting point for the theory of Butler groups of infinite rank. It gives several satisfactory characterizations of countable Butler groups.

THEOREM 7.2. [BS] *For a countable torsion-free group B the following are equivalent:*

(i) *B is finitely Butler;*
(ii) *B is a B_2-group;*
(iii) *B is a B_1-group.* □

Since the property of being finitely Butler is evidently inherited by pure subgroups, we have at once

COROLLARY 7.3. [BS] *Pure subgroups of countable Butler groups are themselves Butler groups.* □

Countable Butler groups need not contain non-trivial balanced subgroups (of course, this can happen only for indecomposable groups). On the other hand, prebalanced subgroups are abundant.

COROLLARY 7.4. *A torsion-free group of countable rank is Butler if and only if all of its finite rank pure subgroups are prebalanced.* □

Another useful consequence is recorded next:

COROLLARY 7.5. *Finite rank pure subgroups of a countable Butler group B are TEP subgroups.* □

The question as to when an arbitrary pure subgroup of a countable Butler group is a TEP-subgroup can also be answered.

COROLLARY 7.6. *Let B be a countable Butler group, and A a pure subgroup of B. The following are equivalent:*

(a) *A is a TEP-subgroup of B;*
(b) *A is prebalanced in B and B/A is a Butler group;*
(c) *A is decent in B.* □

8. B_3-groups

Albrecht-Hill [AH] introduced a third version of Butler groups which they called B_3-groups. A *B_3-group* is defined as a torsion-free group that admits an Axiom-3 family of decent subgroups. They have proved that all B_3-groups are B_2-groups; the gap in their proof that the converse was true was filled by Fuchs-Magidor [FMa].

We say that a subgroup H of G has *property P* if H is a pure subgroup of G, and for each pure subgroup K of G which contains H as a finite corank subgroup, $K = H + B$ holds for some finite rank subgroup B of G. The following lemma plays an essential role in the proof of (8.2).

LEMMA 8.1. [FMa] *If G is the union of a continuous well-ordered ascending chain of pure subgroups, $0 = H_0 < H_1 < \ldots < H_\nu < \ldots (\nu < \mu)$, such that for each $\nu + 1 < \mu$, $H_{\nu+1} = H_\nu + B_\nu$ with a finite rank subgroup B_ν, then G satisfies the 3rd axiom of countability for subgroups of property P.* □

Applying (8.1) to the case in which the finite rank subgroups B_ν in the definition of property P are Butler groups (i.e. property P means being decent in G), one concludes:

THEOREM 8.2. [AH], [FMa] *A torsion-free group is a B_2-group if and only if it is a B_3-group.* □

From the definition it is immediately clear that B_3-groups are finitely Butler. Hence B_2-groups are finitely Butler.

The importance of the last theorem lies in the fact that in certain proofs it is not enough to work with a chain characterizing B_2-groups: often a larger supply of decent subgroups is needed.

COROLLARY 8.3. [FMa] (CH) *Every B_2-group admits a $G(\aleph_1)$-family of balanced subgroups.* □

9. The existence of $\aleph_i$-prebalanced chains

A most relevant question is whether or not there exist $\aleph_0$-prebalanced chains in every torsion-free group; in fact, an answer to this question would imply not only that $\mathrm{Bext}^2(G,T) = 0$ for all torsion-free groups G and torsion groups T, but also that B_1-groups are B_2-groups. However, it is conceivable that B_1-groups admit $\aleph_0$-prebalanced chains, though not all torsion-free groups do.

To establish the existence of $\aleph_0$-prebalanced chains turns out to be quite difficult in groups of large cardinalities. The main obstacle in the transfinite induction lies in handling the cardinals which are cofinal with ω. Dugas-Hill-Rangaswamy [DHR] develop the relevant machinery for all cardinals which are not cofinal with ω, thus they could handle groups up to $\aleph_\omega$. The remaining case was settled by Fuchs-Magidor [FMa]. They proved that ZFC + V = L guarantees that every torsion-free group admits a separative chain. Hence:

THEOREM 9.1. [FMa] *Assuming* V = L, *every torsion-free group admits an* $\aleph_0$-*prebalanced chain.* $\square$

It is not known as yet if V = L is a necessary condition for the existence of such a chain in all torsion-free groups, but CH is certainly necessary; cf. (9.3) below. Recently, Magidor and Shelah have shown (unpublished) that GCH alone is not sufficient to ensure the existence of $\aleph_0$-prebalanced chains. It should be pointed out that the conditions for the existence of separative chains and for $\aleph_0$-prebalanced chains are different; see (9.5) infra.

The proof of (9.1) is based on GCH as well as on the following "Box Principle" by Jensen [Je] ; note that both are consequences of V = L.

$\square_\lambda$: Let λ be a singular cardinal. There exists a family of sets, C_ν, for limit ordinals $\nu < \lambda^+$, such that

(i) C_ν is a cub (i.e. closed and unbounded) in ν;

(ii) the order type of C_ν is $< \lambda$;

(iii) coherence property: if μ is a limit point in C_ν, then $C_\mu = C_\nu \cap \mu$.

If $\mathrm{cf}\,\lambda = \omega$, there is a countable properly ascending chain of regular cardinals $\kappa_n (n < \omega)$ such that $\cup \kappa_n = \lambda$. This is used in the following crucial, but rather technical lemma whose proof relies both on GCH and $\square_\lambda$. We state this key lemma in order to illustrate the idea of proof of (9.1).

LEMMA 9.2. [FMa] (V = L) *Let G be a torsion-free group, A and H_m $(m < \omega)$ pure subgroups of G such that H_m and $A + H_m$ are balanced in G $(H_0 = 0)$. Suppose that $|A| = \lambda^+$ where $\mathrm{cf}\,\lambda = \omega$. There are pure subgroups $A_\alpha (\alpha < \lambda^+)$ in A, pure subgroups A_α^n $(n < \omega)$ in each A_α, and in case $\mathrm{cf}\,\alpha = \omega$, pure subgroups A_α^{nk} $(k < \omega)$ of A_α^n such that*

(a) *for all $\beta < \alpha < \lambda^+$, $A_\beta \leq A_\alpha$; $A_\alpha = \bigcup_{\beta<\alpha} A_\beta$ if α is a limit ordinal; $A = \bigcup_{\alpha<\lambda^+} A_\alpha$;*

(b) *for every $n < \omega$, $A_\alpha^n \leq A_\alpha^{n+1}$ and $A_\alpha = \bigcup_{n<\omega} A_\alpha^n$ for each α;*

(c) *if $\mathrm{cf}\,\alpha = \omega$, $A_\alpha^n = \bigcup_{k<\omega} A_\alpha^{nk}$ where $A_\alpha^{nk} \leq A_\alpha^{nk+1}$ for every $k < \omega$;*

(d) *$|A_\alpha^n| = \kappa_n$ (and hence $|A_\alpha| = \lambda$) for each α, n;*

(e) *if $\mathrm{cf}\,\alpha \neq \omega$, then $A_\alpha^n + H_m$ is balanced in G for all $m, n < \omega$; if $\mathrm{cf}\,\alpha = \omega$, then for each $m, n, k < \omega$, $A_\alpha^{nk} + H_m$ is balanced in G;*

(f) *if $\alpha = \beta + 1$, then in case $\mathrm{cf}\,\beta \neq \omega$, $A_\alpha^n + A_\beta^k + H_m$ is balanced in G for all $n, k, m < \omega$; while in case $\mathrm{cf}\,\beta = \omega, A_\alpha^n + A_\beta^{k\ell} + H_m$ is balanced in G for all $n, k, \ell, m < \omega$.* $\square$

We now present examples of torsion-free groups which fail to admit $\aleph_i$-prebalanced chains for $i \geq 0$.

Let $\mathcal{S} = \{S_\alpha \mid \alpha < \Omega\}$ be a set of almost disjoint countable subsets S_α of the set of all prime numbers; i.e. $S_\alpha \cap S_\beta$ is finite for different $\alpha, \beta < \Omega$; here Ω stands for the initial ordinal of the power of the continuum. (The existence of such an $\mathcal{S}$ is well known; see e.g. Jech [J].) For each $\alpha < \Omega$, let $\mathbf{t}_\alpha$ be the type represented by the characteristic $(n_2, n_3, \ldots, n_p, \ldots)$ where $n_p = 1$ or 0 according as $p \in S_\alpha$ or $p \notin S_\alpha$. It is easily seen that $\mathbf{t}_\alpha \leq \mathbf{t}_{\alpha_1} \cup \ldots \cup \mathbf{t}_{\alpha_k}$ holds if and only if $\alpha \in \{\alpha_1, \ldots, \alpha_k\}$. Consequently, in the lattice of all types, the ideal generated by the $\mathbf{t}_\alpha$ can not be countably generated.

With the indicated choice of types, let A be the direct sum of rational groups A_α with $\mathbf{t}(A_\alpha) = \mathbf{t}_\alpha$ for $\alpha < \Omega$, and G a corank one pure subgroup of A that does not contain any A_α. This group G is homogeneous of type $\mathbb{Z}$, but by (2.2) it is not a B_2-group, so it is not free (see Bican [B2]).

Dugas and Thomé [DT2] improve on the mentioned result on almost disjoint sets by establishing the existence of two collections (both of the power of the continuum) consisting of almost disjoint sets, $\mathcal{S}$ and $\mathcal{S}'$, as above, with the additional property that the intersection $S_\alpha \cap S'_\beta$ is infinite for all $S_\alpha \in \mathcal{S}$, $S'_\beta \in \mathcal{S}'$. (That is, for each α, $\{S_\alpha \cap S'_\beta \mid \beta < \Omega\}$ is a collection of almost disjoint sets.) Using this result, one can verify a generalization of a Dugas-Thomé result:

THEOREM 9.3. [F2] *Let $2^{\aleph_0} \geq \aleph_n$ for some integer $n \geq 1$. Choose an index set I of cardinality $\aleph_n$ and form the completely decomposable group $A = \bigoplus_{\alpha<\omega_n} A_\alpha$ where the set of types of the rank one summands A_α is the union of the sets of types defined by the almost disjoint systems $\mathcal{S}$ and $\mathcal{S}'$ of cardinality $\aleph_n$. Then no corank 1 subgroup G of A that contains none of the A_α admits an $\aleph_{n-2}$-prebalanced chain.* □

From the last theorem it follows readily: if, for some model of ZFC , there is an integer $n \geq 1$, such that every torsion-free group admits an $\aleph_n$-prebalanced chain, then $2^{\aleph_0} \leq \aleph_{n+1}$ holds in this model.

Another noteworthy consequence of the results above is the next theorem.

THEOREM 9.4. [F2] *In any model of* ZFC, *the following conditions are equivalent:*

(i) $\mathrm{Bext}^2(G,T) = 0$ *for all torsion-free groups G and torsion groups T.*
(ii) *Every torsion-free group admits an $\aleph_0$-prebalanced chain.*
(iii) CH *holds and balanced subgroups of completely decomposable groups are B_2-groups.* □

This result explains why the assumption of CH was so relevant in most papers dealing with uncountable Butler groups. That CH is necessary for the vanishing of Bext^2 was first pointed out by Dugas-Thomé [DT2].

We close this section with a comment comparing separativeness and $\aleph_0$-prebalancedness. As pointed out above, it is easy to construct examples of $\aleph_0$-prebalanced subgroups that are not separative; however, it is hard to find groups admitting $\aleph_0$-prebalanced chains but no separative chains. The only such example known so far requires the denial of CH.

LEMMA 9.5. [F2] *($\neg$CH) There is a torsion-free group G of cardinality $\aleph_2$ which admits an $\aleph_0$-prebalanced chain but no separative chains.* □

One of the main results on Butler groups of arbitrary cardinalities is an immediate consequence of (4.1) and (9.1):

THEOREM 9.6. [FMa] *If* V = L, *then*

(a) $\mathrm{Bext}^2(G,T) = 0$ *for all torsion-free G and torsion T;*
(b) *B_1-groups are B_2-groups.* □

10. The groups Bext^n

The derived functors Bext^n of Bext^1 have been discussed in the literature, mostly in connection with the question as to when they vanish. We have already used them in (5.1)-(5.3) in connection with B_2-groups. It is worth while investigating the vanishing of higher Bexts in general.

The first result on Bext^n for $n > 1$ is due to Albrecht and Hill [AH] who proved that $\text{Bext}^2(G, T) = 0$ for all torsion-free groups G of cardinality $\leq \aleph_1$ and for all torsion groups T. In addition, they proved that CH implies $\text{Bext}^3(G, T) = 0$ for all torsion-free groups G. These results have been extended to torsion-free groups of higher cardinalities in Fuchs [F2] provided that for some integer $n > 1$, $\aleph_n$ is assumed to be the continuum.

By induction on the index i it is easy to verify the following lemma:

LEMMA 10.1. [F2] *For all torsion-free groups G of cardinality $\leq \aleph_i$ $(i \geq 1)$ and for all torsion groups T, we have*

$$\text{Bext}^{i+k}(G, T) = 0 \quad \textit{for} \quad k = 1, 2, \dots . \ \square$$

This lemma can be used to derive the following conclusion which yields the above mentioned Albrecht-Hill result in the most important case $n = 1$.

THEOREM 10.2. [F2] *If $2^{\aleph_0} = \aleph_n$ for an integer $n \geq 1$, then*

$$\text{Bext}^{n+k+1}(G, T) = 0 \quad (k = 1, 2, \dots)$$

for all torsion-free groups G and torsion groups T. $\square$

It would of course be most desirable to know how sharp this theorem is. In this connection, we conjecture that if $2^{\aleph_0} = \aleph_n$ for an integer $n \geq 1$, then $\text{Bext}^n(G, T) \neq 0$ must hold for some torsion-free group G and some torsion group T. (This statement is certainly true for $n = 1$ and $n = 2$.) There are no results known similar to (10.2) if the continuum is equal to $\aleph_\alpha$ for some $\alpha \geq \omega$.

11. Subgroups of B_2-groups

We have come to the discussion of the important problem as to when a pure subgroup of a B_2-group is again a B_2-group.

The conditions considered in earlier publications as candidates to ensure that a pure subgroup inherits the B_2-property were in terms of generalized purity. For instance, Hill-Megibben [HM] pointed out that a completely decomposable pure subgroup of a torsion-free group G ought to be separative, though this property was not sufficient even if G itself was completely decomposable. Bican-Fuchs [BF2] have shown that if CH is assumed, then $\aleph_0$-prebalancedness is a necessary condition for a subgroup of a B_2-group to be again a B_2-group. Fortunately, it is possible to establish a necessary and sufficient condition - without additional set-theoretical hypotheses - provided that chains are taken into consideration from the subgroup up to the group itself.

The main result is the following theorem which gives a satisfactory answer to the subgroup problem.

THEOREM 11.1. [F2] *For a pure subgroup A of a B_2-group G the following are equivalent:*

(i) *A is a B_2-group;*
(ii) *there is an $\aleph_0$-prebalanced chain from A to G;*
(iii) *there is a continuous well-ordered ascending chain of B_2-subgroups from A to G with rank 1 factors.* $\square$

Since there is always an $\aleph_0$-prebalanced chain from an $\aleph_0$-prebalanced subgroup of index $\leq \aleph_1$ to the group itself, we see that in any B_2-group, a pure subgroup of index $\leq \aleph_1$ is a B_2-group exactly if it is an $\aleph_0$-prebalanced subgroup.

A noteworthy idea of Albrecht and Hill [AH] is the notion of absolutely separative groups, i.e. groups that are separative in every torsion-free group that contains them as pure subgroups. They show that a torsion-free group which has a so-called balanced cover for its countable subgroups is absolutely separative, and conclude that separative subgroups in direct sums of countable B_2-groups are absolutely separative.

The same idea, applied to $\aleph_0$-prebalancedness, unveils an interesting property of B_2-groups. We say that a torsion-free group A is *absolutely* $\aleph_0$-*prebalanced* if A is $\aleph_0$-prebalanced in every torsion-free group B in which it is contained as a pure subgroup. Clearly, all countable torsion-free groups are absolutely $\aleph_0$-prebalanced. We are interested in a more exciting example of absolutely $\aleph_0$-prebalanced groups:

THEOREM 11.2. [BF2], [R2] *B_2-groups are absolutely $\aleph_0$-prebalanced.* □

The next corollary is an analogue of Theorem 5 by Dugas-Rangaswamy [BR] which follows from the fact that a homogeneous B_2-group is completely decomposable:

COROLLARY 11.3. *A necessary and sufficient condition for a homogeneous pure subgroup A of a B_2-group B to be completely decomposable is that there be an $\aleph_0$-prebalanced chain from A to B.* □

It is an open question whether or not the results of this section are valid if B_2-groups are replaced by B_1-groups.

12. Unions of chains of B_2-groups

Several theorems in abelian group theory are concerned with unions of continuous chains of pure subgroups of free groups. The classical theorem by Pontryagin on countable torsion-free groups has been considerably extended by P. Hill [H1] by showing, inter alia, that the union is itself free whenever the chain is of countable length, no matter what the cardinality of the union is.

It is natural to raise a similar question for B_2-groups. It is not difficult to verify by using an Axiom-3 system of decent subgroups that the answer is analogous:

THEOREM 12.1. [FR3] *If a torsion-free group G is the union of a countable ascending chain $0 = G_0 < G_1 < \ldots < G_n < \ldots$ of pure subgroups, each of which is a B_2-group, then G itself is a B_2-group.* □

The same question for B_1-groups is still open.

A consequence of the last result fills a gap in the proof of Corollary 6.3 by Dugas-Hill-Rangaswamy [DHR].

COROLLARY 12.2. [FR3] *Let λ be a singular cardinal of cofinality ω, and G a torsion-free group of cardinality λ. If every subgroup of G whose cardinality is $< \lambda$ can be embedded in a pure B_2-subgroup of G whose cardinality is $< \lambda$, then G itself is a B_2-group.* □

It should be observed that for chains whose lengths are $> \omega$, (12.1) is no longer valid without additional hypotheses. We refer to Fuchs-Rangaswamy [FR3] for chains of length ω_1.

13. Torsion-free extensions of B_2-groups by p-groups

Turning from structural problems to other questions concerning B_2-groups, let us start mentioning the characterization of those p-groups by which all the torsion-free extensions of B_2-groups are again B_2-groups (Fuchs-Rangaswamy [FR2]).

This problem has its origin in a question investigated for free groups by several authors, most extensively by Dugas and Irwin [DI]. Let G be a torsion-free group, and B a subgroup of G such that $G/B = T$ is a p-group. The problem consists in characterizing those p-groups T for which G is necessarily free whenever B is free. Dugas and Irwin succeeded in describing how this class of p-groups can be built up from groups of smaller cardinalities. It is important to know that the countable p-groups in this class are precisely the reduced ones.

It turns out that a more natural setting for this question is the class of B_2-groups rather than free groups. In fact, the class of p-groups T remains unchanged even if the question is widened from free groups to B_2-groups.

Following Dugas-Irwin [DI], we define a class $\mathcal{G}_p$ of p-groups recursively. Let a countable p-group T belong to $\mathcal{G}_p$ if and only if it is reduced. A p-group T of regular cardinality κ should belong to $\mathcal{G}_p$ if and only if there is a continuous well-ordered ascending chain

$$0 = T_0 < T_1 < \ldots < T_\alpha < T_{\alpha+1} < \ldots \quad (\alpha < \kappa)$$

of subgroups of T such that

(i) $T = \bigcup_{\alpha<\kappa} T_\alpha$;

(ii) for each α, T_α has cardinality $< \kappa$;

(iii) T_β/T_α belongs to $\mathcal{G}_p$ for all $\alpha < \beta < \kappa$.

Finally, a p-group of singular cardinality κ belongs to $\mathcal{G}_p$ if and only if each of its subgroups of cardinality $< \kappa$ belongs to $\mathcal{G}_p$.

From the definition it is readily derived that the class $\mathcal{G}_p$ is closed under the following operations: extensions, arbitrary direct sums, and passage to subgroups. One can also show without much difficulty that the p-groups in $\mathcal{G}_p$ have to be fully starred in the sense of John Irwin, i.e. the basic subgroups in every pure subgroup must have the same cardinality as the subgroup itself.

The next theorem collects the pertinent facts concerning the class in question.

THEOREM 13.1. *For a p-group T, the following properties are equivalent:*

(i) *if G is a torsion-free group and if $G/B \cong T$ for a B_2-subgroup B of G, then G is again a B_2-group;*

(ii) *if G is a torsion-free group and $G/B \cong T$ for a free subgroup B of G, then G is free;*

(iii) *T is a member of $\mathcal{G}_p$.* □

The equivalence of (ii) and (iii) was proved by Dugas-Irwin [DI], while that of (i) and (ii) is due to Fuchs-Rangaswamy [FR2].

It turns out that it is considerably easier to solve the dual problem. In this case, the result reads as follows:

THEOREM 13.2. [FR2] *For a p-group T, the following are equivalent:*

(i) *whenever G is a B_2-group and $G/B \cong T$ for a subgroup B of G, then B is again a B_2-group;*

(ii) *T is reduced.* □

Note that this result gives a sufficient condition under which an essential subgroup of a B_2-group is again such a group.

It should be observed that the last two results generalize immediately to torsion groups with a finite number of p-components, but fail for torsion groups in general - as is shown by easy examples.

14. Indecomposable and superdecomposable B_2-groups

The question of large indecomposable B_2-groups was raised by Fuchs and Metelli [FM1]. Using a result by A.L.S. Corner on fully rigid systems, they show:

THEOREM 14.1. *For every infinite cardinal λ, there is a fully rigid system consisting of 2^λ (pairwise non-quasi-isomorphic) B_2-groups of cardinality λ each of which has endomorphism ring isomorphic to $\mathbb{Z}$.* □

The construction shows that the groups can be chosen to have their typeset contained in a lattice generated by 6 elements. Recently, Arnold and Dugas [AD] succeeded in sharpening this result to 3-generators; they use free groups with distinguished subgroups to derive their result.

The dual problem had been considered by Dugas-Thomé [DT1] and Fuchs-Metelli [FM1] by using different approaches. Recall that a *superdecomposable* group is defined to be a group which has no indecomposable summand $\neq 0$.

THEOREM 14.2. [DT1], [FM1] *There exist superdecomposable Butler groups of countable rank.* □

The mentioned results are superseded by a recent paper [DG] of Dugas-Göbel. They prove e.g.

THEOREM 14.3. [DG] *If the additive group of a ring R is a Butler group which is p-reduced for at least three primes p, then R is the endomorphism ring of a Butler group.* □

In addition, a topological version of this theorem is established.

15. Torsion-free groups by which balanced extensions of a rational group split

Bican-Fuchs [BF1] investigate torsion-free groups A - which they call *R-groups* - such that

$$\mathrm{Bext}^1(A, R) = 0.$$

Here R denotes a fixed subgroup of $\mathbb{Q}$, say of type $\mathbf{t}$. Evidently, all completely decomposable groups are R-groups, but in general there are many more. The most interesting case is the one in which all the non-zero elements of A are of types $\leq \mathbf{t}$; indeed, this case proves to be intimately connected with the theory of Butler groups - a phenomenon for which so far no explanation has been found.

The proofs do not unveil any hidden connection between this problem and Butler groups. The study of R-groups of the mentioned kind bear close resemblance to Whitehead groups. Actually, this is not at all surprising, since the special case $R = \mathbb{Z}$ yields precisely the Whitehead groups.

Let R_0 denote the localization of $\mathbb{Z}$ at the collection of primes p for which $pR \neq R$, and set $\breve{A} = A \otimes R_0$. Thus $\breve{A}$ is the smallest group that contains A and satisfies $p\breve{A} = \breve{A}$ for primes p with $pR = R$; its type is still $\leq \mathbf{t}$.

One can show that a countable group A with elements of types $\leq \mathbf{t}$ is an R-group if and only if $\check{A} = A \otimes R_0$ is a Butler group. For uncountable groups of cardinality $\leq \aleph_\omega$, the following result was established by Bican-Fuchs [BF1]; the cardinality restriction in [BF1] can be removed in view of [FMa].

THEOREM 15.1. [BF1] *Assume* V = L, *and suppose that A is a torsion-free group whose nonzero elements have types $\leq \mathbf{t}$. A is an R-group if and only if $\check{A}$ is a B_2-group.* □

This is, however, not a theorem in ZFC. In fact, if we assume that in our model of ZFC, Shelah's Proper Forcing Axiom (PFA) is valid, then a counterexample to (15.1) can be found even among the $\mathbf{t}$-homogeneous R-groups:

THEOREM 15.2. [BF1] *Assuming* PFA, *there exist $\mathbf{t}$-homogeneous R-groups of any cardinality $\geq \aleph_1$ which are not completely decomposable (and hence they are not B_2-groups).* □

References

[AH] U. Albrecht and P. Hill, *Butler groups of infinite rank and Axiom 3*, Czech. Math. J. **37** (1987), 293-309.

[AD] D.M. Arnold and M. Dugas, *Butler groups with finite typesets and free groups with distinguished subgroups*, Comm. Algebra **21** (1993), 1947-1982.

[AV] D.M. Arnold and C. Vinsonhaler, *Finite rank Butler groups: a survey of recent results*, Abelian Groups, Lecture Notes in Pure Appl. Math. **146** (Marcel Dekker, 1993), 17-41.

[B1] L. Bican, *Splitting in abelian groups*, Czech. Math. J. **28** (1978), 356-364.

[B2] ———————, *Purely finitely generated abelian groups*, Comment. Math. Univ. Carolin. **21** (1980), 209-218.

[BF1] L. Bican and L. Fuchs, *On abelian groups by which balanced extensions of a rational group split*, J. Pure Appl. Algebra **78** (1992), 221-238.

[BF2] ———————, *Subgroups of Butler groups*, Comm. Algebra (to appear).

[BS] L. Bican and L. Salce, *Butler groups of infinite rank*, Abelian Group Theory, Lecture Notes in Math. **1006** (Springer, 1983), 171-189.

[Bu] M.C.R. Butler, *A class of torsion-free abelian groups of finite rank*, Proc. London Math. Soc. **15** (1965), 680-698.

[DG] M. Dugas and R. Göbel, *Every ring with sufficiently reduced additive Butler group is an endomorphism ring of a Butler group* (to appear).

[DHR] M. Dugas, P. Hill and K.M. Rangaswamy, *Infinite rank Butler groups, II*, Trans. Amer. Math. Soc. **320** (1990), 643-664.

[DI] M. Dugas and J. Irwin, *On a class of abelian p-groups*, Forum Math. **4** (1992), 147-158.

[DR] M. Dugas and K.M. Rangaswamy, *Infinite rank Butler groups*, Trans. Amer. Math. Soc. **305** (1988), 129-142.

[DT1] M. Dugas and B. Thomé, *Countable Butler groups and vector spaces with four distinguished subspaces*, J. Algebra **138** (1991), 249-272.

[DT2] ———————, *The functor Bext under the negation of CH*, Forum Math. **3** (1991), 23-33.

[F1] L. Fuchs, *"Infinite Abelian Groups"*, vol. **2**, Academic Press, New York, 1973.

[F2] ———————, *Butler groups of infinite rank*, J. Pure Appl. Algebra (to appear).

[FMa] L. Fuchs and M. Magidor, *Butler groups of arbitrary cardinality*, Israel J. Math. **84** (1993), 239-263.

[FM1] L. Fuchs and C. Metelli, *Indecomposable Butler groups of large cardinalities*, Arch. Math. **57** (1991), 339-344.

[FM2] ———————, *Countable Butler groups*, Contemporary Math. **130** (1992), 133-143.

[FMR] L. Fuchs, C. Metelli and K.M. Rangaswamy, *Corank one subgroups of completely decomposable abelian groups*, Comm. Algebra (to appear).

[FMM] L. Fuchs and E. Monari-Martinez, *Butler modules over valuation domains*, Can. J. Math. **43** (1991), 48-60.

[FR1] L. Fuchs and K.M. Rangaswamy, *Butler groups that are unions of subgroups with countable typesets*, Arch. Math. **61** (1993), 105-110.

[FR2] ——————, *Torsion-free extensions of Butler groups by primary groups* (to appear).

[FR3] ——————, *Unions of chains of Butler groups* (to appear).

[FV] L. Fuchs and G. Viljoen, *Note on the extensions of Butler groups*, Bull. Austral. Math. Soc. **41** (1990), 117-122.

[H1] P. Hill, *New criteria for freeness in abelian groups*, Trans. Amer. Math. Soc. **182** (1973), 201-209 and **196** (1974), 191-202.

[H2] ——————, *The third axiom of countability for abelian groups*, Proc. Amer. Math. Soc. **82** (1981), 347-350.

[HM] P. Hill and C. Megibben, *Torsion-free groups*, Trans. Amer. Math. Soc. **295** (1986), 735-751.

[J] T. Jech, *"Set Theory"*, Academic Press, New York, 1973.

[Je] R.B. Jensen, *The fine structure of the constructible universe*, Annals of Math. Logic **4** (1972), 229-308.

[MV] R. Mines and C. Vinsonhaler, *Butler groups and Bext: A constructive view*, Contemporary Math. 130 (1992), 289-299.

[P] L. Procházka, *Über die Spaltbarkeit der Faktorgruppen torsionsfreier abelscher Gruppen endlichen Ranges*, Czech. Math. J. **11** (1961), 521-557.

[R1] K.M. Rangaswamy, *A homological characterization of Butler groups*, Proc. Amer. Math. Soc. (to appear).

[R2] ——————, *A property of B_2-groups*, Comment. Math. Univ. Carolinae (to appear).

[Ri] F. Richman, *Butler groups, valuated vector spaces and duality*, Rend. Sem. Mat. Univ. Padova **72** (1984), 13-19.

Current address: Department of Mathematics, Tulane University, New Orleans, Louisiana 70118
E-mail address: fuchs@mailhost.tcs.tulane.edu

Contemporary Mathematics
Volume **171**, 1994

Unions of Chains of Butler Groups

L. FUCHS AND K.M. RANGASWAMY

0. Introduction

A torsion-free group G is called a B_2-*group* (a kind of Butler group) if, for some ordinal τ, there is a continuous well-ordered ascending chain of pure subgroups,

$$0 = A_0 < A_1 < \ldots < A_\alpha < \ldots < A_\tau = G = \cup A_\alpha$$

with rank 1 factors (or, equivalently, with finite rank factors) such that, for each $\alpha < \tau$, $A_{\alpha+1} = A_\alpha + B_\alpha$ holds for some finite rank Butler group B_α. Recall that a finite rank Butler group is by definition a pure subgroup of a completely decomposable torsion-free group of finite rank [B].

Suppose that $0 = G_0 < G_1 < \ldots < G_n < \ldots$ is a countable sequence of B_2-groups, each pure in the next one. We emphasize that there is no hypothesis either on the sizes of the groups G_n nor on the embeddings of G_n in G_{n+1} other than purity. Our question is whether or not the union of this chain is again a B_2-group.

This is an analogue of P. Hill's result on free abelian groups [H], replacing freeness by the property of being a B_2-group. An affirmative answer to our question is needed in order to close the gap in the proof of Corollary 6.3 in the paper Dugas-Hill-Rangaswamy [DHR]. We shall give an affirmative answer, and discuss briefly the case of uncountable chains.

Let us recall some additional definitions. A pure subgroup A of a torsion-free group G is called *decent* in G if for all finite rank pure subgroups C/A of G/A we have $C = A + B$ for a suitable finite rank Butler group B; see Albrecht-Hill [AH]. Note that decency is transitive, and a torsion-free group G is a B_2-group whenever it contains a decent subgroup A such that both A and G/A are B_2-groups. A (necessarily pure) subgroup A of G is a TEP-*subgroup* (i.e. has the Torsion Extension Property) if every homomorphism of A into a torsion group T extends to a homomorphism of G into T; see Dugas-Rangaswamy [DR].

1991 *Mathematics Subject Classification*: 20K20.

This paper is in final form, and no version of it will be submitted for publication elsewhere.

A family $\mathcal{C}$ of subgroups of G is called an *Axiom-3 family* if it satisfies: (1) $0, G \in \mathcal{C}$; (2) for any collection $\{H_i \mid i \in I\}$ of subgroups in $\mathcal{C}$, their union $\Sigma\{H_i \mid i \in I\}$ also belongs to $\mathcal{C}$; (3) if $H \in \mathcal{C}$ and if X is a countable subset of G, then there is a $K \in \mathcal{C}$ containing both H and X such that K/H is countable. The family $\mathcal{C}$ is a *$G(\aleph_0)$-family* if a weaker version of (2) holds, viz. only the unions of chains $\{H_i \mid i \in I\}$ of subgroups in $\mathcal{C}$ are required to belong to $\mathcal{C}$. It is an easy consequence that the intersection of two Axiom-3 families (two $G(\aleph_0)$-families) is again an Axiom-3 family (a $G(\aleph_0)$-family).

Recall (see Albrecht-Hill [AH], Fuchs-Magidor [FM]) that a group G is a B_2-group if and only if it is a B_3-group in the sense that it contains an Axiom-3 family $\mathcal{B}$ of decent subgroups. Actually, it suffices for G to admit a $G(\aleph_0)$-family of decent subgroups in order to be a B_2-group. By [DHR, Corollary 3.5] a B_2-group has an Axiom-3 family of TEP subgroups, so by letting $\mathcal{B}$ be the intersection of the two families in G, we may as well assume that the subgroups in a $G(\aleph_0)$-family $\mathcal{B}$ of G are both decent and TEP in G. In particular, this guarantees that all groups in $\mathcal{B}$ are B_2-groups and, whenever the subgroups $C < C'$ belong to $\mathcal{B}$, then C'/C is likewise a B_2-group (cf. Rangaswamy [R, Theorem 4]).

1. Countable chains

Our immediate purpose is to prove the following theorem.

THEOREM 1.1. *If a torsion-free group G is the union of a countable ascending chain $0 = G_0 < G_1 < \ldots < G_n < \ldots$ of pure subgroups, each of which is a B_2-group, then G itself is a B_2-group.*

We require several lemmas. We fix a $G(\aleph_0)$-family $\mathcal{B}_n$ of decent TEP-subgroups in each G_n (to be slightly modified below); these will play a relevant role in our arguments.

LEMMA 1.2. *If $\mathcal{B}_n$ is defined as stated, then the collection*

$$\mathcal{B}'_n = \{A \in \mathcal{B}_n \mid A + G_i \text{ is pure in } G \text{ for each } i < \omega\}$$

is a $G(\aleph_0)$-family.

PROOF: All what we have to check is that the countability condition (3) is satisfied, since (1) and the weaker version of (2) are obvious. Thus let $A_0 \in \mathcal{B}'_n$ and X_0 a countable subgroup of G_n. Suppose that we already have a chain $A_0 < A_1 < \ldots < A_m$ of subgroups in $\mathcal{B}_n$ such that 1) $A_0 + X_0 \leq A_1$; 2) A_{j+1}/A_j is countable for all $j < m$, and in addition, 3) for each $j < m$ and for each $i < \omega, (A_{j+1} + G_i)/(A_0 + G_i)$ contains the purification of $(A_j + G_i)/(A_0 + G_i)$ in $G/(A_0 + G_i)$. To find a next member $A_{m+1} \in \mathcal{B}_n$ of the chain, let $Y_i \subset G_i$ be a complete set of representatives of the purification of $(A_m + G_i)/(A_0 + G_i)$ in $G/(A_0 + G_i)$ modulo $(A_m + G_i)/(A_0 + G_i)$, for each $i < \omega$; these Y_i are countable sets, so $X_{m+1} = \bigcup_{i<\omega} Y_i$ is likewise countable. Therefore, there is an $A_{m+1} \in \mathcal{B}_n$ such that $A_m + X_{m+1} \leq A_{m+1}$ and A_{m+1}/A_m is countable. Then, for each $i < \omega, (A_{m+1} + G_i)/(A_0 + G_i)$ contains the purification of $(A_m + G_i)/(A_0 + G_i)$ in $G/(A_0 + G_i)$. The union A of the chain of the A_m for all $m < \omega$ is a member of $\mathcal{B}_n, A/A_0$ is evidently countable, and our construction guarantees that $(A + G_i)/(A_0 + G_i)$ is pure in $G/(A_0 + G_i)$. Thus $A + G_i$ is pure in G for all i, i.e. $A \in \mathcal{B}'_n$. □

The next step is to build a suitable $G(\aleph_0)$-family for G.

LEMMA 1.3. *The family*

$$\mathcal{B} = \{A \leq G \mid A \cap G_n \in \mathcal{B}'_n \text{ for each } n < \omega\}$$

is a $G(\aleph_0)$-family of subgroups in G.

PROOF: Again, only the countability condition requires a proof. Since there are but countably many indices n to deal with, a usual back-and-forth argument (ω times) suffices to argue that for each $A \in \mathcal{B}$ and for each countable subset X of G, there exists an $A' \in \mathcal{B}$ which contains both A and X such that A'/A is countable. □

To summarize, we have $G(\aleph_0)$-families $\mathcal{B}'_n$ of decent TEP-subgroups in G_n, and a $G(\aleph_0)$-family $\mathcal{B}$ in G satisfying the condition in Lemma 1.3. We will replace $\mathcal{B}'_n$ by the subfamily $\mathcal{B}''_n = \{A \cap G_n \mid A \in \mathcal{B}\}$ which is evidently again a $G(\aleph_0)$-family in G_n. To simplify notation, the new $\mathcal{B}''_n$ will be denoted by $\mathcal{B}_n$.

Before completing the proof of Theorem 1.1, we verify:

LEMMA 1.4. *The subgroups in $\mathcal{B}$ are decent in G.*

PROOF: Let $C \in \mathcal{B}$, and $D > C$ a pure subgroup of G such that D/C is of finite rank. If $F = \{d_1, \dots, d_n\}$ is a maximal independent set in D mod C, then there is an index k such that $F \subset G_k$. Since $C+G_k$ is pure in G, $C+(D\cap G_k) = D\cap(C+G_k)$ is a pure subgroup between C and D. It obviously contains F, so it must be equal to D, i.e. $C + (D \cap G_k) = D$.

By virtue of the construction of $\mathcal{B}$, $C\cap G_k \in \mathcal{B}_k$, thus $C\cap G_k$ is a decent subgroup in G_k. Hence we have $D \cap G_k = (C \cap G_k) + B$ for a suitable finite rank Butler group B. We obtain $D = C + (D \cap G_k) = C + (C \cap G_k) + B = C + B$, proving the assertion. □

We now combine Lemmas 1.3 and 1.4 to conclude that G admits a $G(\aleph_0)$-family of decent subgroups, hence it is a B_2-group. This completes the proof of Theorem 1.1. □

As pointed out in the introduction, our theorem generalizes P. Hill's result on countable unions of free groups [H]. Hill used this theorem to derive that a group of cardinality $\aleph_\omega$ was free whenever all of its subgroups of smaller cardinalities were free. A similar conclusion can be drawn in our case:

COROLLARY 1.5. *Let λ be a singular cardinal of cofinality ω, and G a torsion-free group of cardinality λ. If every subgroup of G whose cardinality is $< \lambda$ can be embedded in a pure B_2-subgroup of cardinality $< \lambda$, then G itself is a B_2-group.* □

Dugas-Hill-Rangaswamy [DHR] utilize a version of Shelah's Singular Compactness Theorem in order to derive that a torsion-free group G of cardinality $\aleph_\omega$ is a B_2-group whenever it admits what they call an $\aleph_\omega$-family of B_2-subgroups. Their proof contains a gap as they tacitly assume that the union of a countable chain of pure B_2-subgroups is again a B_2-group: now this gap is filled by our theorem above. (By the way, their proof works for all singular cardinals.)

2. Uncountable chains

The union of uncountable chains of countable B_2-groups need not be a B_2-group as is shown by Example 3.1 infra. If, however, the embeddings into the next members of the chain are more restrictive and the sizes of the groups are limited to $\aleph_1$, then the conclusion of Theorem 1.1 can be established.

First, we require a simple lemma.

LEMMA 2.1. *Let A and S be pure subgroups of a torsion-free group G. If $A \cap S$ is decent in S, then A is decent in $A + S$.*

PROOF: To check the decency of A in $A+S$, let X be a finite subset of $A+S$. Without loss of generality, $X \subset S$ may be assumed. There is a finite rank Butler group B in S such that $(A \cap S) + B$ is pure in S and contains X. It remains to show that $A+B$ is pure in $A+S$. If for some integer $n > 0$, $n(a+s) = a' + b$ $(a, a' \in A, b \in B, s \in S)$, then $ns - b = a' - na \in A \cap S$, thus $ns \in (A \cap S) + B$. By purity, $s \in (A \cap S) + B$ whence $a + s \in A + B$. □

We can now verify:

THEOREM 2.2. *Let G be the union of a continuous well-ordered ascending chain of decent subgroups,*

$$0 = G_0 < G_1 < \ldots < G_\alpha < \ldots \qquad (\alpha < \omega_1)$$

where each G_α is a B_2-group of cardinality $\leq \aleph_1$. Then G is a B_2-group.

PROOF: For each $\alpha < \omega_1$, let $\mathcal{B}_\alpha$ be an Axiom-3 family of decent subgroups of G_α. Since α is countable, by Lemma 1.2 we may assume that the $\mathcal{B}_\alpha$ are $G(\aleph_0)$-families in G_α such that for each $X \in \mathcal{B}_\alpha$, the subgroup $X + G_\beta$ is pure in G for every $\beta < \alpha$. Write $G = \{g_0, g_1, \ldots, g_\alpha, \ldots \mid \alpha < \omega_1\}$.

Suppose that up to some countable ordinal τ we have constructed a chain of subgroups

$$0 = A_0 < A_1 < \ldots < A_\alpha < \ldots \qquad (\alpha < \tau)$$

with the following properties:

(i) A_α is a countable pure subgroup of G_α for each $\alpha < \tau$;

(ii) $A_{\alpha+1} \in \mathcal{B}_{\alpha+1}$ for all $\alpha + 1 < \tau$;

(iii) if $\beta < \tau$ is a limit ordinal, then $A_\beta = \bigcup_{\alpha<\beta} A_\alpha$;

(iv) for all $\gamma < \alpha < \tau$, $A_\alpha \cap G_\gamma \in \mathcal{B}_\gamma$;

(v) if $g_\gamma \in G_\alpha$ for $\gamma < \alpha + 1 < \tau$, then $g_\gamma \in A_{\alpha+1}$.

If τ happens to be a limit ordinal, then we define $A_\tau = \bigcup_{\alpha<\tau} A_\alpha$. This gives a countable pure subgroup of G_τ such that the chain of the A_α $(\alpha < \tau + 1)$ satisfies conditions (i)-(v) with τ replaced by $\tau + 1$. Of course, (ii) and (v) are now trivial.

If $\tau = \sigma + 1$ for some ordinal σ, then let B be the pure subgroup of $G_{\sigma+1}$ generated by A_σ and by all $g_\gamma \in G_{\sigma+1}$ with $\gamma \leq \sigma$. B is countable, so there is a countable subgroup $C \in \mathcal{B}_{\sigma+1}$ that contains B. We can not set $A_\tau = C$ right away, since this C need not satisfy $C \cap G_\gamma \in \mathcal{B}_\gamma$ for all $\gamma \leq \sigma$. Therefore, we proceed to choose inductively an ascending chain of subgroups $C = C_0 < C_1 < \ldots < C_n < \ldots$ $(n < \omega)$ in $\mathcal{B}_{\sigma+1}$ along with subgroups $S_{\gamma n} \in \mathcal{B}_\gamma$ for each $\gamma \leq \sigma$ and $n < \omega$ such that

$$C_n \cap G_\gamma \leq S_{\gamma n} \leq C_{n+1} \cap G_\gamma.$$

If we set $A_\tau = \bigcup_{n<\omega} C_n$, then conditions (i)-(v) will hold for $\tau + 1$ in place of τ. In this way, we obtain a chain $0 = A_0 < A_1 < \ldots < A_\alpha < \ldots (\alpha < \omega_1)$ with properties (i)-(v) for $\tau = \omega_1$.

Condition (v) guarantees that the union of the chain of the A_α $(\alpha < \omega_1)$ equals G. For a successor ordinal α, (ii) implies that A_α is a B_2-group, whence by Theorem 1.1 the same is true for all limit ordinals $\alpha < \omega_1$. Again in view of (ii), for successor ordinals α, A_α is decent in G_α, so also in G. For a limit ordinal $\alpha < \omega_1$, $A_\alpha \cap G_\gamma$ is decent in G_γ, and so by Lemma 2.1 A_α is decent in $A_\alpha + G_\gamma$. Therefore, A_α is decent in $G_\alpha = \bigcup_{\gamma<\alpha}(A_\alpha + G_\gamma)$, and so in G. We conclude that for each $\alpha < \omega_1, A_\alpha$ is a decent B_2-subgroup of G, and hence of the countable B_2-group $A_{\alpha+1}$. Consequently, there is a countable chain $A_\alpha < A_\alpha + B_{\alpha 1} < \ldots < A_\alpha + B_{\alpha n} < \ldots$ of pure subgroups with union $A_{\alpha+1}$ where the $B_{\alpha n}$ are finite rank Butler groups. It follows that $G = \bigcup_{\alpha<\omega_1} A_\alpha$ itself is a B_2-group. □

It is worth while pointing out that, in the preceding theorem, the size of G_1 is irrelevant; what really matters is that G_1 is of index $\leq \aleph_1$ in G. In fact, we can factor out from G_1 a decent TEP subgroup H of index $\leq \aleph_1$, and apply the theorem to the chain of the G_α/H to conclude that G/H is a B_2-group. Then G itself will be a B_2-group.

If we wish to avoid putting any restriction on the lengths of the chains of the B_2-groups G_α, then we can verify that the union G is likewise a B_2-group provided we are willing to pay a price by imposing the TEP property on the G_α. In fact, we have:

THEOREM 2.3. *The group G is a B_2-group whenever it is the union of a continuous well-ordered ascending chain of TEP subgroups,*

$$0 = G_0 < G_1 < \ldots < G_\alpha < \ldots \qquad (\alpha < \tau)$$

where each G_α is a B_2-group and τ is any ordinal.

PROOF: By [R], if a B_2-group G_α is TEP in a B_2-group $G_{\alpha+1}$, then $G_{\alpha+1}/G_\alpha$ is again a B_2-group. Because of [DHR, Proposition 3.2] G_α is decent in $G_{\alpha+1}$. Hence we have a chain of decent subgroups whose factors are B_2-groups; therefore, the union G is a B_2-group again [DHR, Corollary 3.10]. □

It should be added that the condition of having TEP subgroups in the preceding theorem is not as strong as one might think at the first glance. As a matter of fact, in lots of cases the union G of the chain of the G_α can be a B_2-group only if there is a subchain consisting of TEP subgroups. If $0 = G_0 < G_1 < \ldots < G_\alpha < \ldots (\alpha < \kappa)$ is a filtration of the torsion-free group G with pure B_2-subgroups of cardinality $< \kappa$ for an uncountable regular cardinal κ, then [DHR, Theorem 7.1] shows that G a B_2-group implies that there is a cub C in κ such that G_α is TEP in G for all $\alpha \in C$.

Let us point out that Griffith's theorem that Baer groups are free [G] can easily be derived from the theorem by Rangaswamy [R] mentioned in the proof of the preceding theorem. For, let B be a Baer group and F a free group such that $B \cong F/H$ for some subgoup H of F. Then the standard exact sequence for Hom-Ext shows that H is a TEP-subgroup of F, and so by [R] B is a B_2-group. As it is easily checked, finite rank subgroups of Baer groups are free, so H is decent in F. Thus B is a homogeneous B_2-group of type $\mathbb{Z}$ whence the freeness of B follows at once.

3. Examples

We exhibit two examples in order to show that the conditions on the lengths of the chains in our theorems are relevant.

EXAMPLE 3.1. There exists a group A that is the union of a continuous well-ordered ascending chain $0 = A_0 < A_1 < \ldots < A_\alpha < \ldots (\alpha < \omega_1)$ of countable pure free subgroups A_α such that, for all ordinals $\alpha + 1 < \beta < \omega_1$, the factor groups $A_\beta/A_{\alpha+1}$ are free, but the group A itself is not free. It is not difficult to construct such a group e.g. with $A_{\alpha+1}/A_\alpha \cong \mathbb{Q}$ for limit ordinals $\alpha < \omega_1$. In this example, all the subgroups A_α are B_2-groups, but their union A is not: otherwise A would be free as a homogeneous B_2-group.

EXAMPLE 3.2. We can construct a torsion-free group G as the union of a continuous well-ordered ascending chain $0 = G_0 < G_1 < \ldots < G_\alpha < \ldots (\alpha < \omega_2)$ of subgroups G_α such that

(a) G_α $(\alpha < \omega_2)$ is a free subgroup of cardinality $\leq \aleph_1$, decent in G;

(b) for all ordinals $\alpha + 1 < \beta < \omega_2$, the factor groups $G_\beta/G_{\alpha+1}$ are free;

(c) for limit ordinals $\alpha < \omega_2$, the factor groups $G_{\alpha+1}/G_\alpha \cong$ the group A of Example 3.1.

Now, for every ordinal $\alpha < \omega_2$, the subgroup G_α is decent in $G_{\alpha+1}$, so Theorems 1.1 and 2.2 ensure that all the G_α are B_2-groups. However, the group G is not a B_2-group, as it is homogeneous but not free. It is not free, indeed, since the set $E = \{\alpha < \omega_2 \mid G_{\alpha+1}/G_\alpha \text{ is not free}\}$ is stationary in ω_2.

References

[AH] U. Albrecht and P. Hill, *Butler groups of infinite rank and Axiom 3*, Czech. Math. J. **37** (1987), 293-309.

[B] M.C.R. Butler, *A class of torsion-free abelian groups of finite rank*, Proc. London Math. Soc. **15** (1965), 680-698.

[DHR] M. Dugas, P. Hill and K.M. Rangaswamy, *Infinite rank Butler groups, II*, Trans. Amer. Math. Soc. **320** (1990), 643-664.

[DR] M. Dugas and K.M. Rangaswamy, *Infinite rank Butler groups*, Trans. Amer. Math. Soc. **305** (1988), 129-142.

[F] L. Fuchs, *"Infinite Abelian Groups"*, vol. 2, Academic Press, New York, 1973.

[FM] L. Fuchs and M. Magidor, *Butler groups of arbitrary cardinality*, Israel J. Math. **84** (1993), 239-263.

[G] P. Griffith, *A solution of the splitting mixed group problem of Baer*, Trans. Amer. Math. Soc. **139** (1969), 261-269.

[H] P. Hill, *New criteria for freeness in abelian groups*, Trans. Amer. Math. Soc. **182** (1973), 201-209 and **196** (1974), 191-202.

[R] K.M. Rangaswamy, *A homological characterization of abelian B_2-groups*, to appear.

Current address: Department of Mathematics, Tulane University, New Orleans, Louisiana 70118, USA

Department of Mathematics, University of Colorado, Colorado Springs, Colorado 80933, USA

Contemporary Mathematics
Volume **171**, 1994

Some torsion–free groups arising in measure theory

– R. G. Göbel
– R. M. Shortt

Abstract: A class of torsion-free groups arising in connection with joint extensions of measures is considered. It is shown that any direct limit of a sequence of free groups H_n belongs to the class, so long as the connecting maps $H_{n-1} \longrightarrow H_n$ are represented by row-finite matrices.

§0. Introduction

Let $\mathcal{A}$ be a field (Boolean algebra) of subsets of a non-empty set X and let G be an Abelian group (all groups considered here are Abelian and are written additively). A function $\mu : \mathcal{A} \longrightarrow G$ is a *charge* (finitely additive measure) if $\mu(A_1 \cup A_2) = \mu(A_1) + \mu(A_2)$ for all pairs of disjoint sets A_1, A_2 in $\mathcal{A}$.

Let $\mathcal{A}$ and $\mathcal{B}$ be fields of subsets of a set X and let $\mu : \mathcal{A} \longrightarrow G$ and $\nu : \mathcal{B} \longrightarrow G$ be charges. Then μ and ν are said to be *consistent* if $\mu(C) = \nu(C)$ for all $C \in \mathcal{A} \cap \mathcal{B}$. In the papers [BhS] and [RaR] the question of whether consistent charges admit of a common extension was considered and largely answered: by a common extension we mean a charge $\rho : \mathcal{A} \vee \mathcal{B} \longrightarrow G$ such that $\rho(A) = \mu(A)$ for all $A \in \mathcal{A}$ and $\rho(B) = \nu(B)$ for all $B \in \mathcal{B}$; here, $\mathcal{A} \vee \mathcal{B}$ is the field generated by $\mathcal{A} \cup \mathcal{B}$.

If $\mathcal{A}$ is a field on a set X, we denote by $S(X; \mathcal{A})$ the group of all functions $f : X \longrightarrow \mathbb{Z}$ such that range(f) is finite and such that $f^{-1}(n) \in \mathcal{A}$ for all $n \in \mathbb{Z}$. Then $S(X; \mathcal{A})$ is a "Specker group" [Fu] and so, by a theorem of Nöbeling, is free Abelian. If $\mathcal{A}$ and $\mathcal{B}$ are fields on X, we define

$$H(\mathcal{A}, \mathcal{B}) = \frac{S(X; \mathcal{A} \vee \mathcal{B})}{S(X; \mathcal{A}) + S(X; \mathcal{B})}.$$

Every such group is torsion-free, and the following results have been established in [BhS], [GoS] and [RaR].

1991 Mathematics Subject Classification: Primary 28B10, Secondary 20K20, 20K15

0271-4132/94 $1.00 + $.25 per page

0.1 Theorem: Let $\mathcal{A}$ and $\mathcal{B}$ be fields on a set X and let G be an Abelian group. Then every consistent pair of charges $\mu : \mathcal{A} \longrightarrow G$ and $\nu : \mathcal{B} \longrightarrow G$ has a common extension if and only if $\operatorname{Ext}(H(\mathcal{A}, \mathcal{B}), G) = 0$.

0.2 Corollary: Let G be an Abelian group. Every pair of consistent G-valued charges (on arbitrary fields $\mathcal{A}$ and $\mathcal{B}$) has a common extension if and only if G is a cotorsion group.

To prove the Corollary. it is necessary only to exhibit fields $\mathcal{A}$ and $\mathcal{B}$ such that $H(\mathcal{A}, \mathcal{B})$ contains a copy of $\mathbb{Q}$. In fact, in [GoS], it was shown that every rational group can be realised as some $H(\mathcal{A}, \mathcal{B})$. In this paper, we pursue the question of describing the class of all groups $H(\mathcal{A}, \mathcal{B})$, and it will be shown that all countable torsion-free groups are in this class.

§1. Direct limits

If H is a torsion-free Abelian group, then H can be viewed through the inclusions $F \subseteq H \subseteq F \otimes \mathbb{Q} = V$, where F is a free group $F = \bigoplus_{\alpha<\kappa} e_\alpha \mathbb{Z}$ and V is a vector space over $\mathbb{Q}$. This gives a representation of $H = \bigcup_{n=1}^{\infty} F_n$ as a union of free groups of rank κ, where $F_n = \frac{1}{n!} F \cap H$. The embedding $F_n \subseteq F_{n+1}$ and the choice of free bases give rise to an integer valued embedding matrix M_n having only finitely many non-zero entries in each column (we take images from the left). If the matrix M_n also has only finitely many non-zero entries in each row, then we say that M_n is *row-finite*. This is clearly the case when G has finite rank m, so that each M_n is an $m \times m$ matrix. If H has countable rank, then a small adjustment to this argument applies: we can write $H = \bigcup_{n=1}^{\infty} F_n$, where each F_n has rank $\leq n$. We have the easy

1.1 Observation: If H is countable and torsion-free, then H is a union $H = \bigcup F_n$ of finite rank, free groups F_n, where the embeddings are given by finite (in particular row-finite) matrices M_n.

One way to view a group H as such a union of free groups is to employ the language of direct limits. If $H = \bigcup_\alpha H_\alpha$ is a union of free groups, H_α indexed by an upwardly-directed set of indices α, and $H_\alpha \subseteq H_\beta$ whenever $\alpha \leq \beta$, then $H = \varinjlim H_\alpha$. We are particularly interested in the case where the indices α form a sequence: let $H_0, H_1, \ldots$ be a sequence of groups and let $\varphi_1, \varphi_2, \ldots$ be a sequence of connecting one–one homomorphisms $\varphi_n : H_{n-1} \longrightarrow H_n$. Then a given group H^0 can be recognized as the direct limit $H = \varinjlim H_n$ if there are one–one homomorphisms $\psi_n : H_n \longrightarrow H^0$ such that $\psi_n = \varphi_n \circ \psi_{n-1}$ for all n, and $H^0 = \bigcup \psi_n(H_n)$. If each H_n is isomorphic with a free group $\mathbb{Z}^{(\kappa)} = \bigoplus_{\alpha<\kappa} e_\alpha \mathbb{Z}$ of rank κ, then each φ_n is given by a $\kappa \times \kappa$ matrix M_n as follows,

$$\varphi_n(x) = M_n x = \left(\sum_{\beta<\kappa} M_n(\alpha, \beta) x(\beta) \right)_\alpha .$$

Our aim is to prove the

1.2 Theorem: Let $H = \varinjlim H_n$ be a direct limit of a sequence of free groups of rank κ, where the transition maps $\varphi_n : H_{n-1} \longrightarrow H_n$ are represented by row-finite matrices M_n. Then there are fields $\mathcal{A}$ and $\mathcal{B}$ on a set X such that $H \cong H(\mathcal{A}, \mathcal{B})$.

§2. The combinatorics of zero-one matrices

Let κ be a cardinal number. We call a $\kappa \times \kappa$ matrix N a *staircase matrix* if N is a 0-1 matrix with all row and column sums equal to 2. Such matrices are the incidence matrices of a class of bipartite graphs. If the corresponding graph has κ connected components, each of which is finite, we say that N is a *κ-fold staircase matrix*. Let $n \geq 2$ be an integer. We define a particular $n \times n$, 2-fold staircase matrix N_n as follows:

$$N_n(i,j) = \begin{cases} 1, & \text{if } i = j \quad \text{or} \quad i \equiv j+1 \pmod{n}; \\ 0, & \text{otherwise.} \end{cases}$$

(Here, $0 \leq i, j \leq n-1$, and $n \geq 2$.) It is easy to see that any $n \times n$, 1-fold staircase matrix can, through permutations of rows and columns, be transformed into the standard matrix N_n. Likewise, we see that any κ-fold staircase may be transformed into a concatenated matrix $(N_{n_\alpha})_{\alpha<\kappa}$ with the block form

$$\begin{pmatrix} \ddots & 0 & 0 & 0 \\ 0 & N_{n_\alpha} & 0 & 0 \\ 0 & 0 & N_{n_\alpha} & 0 \\ 0 & 0 & 0 & \ddots \end{pmatrix},$$

with standard matrices N_{n_α} along the diagonal and 0-blocks off the diagonal.

Let π_1 and π_2 be finite partitions of a non-empty set X. We say that the pair (π_1, π_2) is a *κ-fold staircase* if we can write $\pi_1 = \{A_\alpha\}_\alpha$ and $\pi_2 = \{B_\alpha\}_\alpha$ so that

$$\begin{aligned} A_\alpha \cap B_\beta &\neq \emptyset \quad &\text{if} \quad N(\alpha,\beta) = 1 \\ A_\alpha \cap B_\beta &= \emptyset \quad &\text{if} \quad N(\alpha,\beta) = 0 \end{aligned}$$

for a κ-fold staircase matrix N. An elementary calculation proves

2.1 Fact: Let π_1 and π_2 be finite partitions of a set X such that (π_1, π_2) forms a κ-fold staircase. Let $\mathcal{A}$ and $\mathcal{B}$ be the fields on X generated by the π_1 and π_2, respectively. Then $H(\mathcal{A}, \mathcal{B}) \cong \mathbb{Z}^{(\kappa)}$.

§3. The construction

In this section, we prove Theorem 1.2. Given are the groups and connecting homomorphisms $\varphi_n : H_{n-1} \longrightarrow H_n$. We must construct fields $\mathcal{A}^0$ and $\mathcal{B}^0$ on a set X such that $H(\mathcal{A}^0, \mathcal{B}^0) \cong \varinjlim H_n$.

We shall construct a set $X \subseteq (\kappa \times \omega \times \kappa \times \omega)^\omega$, employing an inductive method and defining, for each $n \in \omega$, a set $S_n \subseteq (\kappa \times \omega \times \kappa \times \omega)^{n+1}$; then we set

$$\begin{aligned} U_n &= S_n \times (\kappa \times \omega \times \kappa \times \omega)^{\omega-n-1} \\ X_n &= \bigcap_{m \geq n} U_m \\ X &= \bigcup_{n=1}^{\infty} X_n = \liminf_{n\to\infty} U_n. \end{aligned}$$

Define $p_n : (\kappa \times \omega \times \kappa \times \omega)^\omega \longrightarrow (\kappa \times \omega)^{n+1}$ and $q_n : (\kappa \times \omega \times \kappa \times \omega)^\omega \longrightarrow (\kappa \times \omega)^{n+1}$ as compositions of projections:

$$(\kappa \times \omega \times \kappa \times \omega)^\omega \longrightarrow (\kappa \times \omega \times \kappa \times \omega)^{n+1} \longrightarrow (\kappa \times \omega)^{n+1};$$

in the last projection, p_n is formed by projecting, in each of the $n+1$ copies of $(\kappa \times \omega \times \kappa \times \omega)$, onto the first factor of $\kappa \times \omega$; for q_n, the second factor of $\kappa \times \omega$ is used. Let $\mathcal{A}_n$ [respectively $\mathcal{B}_n$] be the field generated by all sets of the form ${p_n}^{-1}(x)$ [respectively ${q_n}^{-1}(x)$] for $x \in (\kappa \times \omega)^{n+1}$. Put $\mathcal{C}_n = \mathcal{A}_n \vee \mathcal{B}_n$ and

$$\mathcal{A} = \bigvee_n \mathcal{A}_n \qquad \mathcal{B} = \bigvee_n \mathcal{B}_n \qquad \mathcal{C} = \bigvee_n \mathcal{C}_n = \mathcal{A} \vee \mathcal{B}.$$

For $n = 0$, we define $S_0 \subseteq (\kappa \times \omega \times \kappa \times \omega)$ as follows:

$$S_0 = \{(x, y, z, w) : x = z \quad \text{and} \quad 0 \leq y, w \leq 1\}.$$

Then the partitions

$$\Pi_0^1 = \{p_0^{-1}(x) \cap U_0 : x \in \kappa \times \omega,\ p_0^{-1}(x) \cap U_0 \neq \emptyset\}$$
$$\Pi_0^2 = \{q_0^{-1}(y) \cap U_0 : y \in \kappa \times \omega,\ q_0^{-1}(y) \cap U_0 \neq \emptyset\}$$

form a κ-fold staircase.

We assume that the set S_{n-1} has already been constructed so that the pair of partitions $(\Pi_{n-1}^1, \Pi_{n-1}^2)$ of U_{n-1} given by

$$\Pi_{n-1}^1 = \{p_{n-1}^{-1}(x) \cap U_{n-1} : x \in (\kappa \times \omega)^n, \quad p_{n-1}^{-1}(x) \cap U_{n-1} \neq \emptyset\}$$
$$\Pi_{n-1}^2 = \{q_{n-1}^{-1}(y) \cap U_{n-1} : y \in (\kappa \times \omega)^n, \quad q_{n-1}^{-1}(y) \cap U_{n-1} \neq \emptyset\}$$

forms a κ-fold staircase. Thus there are labelings of these partitions

$$\Pi_{n-1}^1 = \{A(\alpha, p) : \alpha < \kappa, \quad p = 0, \ldots, n_\alpha - 1\}$$
$$\Pi_{n-1}^2 = \{B(\alpha, p) : \alpha < \kappa, \quad p = 0, \ldots, n_\alpha - 1\}$$

so that $A(\alpha, p) \cap B(\beta, q) \neq \emptyset$ if and only if $\alpha = \beta$ and $N_{n_\alpha}(p, q) = 1$.

We construct a set $T_n \subseteq S_{n-1} \times (\kappa \times \omega) \times (\kappa \times \omega)$ as a disjoint union of sets T_n^β, $\beta < \kappa$. Fix $\beta < \kappa$. It involves no loss of generality to assume that the βth row of the matrix M_n has a positive entry, so we choose α_0 so that $M_n(\beta, \alpha_0) > 0$. Then define $\mathrm{K}(\beta) = \{\alpha : M_n(\beta, \alpha) \neq 0\} = \{\alpha_0, \alpha_1, \ldots, \alpha_m\}$. We let T_n^β be the union of (singleton) sets of the following forms:

i) $[A(\alpha_i, p) \cap B(\alpha_i, p)] \times \{(\beta, r, \beta, r)\}$, where $i = 0, 1, \ldots, m$ and $1 \leq r \leq |M_n(\beta, \alpha_i)|$ and $0 \leq p \leq n_{\alpha_i} - 1$;

ii) $[A(\alpha_i, p+1) \cap B(\alpha_i, p)] \times \{(\beta, r, \beta, r)\}$, where $i = 0, 1, \ldots, m$ and $1 \leq r \leq |M_n(\beta, \alpha_i)|$ and $0 \leq p \leq n_{\alpha_i} - 1$;

iii) $[A(\alpha_0, 0) \cap B(\alpha_0, n_{\alpha_0} - 1)] \times \{(\beta, r+1, \beta, r)\}$ where $1 \leq r \leq M_n(\alpha_0, \beta)$;

iv) $[A(\alpha_i, 0) \cap B(\alpha_i, n_{\alpha_i} - 1)] \times \{(\beta, r, \beta, r+1)\}$, where $i = 1, 2, \ldots m$ and $1 \leq r \leq |M_n(\beta, \alpha_i)|$.

Then $T_n = \bigcup_{\beta, \alpha} T_n^\beta$.

We now complete the construction of $S_n \supseteq T_n$. Again, S_n will be formed as a disjoint union of sets S_n^β for $\beta < \kappa$. We again fix $\beta < \kappa$ and take $\mathrm{K}(\beta) = \{\alpha_0, \alpha_1, \ldots, \alpha_m\}$ as previously. Choose the point $x \in (\kappa \times \omega)^{n-1}$ such that $A(\alpha_0, 0) = p_{n-1}^{-1}(x) \cap U_{n-1}$ and let $y_0, \ldots, y_m$ be the points of $(\kappa \times \omega)^{n-1}$ such that $B(\alpha_i, n_{\alpha_i} - 1) = q_{n-1}^{-1}(y_i) \cap U_{n-1}$ for $i = 0, \ldots, m$. In passing from T_n^β to $S_n^\beta \supseteq T_n^\beta$, two points are added for each $i = 1, \ldots, m$. For $i = 1, \ldots, m-1$, they are

a) $\Big(x, y_i, \beta, M_n(\alpha_0, \beta) + 2, \beta, 1\Big)$ and $\Big(x, y_i, \beta, M_n(\alpha_0, \beta) + 1, \beta, M_n(\alpha_i, \beta) + 1\Big)$ if$M_n(\alpha_1, \beta) \cdots M_n(\alpha_i, \beta) > 0$ and i is odd, or if $M_n(\alpha_1, \beta) \cdots M_n(\alpha_i, \beta) < 0$ and i is even;

b) $\Big(x, y_i, \beta, M_n(\alpha_0, \beta) + 1, \beta, 1\Big)$ and $\Big(x, y_i, \beta, M_n(\alpha_0, \beta) + 2, \beta, M_n(\alpha_i, \beta) + 1\Big)$ if $M_n(\alpha_1, \beta) \cdots M_n(\alpha_i, \beta) < 0$ and i is odd, or if $M_n(\alpha_1, \beta) \cdots M_n(\alpha_i, \beta) > 0$ and i is even.

For $i = m$, the points to be added are

c) $\big(x, y_m, \beta, 1, \beta, 1\big)$ and $\Big(x, y_n, \beta, M_n(\alpha_0, \beta) + 1, \beta, M_n(\alpha_m, \beta) + 1\Big)$ if $M_n(\alpha_1, \beta) \cdots M_n(\alpha_m, \beta) > 0$ and m is odd, or if $M_n(\alpha_1, \beta) \cdots M_n(\alpha_m, \beta) < 0$ and m is even;

d) $\big(x, y_m, \beta, 1, \beta, 1\big)$ and $\Big(x, y_m, \beta, M_n(\alpha_0, \beta) + 2, \beta, M_n(\alpha_m, \beta) + 1\Big)$ if $M_n(\alpha_1, \beta) \cdots M_n(\alpha_m, \beta) < 0$ and m is odd, or if $M_n(\alpha_1, \beta) \cdots M_n(\alpha_m, \beta) > 0$ and m is even.

If $m = 0$, then we add the points $\Big(x, y_0, \beta, M_n(\alpha_0, \beta) + 1, \beta, M_n(\alpha_0, \beta) + 1\Big)$ and $\Big(x, y_0, \beta, 1, \beta, M_n(\alpha_0, \beta) + 1\Big)$.

3.1 Remark: We note that although the passage from $S_{n-1} \times (\kappa \times \omega)^2$ to T_n involves the loss of some points, and that from T_n to S_n the addition of others, these transitions are relatively conservative. For example, suppose that $x \in (\kappa \times \omega)^n$ and $y \in (\kappa \times \omega)^n$ are such that $(x, y) \in S_{n-1}$; an easy compactness argument shows that there is some $s \in X_{n-1}$ such that $p_{n-1}(s) = x$ and $q_{n-1}(s) = y$.

This completes the construction.

We now consider the partitions of U_n given by

$$\Pi_n^1 = \{p_n^{-1}(x) \cap U_n : x \in (\kappa \times \omega)^{n+1}, \quad p_n^{-1}(x) \cap U_n \neq \emptyset\}$$
$$\Pi_n^2 = \{q_n^{-1}(y) \cap U_n : y \in (\kappa \times \omega)^{n+1}, \quad q_n^{-1}(y) \cap U_n \neq \emptyset\}.$$

It is not hard, although admittedly awkward, to verify that the pair $\big(\Pi_n^1, \Pi_n^2\big)$ forms a κ-fold staircase. In general, the flights of the staircase are given by $S_n^\beta \times (\kappa \times \omega \times \kappa \times \omega)^{\omega - n - 1}$ for $\beta < \kappa$.

We now define the groups and homomorphisms relevant to our direct limit construction. If $\mathcal{D}$ is a field on a set X, and $A \subseteq X$, then $\mathcal{D}(A) = \{A \cap D : D \in \mathcal{D}\}$ is the *trace* of $\mathcal{D}$ on A. We put

$$H_n^\circ = H\Big(\mathcal{A}_n(U_n), \mathcal{B}_n(U_n)\Big)$$
$$G_n = H\Big(\mathcal{A}_{n-1}(U_n), \mathcal{B}_{n-1}(U_n)\Big)$$
$$K_n = H\Big(\mathcal{A}_n(X_n), \mathcal{B}_n(X_n)\Big).$$

Referring to the construction of U_n and X_n, we see that since the pair of partitions (Π_n^1, Π_n^2) of U_n form a κ-fold staircase, we have $H_n^\circ \cong H_n \cong \mathbb{Z}^{(\kappa)}$. (Thus we are free to drop the little circle and write simply H_n.) One set of generators for H_n is given by the cosets of the indicator functions of the sets

$$[A(\alpha_0, 0) \cap B(\alpha_0, 0)] \times \{(\beta, r, \beta, r)\} \times (\kappa \times \omega \times \kappa \times \omega)^{\omega - n - 1}$$

for $\mathrm{K}(\beta) = \{\alpha_0, \ldots \alpha_m\}$ and $1 \leq r \leq |M_n(\beta, \alpha_0)|$. For each fixed $\beta < \kappa$, these $|M_n(\beta, \alpha_0)|$ cosets all represent the same generator in $\mathbb{Z}^{(\kappa)}$, say e_β, where

$$e_\beta(\gamma) = \begin{cases} 1, & \gamma = \beta \\ 0, & \gamma \neq \beta. \end{cases}$$

Likewise, if we fix both $\beta < \kappa$ and $0 \leq i \leq m$, we find that the $|M_n(\beta, \alpha_i)|$ sets

$$[A(\alpha_i, 0) \cap B(\alpha_i, 0)] \times \{(\beta, r, \beta, r)\} \times (\kappa \times \omega \times \kappa \times \omega)^{\omega - n}$$

represent either e_β or $-e_\beta$ according as $M_n(\beta, \alpha_i)$ is positive or negative.

We turn our attention to the groups G_n. The pair of partitions $(\Sigma_{n-1}^1, \Sigma_{n-1}^2)$ of U_n given by

$$\Sigma_{n-1}^1 = \{p_{n-1}^{-1}(x) \cap U_n : x \in (\kappa \times \omega)^n, \quad p_{n-1}^{-1}(x) \cap U_n \neq \emptyset\}$$
$$\Sigma_{n-1}^2 = \{q_{n-1}^{-1}(y) \cap U_n : y \in (\kappa \times \omega)^n, \quad q_{n-1}^{-1}(y) \cap U_n \neq \emptyset\}$$

is the κ-fold staircase $(\Pi_{n-1}^1, \Pi_{n-1}^2)$ augmented by the addition of the "links"

$$L^i = [A(\alpha_0, 0) \cap B(\alpha_i, n_{\alpha_i} - 1)] \times (\kappa \times \omega \times \kappa \times \omega)^{\omega - n - 1},$$

where $i = 1, \ldots, m$. For each $\beta < \kappa$, we have the following schematic:

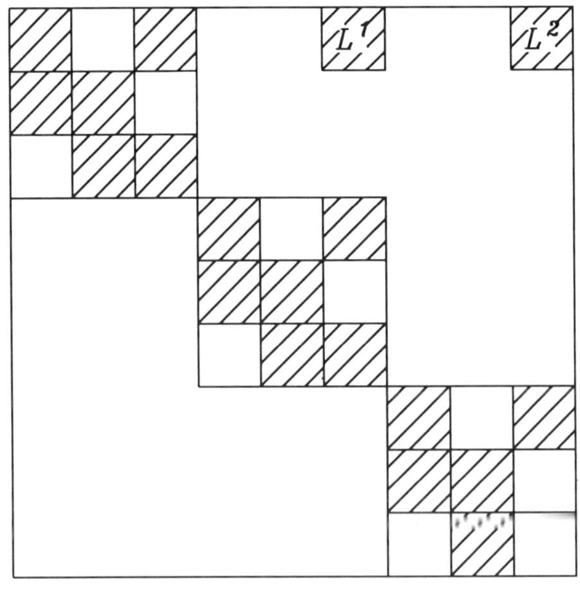

There is one such schematic for each $\beta < \kappa$, and, by a suitable permutation of rows and columns, they can be arranged along the diagonal of an otherwise empty matrix. An elementary calculation shows that $G_n \cong \mathbb{Z}^{(\kappa)} \oplus F$, where F is a free group whose generators are given by the (coset of) indicator functions of the links L^i, and the generators of $\mathbb{Z}^{(\kappa)}$ are those of H_{n-1}.

Remark 3.1 implies that $K_n \cong H_n$ for each n. The κ generators of K_n are given by the sets

$$\{[A(\alpha_0, 0) \cap B(\alpha_0, 0)] \times \{(\beta, r, \beta, r)\} \times (\kappa \times \omega)^{\omega - n}\} \cap X_n.$$

The isomorphism $h_n : H_n \longrightarrow K_n$ is given by the formula $h_n([f]) = [f_0]$, where $[f]$ denotes the coset of $f : U_n \longrightarrow \mathbb{Z}$ in H_n, $f_0 : X_n \longrightarrow \mathbb{Z}$ is the restriction of f, and $[f_0]$ is its coset in K_n.

We define a homomorphism $i_n : H_{n-1} \longrightarrow G_n$ by the formula $i_n([f]) = [f_{00}]$. Here, $[f]$ is the coset of $f : U_{n-1} \longrightarrow \mathbb{Z}$ in H_{n-1}, and $[f_{00}]$ is the coset in G_n of the function $f_{00} : U_n \longrightarrow \mathbb{Z}$ defined by

$$f_{00}(x) = \begin{cases} f(x) & x \in U_{n-1} \\ 0 & x \in U_n - U_{n-1}. \end{cases}$$

Inspection of generators shows that i_n is the canonical embedding of $\mathbb{Z}^{(\kappa)}$ in $\mathbb{Z}^{(\kappa)} \oplus F$.

Next, we define a homomorphism $j_n : G_n \longrightarrow H_n$ by setting $j_n([g]) = [g]$. Here, on the left, $[g]$ represents the coset of $g : U_n \longrightarrow \mathbb{Z}$ in G_n, and on the right, the coset of the same g in H_n. We see, again by inspection of generators, that j_n maps the factor $F \subseteq G_n$ to zero, and that the composition $j_n \circ i_n$ is precisely the transition homomorphism φ_n specified in the direct limit we started with.

We define $\mathcal{A}^\circ$ to be the field on X generated by the union of the collections $\mathcal{A}_n(X_n)$; likewise, $\mathcal{B}^\circ$ is the field generated by the union of all $\mathcal{B}_n(X_n)$; put $\mathcal{C}^\circ = \mathcal{A}^\circ \vee \mathcal{B}^\circ$ and define $H^\circ = H\big(\mathcal{A}^\circ(X), \mathcal{B}^\circ(X)\big)$. Our purpose is to show that $H^\circ \cong H = \varinjlim H_n$. We define a homomorphism $k_n : H_n \longrightarrow H^\circ$ by the formula $k_n([h]) = [h^\circ]$, where $[h]$ is the coset in H_n of the function $h : X_n \longrightarrow \mathbb{Z}$ and $[h^\circ]$ is the coset in H° of the function $h^\circ : X \longrightarrow \mathbb{Z}$ defined by

$$h^\circ(x) = \begin{cases} h(x) & x \in X_n \\ 0 & x \in X - X_n. \end{cases}$$

The diagram

$$\begin{array}{ccccccccccc}
\cdots \longrightarrow & H_{n-1} & \xrightarrow{i_n} & G_n & \xrightarrow{j_n} & H_n & \xrightarrow{i_{n+1}} & G_{n+1} & \xrightarrow{j_{n+1}} & H_{n+1} & \longrightarrow \cdots \\
& & & & & \downarrow{\scriptstyle h_n} & & & & & \\
& \downarrow{\scriptstyle h_{n-1}} & & & & K_n & & & & \downarrow{\scriptstyle h_{n+1}} & \\
& & & & & \downarrow{\scriptstyle k_n} & & & & & \\
& K_{n-1} & & \xrightarrow{\quad k_{n-1} \quad} & & H^\circ & & \xleftarrow{\quad k_{n+1} \quad} & & K_{n+1} &
\end{array}$$

commutes.

Scholium: The homomorphism $\psi_n : H_n \longrightarrow H^\circ$ defined by $\psi_n = k_n \circ h_n$ is one–one, and $H^\circ = \bigcup \psi_n(H_n)$.

Proof of scholium: Since $\psi_{n-1} = \psi_n \circ \varphi_n$, and φ_n is an isomorphism, it suffices to prove that ψ_0 is one–one, and, since h_0 is an isomorphism, it is enough to check that k_0 is one–one. Suppose that $h : X_0 \longrightarrow \mathbb{Z}$ is $\mathcal{C}_0$-measurable and $h^\circ : X \longrightarrow \mathbb{Z}$ is the "zero-extension" of h defined above. Further, suppose that $h^\circ = f + g$ for $f \in \mathcal{A}_n(X_n)$ and $G \in \mathcal{B}_n(X_n)$. We may assume that $f = g = 0$ on $X - X_0$. Then f is measurable for the trace of $\mathcal{A}_n$ on X_0, which is $\mathcal{A}_0$, and similarly for g. It follows that $[h] = 0$.

The second half of the scholium is easy, since each $\mathcal{C}^\circ$-measurable function is $\mathcal{C}_n$-measurable for some finite n.

Thus we see that $H\left(\mathcal{A}^\circ(X), \mathcal{B}^\circ(X)\right)$ is isomorphic with the direct limit $H = \varinjlim H_n$, as required.

§4. Illustration

We provide here a diagram to help illustrate the key points of the construction. One stage of the process is pictured. We take $\kappa = 1$ (rank 2) and

$$M_n = \begin{pmatrix} 3 & -2 \\ 1 & 2 \end{pmatrix}.$$

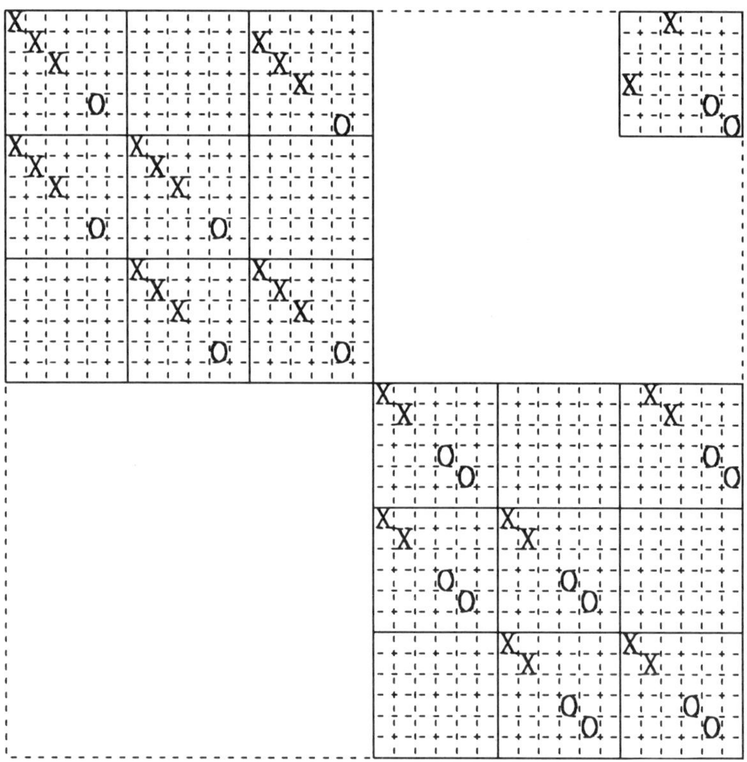

The X's and O's represent generators of $\mathbb{Z} \oplus \mathbb{Z}$ (or their negatives). The upper left block, representing one generator of $\mathbb{Z} \oplus \mathbb{Z}$ (say $(1,0)$), is mapped to $(3,1)$ (or three X's and one O).

§5. More direct limits

By Observation 1.1 and Theorem 1.2, we see that every countable torsion-free group occurs as one of the group $H(\mathcal{A}, \mathcal{B})$. But Theorem 1.2 actually allows for the construction of certain uncountable groups, *e.g.*, all sums of countable groups. There arises the

5.1 Question: Is the class of all groups $H(\mathcal{A}, \mathcal{B})$ coextensive with the class of all torsion-free Abelian groups?

Our Theorem 1.2 does not serve to provide an easy answer in the positive, to wit, inspection of Shelah's indecomposable groups [Sh] (see also [FrG] for their generators) shows that the embedding matrices given by our representation in §1 have rows with κ many entries -1 and so are not row-finite. But perhaps our representation is simply not well chosen and could be made row-finite *via* a change of basis.

So we ask whether there are groups having no row-finite representation as the direct limit of free groups. In fact, we can determine the class of groups representable under these or more general conditions. Let ρ and λ be infinite cardinals. We say that a group H has a *free ρ-λ-representation* if H is the union of a chain $\{F_\alpha : \alpha < \lambda\}$ of free subgroups F_α with sets B_α of basis elements such that the embedding $F_\alpha \subseteq F_{\alpha+1}$ is given by an integer matrix M_α with only $< \rho$ non-zero entries in each row. We also say that H is *ρ-λ-represented by* $\{B_\alpha : \alpha < \lambda\}$.

5.2 Theorem: If H is a torsion-free Abelian group of rank κ having a free ρ-λ-representation with $\rho, \lambda < \kappa$ and κ regular, then H is a direct sum of κ non-trivial summands of size $\leq \rho \cdot \lambda$.

Applying this result to the case $\rho = \lambda = \omega$, we see that indecomposable groups of regular cardinality, *e.g.*, Shelah's examples, can never be represented by a countable chain of free groups with row-finite embedding matrices.

5.3 Corollary: A torsion-free group of regular cardinality is a direct sum of countable groups if and only if it has a free ω-ω-representation.

Proof of the theorem: Let H be ρ-λ-represented by $\{B_\alpha : \alpha < \lambda\}$. The proof is by transfinite induction, constructing

(τ) $H = \left(\bigoplus_{\beta<\tau} D_\beta\right) \oplus C_\tau$ for each $\tau < \kappa$ such that $1 \neq \text{card}(D_\beta) \leq \rho \cdot \lambda$ and C_τ such that $\{C_\gamma : \gamma \leq \tau\}$ is a continuous descending chain of groups C_γ ρ-λ-represented by $\{B_\beta \cup C_\gamma : \beta < \lambda\}$.

Suppose that $\alpha < \kappa$ and that condition (τ) holds for all $\tau < \alpha$. If $\alpha = 0$, then $H = C_0$ and condition (0) holds. If α is a limit ordinal, take $C_\alpha = \bigcap_{\tau<\alpha} C_\tau$; condition ($\alpha$) holds. It remains to construct D_α amd $C_{\alpha+1}$ so that ($\alpha + 1$) holds. Changing names, we may assume $H = C_\alpha$, and we must write $H = D \oplus C$ such that $D = D_{\alpha+1}$ and $C = C_{\alpha+1}$. Since H is ρ-λ-representable by $\{B_\alpha : \alpha < \lambda\}$, let $M_\alpha : F_\alpha \longrightarrow F_{\alpha+1}$ be the corresponding embedding matrices, which are all bounded by ρ in rows. If $\gamma < \kappa$, put $r_\gamma = \sup\{\eta : M_\alpha(\delta, \eta) \neq 0, \quad \alpha < \lambda, \quad \eta < \kappa, \quad \delta < \gamma\}$. Clearly $r_\gamma < \kappa$, since κ is regular, and $\left|\{\eta : M_\alpha(\delta, \eta) \neq 0, \quad \alpha < \lambda, \quad \eta < \kappa, \quad \delta < \gamma\}\right| \leq \lambda\rho|\gamma| < \text{cf}(\kappa) = \kappa$, where $|\gamma|$ denotes the cardinality of the ordinal γ. Similarly define c_γ for all columns of M_α. Now it is easy to choose a countable sequence of ordinals $\gamma_n < \kappa$ such that $|\gamma_n| \leq \rho\lambda$ and

$$r_\gamma < \max\{r_{\gamma_1}, c_{\gamma_2}\} < \max\{r_{\gamma_1}, c_{\gamma_2}, r_{\gamma_3}\} < \cdots.$$

If $\sigma = \sup_n \gamma_n$, then each $M_\alpha(\delta, \eta)$ has 0 entries for $\delta < \sigma \leq \eta$ and $\eta < \sigma \leq \delta$. Hence, M_α decomposes into squares for each $\alpha < \lambda$. We have a uniform decomposition $F_\alpha = F_\alpha^1 \oplus F_\alpha^2$ such that $F_\alpha^1 \subseteq F_{\alpha+1}^1$ and $F_\alpha^2 \subseteq F_{\alpha+1}^2$ for all $\alpha < \lambda$. Moreover, $\sup_{\alpha<\lambda} |F_\alpha^1| \leq \rho\lambda$. We derive $H = \bigcup_\alpha F_\alpha = \bigcup_\alpha (F_\alpha^1 \oplus F_\alpha^2) = \bigcup_\alpha F_\alpha^1 \oplus \bigcup_\alpha F_\alpha^2$; then $D = \bigcup_{\alpha<\lambda} F_\alpha^1$, with $|D| \leq \rho\lambda$, and $C = \bigcup_\alpha F_\alpha^2$ are the direct summands.

In [GoS] we considered rational cotorsion theories and their connection with the common extension of measures. In fact, any rational group $S \subseteq \mathbb{Q}$ was realized as $H(\mathcal{A}, \mathcal{B}) = S$ in [GoS]. The larger class of all direct sums of countable, torsion-free Abelian groups realised as $H(\mathcal{A}, \mathcal{B})$ in this paper also leads to cotorsion theories both different from rational cotorsion theories and giving new interesting examples for common extensions of measures. This will be discussed now.

Using [Fa] we can find an E-ring E of finite rank, such that E^+ is of type 0 and different from $\mathbb{Z}$. Recall that a ring E is an E-ring if $E \cong \mathrm{End}E^+$ with the isomorphism induced by left scalar multiplication on E^+. Let G be any countable, torsion-free Abelian group of type 0 with $\mathrm{End}G \cong E$, *e.g.*, $E^+ = G$ or any group constructed in [GoN]. If $\chi = (r_p)$ denotes the characteristic of a rational group $S \not\cong \mathbb{Z}$, then either $r_p = \infty$ for some p or $r_p > 0$ for an infinite number of primes. We have $\mathrm{Ext}(Z(p^{r_p}), G) \neq 0$ for these primes and $|\mathrm{Ext}(Z(p^\infty), G)| \geq 2^{\aleph_0}$ in the first case. Hence $|\Pi_p \mathrm{Ext}(Z(p^{r_p}, G)| \geq 2^{\aleph_0}$ in any case and this product cannot be an epimorphic image of a countable group. From Salce [Sa], Theorem 3.5 we derive that G cannot be cotorsion with respect to any non-trivial rational cotorsion theory. Recall that in the trivial cotorsion-theory cogenerated by $\mathbb{Z}$, all groups are cotorsion. Obviously E^+ is free over $\mathrm{End}G \cong E^+$; hence $\mathrm{Ext}_E(E, G) = 0$ and the following easy homological argument gives $\mathrm{Ext}_{\mathbb{Z}}(E, G) \cong \mathrm{Ext}_E(E, G)$.

If $0 \longrightarrow F_1 \longrightarrow F_2 \longrightarrow E^+ \longrightarrow 0$ is a free $\mathbb{Z}$-resolution, then we have a commuting diagram

$$\begin{array}{ccccccc}
\mathrm{Hom}_{\mathbb{Z}}(F_2, G) & \longrightarrow & \mathrm{Hom}_{\mathbb{Z}}(F_1, G) & \longrightarrow & \mathrm{Ext}_{\mathbb{Z}}(E, G) & \longrightarrow & 0 \\
\downarrow\iota & & \downarrow\iota & & & & \\
\mathrm{Hom}_E(E \otimes F_2, G) & \longrightarrow & \mathrm{Hom}_E(E \otimes F_1, G) & \longrightarrow & \mathrm{Ext}_E(E, G) & \longrightarrow & 0
\end{array}$$

giving the isomorphism between $\mathrm{Ext}_{\mathbb{Z}}$ and Ext_E. Therefore $\mathrm{Ext}_{\mathbb{Z}}(E, G) = 0$ and G is cotorsion with respect to E. Finally recall that E is one of the group realized as $H(\mathcal{A}, \mathcal{B})$ in the first part of this paper.

§6. References

[BhS] Bhaskara Rao, K. P. S. and Shortt, R. M., Group-values charges: common extensions and the infinite Chinese remainder property, *Proc. AMS* **113** (1991) 965–972

[Fa] Faticoni, T. G., Each countable reduced torsion-free commutative ring is a pure subring of an E-ring, *Comm. in Algebra* **15** (1987) 2545–2564

[FrG] Franzen, B. and Göbel, R., The Brenner-Butler-Corner theorem and its application to modules, in *Abelian Group Theory*, Gordon and Breach, London 1987, pp. 209–227

[Fu] Fuchs, L., *Infinite Abelian Groups, Vols. I and II*, Academic Press, New York 1973

[GoN] Göbel, R. and Nay, W., Independence in completions and endomorphism algebras, *Forum mathematicum* **1** (1989) 215–226

[GoS] Göbel, R. and Shortt, R. M., Algebraic ramifications of the common extension problem for group-values measures, *Fundamenta Math.*, to appear

[RaR] Rangaswamy, K. M. and Reid, J. D., Common extensions of finitely additive measures and a characterization of cotorsion Abelian groups, in *Abelian Groups, Proc. of 1991 Curaçao Conference*, M. Dekker, New York 1993, pp. 231–239

[Sa] Salce, L., Cotorsion theories for abelian groups, *Symposia Mathematica* **23** (1979) 11–32

[Sh] Shelah, S., Infinite abelian groups, Whitehead problem and some constructions, *Israel Journal of Math.* **18** (1974) 243–256

R. Göbel
University of Essen, Essen, Germany

R. M. Shortt
Wesleyan University, Middletown, Connecticut, USA

Contemporary Mathematics
Volume **171**, 1994

Numerical Invariants for a Class of Butler Groups

H. PAT GOETERS, WILLIAM ULLERY AND CHARLES VINSONHALER

Dedicated to the Memory of R. S. Pierce

Suppose $A_1, A_2, \ldots, A_n$ $(n \geq 2)$ are nonzero subgroups of the additive group of rationals $\mathbb{Q}$. The cokernel of the diagonal embedding $\bigcap_{1\leq i\leq n} A_i \to \bigoplus_{1\leq i\leq n} A_i$ is a rank $n-1$ Butler group which we denote by $\mathcal{G}[A_1, A_2, ..., A_n]$. Groups of the form $\mathcal{G}[A_1, A_2, ..., A_n]$ are a subclass of the $\mathcal{B}^{(1)}$-groups of [**FM**]. We shall adopt a more colloquial terminology and call such a group a *bracket group*.

There is an extensive literature on the quasi-isomorphism problem for bracket groups. Most notable, at least in terms of numerical invariants, is the approach that culminated in [**AV3**]. If G is a torsion-free abelian group and if M is a nonempty subset of the lattice of types, define

$$r_G(M) = \text{rank} \sum_{\tau \epsilon M} G(\tau)$$

where $G(\tau)$ is the τ-socle of G. In [**AV3**], it is shown that two bracket groups $G = \mathcal{G}[A_1, A_2, ..., A_n]$ and $H = \mathcal{G}[B_1, B_2, ..., B_m]$ are quasi-isomorphic if and only if $r_G(M) = r_H(M)$ for each finite subset M of the finite lattice generated by typeset $G \cup$ typeset H. In this paper we show that it suffices to consider only those subsets M which are singletons. Specifically, we prove

MAIN THEOREM. *Let $G = \mathcal{G}[A_1, A_2, ..., A_n]$ and $H = \mathcal{G}[B_1, B_2, ..., B_m]$ be bracket groups. Then G is quasi-isomorphic to H if and only if* rank $G(\tau) =$ rank $H(\tau)$ *for each type τ in the lattice generated by* typeset G $\cup$ typeset H.

At this point it is convenient to introduce notation which will be used frequently in the sequel. If n is a positive integer, we write $\bar{n}$ for the set $\{1, 2, ..., n\}$. If $A_1, \ldots, A_n$ and $B_1, \ldots, B_n$ are nonzero subgroups of $\mathbb{Q}$ with $A_i \cong B_i$ for each $i \in \bar{n}$, it is easily seen that $\mathcal{G}[A_1, \ldots, A_n]$ is quasi-isomorphic to $\mathcal{G}[B_1, \ldots, B_n]$.

1991 *Mathematics Subject Classification.* 20K15.

Thus, $G = \mathcal{G}[A_1, ..., A_n]$ is determined up to quasi-isomorphism by the types $\tau_1, \tau_2, \ldots, \tau_n$, where $\tau_i =$ type(A_i) for $i \in \bar{n}$. In the future we denote the bracket group G by $G = [\tau_1, ..., \tau_n]$, keeping in mind that this notation defines G only up to quasi-isomorphism. If I is a nonempty subset of $\bar{n}$, say $I = \{i_1, \ldots, i_k\}$, we write $[\tau(I)]$ for $[\tau_{i_1}, \ldots, \tau_{i_k}]$, and τ_I for the meet $\tau_{i_1} \wedge \ldots \wedge \tau_{i_k}$. If, in addition, $\gamma_1, \ldots, \gamma_m$ are types, $[\tau(I), \gamma_1, ..., \gamma_m]$ means $[\tau_{i_1}, ..., \tau_{i_k}, \gamma_1, \ldots, \gamma_m]$.

For a torsion-free abelian group G, we write $T_G = \{\tau \mid \tau = \text{type}(a)$ for some nonzero $a \in G\}$ for the typeset of G, $\Delta_G = \{\tau \in T_G \mid \text{rank } G(\tau) = 1\}$ and $\text{Max}(G)$ for the set of maximal elements in T_G. Observe that $\Delta_G \subseteq \text{Max}(G) \subseteq T_G$ and if G is a Butler group, then $\text{Max}(G)$ generates T_G under meets, and hence T_G is closed under meets (see for example, [**A1**]). If H is another group, we write $G \simeq H$ to mean that G and H are quasi-isomorphic and $G \doteq H$ to mean that G is quasi-equal to H. In addition, we shall frequently abbreviate rank $G(\tau)$ as $r_G(\tau)$. All notation and terminology not defined herein should be in agreement with [**A1**], [**AV1**] and [**F**]. In particular, "group" always means "abelian group". We begin by quoting several known results for future reference. The first appears in [**GU3**] and is a modification of [**AV3**, Proposition 5].

PROPOSITION 1. ([**AV3**], [**GU3**]). *Suppose $G = [\tau_1, \ldots, \tau_n]$ is quasi-decomposable. Then, there exist nonempty subsets $I_1, I_2, ..., I_k$ $(k \geq 2)$ of $\bar{n}$ that satisfy the following conditions.*

(a) *$|I_i| \geq 2$ for each $i \in \bar{k}$, $\bigcup_{i \in \bar{k}} I_i = \bar{n}$, and $|I_i \cap I_j| \leq 1$ whenever $i \neq j$.*
(b) *$G \simeq [\tau(I_1)] \oplus \ldots \oplus [\tau(I_k)]$ and each $[\tau(I_i)]$ is strongly indecomposable.*
(c) *If $i \neq j$ and $G_i = [\tau(I_i)]$, then* rank *$G_i(\tau_{I_j}) \leq 1$.*
(d) *If $i \neq j$, there exist $i_0 \in I_i$ and $j_0 \in I_j$ such that $\tau_{i_0} \vee \tau_{j_0} = \tau_{I_i} \vee \tau_{I_j}$.*

A quasi-decomposition $G \simeq [\tau(I_1)] \oplus \ldots \oplus [\tau(I_k)]$ of $G = [\tau_1, ..., \tau_n]$ that satisfies all the conditions of Proposition 1 is called a *canonical decomposition* of G. One consequence of the next result is that any socle of a bracket group is (quasi-isomorphic to) a bracket group. This is a key fact needed in an inductive proof of the Main Theorem.

PROPOSITION 2. [**GM**, Theorem 1.8] *If $G = [\tau_1, ..., \tau_n]$ and $\sigma \in T_G$, there exists a partition $E_1, E_2, \ldots, E_k$ $(k \geq 2)$ of $\bar{n}$ into nonempty proper subsets such that*

(a) *$\sigma = \bigwedge_{i \in \bar{k}} (\tau_{E_i} \vee \tau_{E_i'})$, where $E_i' = \bar{n} \setminus E_i$ for all $i \in \bar{k}$.*
(b) *For an arbitrary partition $J_1, \ldots, J_s$ of $\bar{n}$, $\sigma \leq \bigwedge_{\ell \in \bar{s}} (\tau_{J_\ell} \vee \tau_{J_\ell'})$ if and only if $J_\ell = \bigcup \{E_i \mid E_i \cap J_\ell \neq \emptyset\}$ for all $\ell \in \bar{s}$.*
(c) *$G(\sigma) \simeq [\tau_{E_1} \vee \tau_{E_1'}, \ldots, \tau_{E_k} \vee \tau_{E_k'}]$.*

In the sequel, the (unique) partition $E_1, \ldots, E_k$ of $\bar{n}$ described in Proposition 2 will be referred to as the *canonical partition* associated with σ.

LEMMA 1. [**GU3**, Proposition 4] *Suppose G is a subgroup of a Butler group H. If $r_G(\tau) = r_H(\tau)$ for all types τ, then $G \doteq H$.*

Our application of Lemma 1 will be in the situation where G and H are bracket groups such that G embeds in H. In this case, if $r_G(\tau) = r_H(\tau)$ for all τ, one concludes that $G \doteq H$.

Call a collection of types $\tau_1, \dots, \tau_n$ *cotrimmed* if $\tau_i \geq \bigwedge_{j \neq i} \tau_j$ for all $i \in \bar{n}$. If $\tau_1, \dots, \tau_n$ is cotrimmed and $G = [\tau_1, \dots, \tau_n]$, it is easily seen that each $\tau_i \in T_G$. If $\tau_1, ..., \tau_n$ are types, then

$$\tau_1' = \tau_1 \vee (\bigwedge{}_{j \neq 1} \tau_j), \dots, \tau_n' = \tau_n \vee (\bigwedge{}_{j \neq n} \tau_j)$$

is a cotrimmed collection of types which we call the *cotrimmed version* of $\tau_1, \dots \tau_n$. It is not difficult to show that $[\tau_1, \dots, \tau_n] \doteq [\tau_1', \dots, \tau_n']$. These notions were apparently first worked out in detail in [**L**]. If $G = [\tau_1, \dots, \tau_n]$, it follows from results in [**FM**] or [**GU1**] that $\tau_I \vee \tau_{I'} \in T_G$ for every nonempty proper subset I of $\bar{n}$, where $I' = \bar{n} \setminus I$. Thus, from Proposition 2 (or [**FM**] and [**GU1**]), if $\sigma \in \operatorname{Max}(G)$, there exists a nonempty proper subset I of $\bar{n}$ such that $\sigma = \tau_I \vee \tau_{I'}$. We collect some needed facts concerning $\operatorname{Max}(G)$ in the following two lemmas.

LEMMA 2. *Suppose $G = [\tau_1, \dots, \tau_n]$ and $\sigma = \tau_I \vee \tau_{I'}$ is in* $\operatorname{Max}(G)$ *with $|I|$ and $|I'|$ both at least 2. If G has no rank-1 quasi-summands, then both $G(\tau_I)$ and $G(\tau_I')$ are proper subgroups of G.*

PROOF. If $G(\tau_I) = G$, then $G(\tau_{I'}) = G(\tau_I) \cap G(\tau_{I'}) = G(\tau_I \vee \tau_{I'}) = G(\sigma)$. Thus, σ is the smallest element of T_G with the property that $\sigma \geq \tau_{I'}$. If $\tau_1', \dots, \tau_n'$ is the cotrimmed version of $\tau_1, \dots, \tau_n$, there exist distinct elements i and j of I' such that $\tau_{I'} \leq \tau_i'$ and $\tau_{I'} \leq \tau_j'$. Since $\tau_i', \tau_j' \in T_G$ and $\sigma \in T_G$ is maximal, we conclude that $\tau_i' = \sigma = \tau_j'$. But this contradicts the hypothesis that G has no rank-1 quasi-summands (see, for example, [**AV3**]). Therefore $G(\tau_I) \neq G$. Similarly, $G(\tau_{I'}) \neq G$. $\square$

In preparation for the proof of our next result, we recall from [**GU4**] the following characterization of bracket groups with rank-1 quasi-summands. Namely, $G = [\tau_1, \tau_2, \dots, \tau_n]$ has a rank-1 quasi-summand if and only if there exist distinct $i, j \in \bar{n}$ and a partition $\bar{n} = I \cup J$, with $i \in I$ and $j \in J$, such that $\tau_i \vee \tau_j = \tau_I \vee \tau_J$. In particular, if G has a rank-1 quasi-summand, there exist distinct $i, j \in \bar{n}$ with $\tau_i \vee \tau_j \in T_G$.

LEMMA 3. *Suppose $G = [\tau, \dots, \tau_n]$ with $\tau_1, ..., \tau_n$ cotrimmed and $\sigma = \tau_I \vee \tau_{I'} \in$* $\operatorname{Max}(G)$ *for some nonempty proper subset I of $\bar{n}$. Then*

$$G(\tau_I) \doteq [\tau(I), \sigma, \dots, \sigma]$$

with σ repeated rank $G(\tau_I) - |I| + 1$ *times. Moreover, if G is strongly indecomposable and if $B = [\tau(I), \sigma]$ has a rank-1 quasi-summand, then $\sigma = \tau_i$ for some $i \in I$.*

PROOF. There is no harm in assuming that $I = \bar{k} = \{1, 2, \ldots, k\}$ with $1 \leq k < n$. Observe that $\tau_I \in T_G$, since T_G is closed under meets, and that the canonical partition of $\bar{n}$ associated with τ_I is of the form

$$\{1\}, \{2\}, \ldots, \{k\}, J_1, \ldots, J_s$$

where $J_1, \ldots, J_s$ is a partition of $I' = \{k+1, \ldots, n\}$. Therefore, if we set $\gamma_i = \tau_{J_i} \vee \tau_{J_i'}$ for each $i \in \bar{s}$, it follows from Proposition 2 and cotrimming that $\tau_I = \tau_1 \wedge \tau_2 \wedge \ldots \wedge \tau_k \wedge \gamma_1 \wedge \ldots \wedge \gamma_s$ and $G(\tau_I) \doteq [\tau_1, \ldots, \tau_k, \gamma_1, \ldots, \gamma_s]$. Next observe that $\sigma = \tau_I \vee \tau_{I'} \leq \tau_{J_i} \vee \tau_{J_i'} = \gamma_i \in T_G$ for each $i \in \bar{s}$. By the maximality of σ, $\sigma = \gamma_i$ for all i and $G(\tau_I) \doteq [\tau(I), \sigma, \ldots, \sigma]$.

Now suppose that G is strongly indecomposable and $B = [\tau(I), \sigma]$ has a rank-1 quasi-summand. If i, j are distinct elements of I, then G strongly indecomposable implies that $0 = \text{rank } G(\tau_i \vee \tau_j) \geq \text{rank } B(\tau_i \vee \tau_j)$. Therefore, by the remarks preceding the proof, there exist $i \in I$ and a partition $I = J_1 \cup J_2$ of I such that $i \in J_2$ and $\sigma \vee \tau_i = (\tau_{J_1} \wedge \sigma) \vee \tau_{J_2}$. Thus, $\sigma \vee \tau_i = (\tau_{J_1} \wedge (\tau_I \vee \tau_{I'})) \vee \tau_{J_2} = \tau_I \vee \tau_{J_1 \cup I'} \vee \tau_{J_2} = \tau_{J_1 \cup I'} \vee \tau_{J_2} \in T_G$. Since $\sigma \in \text{Max}(G)$ and since G strongly indecomposable implies that $\tau_i \in \text{Max}(G)$, $\sigma = \tau_i$. □

A surprising feature of bracket groups is that the possible presence of large numbers of rank-1 quasi-summands complicates their analysis. Prime examples of this appear in the papers [**AV3**] and [**GU3**]. Our next result will be used to gain control over this difficult case.

LEMMA 4. *Suppose $G = [\tau_1, \ldots, \tau_n]$ is quasi-decomposable with $\tau_1, \ldots, \tau_n$ cotrimmed. Let $G \doteq [\tau(I_1)] \oplus \ldots \oplus [\tau(I_k)]$ be a canonical decomposition of G. If $G(\tau_{I_1}) = G$, then for each $j \in \bar{n} \setminus I_1$ there exists $i \in I_1$ such that $\tau_i \leq \tau_j$.*

PROOF. Since τ_{I_1} and each τ_j are in T_G, $G(\tau_{I_1}) = G$ implies that $\tau_{I_1} \leq \tau_j$ for all $j \in \bar{n}$. If $j \in \bar{n} \setminus I_1$, and since $\bar{n} = \cup_{i \in \bar{k}} I_i$ by Proposition 1(a), there is no harm in assuming that $j \in I_2$. Then, Proposition 1(d) provides an $i \in I_1$ such that $\tau_i \leq \tau_{I_1} \vee \tau_{I_2}$. Consequently, $\tau_{I_1} \vee \tau_{I_2} \leq \tau_{I_1} \vee \tau_j = \tau_j$ gives $\tau_i \leq \tau_j$. □

Call a bracket group $G = [\tau_1, \ldots, \tau_n]$ *type I* if it satisfies the following three conditions.

(i) G is not strongly indecomposable.
(ii) $\tau_1, \ldots, \tau_n$ is cotrimmed.
(iii) If $G \doteq [\tau(I_1)] \oplus \ldots \oplus [\tau(I_k)]$ is a canonical decomposition of G, there exists $i \in \bar{k}$ such that $G(\tau_{I_i}) = G$ and rank $[\tau(I_i)] \geq 2$.

In view of Proposition 1(c), a type I bracket group G has the form $G \doteq G_1 \oplus C$ where G_1 is strongly indecomposable of rank at least 2 and C is completely decomposable. Also note that the group G of Lemma 4 is type I, provided that it is not almost completely decomposable. We next deal with this exceptional case.

Here and in the sequel we use without further mention the main theorem of [**GU2**]; namely, any quasi-summand of a bracket group is quasi-isomorphic to a bracket group.

LEMMA 5. *Suppose* $G = [\tau_1, ..., \tau_n]$ *is almost completely decomposable and* $H = [\sigma_1, \ldots, \sigma_n]$. *If* $r_G(\tau) = r_H(\tau)$ *for all types* τ, *then* $G \doteq H$.

PROOF. Induct on n. The case $n = 2$ being trivial, we may assume $n \geq 3$. Moreover, we may assume that $\sigma_1, \ldots, \sigma_n$ is cotrimmed. Since rank $G = n-1$ and G is almost completely decomposable, $|\mathrm{Max}(G)| \leq n-1$, while if H is strongly indecomposable, then $\sigma_i \in \Delta_H = \Delta_G \subsetneq \mathrm{Max}(G)$ for all $i \in \bar{n}$ (see, for example, [**AV3**]). Consequently, H cannot be strongly indecomposable.

Let $H \doteq [\sigma(J_1)] \oplus \ldots \oplus [\sigma(J_\ell)]$ be a canonical decomposition of H ($\ell \geq 2$). For each $i \in \bar{\ell}$, set $H_i = [\sigma(J_i)]$ and $\mu_i = \sigma_{J_i}$. Thus, for each i, $H(\mu_i) \doteq H_i \oplus F_i$, where $F_i = \oplus_{j \neq i} H_j(\mu_i)$ is completely decomposable by Proposition 1(c).

If $H(\mu_i) \neq H$ for all $i \in \bar{\ell}$, induction applies to each $H(\mu_i)$ and we easily conclude that $G \doteq H$. Otherwise, we may suppose that $H(\mu_1) = H$. In this case, $H_2, \ldots, H_\ell$ are all rank-1 by Proposition 1(c). Now, either H_1 is also rank-1, in which case the result is clear, or rank $H_1 \geq 2$ and $H(\mu_2) \neq H$. In the latter case, induction applies to $H(\mu_2)$ to show that H_2 is isomorphic to a strongly indecomposable quasi-summand of G and $G \doteq H$. Consequently, rank $H_1 = 1$. □

In final preparation for the proof of the Main Theorem, we need the following result of [**G**], which we state without proof.

THEOREM 1. ([**G**]). *Let G and H be torsion-free abelian groups of finite rank. If G embeds in a direct sum of copies of H and if H embeds in a direct sum of copies of G, then G and H have a common quasi-summand.*

In order to put the proof of the Main Theorem in a form that is more easily digested, we separate key parts of the argument in Lemmas 6 through 11 below. To facilitate the statement of these lemmas, we shall at times assume the following.

STANDING HYPOTHESIS. *G and H are bracket groups of equal ranks and if τ is a type with* rank $G(\tau) <$ rank G, *then* $G(\tau) \doteq H(\tau)$.

Those results which depend on the Standing Hypothesis will be clearly indicated. Also, we shall frequently use the following fact without further mention.

If B is a strongly indecomposable bracket group and if A is a torsion-free group of finite rank such that there is a nonzero composition $B \to A \to B$, then B is a quasi-summand of A. This is easily seen since $\mathrm{End}(B)$ is isomorphic to a subring of $\mathbb{Q}$ (see, for example, [**AV3**], [**FM**] or [**GU1**]).

The next result will prove to be our main tool in constructing nontrivial homomorphisms between bracket groups.

LEMMA 6. (Standing Hypothesis). *Suppose* $G = G_1 \oplus C$ *is type* I *with* G_1 *strongly indecomposable of rank at least 2 and* C *completely decomposable. Then,* C *embeds in a direct sum of copies of* H *and* $\mathrm{Hom}(G_1, H) \neq 0$.

PROOF. The fact that C embeds in a direct sum of copies of H follows from Proposition 1(c) and the fact that $G(\tau)$ and $H(\tau)$ are quasi-isomorphic whenever rank $G(\tau) <$ rank G. To show that $\mathrm{Hom}(G_1, H) \neq 0$, we introduce an additional definition. A set of types T with $|T| \geq 3$ is called a *minimal dependency set for* H if there is a set of elements $\{h_\tau \mid \tau \in T\}$ satisfying the following three conditions.

6(a) $h_\tau \in H(\tau)$ for each $\tau \in T$.

6(b) $\sum_{\tau \in T} h_\tau = 0$

6(c) If S is nonempty proper subset of T, then $\{h_\tau \mid \tau \in S\}$ is $\mathbb{Z}$-independent in H.

By a standard argument as in [**AV3**] or [**GU3**], it is seen that if $\{\tau_1, \dots, \tau_r\}$ is a minimal dependency set for H, there is an embedding of the bracket group $[\tau_1, \dots, \tau_r]$ into H. Furthermore, if T is a set of types with $|T| \geq 3$, then T contains a minimal dependency set for H if the following conditions hold.

6(d) $\sum_{\tau \in T} \mathrm{rank}\ H(\tau) > \mathrm{rank}\ H$; and

6(e) $H(\sigma) \not\subseteq H(\tau)$ for distinct $\sigma, \tau \in T$.

In our standard notation for type I groups, let $G = [\tau_1, \dots, \tau_n] \doteq G_1 \oplus C$ where $\tau_1, \dots, \tau_n$ is cotrimmed, $G_1 = [\tau_1, \dots, \tau_k]$ is strongly indecomposable of rank at least 2 and C is completely decomposable. By Lemma 4, for each $j > k$ there exists $i \leq k$ such that $\tau_i \leq \tau_j$. As a consequence $\sum_{i=1}^{k} \mathrm{rank}\ H(\tau_i) > \mathrm{rank}\ H$. Therefore 6(d) holds for $T = \{\tau_1, \dots, \tau_k\}$. Observe that 6(e) holds as well since each $\tau \in T$ is in T_G and G_1 is strongly indecomposable. Therefore, $\{\tau_1, \dots, \tau_k\}$ contains a minimal dependency set for H.

After reindexing as necessary, we may assume that $\{\tau_1, \dots, \tau_r\}$ is a minimal dependency set for H with $3 \leq r \leq k$. Observe that there is a natural nonzero homomorphism $[\tau_1, \dots, \tau_k] \to [\tau_1, \dots, \tau_r]$ and, as observed above, $[\tau_1, \dots, \tau_r]$ embeds in H. Consequently there is a nonzero composition

$$G_1 = [\tau_1, \dots, \tau_k] \to [\tau_1, \dots, \tau_r] \to H$$

and $\mathrm{Hom}(G_1, H) \neq 0$. □

LEMMA 7. (Standing Hypothesis.) *Suppose* G *is neither almost completely decomposable nor strongly indecomposable. If* G *is not type* I, *then* G *embeds in a direct sum of copies of* H.

PROOF. Under the stated hypotheses, there are types $\delta_1, \dots, \delta_k$ such that $G(\delta_i) \doteq H(\delta_i)$ and each strongly indecomposable quasi-summand of G is contained in some $G(\delta_i)$. □

LEMMA 8. (Standing Hypothesis.) *Suppose $G \doteq G_1 \oplus C$ is type* I *with G_1 strongly indecomposable of rank at least 2 and C completely decomposable. Suppose further that H is neither almost completely decomposable nor strongly indecomposable. Let X be a summand of C of maximal type τ such that* $\operatorname{Hom}(G_1, X) \neq 0$. *If H is not type* I, *then X is a quasi-summand of H.*

PROOF. Under the Standing Hypothesis, $G(\tau) \doteq H(\tau)$. Thus, there is an embedding $X \to H_0$, where H_0 is a strongly indecomposable quasi-summand of H. By Lemma 7 applied to H, there is an embedding of H_0 into a direct sum of copies of G. Consequently, we obtain a nonzero composition $X \to H_0 \to G$. Observe that this composite followed by projection of G onto G_1 must be 0, since otherwise the condition $\operatorname{Hom}(G_1, X) \neq 0$ would imply that rank $G_1 = 1$, contrary to hypothesis. Therefore, since τ is a maximal type in C, there is a nonzero composition

$$X \to H_0 \to X,$$

and so X is a quasi-summand of H_0 and hence one of H as well. □

Our next result does not depend on the Standing Hypothesis.

LEMMA 9. *Suppose G and H are torsion-free groups of finite rank with G_1 and G_2 (respectively H_1 and H_2) pure subgroups of G (respectively H). Set $G_0 = G_1 \cap G_2$ and $H_0 = H_1 \cap H_2$. Suppose further that $G_i \doteq H_i$ for all $i \in \{0, 1, 2\}$ and if $f \in \operatorname{Hom}(G_i, H_i)$ for $i \in \{1, 2\}$, then $f(G_0) \subseteq H_0$. If* rank *$G_0 = 1$, then*

$$G_1 + G_2 \doteq H_1 + H_2.$$

PROOF. Define mappings $\theta : G_1 \oplus G_2 \to G_1 + G_2$ and $\varphi : H_1 \oplus H_2 \to H_1 + H_2$ by $\theta(g_1, g_2) = g_1 - g_2$ and $\varphi(h_1, h_2) = h_1 - h_2$. Then θ and φ are epimorphisms with respective kernels $d(G_0)$ and $d'(H_0)$, where $d : G_0 \to G_1 \oplus G_2$ and $d' : H_0 \to H_1 \oplus H_2$ are the diagonal embeddings. Select quasi-isomorphisms $f : G_1 \to H_1$ and $g : G_2 \to H_2$. Since $f|G_0$ and $g|G_0$ are both nonzero elements of $\operatorname{Hom}(G_0, H_0)$ and rank $\operatorname{Hom}(G_0, H_0) = 1$, we can multiply f and g by suitable nonzero integers and assume that $f|G_0 = g|G_0$. Thus, we obtain a commutative diagram

$$\begin{array}{ccccccccc} 0 & \longrightarrow & G_0 & \xrightarrow{d} & G_1 \oplus G_2 & \xrightarrow{\theta} & G_1 + G_2 & \longrightarrow & 0 \\ & & \downarrow{\scriptstyle f|G_0 = g|G_0} & & \downarrow{\scriptstyle f \oplus g} & & \downarrow & & \\ 0 & \longrightarrow & H_0 & \xrightarrow[d']{} & H_1 \oplus H_2 & \xrightarrow[\varphi]{} & H_1 + H_2 & \longrightarrow & 0 \end{array}$$

with exact rows, which induces a quasi-isomorphism $G_1 + G_2 \doteq H_1 + H_2$. □

Two more lemmas under the Standing Hypothesis precede the proof of the Main Theorem.

LEMMA 10. (Standing Hypothesis). *Suppose* $G = [\tau_1, \ldots, \tau_n]$ *with* $\tau_1, \ldots, \tau_n$ *cotrimmed. If* $\Delta_G \setminus \{\tau_1, \ldots, \tau_n\}$ *is nonempty and* $r_G(\tau) = r_H(\tau)$ *for all types* τ, *then* $G \doteq H$.

PROOF. Suppose $\sigma \in \Delta_G \setminus \{\tau_1, \ldots, \tau_n\}$. Then, $\sigma = \tau_I \vee \tau_{I'}$ where I is a nonempty proper subset of $\bar{n}$. Clearly $|I|$ and $|I'|$ each have at least two elements since $\tau_1 \ldots, \tau_n$ is cotrimmed. Further observe that both $G(\tau_I)$ and $G(\tau_{I'})$ are proper subgroups of G by an argument similar to that in the proof of Lemma 2. Under the Standing Hypothesis, we have $G(\tau_I) \doteq H(\tau_I)$, $G(\tau_{I'}) \doteq H(\tau_{I'})$ and $G(\tau_I) \cap G(\tau_{I'}) = G(\sigma) \doteq H(\sigma) = H(\tau_I) \cap H(\tau_{I'})$. Since rank $G(\sigma) = 1$, Lemma 9 implies that $G = G(\tau_I) + G(\tau_{I'}) \doteq H(\tau_I) + H(\tau_{I'})$. Therefore, G embeds in H and Lemma 3 completes the proof. $\square$

LEMMA 11. (Standing Hypothesis). *Suppose* $G \doteq G_1 \oplus C$ *and* $H \doteq H_1 \oplus D$ *are type* I *with* G_1 *and* H_1 *strongly indecomposable of rank at least* 2 *and* C *and* D *completely decomposable. If* G_1 *embeds in* H_1, *then* G *and* H *have a common quasi-summand.*

PROOF. Let $\varphi : G_1 \to H_1$ be an embedding. By Lemma 6, there exists a nonzero map $\psi : H_1 \to G_1 \oplus C$. If $\psi(H_1) \not\subseteq C$, there is a nonzero composition

$$H_1 \overset{\psi'}{\to} G_1 \overset{\varphi}{\to} H_1$$

where ψ' is ψ followed by projection onto G_1. Thus, H_1 is a quasi-summand of G_1 and $H_1 \doteq G_1$, since G_1 is strongly indecomposable.

On the other hand, if $\psi(H_1) \subseteq C$, there is a nonzero map $H_1 \to G_0$, where G_0 is a rank-1 quasi-summand of C. By Lemma 6, G_0 embeds in a strongly indecomposable quasi-summand H_0 of H. If $H_0 \doteq H_1$, we obtain a nonzero composition $H_1 \to G_0 \to H_1$ contradicting rank $G_0 = 1 <$ rank H_1. Consequently, H_0 is a rank-1 quasi-summand of D. By Proposition 1, there is a type τ such that $G(\tau) \doteq G_0 \oplus G_1(\tau) \oplus C'$ where $G_1(\tau) \oplus C'$ is completely decomposable. Since rank $G(\tau) <$ rank G, $G(\tau) \doteq H(\tau)$ by the Standing Hypothesis. This, in conjunction with type $H_0 \geq$ type $G_0 \geq \tau$, gives

$$G_0 \oplus G_1(\tau) \oplus C' \doteq H_0 \oplus H_1(\tau) \oplus D'$$

where $H_1(\tau) \oplus D'$ is completely decomposable and D' is a quasi-summand of H. Since G_0 does not embed in H_1, it follows that G_0 is isomorphic to some rank-1 quasi-summand of H. $\square$

We are now ready for the proof proper of our principal result.

PROOF OF THE MAIN THEOREM. If $G \doteq H$, it is clearly necessary that $r_G(\tau) = r_H(\tau)$ for all types τ. To see that the condition is sufficient, we assume that $r_G(\tau) = r_H(\tau)$ for all types τ and we induct on rank $G =$ rank H. If rank $G = 1$, the result is trivial, so we may assume rank $G \geq 2$. Observe that the condition $r_G(\tau) = r_H(\tau)$ for all τ implies $T_G = T_H$ (see, for example, [**GU3**]). Moreover, the induction hypothesis implies that we can invoke the Standing Hypothesis. For the remainder of the proof we set the following notation: $G = [\tau_1, \dots, \tau_n]$, $H = [\sigma_1, \dots, \sigma_n]$ and both $\tau_1, \dots, \tau_n$ and $\sigma_1, \dots, \sigma_n$ are cotrimmed.

First observe that Lemma 5 says that we may assume that neither G nor H is almost completely decomposable. Set $\Delta = \Delta_G = \Delta_H$. Next observe that Lemma 10 implies that we may also assume

$$\Delta \subseteq \{\tau_1, \dots, \tau_n\} \cap \{\sigma_1, \dots, \sigma_n\}. \tag{1}$$

If either G or H is strongly indecomposable, then (1) yields $\{\tau_1, \dots, \tau_n\} = \Delta = \{\sigma_1, \dots, \sigma_n\}$ in which case it is obvious that $G \doteq H$.

To summarize what we have done so far, we may assume that G is neither almost completely decomposable nor strongly indecomposable. Of course, these same assumptions can be imposed on H as well. We now consider three cases.

CASE 1. If G and H are both not type I, then Lemma 7 and Theorem 1 imply G and H have a common quasi-summand. Therefore $G \doteq H$ by induction.

CASE 2. Suppose now that G is type I but H is not type I. Write $G \doteq G_1 \oplus C$ where G_1 is strongly indecomposable of rank at least 2 and C is completely decomposable. By Lemma 6, there is a nonzero map $G_1 \to H$. Since H is not type I, we can interchange the roles of G and H in Lemma 7 to obtain a nonzero composition $\Theta : G_1 \to H_0 \to G_1 \oplus C$ where H_0 is a strongly indecomposable quasi-summand of H.

If $\Theta(G_1) \not\subseteq G_1$, then $\operatorname{Hom}(G_1, X) \neq 0$ for some summand X of C of maximal type. Therefore, by Lemma 8, X would be a common quasi-summand of G and H and $G \doteq H$, by induction. On the other hand, if $\Theta(G_1) \subseteq G_1$, there is a nonzero composition $G_1 \to H_0 \to G_1$. We conclude that $G_1 \doteq H_0$, and again $G \doteq H$ by induction. We now have reduced to the troublesome case.

CASE 3. G and H are both type I. In this case

$$G \doteq G_1 \oplus C \text{ and } H \doteq H_1 \oplus D \tag{2}$$

where G_1 and H_1 are strongly indecomposable of ranks at least 2 and C and D are completely decomposable. From Proposition 1, after a suitable reindexing, we have $G_1 = [\tau_1, \dots, \tau_r]$ and $H_1 = [\sigma_1, \dots, \sigma_s]$ with $r \geq 3$ and $s \geq 3$.

We first claim that we can assume that both $\tau_1, ..., \tau_r$ and $\sigma_1, \dots, \sigma_s$ are cotrimmed. Indeed, suppose that $\tau_1, \dots, \tau_r$ is not cotrimmed. Then, by Lemma 4, we have

$$\rho = \bigwedge\{\tau_i \mid i \neq i_0, 1 \leq i \leq r\} \not\leq \bigwedge\{\tau_i \mid r < i \leq n\}$$

for some $i_0 \in \bar{r}$. Note $\rho \not\leq \bigwedge\{\tau_i \mid r < i \leq n\}$ implies that rank $G(\rho) <$ rank G, since $C(\rho) \subsetneq C$. Thus $G(\rho) \doteq H(\rho)$ and we conclude that G_1 embeds in H_1. By Lemma 11, G and H have a common quasi-summand and we have $G \doteq H$ by induction. Therefore, we may assume that $\tau_1, \dots, \tau_r$ is cotrimmed.

Similarly, we may also assume that $\sigma_1, \dots, \sigma_s$ is cotrimmed. Also, we will need the observation that if γ and δ are types of rank-1 quasi-summands of C and D, respectively, then Lemma 4 implies that there exist $i \in \bar{r}$ and $j \in \bar{s}$ such that $\tau_i \leq \gamma$ and $\sigma_j \leq \delta$.

Suppose $\sigma \in \operatorname{Max}(C)$. Then, σ is the type of a rank-1 quasi-summand of C. By the observation just made above, and the fact that G_1 is strongly indecomposable, $\sigma \in \operatorname{Max}(G) = \operatorname{Max}(H)$. If $\sigma \in \operatorname{Max}(D)$, then C and D have a common rank-1 quasi-summand of type σ and $G \doteq H$ by induction. Therefore, we may assume that $\sigma \in \operatorname{Max}(H_1)$.

We first consider the possibility that $\sigma \notin \{\sigma_1, \dots, \sigma_s\}$. Then $\sigma = \sigma_I \vee \sigma_{I'}$ for some nonempty proper subset I of $\bar{s}$. Since $\sigma_1, \dots, \sigma_s$ is cotrimmed, $|I| \geq 2$ and $|I'| \geq 2$. By Lemma 2 (applied to H_1) and induction we now have

$$G(\sigma_I) \doteq G_1(\sigma_I) \oplus C(\sigma_I) \doteq H(\sigma_I) \doteq H_1(\sigma_I) \oplus D(\sigma_I), \text{ and}$$
$$G(\sigma_{I'}) \doteq G_1(\sigma_{I'}) \oplus C(\sigma_{I'}) \doteq H(\sigma_{I'}) \doteq H_1(\sigma_{I'}) \oplus D(\sigma_{I'}).$$

Observe that either $C(\sigma_I)$ and $D(\sigma_I)$ share a summand (in which case G and H share a summand and $G \doteq H$ by induction), or else $H_1(\sigma_I) \doteq C(\sigma_I) \oplus K_1$ for some quasi-summand K_1 of $G_1(\sigma_I)$. Thus we may assume the latter case and obtain $G_1(\sigma_I) \doteq K_1 \oplus D(\sigma_I)$. By a similar argument, we may also assume that $H_1(\sigma_{I'}) \doteq C(\sigma_{I'}) \oplus K_1'$ for some quasi-summand K_1' of $G_1(\sigma_{I'})$ with $G_1(\sigma_{I'}) \doteq K_1' \oplus D(\sigma_{I'})$.

Since $\sigma_1, \dots, \sigma_s$ is cotrimmed, Lemma 3 applies to H_1 and we obtain $H_1(\sigma_I) \doteq B \oplus X$, where $B = [\sigma(I), \sigma]$ and X is homogeneous completely decomposable of type σ. If B had a rank-1 quasi-summand, Lemma 3 would imply that $\sigma = \sigma_i$ for some $i \in I$. But $\sigma \not\leq \sigma_i$ for each $i \in I$. Consequently, if B quasi-decomposes, it decomposes into strongly indecomposable quasi-summands with each of rank at least 2. Recalling that

$$H_1(\sigma_I) \doteq C(\sigma_I) \oplus K_1 \doteq B \oplus X, \tag{3}$$

with $C(\sigma_I)$ completely decomposable, it follows that B is a quasi-summand of K_1.

We claim that in fact $B \doteq K_1$. To see this, recall that Lemma 4 implies that $\sigma \geq \tau_j$ for some $j \in \bar{r}$. Say $\sigma \geq \tau_r$. Then, $1 \leq \operatorname{rank} B(\sigma) \leq \operatorname{rank} K_1(\sigma) \leq \operatorname{rank} G_1(\sigma_I)(\sigma) = \operatorname{rank} G_1(\sigma) \leq \operatorname{rank} G_1(\tau_r) = 1$. From this we obtain

$$\operatorname{rank} B(\sigma) = \operatorname{rank} K_1(\sigma) = \operatorname{rank} G_1(\sigma) = 1 \tag{4}$$

If we consider (3) and (4) in conjunction with the facts that $C(\sigma_I)$ is completely decomposable, X is homogeneous completely decomposable of type σ and B is a quasi-summand of K_1, we conclude that B is quasi-isomorphic to K_1, as claimed. Similarly, $B' = [\sigma(I'), \sigma] \doteq K_1'$.

Consider the map $\phi : B \oplus B' \to H_1$, given by $\phi(b, b') = b - b'$. It is easily seen that ϕ maps onto H_1. By comparing ranks, we also see that $1 = \operatorname{rank}(\operatorname{Ker} \phi) =$

rank $(B \cap B')$. So, $B \cap B'$ is a rank-1 pure subgroup of $H_1(\sigma)$, which is homogeneous completely decomposable of type σ. Thus, type $(B \cap B') = \sigma$. We conclude that $B \cap B' \cong G_1(\sigma) = K_1 \cap K_1'$, and any homomorphisms $f : B \to K_1$ and $g : B' \to K_1'$ map $B \cap B'$ into $K_1 \cap K_1'$. So, all hypotheses of Lemma 9 are satisfied and we conclude that H_1 embeds in G_1. Therefore, G and H have a common quasi-summand by Lemma 11 and $G \doteq H$ by induction.

We now have reduced to the case $\mathrm{Max}(C) \subseteq \{\sigma_1, \dots, \sigma_s\}$ and (by an argument similar to the above) $\mathrm{Max}(D) \subseteq \{\tau_1, \dots, \tau_r\}$. In this case, observe that if γ is the type of a rank-1 quasi-summand of C, then $\gamma \le \sigma_j$ for some $j \in \bar{s}$. Likewise, if δ is the type of a rank-1 quasi-summand of D, $\delta \le \tau_i$ for some $i \in \bar{r}$. By (1) and (2) and the observations just made, we may reindex and obtain

$$G \doteq [\tau_1, \dots, \tau_t, \tau_{t+1}, \dots, \tau_r] \oplus C \text{ and } H \doteq [\tau_1, \dots, \tau_t, \sigma_{t+1}, \dots, \sigma_s] \oplus D,$$

where $\tau_1, \dots, \tau_t \in \Delta \subseteq \{\tau_1, \dots, \tau_r\} \cap \{\sigma_1 \dots, \sigma_s\}$; for $t+1 \le i \le r$, there is $t+1 \le j \le s$ with $\tau_i \le \sigma_j$; and for $t+1 \le u \le s$, there is $t+1 \le v \le s$ with $\sigma_u \le \tau_v$. Since $\tau_{t+1}, ..., \tau_r \in \mathrm{Max}(G_1)$ are distinct and $\sigma_{t+1}, \dots, \sigma_s \in \mathrm{Max}(H_1)$ are distinct, it now follows that $r = s$, and after reindexing, $\sigma_i = \tau_i$ for all $i \in \bar{r}$. Therefore $G_1 \doteq H_1$ and induction applies to complete the proof. □

In conclusion, we present several applications of the Main Theorem. The first deals with a certain connection between representations of posets and bracket groups.

Suppose $(T, \le)$ is a finite poset and F is a field. An *F-representation of T* is a family $(V, V_i\ ; i \in T)$, where V is a finite-dimensional F-space, each V_i is a subspace of V and $V_i \subseteq V_j$ whenever $i \le j$. The *dimension vector* of the F-representation $\mathcal{V} = (V, V_i\ ;\ i \in T)$ is defined by $\dim \mathcal{V} = (\dim_F V, \dim_F V_i; i \in T)$, a $(|T| + 1)$-tuple with nonnegative integer entries. If $\mathcal{U} = (U, U_i\ ;\ i \in T)$ is another such F-representation of T, then $\mathcal{U}$ and $\mathcal{V}$ are *isomorphic* if there is an F-isomorphism $\varphi : U \to V$ such that $\varphi(U_i) = V_i$ for all $i \in T$. Of course if $\mathcal{U}$ and $\mathcal{V}$ are isomorphic, then $\dim \mathcal{U} = \dim \mathcal{V}$. However in general, the converse does not hold.

If G is a bracket group with typeset T_G, we consider two representations of the finite poset T_G^{opp}, the poset consisting of the elements of T_G with the reverse ordering. One of these representations, given in [**GM**], is over the field $\mathbb{Z}_2 = \mathbb{Z}/2\mathbb{Z}$, and the other is over the rational field $\mathbb{Q}$, due essentially to M.C.R. Butler [**B**] as reported in [**A2**].

First, if $G = [\tau_1, \dots, \tau_n]$ and $2^{\bar{n}}$ is the Boolean ring of subsets of $\bar{n}$, then $2^{\bar{n}}$ may be regarded as a vector space over $\mathbb{Z}_2$ (where, as usual, addition in $2^{\bar{n}}$ is given by the symmetric difference). For each $\sigma \in T_G$, define $\bar{n}_G(\sigma) = \{E \in 2^{\bar{n}} \mid \tau_E \vee \tau_{E'} \ge \sigma\}$, with the understanding that if E is the empty set, then $\tau_E =$ type $(\mathbb{Q})$, the type determined by the characteristic $(\infty, \infty, \infty, ...)$. It is routine to verify that each $\bar{n}_G(\sigma)$ is a subspace of $2^{\bar{n}}$ and that $\mathcal{R}_G = (2^{\bar{n}}, \bar{n}_G(\sigma)\ ;\ \sigma \in T_G^{opp})$ is a $\mathbb{Z}_2$-representation of T_G^{opp}. In [**GM**] it is shown that two bracket groups G

and H with equal typesets are quasi-isomorphic if and only if the respective $\mathbb{Z}_2$-representations $\mathcal{R}_G$ and $\mathcal{R}_H$ are isomorphic.

The $\mathbb{Q}$-representation of T_G^{opp} that we consider is given by

$$\mathcal{Q}_G = (\mathbb{Q}G, \mathbb{Q}G(\sigma) \ ; \ \sigma \in T_G^{opp})$$

where $\mathbb{Q}G = \mathbb{Q} \bigotimes G$ and $\mathbb{Q}G(\sigma) = \mathbb{Q} \bigotimes G(\sigma)$. Observe that $\mathcal{Q}_G$ is a variant of the usual construction, where T_G is replaced by the set of join-irreducible elements in the lattice generated by $\tau_1, \dots, \tau_n$.

The content of our first corollary is that $\dim \mathcal{R}_G$ and $\dim \mathcal{Q}_G$ each determine G up to quasi-isomorphism and both $\mathcal{R}_G$ and $\mathcal{Q}_G$ up to isomorphism.

COROLLARY 1. *For two bracket groups G and H with $T_G = T_H$, the following statements are equivalent.*

(a) $G \doteq H$.

(b) *The $\mathbb{Z}_2$-representations $\mathcal{R}_G$ and $\mathcal{R}_H$ are isomorphic.*

(c) $\dim \mathcal{R}_G = \dim \mathcal{R}_H$.

(d) *The $\mathbb{Q}$-representations $\mathcal{Q}_G$ and $\mathcal{Q}_H$ are isomorphic.*

(e) $\dim \mathcal{Q}_G = \dim \mathcal{Q}_H$.

PROOF. Since $\dim \bar{n}_G(\sigma) = \text{rank } G(\sigma) + 1$ by a result of [**GM**], and $\dim \mathbb{Q}G(\sigma) = \text{rank } G(\sigma)$, the equivalence of (a), (c) and (e) follows from the Main Theorem. The equivalence of (a) and (b) is the main result of [**GM**], and the equivalence of (a) and (d) is a consequence of Theorem 2.2 of [**A2**]. □

If G is a torsion-free group and if X is a rank-1 group of type τ, recall that the *τ-radical* of G is defined by $G[\tau] = \bigcap\{ \text{ Ker } f \mid f \in \text{ Hom}(G, X)\}$. For a set of types M, set

$$r_G[M] = \text{rank } \bigcap_{\tau \in M} G[\tau].$$

In a certain sense, $r_G[M]$ is the dual of the invariant $r_G(M)$ mentioned earlier. By Proposition 1.9 of [**AV1**], for a type σ, $G(\sigma) = \bigcap\{G[\tau] \mid \sigma \not\leq \tau\}$. Therefore, from our Main Theorem, we immediately have

COROLLARY 2. *Suppose G and H are bracket groups of equal ranks. Then $G \doteq H$ if and only if $r_G[M] = r_H[M]$ for all finite subsets M of the lattice-generated by $T_G \cup T_H$.*

For nonzero subgroups $A_1, \dots, A_n$ of $\mathbb{Q}$, define $\mathcal{G}(A_1, \dots, A_n)$ to be the kernel of the codiagonal map $A_1 \oplus \dots \oplus A_n \to \mathbb{Q}$ given by $(a_1, \dots, a_n) \mapsto \sum a_i$. Let $\mathcal{C}^{(\ )}$ denote the class of all groups of the form $\mathcal{G}(A_1, \dots, A_n)$, and let $\mathcal{C}^{[\ ]}$ denote the class of bracket groups. The classes $\mathcal{C}^{(\ )}$ and $\mathcal{C}^{[\ ]}$ are each classified separately in [**AV3**]. However, using Corollary 2, we can unify these classifications.

COROLLARY 3. *Suppose G and H are torsion-free groups of equal ranks in the class $\mathcal{C}^{[\ ]} \cup \mathcal{C}^{(\ )}$. Then, $G \doteq H$ if and only if $r_G[M] = r_H[M]$ for all finite subsets M of the lattice generated by $T_G \cup T_H$.*

PROOF. By Corollary 2 and by Theorem 7 and Corollary 8 of [**AV3**], the invariants $r_*[M]$ serve to classify $\mathcal{C}^{[\,]}$ and $\mathcal{C}^{(\,)}$, respectively. Moreover, by Corollary 10 of [**GU3**], if $G \in \mathcal{C}^{(\,)}$ and if H is any finite rank Butler group, then $r_G[M] = r_H[M]$ for all M implies that $G \doteq H$. The result now follows. □

Finally, applying the Butler duality of [**AV2**], we obtain the following duals of the Main Theorem and Corollary 3.

COROLLARY 4. (Dual version of the Main Theorem). *$G = \mathcal{G}(A_1, \dots, A_n)$ and $H = \mathcal{G}(B_1, \dots, B_n)$ are quasi-isomorphic if and only if* rank $G[\tau]$ = rank $H[\tau]$ *for all types τ in the lattice generated by $T_G \cup T_H$.*

COROLLARY 5. (Dual version of Corollary 3). *Suppose G and H are torsion-free groups in the class $\mathcal{C}^{[\,]} \cup \mathcal{C}^{(\,)}$. Then, $G \doteq H$ if and only if $r_G(M) = r_H(M)$ for all finite subsets M of the lattice generated by $T_G \cup T_H$.*

REFERENCES

[A1] D. Arnold, Finite Rank Torsion-Free Abelian Groups and Rings, Lecture Notes in Math., vol. 931, Springer-Verlag, New York, 1982.

[A2] D. Arnold, *Representations of partially ordered sets and abelian groups*, Abelian Group Theory, Contemporary Math., vol. 87, Amer. Math. Soc., Providence, 1989, 91-109.

[AV1] D. Arnold and C. Vinsonhaler, *Pure subgroups of finite rank completely decomposable groups II*, Abelian Group Theory, Lecture Notes in Math., vol. 1006, Springer-Verlag, New York, 1983, 97-143.

[AV2] D. Arnold and C. Vinsonhaler, *Duality and invariants for Butler groups*, Pacific J. Math. 148 (1991), 1-10.

[AV3] D. Arnold and C. Vinsonhaler, *Quasi-isomorphism invariants for two classes of finite rank Butler groups*, Proc. Amer. Math. Soc. 118 (1993), 19-26.

[B] M. C. R. Butler, *Torsion-free modules and diagrams of vector spaces*, Proc. London Math. Soc. 15 (1968), 635-652.

[F] L. Fuchs, *Infinite Abelian Groups*, vol. II, Academic Press, New York, 1973.

[FM] L. Fuchs and C. Metelli, *On a class of Butler groups*, Manuscripte Math. 71 (1991), 1-28.

[G] H. P. Goeters, *Torsion-free abelian groups with finite rank endomorphism rings*, Questiones Math. 14 (1991), 111-115.

[GM] H. P. Goeters and C. Megibben, *Quasi-isomorphism and $\mathbb{Z}_2$-representations for a class of Butler groups*, preprint.

[GU1] H. P. Goeters and W. Ullery, *Butler groups and lattices of types*, Comment. Math. Univ. Carolinae 31 (1990), 613-619.

[GU2] H. P. Goeters and W. Ullery, *Quasi-summands of a certain class of Butler groups*, Abelian Groups: Proceedings of the 1991 Curacao Conference, Marcel Dekker, New York, 1993, 167-174.

[GU3] H. P. Goeters and W. Ullery, *The Arnold-Vinsonhaler invariants*, Houston J. Math., to appear.

[GU4] H. P. Goeters and W. Ullery, *On quasi-decompositions of $\mathcal{B}^{(1)}$-groups*, preprint.

[L] W. Y. Lee, *Codiagonal Butler groups*, Chinese J. Math. 17 (1989), 259-271.

DEPARTMENT OF MATHEMATICS, AUBURN UNIVERSITY, AUBURN, AL
E-mail address: goetehp@mail.auburn.edu

DEPARTMENT OF MATHEMATICS, AUBURN UNIVERSITY, AUBURN, AL
E-mail address: ullery@mail.auburn.edu

DEPARTMENT OF MATHEMATICS, UNIVERSITY OF CONNECTICUT, STORRS, CT
E-mail address: vinson@uconnvm.bitnet

Contemporary Mathematics
Volume **171**, 1994

K_0 of Regular Rings with Bounded Index of Nilpotence

K. R. GOODEARL

Dedicated to the memory of Dick Pierce

ABSTRACT. A complete description of K_0 of any (von Neumann) regular ring R with bounded index of nilpotence is given, in which $K_0(R)$ is represented as a partially ordered group of locally constant rational-valued functions on a boolean space subject to certain constraints. This representation is derived from a corresponding representation of an arbitrary Riesz group with an order-unit of finite index. The problem of realizing any such group as K_0 of a regular ring with bounded index is discussed, and some topological difficulties are pinpointed.

Introduction

The structure of regular rings with bounded index was first systematically investigated by Kaplansky in 1950 [**11**] (cf. [**5**, Chapter 7]); for a sample of more recent work see [**1**, **2**, **3**, **4**, **8**, **10**, **12**]. In the biregular case, the Pierce sheaf [**14**] provides a very useful tool, since then the stalks are simple artinian rings. Using this, Burgess and Goursaud gave a complete description of K_0 of any regular, biregular ring of bounded index [**1**].

In general, however, the stalks of Pierce sheaves can be almost arbitrarily complicated, and consequently the Pierce sheaf is a less useful tool than in the biregular case. For example, if F is a field, $\{R_i\}$ a collection of copies of some fixed matrix ring $M_n(F)$, and S a regular subring of $(\prod R_i)/(\bigoplus R_i)$ containing no central idempotents of $(\prod R_i)/(\bigoplus R_i)$ other than 0 and 1, then S is a Pierce

1991 *Mathematics Subject Classification.* 16E50, 16E20, 16S60, 19K14, 06F20.

This research was partially supported by research grants from the National Science Foundation.

This paper is in final form, and no version of it will be submitted for publication elsewhere.

stalk of a regular ring with index n. In particular, since calculating K_0 of a ring via its Pierce sheaf is necessarily at least as complicated as calculating K_0 of the Pierce stalks, Pierce sheaf methods do not generally yield manageable descriptions of K_0.

The main goal of this paper is to develop a precise description of $K_0(R)$ for an arbitrary regular ring R with bounded index, in terms of certain locally constant functions from a boolean space X to the rational field. (In fact, if R has index n, these functions will take values in $(1/n!)\mathbb{Z}$.) When R is biregular, X will just be the maximal ideal space of the boolean algebra of central idempotents in R (i.e., the base space for the Pierce sheaf), and our description of $K_0(R)$ will coincide with that of Burgess and Goursaud. In the general case, X corresponds to a collection of certain ideals I of R for which the factor rings R/I are semisimple artinian. However, the topology on X is not a hull-kernel topology, but is derived from the topology on the space $\mathbb{P}(R)$ of pseudo-rank functions on R; in fact, X is best described as a compact subset of $\mathbb{P}(R)$.

We also address the problem of which partially ordered abelian groups can be realized as K_0 of regular rings with bounded index. It is easily seen that such a group G must be a Riesz group (cf. [**7**]), and that G must have an order-unit which has finite index in a suitable sense (see Section 1). We show that any such G can be completely described in terms of suitable locally constant functions from a boolean space X to $\mathbb{Q}$; in fact, our description of K_0 of any regular ring with bounded index is derived from this ordered group representation. Using the Burgess-Goursaud result, we show that any lattice-ordered abelian group with an order-unit of finite index is isomorphic to K_0 of a regular, biregular ring with bounded index. The general case, however, is less tractable, for topological difficulties arise even in index 2. We discuss these difficulties in the final section of the paper, and show how to realize a large class of Riesz groups with order-units of index 2 as K_0 of regular rings with index 2.

All rings in this paper are associative with unit, and all modules are unital. We refer the reader to [**5**] and [**7**] for the general theories of regular rings and partially ordered abelian groups. We follow the notation of these books; in particular, $\mathbb{P}(R)$ will denote the compact convex set of all pseudo-rank functions on a regular ring R, and $S(G,u)$ will denote the compact convex set of all states on a pre-ordered abelian group G with an order-unit u. The symbol $\partial_e K$ denotes the set of extreme points of a convex set K, and $\operatorname{conv} X$ denotes the convex hull of a subset $X \subseteq K$. The *category of pre-ordered abelian groups with order-unit* has as objects all pairs (G,u) where G is a pre-ordered abelian group and u is an order-unit in G; the morphisms $(G,u) \to (H,v)$ in this category are the positive homomorphisms $f : G \to H$ such that $f(u) = v$. Such maps will be referred to as *morphisms* without repeating the name of the category. We view K_0 as a functor from the category of rings to the category of pre-ordered abelian groups with order-unit, under which a ring R is sent to the object $(K_0(R), [R])$.

1. Background and basics

Recall that any regular ring R with bounded index is unit-regular [**5**, Corollary 7.11]. Consequently, $K_0(R)$ is a *Riesz group*, i.e., a directed, partially ordered abelian group satisfying the Riesz interpolation and decomposition properties [**9**, Propositions II.10.2, II.10.3]. Moreover, it follows from [**5**, Theorem 15.12] that $K_0(R)$ is also unperforated, and therefore it is a *dimension group* (cf. [**7**]). In addition, R is Hausdorff and complete with respect to its N^*-metric [**6**, Theorem 1.3], from which it follows that $K_0(R)$ is archimedean and complete with respect to its order-unit norm [**6**, Theorem 2.11]. These properties in fact hold for any Riesz group with an order-unit of finite index, as we show in the following section.

Since bounded index for regular rings is equivalent to a condition on direct sums of (principal) right ideals [**5**, Theorem 7.2], we can easily translate this property into a natural condition in K_0, as follows.

DEFINITION. Given a positive element u in a partially ordered abelian group G, let us define the *index (of nilpotence)* of u to be the least positive integer n such that $(n+1)x \nleq u$ for all nonzero $x \in G^+$, provided such n exists. If no such n exists, we set the index of u equal to ∞.

PROPOSITION 1.1. *The index of a unit-regular ring R equals the index of the element $[R]$ in $K_0(R)$.*

PROOF. The elements $x \in K_0(R)$ satisfying $0 \leq x \leq [R]$ are precisely the classes $[A]$ for principal right ideals A of R. Hence, $[R]$ has index greater than a given positive integer n if and only if R contains a direct sum of $n+1$ nonzero pairwise isomorphic principal right ideals. By [**5**, Theorem 7.2], the latter condition occurs if and only if R has index greater than n. $\square$

Recall that a ring R is *biregular* provided that each principal two-sided ideal RaR in R is generated by a central idempotent. Among regular rings with bounded index, biregularity can be detected in K_0 as follows.

THEOREM 1.2. *Let R be a regular ring with bounded index. Then R is biregular if and only if $\partial_e S(K_0(R),[R])$ is compact, if and only if $K_0(R)$ is lattice-ordered.*

PROOF. By [**8**, Theorem 17], R is biregular if and only if $\partial_e \mathbb{P}(R)$ is compact. Since $\mathbb{P}(R)$ is affinely homeomorphic to $S(K_0(R),[R])$ [**5**, Proposition 17.12], the first equivalence of the theorem is clear. As noted above, $K_0(R)$ is an archimedean norm-complete dimension group. Therefore by [**7**, Corollary 15.10], $\partial_e S(K_0(R),[R])$ is compact if and only if $K_0(R)$ is lattice-ordered. $\square$

Note that any lattice-ordered abelian group is a Riesz group, since directedness and Riesz interpolation are clear. (In fact, lattice-ordered abelian groups are also unperforated [**7**, Proposition 1.24].) We show in Section 4 that every lattice-ordered abelian group with an order-unit of finite index can be realized as K_0 of a regular, biregular ring with bounded index.

2. Riesz groups with order-units of finite index, Part I

We start by deriving some general results about Riesz groups with order-units of finite index; in particular, we prove that any such group is unperforated, archimedean, and norm-complete. A structure theory for these groups is developed in Section 5.

DEFINITION. An element x in a partially ordered set P is *meet-irreducible* provided the set $\{y \in P \mid y > x\}$ is downward directed, i.e., given any elements $y, y' > x$ in P, there exists an element $z \in P$ such that $y, y' \geq z > x$. An ideal H in a partially ordered abelian group G is *prime* provided H is proper and the zero element of G/H is meet-irreducible.

LEMMA 2.1. *Let G be an interpolation group. Then 0 is meet-irreducible in G if and only if the intersection of any two nonzero ideals of G is nonzero.*

PROOF. Assume first that 0 is meet-irreducible, and let I and J be nonzero ideals in G. Then there exist nonzero positive elements $x \in I$ and $y \in J$. Since 0 is meet-irreducible, there exists $z \in G$ such that $x, y \geq z > 0$. Hence, $z \in I \cap J$, and therefore $I \cap J$ is nonzero.

Conversely, assume that the intersection of any two nonzero ideals of G is nonzero, and consider elements $x, y > 0$ in G. Let I be the ideal generated by x and J the ideal generated by y; by assumption, $I \cap J$ is nonzero. Since $I \cap J$ is an ideal of G [**7**, Proposition 2.4], there must exist a strictly positive element $w \in I \cap J$. Then $w \leq mx, my$ for some $m \in \mathbb{N}$. By Riesz decomposition, $w = w_1 + \cdots + w_m$ for some $w_i \in G^+$ such that each $w_i \leq x$. Since $w > 0$, some $w_i > 0$, say $w_1 > 0$. Now $w_1 \leq w \leq my$, and so $w_1 = z_1 + \cdots + z_m$ for some $z_j \in G^+$ such that each $z_j \leq y$. Some $z_j > 0$, say $z_1 > 0$. Since $z_1 \leq y$ and $z_1 \leq w_1 \leq x$, this proves that 0 is meet-irreducible. □

PROPOSITION 2.2. *Let G be a Riesz group and $x \in G$. If $x + H \in (G/H)^+$ for all prime ideals H of G, then $x \in G^+$.*

PROOF. Suppose that $x \notin G^+$. Recall that for any ideal I of G, we have $x + I \in (G/I)^+$ if and only if $x + a \geq 0$ for some $a \in I^+$. Thus by Zorn's Lemma there exists an ideal K of G maximal with respect to the property $x + K \notin (G/K)^+$. Now pass to G/K. Therefore we may assume, without loss of generality, that $x \notin G^+$ but $x + I \in (G/I)^+$ for all nonzero ideals I of G.

Since $x \notin G^+$, the zero ideal of G is not prime. By Lemma 2.1, there exist nonzero ideals I and J in G whose intersection is zero. Then $x + I \in (G/I)^+$ and $x + J \in (G/J)^+$, and hence $x + a \geq 0$ and $x + b \geq 0$ for some $a \in I^+$ and $b \in J^+$.

Now $x = u - v$ for some $u, v \in G^+$. Since $u - v + a = x + a \geq 0$, we have $v \leq u + a$, and so $v = v_1 + v_2$ for some $v_1, v_2 \in G^+$ such that $v_1 \leq u$ and $v_2 \leq a$. Set $u_2 = u - v_1 \in G^+$, and observe that $u_2 - v_2 = u - v = x$, whence $u_2 - v_2 + b \geq 0$. Then $v_2 \leq u_2 + b$, and so $v_2 = w_1 + w_2$ for some $w_1, w_2 \in G^+$ such that $w_1 \leq u_2$ and $w_2 \leq b$.

At this point, we have $0 \le w_2 \le b$, whence $w_2 \in J$. In addition, $0 \le w_2 \le v_2 \le a$, whence $w_2 \in I$. Since $I \cap J$ is zero, we conclude that $w_2 = 0$, and so $v_2 = w_1 \le u_2$. But then $x = u_2 - v_2 \ge 0$, contradicting our initial assumption. $\square$

LEMMA 2.3. *Let (G, u) be a Riesz group with order-unit, and assume that there is a positive integer n such that G^+ contains no descending chains of the form*

$$u \ge b_0 > b_1 > \cdots > b_{n+1} \ge 0.$$

Then $G \cong \mathbb{Z}^k$ (with the product ordering) for some nonnegative integer $k \le n$.

PROOF. It follows easily from the hypotheses that u is a sum of atoms of G^+. Thus we can write $u = m_1x_1 + \cdots + m_kx_k$ for some positive integers m_i and distinct atoms x_i of G^+. In particular,

$$u > x_1 + \cdots + x_{k-1} > x_1 + \cdots + x_{k-2} > \cdots > x_1 > 0,$$

and consequently $k \le n$. By Riesz decomposition, every element of G^+ is a $\mathbb{Z}^+$-linear combination of the x_i, whence the x_i generate G as a group. Therefore $G \cong \mathbb{Z}^k$ by [**7**, Proposition 3.12]. $\square$

PROPOSITION 2.4. *Let (G, u) be a Riesz group with order-unit, and assume that u has finite index n. Let H be an ideal of G.*

(a) *The order-unit $u + H \in G/H$ has index at most n.*

(b) *If H is a prime ideal, then $(G/H, u + H) \cong (\mathbb{Z}, m)$ for some positive integer $m \le n$.*

PROOF. (a) Suppose that $(n + 1)(x + H) \le u + H$ for some $x \in G^+$. Then $(n+1)x \le u+v$ for some $v \in H^+$, and so there exist decompositions $x = x_{i1}+x_{i2}$ for $i = 1, \ldots, n + 1$, with all $x_{ij} \in G^+$, such that $x_{11} + \cdots + x_{n+1,1} \le u$ and $x_{12} + \cdots + x_{n+1,2} \le v$. Note in particular that $x_{i2} \le x, v$ and $x_{i2} \in H$ for all i. By interpolation, there exists $w \in G$ such that $x_{i2} \le w \le x, v$ for all i. Since $x_{12}, v \in H$, it follows that $w \in H$. Further, $0 \le x - w \le x - x_{i2} = x_{i1}$ for all i, whence $(n + 1)(x - w) \le x_{11} + \cdots + x_{n+1,1} \le u$. Since u has index n, we must have $x - w = 0$, whence $x \in H$ and so $x + H = 0 + H$. Therefore $u + H$ has index at most n.

(b) After passing to G/H and reducing n if necessary, we may assume that $H = \{0\}$. We claim that there cannot exist a descending chain of the form

$$u \ge b_0 > b_1 > \cdots > b_{n+1} \ge 0$$

in G^+. If such a chain did exist, then since 0 is meet-irreducible there would be an element $b > 0$ in G such that $b \le b_i - b_{i+1}$ for $i = 0, \ldots, n$. But then

$$(n + 1)b \le (b_0 - b_1) + (b_1 - b_2) + \cdots + (b_n - b_{n+1}) = b_0 - b_{n+1} \le u,$$

contradicting the assumption that u has index n. Thus the claim holds.

By Lemma 2.3, $G \cong \mathbb{Z}^k$ (with the product ordering) for some nonnegative integer k. Since $\{0\}$ is assumed to be a prime ideal of G, we must have $k = 1$.

Therefore $(G, u) \cong (\mathbb{Z}, m)$ for some positive integer m. Then m has index n as an order-unit in $\mathbb{Z}$, and so we conclude that $m = n$. $\square$

COROLLARY 2.5. *If (G, u) is a Riesz group with an order-unit of finite index, then G is unperforated.*

PROOF. Consider $x \in G$ such that $mx \geq 0$ for some positive integer m. If H is any prime ideal of G, then $G/H \cong \mathbb{Z}$ by Proposition 2.4, and so the inequality $m(x + H) \geq 0$ in G/H implies $x + H \geq 0$. Therefore $x \geq 0$ by Proposition 2.2. $\square$

Corollary 2.5 shows that every Riesz group with an order-unit of finite index is a dimension group. We will henceforth use this fact without explicit mention.

LEMMA 2.6. *Let G be a dimension group and u an element of G^+ with finite index n.*

(a) *If $x, y \in G^+$ and $mx \leq my + u$ for some integer $m > n$, then $x \leq y$.*

(b) *If $a \in G$ and $ma \leq u$ for some integer $m > n$, then $a \leq 0$.*

PROOF. (a) By [**7**, Proposition 3.22], $x = v + w$ for some $v, w \in G^+$ such that $v \leq y$ and $mw \leq u$. Since u has index n, we obtain $w = 0$, and hence $x \leq y$.

(b) This is immediate from part (a) since $a = x - y$ for some $x, y \in G^+$. $\square$

THEOREM 2.7. *Let (G, u) be a Riesz group with order-unit, and assume that u has finite index n. Then G is archimedean and norm-complete. Moreover, for each extremal state $s \in S(G, u)$, there is a positive integer $m_s \leq n$ such that $s(G) = (1/m_s)\mathbb{Z}$.*

PROOF. First consider $a, b \in G$ such that $ma \leq b$ for all $m \in \mathbb{N}$. Then $b \leq ku$ for some $k \in \mathbb{N}$, whence $(n + 1)ka \leq ku$ and so $(n + 1)a \leq u$. Hence, $a \leq 0$ by Lemma 2.6. Therefore G is archimedean.

Next, consider an extremal state $s \in S(G, u)$, and suppose there exists $a \in G$ such that $0 < s(a) \leq 1/(n+1)$. Write $a = x - y$ for some $x, y \in G^+$, and observe that $s((n+1)x) \leq s((n+1)y + u)$. By [**7**, Corollary 12.17], there is some $z \in G^+$ such that $z \leq x$ and $(n + 1)z \leq (n + 1)y + u$, while $s(z) > s(x) - s(a) = s(y)$. By Lemma 2.6, $z \leq y$. But this is incompatible with the inequality $s(z) > s(y)$. Therefore there does not exist $a \in G$ such that $0 < s(a) \leq 1/(n + 1)$. Hence, $s(G)$ is a discrete subgroup of $\mathbb{R}$. Since $1 \in s(G)$, it follows that $s(G) = (1/m_s)\mathbb{Z}$ for some positive integer $m_s \leq n$.

Given any nonzero element $c \in G$, we have $\|c\| = \max\{|s(c)| : s \in S(G, u)\}$ [**7**, Proposition 7.12], and $\|c\| > 0$ because G is archimedean [**7**, Theorem 4.14]. It follows from the Krein-Mil'man Theorem that the values $\max\{s(c) \mid s \in S(G, u)\}$ and $\min\{s(c) \mid s \in S(G, u)\}$ both occur at extremal states [**7**, Corollary 5.19]. In view of the previous paragraph, we conclude that $\|c\| \in (1/m)\mathbb{Z}$ for some positive integer $m \leq n$, and thus $\|c\| \geq 1/n$.

Therefore the order-unit norm on G is discrete, and hence G is norm-complete. $\square$

3. Lattice-ordered abelian groups with order-units of finite index

With the help of Section 2 and other known results, we can now easily derive a structure theory for lattice-ordered abelian groups with order-units of finite index.

THEOREM 3.1. *Let (G, u) be a Riesz group with order-unit, and assume that u has finite index. Then G is lattice-ordered if and only if $\partial_e S(G, u)$ is compact.*

PROOF. Theorem 2.7 and [**7**, Corollary 15.10]. □

As a consequence of Theorems 2.7 and 3.1, we can give a complete representation of any lattice-ordered abelian group with an order-unit of finite index in terms of $\mathbb{Q}$-valued functions on a boolean space. As in [**7**], all groups of real-valued functions that appear in this paper are assumed to be equipped with the *pointwise ordering*. (That is, the relation $f \leq g$ for functions f and g on a set X means that $f(x) \leq g(x)$ for all $x \in X$; the relation $f < g$ means only that $f \leq g$ and $f \neq g$.)

COROLLARY 3.2. *Let (G, u) be a lattice-ordered abelian group with order-unit, and assume that u has finite index. Then the space $X = \partial_e S(G, u)$ is compact and totally disconnected, and for each $s \in X$ there is a positive integer m_s such that $s(G) = (1/m_s)\mathbb{Z}$. If L denotes the ordered group of those locally constant functions $p : X \to \mathbb{Q}$ such that $p(s) \in (1/m_s)\mathbb{Z}$ for all $s \in X$, then $(G, u) \cong (L, 1)$.*

PROOF. The existence of the m_s is given by Theorem 2.7, and X is compact by Theorem 3.1. By [**7**, Theorem 15.9], $(G, u) \cong (B, 1)$ where B is the ordered group of those continuous functions $p : X \to \mathbb{R}$ such that $p(s) \in (1/m_s)\mathbb{Z}$ for all $s \in X$. If n is the index of u, then all the functions in B take values in the discrete group $(1/n!)\mathbb{Z}$. Therefore $B = L$.

Given any distinct states $s, t \in X$, there is some $x \in G$ for which $s(x) \neq t(x)$. Since evaluation at x is locally constant on X, the set $\{s' \in X \mid s'(x) = s(x)\}$ is a (relatively) clopen subset of X that contains s but not t. Therefore X is totally disconnected. □

In the case of index 1, the group L in Corollary 3.2 consists of all locally constant integer-valued functions on X. Moreover, the lattice conditions can be obtained from other assumptions, as we see using the following lemma.

LEMMA 3.3. *Let (G, u) be a Riesz group with order-unit, and $s \in S(G, u)$. If $s(G) = \mathbb{Z}$, then s is extremal.*

PROOF. Let H be the subgroup of G generated by $\ker(s)^+$. By [**7**, Proposition 4.22], H is an ideal of G and G/H is simplicial. Now s is induced from some state $\overline{s}$ in $S(G/H, u + H)$, and if $\overline{s}$ is extremal then so is s [**7**, Proposition 6.12]. Hence, there is no loss of generality in assuming that G is simplicial and that $\ker(s)^+ = \{0\}$.

We may now assume that $G = \mathbb{Z}^k$ (with the product ordering) for some positive integer k, and that $u = (u_1, \dots, u_k)$ for some positive integers u_i. The extreme points in $S(G,u)$ are the states of the form $s_i = (1/u_i)\pi_i$ where $\pi_i : G \to \mathbb{Z}$ is the projection onto the i-th factor, and s is a positive convex combination of the s_i, say $s = \alpha_1 s_1 + \cdots + \alpha_k s_k$. If $x = (1,0,0,\ldots,0) \in G^+$, then $s(x) = \alpha_1/u_1$. Since $\alpha_1 \leq 1 \leq u_1$ and $0 < s(x) \in \mathbb{Z}$, we must have $\alpha_1 = 1$. Therefore $s = s_1$, and hence s is extremal. □

COROLLARY 3.4. *Let (G,u) be a Riesz group with order-unit, and suppose that u has index 1. Then the space $X = \partial_e S(G,u)$ is compact, and $(G,u) \cong (L,1)$ where L is the ordered group of all locally constant integer-valued functions on X.*

PROOF. We have $X = \{s \in S(G,u) \mid s(G) = \mathbb{Z}\}$ in view of Theorem 2.7 and Lemma 3.3. It follows that X is closed in $S(G,u)$, whence X is compact. Theorem 3.1 and Corollary 3.2 now yield the desired conclusions. □

In general, the extreme boundary of the state space of a Riesz group with order-unit need not be compact, even if the order-unit has finite index. Here is the easiest example. Let G be the group of those eventually constant integer sequences $x = \{x_n\}_{n=1}^{\infty}$ such that $x_n = x_1 + x_2$ for all sufficiently large n. Equip G with the pointwise ordering and the order-unit $u = (1,1,2,2,2,\ldots)$. It is easily checked that G is a Riesz group and that u has index 2. The extreme points in $S(G,u)$ are π_1, π_2, and $s_i = \frac{1}{2}\pi_i$ for $i \geq 3$, where $\pi_i : G \to \mathbb{Z}$ is the projection on the i-th component. Since $s_i \to \frac{1}{2}\pi_1 + \frac{1}{2}\pi_2$, we conclude that $\partial_e S(G,u)$ is not closed in $S(G,u)$, and thus is not compact.

In this example, the closure of $\partial_e S(G,u)$ is just $\partial_e S(G,u) \cup \{\frac{1}{2}\pi_1 + \frac{1}{2}\pi_2\}$. In particular, $\overline{\partial_e S(G,u)}$ is contained in the convex hull of $\partial_e S(G,u)$. This occurs for any Riesz group with an order-unit of finite index, as we prove in Section 5. As a consequence, we obtain a locally constant function representation for any such Riesz group.

4. K_0 of regular, biregular rings with bounded index

One of the key advantages of biregularity is that all the stalks of the Pierce sheaf of a biregular ring R are simple rings; in case R is also a regular ring with bounded index, its Pierce stalks are all simple artinian. Under these assumptions, a complete description of $K_0(R)$ in terms of locally constant functions on a boolean space was obtained in [**1**]. The K_0 calculation in question is actually somewhat more general than needed here; we state the bounded index case in the following theorem, where as in [**5**] $BS(R)$ denotes the *boolean spectrum* of R, namely the maximal ideal space of the boolean algebra $B(R)$ of central idempotents in R. By the *rank* of a simple artinian ring A we mean the uniform rank of A as a right or left A-module; this equals the integer n that gives the matrix size when A is expressed in the form $M_n(D)$ for some division ring D.

THEOREM 4.1. [Burgess-Goursaud] *Let R be a regular, biregular ring with bounded index, and for $x \in BS(R)$ let n_x denote the rank of the simple artinian ring R/Rx. Then $(K_0(R),[R]) \cong (L,1)$ where L denotes the ordered group of* *those locally constant functions* $p : BS(R) \to \mathbb{Q}$ *such that* $p(x) \in (1/n_x)\mathbb{Z}$ *for all* $x \in BS(R)$.

PROOF. [**1**, Théorème 1]. $\square$

Comparison of Theorem 4.1 with Corollary 3.2 suggests that any lattice-ordered abelian group G with an order-unit u of finite index might be realizable as K_0 of some regular, biregular ring with bounded index. To achieve this, we pass to the group L of locally constant functions on the space X as in Corollary 3.2, and we construct an algebra of locally constant matrix-valued functions on X, such that the values of these functions at each point $s \in X$ are $m_s \times m_s$ matrices over a ground field F. The resulting algebra is *locally matricial* in the sense that each finite subset is contained in a matricial subalgebra. (Recall that a *matricial F-algebra* is any F-algebra isomorphic to $M_{n_1}(F) \times \cdots \times M_{n_t}(F)$ for some positive integers n_i.) Note that all locally matricial algebras are unit-regular.

LEMMA 4.2. *Let F be a ring, $t \in \mathbb{N}$, and D the set of positive divisors of t. For $d \in D$, let B_d denote the image of the unital block diagonal ring homomorphism $M_d(F) \to M_t(F)$. Then $B_m \cap B_n = B_{(m,n)}$ for any $m, n \in D$, where (m,n) denotes the greatest common divisor of m and n.*

PROOF. Fix $m, n \in D$, and set $r = (m,n)$. If we identify $M_t(F)$ with $M_{t/r}(M_r(F))$, then B_d (for $d = m,n,r$) is the image of the block diagonal map $M_{d/r}(M_r(F)) \to M_{t/r}(M_r(F))$. Hence, after replacing F by $M_r(F)$, we may assume that $r = 1$. Note that now $m + n\mathbb{Z}$ is a unit in $\mathbb{Z}/n\mathbb{Z}$.

It is clear that $B_1 \subseteq B_m \cap B_n$. Now consider a matrix $a \in B_m \cap B_n$, and write

$$a = \begin{pmatrix} b & & & \\ & b & & \\ & & \ddots & \\ & & & b \end{pmatrix} = \begin{pmatrix} c & & & \\ & c & & \\ & & \ddots & \\ & & & c \end{pmatrix}$$

for some $b \in M_m(F)$ and $c \in M_n(F)$. To show that $a \in B_1$, it suffices to show that c is a scalar matrix.

Consider an integer $i \in \{1, 2, \dots, n\}$. Then there exists $q \in \{1, 2, \dots, n\}$ such that $qm \equiv i \pmod{n}$. Note that $qm \le mn \le t$; we examine the entries in the qm-th row of a, starting with the qm, qm-entry. Since b is $m \times m$, we see that the entries in question are $b_{mm}, 0, 0, \dots, 0$. On the other hand, since c is $n \times n$, these entries are $c_{ii}, c_{i,i+1}, \dots, c_{in}, 0, 0, \dots, 0$. Thus $c_{ii} = b_{mm}$ and $c_{ij} = 0$ for $j > i$.

A symmetric argument shows that $c_{ji} = 0$ for $j > i$. Therefore c equals the diagonal matrix $\operatorname{diag}(b_{mm}, \dots, b_{mm})$. $\square$

THEOREM 4.3. *Let (G,u) be a lattice-ordered abelian group with order-unit, and assume that u has finite index. Given any field F, there exists a biregular, locally matricial F-algebra R such that $(K_0(R), [R]) \cong (G,u)$.*

PROOF. We have $(G,u) \cong (L,1)$ where L is the group of locally constant functions described in Corollary 3.2. If n denotes the index of u, then each

$m_s \leq n$ by Theorem 2.7. For any $s \in X$, there exists an element $x \in G$ such that $s(x) = 1/m_s$, and so there exists a function $p \in L$ such that $p(s) = 1/m_s$. Hence, s has a neighborhood V on which p takes the value $1/m_s$, and so $m_s \mid m_t$ for all $t \in V$.

Set $T = M_{n!}(F)$, and for $m = 1, \ldots, n$ let $T(m)$ denote the image of the unital block diagonal homomorphism $M_m(F) \to T$. Let R be the set of those locally constant functions $r : X \to T$ such that $r(s) \in T(m_s)$ for all $s \in X$, and observe that R is an F-subalgebra of T^X, with index at most n. Given $r_1, \ldots, r_t \in R$, we can express X as a disjoint union of clopen subsets $V_1, \ldots, V_k$ on which each r_i is constant. Note from Lemma 4.2 that the algebras $T_j = \bigcap\{T(m_s) \mid s \in V_j\}$ are isomorphic to full matrix algebras over F, whence the direct product $T_1 \times \cdots \times T_k$ is a matricial F-algebra. Since $r_1, \ldots, r_t$ all lie in the subalgebra

$$\{r \in R \mid r \text{ is constant on } V_j \text{ for } j = 1, \ldots, k\} \cong T_1 \times \cdots \times T_k,$$

we thus see that R is locally matricial. Further, since any central idempotent in a subalgebra of the form just described remains central in R, it follows that R is biregular.

Next, set $I_s = \{r \in R \mid r(s) = 0\}$ for $s \in X$. As observed above, s has a neighborhood V such that $m_s \mid m_t$ for all $t \in V$, and there is no loss of generality in assuming that V is clopen. Then $T(m_s) \subseteq T(m_t)$ for all $t \in V$, and so any constant function from V to $T(m_s)$ extends to a function in R. This shows that evaluation at s induces an isomorphism of R/I_s onto $T(m_s)$. Thus R/I_s is a simple artinian ring of rank m_s.

Observe that the central idempotents in R are precisely the functions whose only values are 0 and 1. Thus $B(R)$ is isomorphic to the boolean algebra of clopen subsets of X, and there is a homeomorphism $\tau : X \to BS(R)$ such that

$$\tau(s) = \{e \in B(R) \mid e(s) = 0\}$$

for all $s \in X$. In particular, $R\tau(s)$ is a maximal ideal (by biregularity) contained in I_s, whence $R\tau(s) = I_s$ and so the Pierce stalk $R/R\tau(s)$ is a simple artinian ring of rank m_s. Therefore we conclude from Theorem 4.1 that

$$(K_0(R), [R]) \cong (L, 1) \cong (G, u). \quad \square$$

We now have a complete characterization of the partially ordered abelian groups that appear as K_0 of regular, biregular rings with bounded index, as follows.

THEOREM 4.4. *Let (G, u) be a partially ordered abelian group with order-unit. Then $(G, u) \cong (K_0(R), [R])$ for some regular, biregular ring R with bounded index if and only if G is lattice-ordered and u has finite index.*

PROOF. Proposition 1.1 and Theorems 1.2, 4.3. $\square$

5. Riesz groups with order-units of finite index, Part II

In this section, we give an explicit description of an arbitrary Riesz group G with an order-unit u of finite index in terms of a group L of locally constant $\mathbb{Q}$-valued functions on a boolean space X. This description is similar to the one given in Corollary 3.2 in that X is a subset of $S(G,u)$ and the functions $p \in L$ are required to satisfy $p(s) \in s(G)$ for all $s \in X$. However, we cannot in general take $X = \partial_e S(G,u)$ since the latter set need not be compact. Instead, we take a compact set containing the closure of $\partial_e S(G,u)$; this choice of X requires the functions in L to preserve certain convex combinations.

As in Section 3, known results provide us with an isomorphism of G onto a subgroup A of $\operatorname{Aff} S(G,u)$. Since $X \supseteq \partial_e S(G,u)$, it is immediate from the Krein-Mil'man Theorem that restriction from $S(G,u)$ to X provides an order-embedding of A into L; the difficult part is to show that every function in L extends to a function in A. We prove this in two stages, first considering locally constant functions on a compact subset of $S(G,u)$ somewhat larger than X.

DEFINITION. Let (G,u) be a partially ordered abelian group with order-unit, $s \in S(G,u)$, and H the subgroup of G generated by $\ker(s)^+$. By [**7**, Proposition 4.22], H is an ideal of G. Let us define the *index* of s to be the index of the order-unit $u+H$ in G/H.

LEMMA 5.1. *Let (G,u) be a dimension group with order-unit, $s \in S(G,u)$, and $n \in \mathbb{N}$.*

(a) $\operatorname{index}(s) \leq n$ *if and only if there does not exist $x \in G^+$ such that $(n+1)x \leq u$ and $s(x) > 0$.*

(b) *Suppose that s is extremal. Then* $\operatorname{index}(s) \leq n$ *if and only if there do not exist $x_1, \dots, x_{n+1} \in G^+$ such that $x_1 + \cdots + x_{n+1} \leq u$ and $s(x_i) > 0$ for all i.*

(c) *Suppose that s is extremal. Then* $\operatorname{index}(s) = n$ *if and only if $s(G) = (1/n)\mathbb{Z}$.*

(d) *s is a convex combination of extremal states $s_1, \dots, s_k$ with* $\sum \operatorname{index}(s_j) \leq n$ *if and only if there do not exist $x_1, \dots, x_{n+1} \in G^+$ such that $x_1 + \cdots + x_{n+1} \leq u$ and $s(x_i) > 0$ for all i.*

PROOF. Let H be the subgroup of G generated by $\ker(s)^+$, and note that $H^+ = \ker(s)^+$.

(a) If there exists an element $x \in G^+$ with the given properties, then $x+H$ is a strictly positive element in G/H and $(n+1)(x+H) \leq u+H$. Hence, $u+H$ has index greater than n.

Conversely, if $u+H$ has index greater than n, then there exists an element $y \in G^+$ such that $y+H > 0$ and $(n+1)(y+H) \leq u+H$. As a result, $s(y) > 0$ and $(n+1)y \leq u+z$ for some $z \in H^+$. Since $(n+1)y \leq (n+1)z + u$, it follows from [**7**, Proposition 3.22] that $y = v+x$ for some $v, x \in G^+$ such that $v \leq z$ and $(n+1)x \leq u$. Now $v \in H$ because H is an ideal of G, and thus $s(x) = s(y) > 0$.

(b) The implication ($\Longleftarrow$) is clear from part (a). Conversely, assume that there exist elements $x_1, \dots, x_{n+1} \in G^+$ with the given properties. Since s is extremal, there exists $x \in G^+$ such that $s(x) > 0$ and $x \le x_i$ for all i [**7**, Theorem 12.14]. Then $(n+1)x \le u$, and thus $\operatorname{index}(s) > n$ by part (a).

(c) If $s(G) = (1/n)\mathbb{Z}$, then $(G/H, u+H) \cong ((1/n)\mathbb{Z}, 1)$ by [**7**, Proposition 6.19]. In this case, it is clear that $u+H$ has index n.

Conversely, assume that $\operatorname{index}(s) = n$. Since s is extremal, it is induced from an extremal state $\overline{s}$ on $(G/H, u+H)$, and $\overline{s}$ is discrete by Theorem 2.7. Hence, $s(G) = \overline{s}(G/H) = (1/m)\mathbb{Z}$ for some $m \in \mathbb{N}$. Moreover, $m = \operatorname{index}(s) = n$ by what we have just proved.

(d) Assume first that s is a convex combination of extremal states s_j with $\sum \operatorname{index}(s_j) \le n$. Consider $x_1, \dots, x_{n+1} \in G^+$ with $\sum x_i \le u$. For each $j = 1, \dots, k$, we have $\operatorname{index}(s_j) < \infty$ and so $s_j(x_i)$ can be nonzero for at most $\operatorname{index}(s_j)$ choices of i, by part (b). Since $\sum \operatorname{index}(s_j) \le n$, there must be an index i such that $s_j(x_i) = 0$ for all j, and thus $s(x_i) = 0$.

Conversely, assume that there do not exist $x_1, \dots, x_{n+1} \in G^+$ such that $\sum x_i \le u$ and $s(x_i) > 0$ for all i. We claim that there do not exist $n+1$ elements $a_i > 0$ in G/H such that $a_1 + \cdots + a_{n+1} \le u + H$. Given $y_1, \dots, y_{n+1} \in G^+$ with $\sum (y_i + H) \le u + H$, we obtain $\sum y_i \le u + v$ for some $v \in H^+$. By Riesz decomposition, each $y_i = u_i + v_i$ for some $u_i, v_i \in G^+$ such that $\sum u_i \le u$ and $\sum v_i \le v$. By assumption, $s(u_j) = 0$ for some j, whence $u_j \in H$. Since $0 \le v_j \le v \in H$, we also have $v_j \in H$, and so $y_j \in H$. Hence, $y_j + H = 0 + H$, verifying the claim.

In particular, the claim shows that there cannot exist a descending chain of the form

$$u + H \ge b_0 > b_1 > \cdots > b_{n+1} \ge 0$$

in $(G/H)^+$. By Lemma 2.3, $(G/H, u+H) \cong (\mathbb{Z}^k, v)$ for some positive integer k, where $\mathbb{Z}^k$ has the product ordering. Further, $v = (v_1, \dots, v_k)$ for some positive integers v_j such that $\sum v_j \le n$. The extremal states on $(\mathbb{Z}^k, v)$ are those of the form $(1/v_j)\pi_j$ where $\pi_j : \mathbb{Z}^k \to \mathbb{Z}$ is the projection onto the j-th component. Hence, the state $\overline{s}$ on $(G/H, u+H)$ induced by s is a convex combination of extremal states $\overline{s_1}, \dots, \overline{s_k}$ such that $\overline{s_j}(G/H) = (1/v_j)\mathbb{Z}$. Each of the states s_j on (G, u) obtained by composing $\overline{s_j}$ with the quotient map $G \to G/H$ is extremal [**7**, Proposition 6.12], and s is a convex combination of $s_1, \dots, s_k$. Moreover, $s_j(G) = (1/v_j)\mathbb{Z}$, and thus $\operatorname{index}(s_j) = v_j$ by the portion of part (c) proved above. Therefore $\sum \operatorname{index}(s_j) \le n$. □

DEFINITION. Let X be a closed subset of a compact convex set K. We shall call a real-valued function f on X *affine* provided f preserves all convex combinations occurring within X, that is, $f(x) = \sum \alpha_i f(x_i)$ for any convex combination $x = \sum \alpha_i x_i$ such that x and the x_i all lie in X. Set $\operatorname{Aff} X$ equal to the set of all affine, continuous, real-valued functions on X. Then $\operatorname{Aff} X$ is a closed subspace of $C(X, \mathbb{R})$, and so it is a partially ordered real Banach space with respect to the pointwise ordering and the supremum norm.

DEFINITION. Let (G, u) be a Riesz group with order-unit and F a nonempty finite dimensional face of $S(G, u)$; then F has only finitely many extreme points, and $F = \operatorname{conv} \partial_e F$. We define the *total index* of F to be the sum of the indices of the extreme points of F.

PROPOSITION 5.2. *Let (G, u) be a Riesz group with order-unit, and assume that u has finite index n. Let Y denote the union of those nonempty finite dimensional faces of $S(G, u)$ that have total index at most n.*

(a) *Y is closed in $S(G, u)$, and $\partial_e S(G, u) \subseteq Y \subseteq \operatorname{conv} \partial_e S(G, u)$.*

(b) *The restriction map $\rho : \operatorname{Aff} S(G, u) \to \operatorname{Aff} Y$ is an isometric isomorphism of partially ordered real Banach spaces.*

PROOF. (a) Since $S(G, u)$ is a Choquet simplex, its extreme points are affinely independent [**7**, Corollary 10.8], and the convex hull of any set of extreme points is a face [**7**, Proposition 10.10]. Hence, the finite dimensional faces of $S(G, u)$ are just the convex hulls of finite subsets of $\partial_e S(G, u)$. In view of Lemma 5.1, it follows that a state $s \in S(G, u)$ belongs to Y if and only if there do not exist $x_1, \dots, x_{n+1} \in G^+$ such that $x_1 + \cdots + x_{n+1} \leq u$ and $s(x_i) > 0$ for all i. Hence, Y equals the intersection of the sets

$$\bigcup_{i=1}^{n+1} \{s \in S(G, u) \mid s(x_i) = 0\}$$

as $(x_1, \dots, x_{n+1})$ runs over all the $(n+1)$-tuples of elements of G^+ satisfying $x_1 + \cdots + x_{n+1} \leq u$. Thus Y is closed in $S(G, u)$.

By definition, $Y \subseteq \operatorname{conv} \partial_e S(G, u)$. If s is any state in $S(G, u)$, then $\operatorname{index}(s) \leq n$ by Proposition 2.4. Thus $\partial_e S(G, u) \subseteq Y$.

(b) Standard applications of the Krein-Mil'man Theorem show that any function in $\operatorname{Aff} S(G, u)$ attains its maximum and minimum values on $\partial_e S(G, u)$ [**7**, Corollary 5.19]. Since $\partial_e S(G, u) \subseteq Y$, it follows immediately that ρ is an isometry and an order-embedding. Hence, it only remains to show that ρ is surjective.

Given $g \in \operatorname{Aff} Y$, set $m = \min g(Y)$ and $M = \max g(Y)$ and define real-valued functions f and h on $S(G, u)$ by

$$f(s) = \begin{cases} g(s) & \text{for } s \in Y \\ m & \text{for } s \notin Y; \end{cases} \qquad h(s) = \begin{cases} g(s) & \text{for } s \in Y \\ M & \text{for } s \notin Y. \end{cases}$$

Since g is continuous and Y is closed in $S(G, u)$, it is easily checked that f is upper semicontinuous and h is lower semicontinuous. We claim that f is convex and h is concave.

Consider a positive convex combination $s = \alpha_1 s_1 + \alpha_2 s_2$ in $S(G, u)$. If $s \notin Y$, then

$$f(s) = m = \alpha_1 m + \alpha_2 m \leq \alpha_1 f(s_1) + \alpha_2 f(s_2).$$

If $s \in Y$, then since Y is a union of faces of $S(G, u)$, we must have $s_1, s_2 \in Y$. In this case,

$$f(s) = g(s) = \alpha_1 g(s_1) + \alpha_2 g(s_2) = \alpha_1 f(s_1) + \alpha_2 f(s_2)$$

because g is affine. Therefore f is convex, as claimed. A symmetric argument shows that h is concave.

By Edwards' Separation Theorem [**7**, Theorem 11.13], there exists a function $g^* \in \operatorname{Aff} S(G,u)$ such that $f \leq g^* \leq h$. Since $f|_Y = h|_Y = g$, we conclude that $g^*|_Y = g$. Therefore ρ is surjective. $\square$

PROPOSITION 5.3. *Let (G,u) be a Riesz group with order-unit, and assume that u has finite index n. Set*

$$X = \{s \in S(G,u) \mid s(G) = (1/m)\mathbb{Z} \text{ for some } m \in \{1,2,\ldots,n\}\}.$$

(a) *The space X is compact and totally disconnected, and $\partial_e S(G,u) \subseteq X \subseteq \operatorname{conv} \partial_e S(G,u)$.*

(b) *The restriction map $\operatorname{Aff} S(G,u) \to \operatorname{Aff} X$ is an isometric isomorphism of partially ordered real Banach spaces.*

PROOF. (a) Clearly the set $X_m = \{s \in S(G,u) \mid s(G) \subseteq (1/m)\mathbb{Z}\}$ is closed in $S(G,u)$ for each m. Further, if $s \in X_m$ then $s(G) = (1/m')\mathbb{Z}$ for some positive divisor m' of m. Thus $X = X_1 \cup \cdots \cup X_n$, whence X is closed in $S(G,u)$ and thus compact. That $\partial_e S(G,u) \subseteq X$ is shown in Theorem 2.7.

Note that if $s \in X$ and $x \in G$ with $s(x) > 0$, then $s(x) \geq 1/n$. Hence, there cannot exist $x_1, \ldots, x_{n+1} \in G^+$ such that $x_1 + \cdots + x_{n+1} \leq u$ and $s(x_i) > 0$ for all i. Thus $s \in \operatorname{conv} \partial_e S(G,u)$ by Lemma 5.1.

Given distinct $s,t \in X$, there is some $x \in G$ such that $s(x) \neq t(x)$. Evaluation at x then gives a continuous function $f : X \to \mathbb{R}$ such that $f(s) \neq f(t)$. Since f maps X into the discrete space $(1/n!)\mathbb{Z}$ (by definition of X), the set $f^{-1}(f(s))$ is a clopen subset of X containing s but not t. Therefore X is totally disconnected.

(b) Since $\partial_e S(G,u) \subseteq X$, we see as in the proof of Proposition 5.2 that the given restriction map is an isometry and an order-embedding. It only remains to show that any function $g \in \operatorname{Aff} X$ can be extended to a function in $\operatorname{Aff} S(G,u)$.

Define Y as in Proposition 5.2. As noted above, for any $s \in X$ there do not exist $x_1, \ldots, x_{n+1} \in G^+$ such that $x_1 + \cdots + x_{n+1} \leq u$ and $s(x_i) > 0$ for all i, and so $s \in Y$ by Lemma 5.1. Thus $X \subseteq Y$. In view of Proposition 5.2, it now suffices to extend g to a function in $\operatorname{Aff} Y$.

Since the extreme points of $S(G,u)$ are affinely independent [**7**, Corollary 10.8], there is a unique affine real-valued function g' on $\operatorname{conv} \partial_e S(G,u)$ that agrees with g on $\partial_e S(G,u)$. Since $\partial_e S(G,u) \subseteq X \subseteq \operatorname{conv} \partial_e S(G,u)$ and g is affine, it follows that g' is an extension of g. On the other hand, since $Y \subseteq \operatorname{conv} \partial_e S(G,u)$, we can restrict g' to a function g'' on Y. Thus g'' is an affine real-valued function on Y that extends g, and we will be done if g'' is continuous.

Let Δ be the standard $(n-1)$-simplex in $\mathbb{R}^n$, that is,

$$\Delta = \{(\alpha_1, \ldots, \alpha_n) \in \mathbb{R}^n \mid \alpha_1 + \cdots + \alpha_n = 1 \text{ and all } \alpha_i \geq 0\}.$$

Define a function $f : \Delta \times X^n \to S(G,u)$ by the rule

$$f((\alpha_1, \ldots, \alpha_n), s_1, \ldots, s_n) = \alpha_1 s_1 + \cdots + \alpha_n s_n,$$

and observe that f is continuous. Since X is contained in $\operatorname{conv} \partial_e S(G,u)$, so is the image of f, and hence the composition $g'f$ is defined. Since g' is affine,

$$g'f((\alpha_1,\dots,\alpha_n),s_1,\dots,s_n) = \alpha_1 g'(s_1)+\dots+\alpha_n g'(s_n) = \alpha_1 g(s_1)+\dots+\alpha_n g(s_n)$$

for all $((\alpha_1,\dots,\alpha_n),s_1,\dots,s_n) \in \Delta \times X^n$. In other words, $g'f$ equals the composition of the map $1\times g\times\dots\times g : \Delta\times X^n \to \Delta\times\mathbb{R}^n$ with the map $\Delta\times\mathbb{R}^n \to \mathbb{R}$ given by forming convex combinations. Therefore $g'f$ is continuous. Since f is continuous and $\Delta \times X^n$ is compact, it follows that the restriction of g' to $f(\Delta \times X^n)$ is continuous.

Any $s \in Y$ is a convex combination of extremal states $s_1,\dots,s_k$ such that $\sum \operatorname{index}(s_j) \le n$. In particular, $k \le n$. On the other hand, by Lemma 5.1 each $s_j(G) = (1/m_j)\mathbb{Z}$ where $m_j = \operatorname{index}(s_j) \le n$, whence $s_j \in X$. Thus $s \in f(\Delta \times X^n)$.

Since $Y \subseteq f(\Delta\times X^n)$, we conclude that the restriction of g' to Y, namely g'', is continuous, as desired. □

Part (b) of Proposition 5.3 is actually a special case of a more general result about Choquet simplices. Namely, if K is a Choquet simplex and X is a compact subset of K such that $\partial_e K \subseteq X \subseteq \operatorname{conv} \partial_e K$, then the restriction map $\operatorname{Aff} K \to \operatorname{Aff} X$ is an isometric isomorphism of partially ordered real Banach spaces. We do not prove this general result here because it seems to require the full machinery of Choquet theory.

THEOREM 5.4. *Let (G,u) be a Riesz group with order-unit, and assume that u has finite index n. Set*

$$X = \{s \in S(G,u) \mid s(G) = (1/m)\mathbb{Z} \text{ for some } m \in \{1,2,\dots,n\}\},$$

and let L denote the ordered group of those locally constant affine functions $p : X \to \mathbb{Q}$ such that $p(s) \in s(G)$ for all $s \in X$. Then the evaluation map $G \to C(X,\mathbb{R})$ provides an isomorphism of (G,u) onto $(L,1)$.

PROOF. By [**7**, Theorem 15.7], $(G,u) \cong (A,1)$ via the evaluation map $\phi : G \to \operatorname{Aff} S(G,u)$, where A is the ordered group of those functions $p \in \operatorname{Aff} S(G,u)$ such that $p(s) \in s(G)$ for all $s \in \partial_e S(G,u)$. Now any $p \in A$ is given by evaluation at some element of G, whence $p(s) \in s(G)$ for all $s \in X$ (not just for $s \in \partial_e S(G,u)$). Hence, p maps X into the discrete subgroup $(1/n!)\mathbb{Z}$ of $\mathbb{Q}$, and so the restriction of p to X is locally constant.

We conclude from Proposition 5.3 that the restriction map $A \to \operatorname{Aff} X$ gives an isomorphism $\rho : (A,1) \to (L,1)$. Therefore the evaluation map $\rho\phi : (G,u) \to (L,1)$ is an isomorphism. □

6. K_0 of regular rings with bounded index

Theorem 5.4 provides a precise description, in terms of locally constant functions on a boolean space, of K_0 of any regular ring R with bounded index. In order to relate this description more directly to the structure of R, we restate it in terms of locally constant functions on a compact subset X of $\mathbb{P}(R)$. The following lemmas help effect the translation, and also show how X is determined by the ideal structure of R.

LEMMA 6.1. *Let (G, u) be a dimension group with order-unit, $s \in S(G, u)$, and $m \in \mathbb{N}$. Then the following conditions are equivalent:*

(a) $s([0, u]) = \{0, 1/m, 2/m, \ldots, 1\}$.

(b) $s(G) = (1/m)\mathbb{Z}$.

(c) $s = (a_1/m)s_1 + \cdots + (a_k/m)s_k$ *for some positive integers* $a_1, \ldots, a_k$ *and some distinct extremal states* $s_1, \ldots, s_k \in \partial_e S(G, u)$ *such that* $a_1 + \cdots + a_k = m$ *and* $s_j(G) = (1/a_j)\mathbb{Z}$ *for all* j. *In particular, each* s_j *has index* a_j.

PROOF. (a)$\Longrightarrow$(b): Since G is directed and has Riesz decomposition, it is generated as a group by $[0, u]$, and hence $s(G) \subseteq (1/m)\mathbb{Z}$. The reverse inclusion is clear since $s(G)$ is a subgroup of $\mathbb{R}$ containing $1/m$.

(b)$\Longrightarrow$(c): By the proof of [**7**, Proposition 6.22], $s = (a_1/m)s_1 + \cdots + (a_k/m)s_k$ for some $a_j \in \mathbb{N}$ and some distinct $s_j \in \partial_e S(G, u)$ such that $\sum a_j = m$ and $s_j(G) = (1/a_j)\mathbb{Z}$ for all j. Then $\operatorname{index}(s_j) = a_j$ by Lemma 5.1.

(c)$\Longrightarrow$(a): Obviously $s(G) \subseteq (1/m)\mathbb{Z}$, whence s is discrete. By [**7**, Proposition 4.22], the subgroup H of G generated by $\ker(s)^+$ is an ideal of G, and the quotient G/H is simplicial. Then $s, s_1, \ldots, s_k$ are induced from states $\overline{s}, \overline{s_1}, \ldots, \overline{s_k}$ on $(G/H, u + H)$ satisfying the assumptions of part (c), and $s([0, u]) \subseteq \overline{s}([0, u + H])$. Any coset in $[0, u + H]$ can be written in the form $x + H$ for some $x \in G^+$, and $x \leq u + h$ for some $h \in H^+$. Then $x = y + z$ for some $y, z \in G^+$ such that $y \leq u$ and $z \leq h$, whence $y \in [0, u]$ and $z \in H$. As a result, $\overline{s}(x + H) = s(x) = s(y) \in s([0, u])$. Thus $s([0, u]) = \overline{s}([0, u + H])$, and so there is no loss of generality in passing to G/H.

Therefore we may assume that $\ker(s)^+ = \{0\}$ and that G is simplicial. It follows from the first condition that the s_j are the only extremal states on (G, u), and so $G \cong \mathbb{Z}^k$. Hence, we may assume that $G = \mathbb{Z}^k$ and that each s_j is a multiple of the j-th coordinate projection $\pi_j : \mathbb{Z}^k \to \mathbb{Z}$. Since $s_j(G) = (1/a_j)\mathbb{Z}$, we must have $s_j = (1/a_j)\pi_j$ and $\pi_j(u) = a_j$. Thus $u = (a_1, \ldots, a_k)$ and $s = (1/m)(\pi_1 + \cdots + \pi_k)$. Since $a_1 + \cdots + a_k = m$, we conclude that $s([0, u]) = \{0, 1/m, 2/m, \ldots, 1\}$, as desired. □

COROLLARY 6.2. *Let R be a unit-regular ring, $N \in \mathbb{P}(R)$, and $m \in \mathbb{N}$. Then $N(R) = \{0, 1/m, 2/m, \ldots, 1\}$ if and only if $N = (a_1/m)N_1 + \cdots + (a_k/m)N_k$ for some positive integers $a_1, \ldots, a_k$ such that $a_1 + \cdots + a_k = m$ and some distinct pseudo-rank functions $N_1, \ldots, N_k$ on R such that each $R/\ker(N_j)$ is a simple artinian ring of rank a_j.*

PROOF. This follows from Lemma 6.1 together with the canonical affine homeomorphism of $S(K_0(R), [R])$ onto $\mathbb{P}(R)$ [**5**, Proposition 17.12] and the observation that a pseudo-rank function $P \in \mathbb{P}(R)$ corresponds to a discrete extremal state $s \in S(K_0(R), [R])$ with $s(K_0(R)) = (1/b)\mathbb{Z}$ if and only if $R/\ker(P)$ is simple artinian with rank b (cf. [**13**, Lemma 1.13]). □

THEOREM 6.3. *Let R be a regular ring with bounded index n and X the set of those pseudo-rank functions $N \in \mathbb{P}(R)$ such that $N(R) = \{0, 1/m, 2/m, \ldots, 1\}$ for some $m \in \{1, 2, \ldots, n\}$. For $N \in X$, let G_N denote the additive subgroup of $\mathbb{Q}$ generated by $N(R)$. Let L denote the ordered group of those locally constant affine functions $p : X \to \mathbb{Q}$ such that $p(N) \in G_N$ for all $N \in X$. Then X is compact and totally disconnected, and there exists an isomorphism $\psi : (K_0(R), [R]) \to (L, 1)$ such that $\psi([aR])(N) = N(a)$ for all $a \in R$ and $N \in X$.*

PROOF. Recall from Section 1 that R is unit-regular and that $K_0(R)$ is a dimension group. Further, $[R]$ has index n in $K_0(R)$ by Proposition 1.1. Now set

$$X' = \{s \in S(K_0(R), [R]) \mid s(K_0(R)) = (1/m)\mathbb{Z} \text{ for some } m \in \{1, 2, \ldots, n\}\},$$

and let L' denote the ordered group of those locally constant affine functions $p : X' \to \mathbb{Q}$ such that $p(s) \in s(K_0(R))$ for all $s \in X'$. By Proposition 5.3 and Theorem 5.4, X' is compact and totally disconnected, and the evaluation map $K_0(R) \to C(X', \mathbb{R})$ provides an isomorphism $\phi : (K_0(R), [R]) \to (L', 1)$.

By [**5**, Proposition 17.12], there is an affine homeomorphism $\eta : \mathbb{P}(R) \to S(K_0(R), [R])$ such that $\eta(N)([aR]) = N(a)$ for all $N \in \mathbb{P}(R)$ and $a \in R$. In view of Lemma 6.1, η restricts to a homeomorphism χ of X onto X', and $\chi(N)(K_0(R)) = G_N$ for all $N \in X$. Then χ induces an isomorphism $\chi^* : (L', 1) \to (L, 1)$, and the composition $\psi = \chi^*\phi$ is the desired isomorphism of $(K_0(R), [R])$ onto $(L, 1)$. □

The affineness conditions required of the functions in L in Theorem 6.3 are easily specified because $X \subseteq \operatorname{conv} \partial_e \mathbb{P}(R)$. Namely, each $N \in X$ can be written as a convex combination

$$N = \alpha_{N,1} P_{N,1} + \cdots + \alpha_{N,t(N)} P_{N,t(N)}$$

for some extremal pseudo-rank functions $P_{N,j}$, and a real-valued function f on X is affine if and only if $f(N) = \alpha_{N,1} f(P_{N,1}) + \cdots + \alpha_{N,t(N)} f(P_{N,t(N)})$ for all $N \in X$.

7. Regular rings of index 2

The problem of whether every Riesz group with an order-unit of finite index is realizable as K_0 of a regular ring with bounded index appears to be much more difficult than the corresponding problem for lattice-ordered abelian groups (Theorem 4.3), even in the case of index 2. We discuss the form of such Riesz groups in this section, and show that at least a large class of them are realizable. Note that this problem is only open for uncountable Riesz groups, since every countable dimension group with order-unit can be realized as K_0 of a unit-regular ring (in fact, as K_0 of an ultramatricial algebra) – e.g., combine [**7**, Corollary 3.18] with [**5**, Theorem 15.24] (see also [**7**, pp. xvi-xviii] for a historical discussion).

Given a regular ring R with index 2, Theorem 6.3 provides a description of $K_0(R)$ in terms of functions on the boolean space

$$X = \{N \in \mathbb{P}(R) \mid N(R) \subseteq \{0, 1/2, 1\}\}.$$

For $N \in X$, the ring $R/\ker(N)$ is one of the following: a division ring, a direct product of two division rings, or a simple artinian ring of rank 2. The first two cases are precisely those in which $R/\ker(N)$ has index 1, while the second two are those in which $N(R) = \{0, 1/2, 1\}$. Let us write $X = X_1 \cup X_2$ and $X_1 = X_{11} \cup X_{12}$ where

$$\begin{aligned} X_i &= \{N \in X \mid R/\ker(N) \text{ has index } i\} \\ X_{11} &= \{N \in X \mid N(R) = \{0, 1\}\} \\ X_{12} &= \{N \in X_1 \mid N(R) = \{0, 1/2, 1\}\}. \end{aligned}$$

Note that X_{11} is closed in X. As in Lemma 5.1, it is not hard to see that

$$X_1 = \{N \in X \mid N(e) = 0 \text{ for all } e \in R \text{ such that } 2(eR) \lesssim R\}$$

(use [**5**, Theorem 7.2 and Proposition 2.19]). Hence, X_1 is closed in X.

Easy examples show that neither X_1 nor X_{11} is necessarily open in X. For instance, choose a field F and let

$$R = \{(r_1, r_2, \ldots, r_n, r, r, r, \ldots) \in M_2(F)^{\mathbb{N}} \mid r \in \left(\begin{smallmatrix} F & 0 \\ 0 & F \end{smallmatrix}\right)\}.$$

There is an extremal pseudo-rank function N_k corresponding to each coordinate projection $R \to M_2(F)$ (given by the rule $N_k(r) = \frac{1}{2} \operatorname{rank} r_k$), there are two other extremal pseudo-rank functions N_0 and N_{-1} corresponding to the fact that $R/\operatorname{soc}(R) \cong F \times F$, and the N_k converge to the average $P = \frac{1}{2}N_0 + \frac{1}{2}N_{-1}$. (Cf. [**5**, Examples 17.7, 17.15].) Hence, $X_{11} = \{N_0, N_{-1}\}$ and $X_{12} = \{P\}$. In this example, X_{11} is open in X but $X_1 = \{N_0, N_{-1}, P\}$ is not. For a second example, we take

$$R = \{(r_1, r_2, \ldots, r_n, r, r, r, \ldots) \in M_2(F)^{\mathbb{N}} \mid r \in F \cdot \left(\begin{smallmatrix} 1 & 0 \\ 0 & 1 \end{smallmatrix}\right)\}.$$

There are again extremal pseudo-rank functions N_k corresponding to the coordinate projections, but only one other extremal pseudo-rank function N_0, corresponding to the fact that $R/\operatorname{soc}(R) \cong F$, and $N_k \longrightarrow N_0$. In this example, $X_{11} = \{N_0\}$, which is not open in X, while X_{12} is empty.

The only nontrivial convex combinations in X are those expressing elements of X_{12} as averages of elements of X_{11}. We may choose functions $a, b : X_{12} \to X_{11}$ such that $N = \frac{1}{2}a(N) + \frac{1}{2}b(N)$ for all $N \in X_{12}$. Then the condition for a real-valued function p on X to be affine is just that $p(N) = \frac{1}{2}pa(N) + \frac{1}{2}pb(N)$ for all $N \in X_{12}$. To summarize, we rewrite the index 2 case of Theorem 6.3 as follows.

THEOREM 7.1. *Let R be a regular ring of index 2. Then there exist a boolean space X, closed subsets $X_1 \supseteq X_{11}$ in X, and functions $a, b : X_1 \setminus X_{11} \to X_{11}$ such that $(K_0(R), [R]) \cong (L, 1)$ where L is the ordered group of those locally constant functions $p : X \to (1/2)\mathbb{Z}$ such that $p(X_{11}) \subseteq \mathbb{Z}$ and $p = \frac{1}{2}(pa + pb)$ on $X_1 \setminus X_{11}$.* □

While the functions a and b in Theorem 7.1 need not be continuous, the pair $\{a, b\}$, viewed as a function from $X_1 \setminus X_{11}$ to the power set of X_{11}, is continuous in a suitable topology. In discussing this, it is convenient to extend a and b to X_1 by setting them equal to the identity on X_{11}. Note that all functions $p : X \to \mathbb{R}$ automatically satisfy $p = \frac{1}{2}(pa + pb)$ on X_{11}.

DEFINITION. For any topological space Y, let $\mathcal{P}_{1,2}(Y)$ denote the collection of all one- and two-point subsets of Y. There is a natural surjection $\xi_Y : Y \times Y \to \mathcal{P}_{1,2}(Y)$, given by the rule $\xi_Y(y, z) = \{y, z\}$, and we give $\mathcal{P}_{1,2}(Y)$ the quotient topology induced from the product topology on $Y \times Y$ via the map ξ_Y. This topology has a basis of open sets of the form

$$U(A, B) = \{\{y, z\} \in \mathcal{P}_{1,2}(Y) \mid (y, z) \in (A \times B) \cup (B \times A)\}$$

where A and B are arbitrary open sets in Y. It is easily checked that if Y is a boolean space, then so is $\mathcal{P}_{1,2}(Y)$.

Returning to the setting of Theorem 7.1, the pair (a, b) gives us a function from X_1 to $X_{11} \times X_{11}$, where as above $a(N) = b(N) = N$ for $N \in X_{11}$, and the composition $\xi_Y \circ (a, b)$ gives us a function from X_1 to $\mathcal{P}_{1,2}(X_{11})$. This latter function is continuous, as we now show. In fact, the corresponding function that arises when any Riesz group with an order-unit of index 2 is presented in the format of Theorem 7.1 is necessarily continuous, as follows.

THEOREM 7.2. *Let (G, u) be a Riesz group with order-unit, and assume that u has index 2. Set*

$$\begin{aligned} X &= \{s \in S(G, u) \mid s(G) \subseteq (1/2)\mathbb{Z}\} \\ X_1 &= \{s \in X \mid \text{index}(s) = 1\} \\ X_{11} &= \{s \in X \mid s(G) = \mathbb{Z}\}. \end{aligned}$$

(a) *The space X is boolean, the subsets $X_1 \supseteq X_{11}$ are closed in X, and there exist functions $a, b : X_1 \to X_{11}$ such that $a = b =$ identity on X_{11} and $s = \frac{1}{2}a(s) + \frac{1}{2}b(s)$ for all $s \in X_1$.*

(b) *For any choice of the functions a and b, the composition $\xi_{X_{11}} \circ (a, b) : X_1 \to \mathcal{P}_{1,2}(X_{11})$ is continuous.*

(c) *$(G, u) \cong (L, 1)$ where L is the ordered group of those locally constant functions $p : X \to (1/2)\mathbb{Z}$ such that $p(X_{11}) \subseteq \mathbb{Z}$ and $p = \frac{1}{2}(pa + pb)$ on X_1.*

PROOF. Let L' denote the ordered group of those locally constant affine functions $p : X \to \mathbb{Q}$ such that $p(s) \in s(G)$ for all $s \in X$. Then X is boolean by Proposition 5.3, and $(G, u) \cong (L', 1)$ by Theorem 5.4. Since u has index 2, it follows from Proposition 2.4 that every state in $S(G,u)$ has index 1 or 2. Now $X = X_1 \cup X_2$ and $X_1 = X_{11} \cup X_{12}$ where

$$X_2 = \{s \in X \mid \text{index}(s) = 2\}$$
$$X_{12} = \{s \in X_1 \mid s(G) = (1/2)\mathbb{Z}\}.$$

Obviously X_{11} is closed in X. In view of Lemma 5.1,

$$X_1 = \{s \in X \mid s(x) = 0 \text{ for all } x \in G^+ \text{ such that } 2x \leq u\},$$

from which it is clear that X_1 is closed in X.

States in X_{11} are extremal by Lemma 3.3. For $s \in X \setminus X_{11}$, we have $s(G) = (1/2)\mathbb{Z}$, and so by Lemma 6.1, $s = (a_1/2)s_1 + \cdots + (a_k/2)s_k$ for some $a_j \in \mathbb{N}$ and $s_j \in \partial_e S(G,u)$ such that $a_1 + \cdots + a_k = 2$ and each $s_j(G) = (1/a_j)\mathbb{Z}$. If $k = 1$, then s is extremal; in this case, $\text{index}(s) = 2$ by Lemma 5.1 and so $s \in X_2$. Otherwise, $s = \frac{1}{2}s_1 + \frac{1}{2}s_2$ and each $s_j(G) = \mathbb{Z}$; in this case, each $s_j \in X_{11}$. It follows from Lemma 5.1 that $\text{index}(s) = 1$, and so $s \in X_{12}$.

Thus $X_{11} \cup X_2 \subseteq \partial_e S(G,u)$ and each state in X_{12} is an average of two states from X_{11}. Therefore there exist functions $a, b : X_1 \to X_{11}$ such that $a = b =$ identity on X_{11} and $s = \frac{1}{2}a(s) + \frac{1}{2}b(s)$ for all $s \in X_{12}$. Since the latter relations are the only nontrivial convex combinations among elements of X, we conclude that $L' = L$.

Parts (a) and (c) are now proved, and it only remains to show that $\xi_{X_{11}} \circ (a,b)$ is continuous. Recall that there is a subbasis of open neighborhoods for any state $t \in S(G,u)$ consisting of the sets

$$V(t,x,\epsilon) = \{s \in S(G,u) : |s(x) - t(x)| < \epsilon\}$$

for $x \in G$ and $\epsilon > 0$. By Riesz decomposition, any $x \in G$ can be written in the form

$$x = x_1 + \cdots + x_n - x_{n+1} - \cdots - x_{2n}$$

for some elements x_i in the interval $[0,u]$, and

$$V(t,x,\epsilon) \supseteq V(t,x_1,\epsilon/2n) \cap \cdots \cap V(t,x_{2n},\epsilon/2n).$$

Thus the sets $V(t,x,\epsilon)$ with $x \in [0,u]$ and $\epsilon > 0$ constitute a subbasis of open neighborhoods of t.

Now consider a state $t \in X_1$ and an open subset U of $\mathcal{P}_{1,2}(X_{11})$ containing $\xi_{X_{11}}(a(t),b(t))$. Choose relative open sets $A, B \subseteq X_{11}$ such that $\xi_{X_{11}}(a(t),b(t)) \in U(A,B) \subseteq U$. Without loss of generality, we may assume that $(a(t),b(t)) \in A \times B$ and that

$$A = X_{11} \cap V(a(t),x_1,\epsilon) \cap \cdots \cap V(a(t),x_n,\epsilon)$$
$$B = X_{11} \cap V(b(t),x_1,\epsilon) \cap \cdots \cap V(b(t),x_n,\epsilon)$$

for some $x_1, \ldots, x_n \in [0, u]$ and some real number $\epsilon \in (0,1)$. Since states in X_{11} take values in $\mathbb{Z}$, we thus have

$$A = \{s \in X_{11} \mid s(x_i) = a(t)(x_i) \text{ for all } i = 1, \ldots, n\}$$
$$B = \{s \in X_{11} \mid s(x_i) = b(t)(x_i) \text{ for all } i = 1, \ldots, n\}.$$

Since any x_i can be replaced by $u - x_i$ if desired, we may also assume that $a(t)(x_i) \leq b(t)(x_i)$ for all i.

For each i, we have $0 \leq x_i \leq u$ and $a(t)(G) = b(t)(G) = \mathbb{Z}$, whence the values $a(t)(x_i)$ and $b(t)(x_i)$ can only be 0 or 1. After renumbering the x_i if necessary, we may assume that there is an index m such that $a(t)(x_i) = 0$ for $i \leq m$ and $a(t)(x_i) = 1$ for $i > m$. Hence, the elements $x_1, \ldots, x_m, u - x_{m+1}, \ldots, u - x_n$ all lie in $\ker a(t)$. Since $a(t)$ is a discrete extremal state, its kernel is an ideal of G [**7**, Proposition 6.19], and so there exists $x \in \ker a(t)$ such that $x_i \leq x$ for all $i \leq m$ and $u - x_i \leq x$ for all $i > m$. By Riesz interpolation, there is an element $x' \in G$ such that

$$x_1, \ldots, x_m, u - x_{m+1}, \ldots, u - x_n \leq x' \leq x, u.$$

Hence, after replacing x by x', we may assume that $0 \leq x \leq u$.

The set $V = X_1 \cap V(t, x, \frac{1}{2}) \cap V(t, x_1, \frac{1}{2}) \cap \cdots \cap V(t, x_n, \frac{1}{2})$ is an open neighborhood of t in X_1. Since states in X_1 take values in $(1/2)\mathbb{Z}$, we have

$$V = \{s \in X_1 \mid s(x) = t(x) \text{ and } s(x_i) = t(x_i) \text{ for } i = 1, \ldots, n\}.$$

Consider a state $s \in V$, and observe (as in the preceding paragraph) that $a(s)(x) \in \{0, 1\}$.

If $a(s)(x) = 0$, then $a(s)(x_i) = 0$ for $i \leq m$ and $a(s)(u - x_i) = 0$ for $i > m$, whence $a(s) \in A$. Further,

$$a(s)(x_i) + b(s)(x_i) = 2s(x_i) = 2t(x_i) = a(t)(x_i) + b(t)(x_i)$$

for all i, from which we see that $b(s) \in B$, and so $(a(s), b(s)) \in A \times B$. On the other hand, if $a(s)(x) = 1$, then

$$1 + b(s)(x) = 2s(x) = 2t(x) = a(t)(x) + b(t)(x) = b(t)(x) \leq 1$$

and so $b(s)(x) = 0$. In this case, we find that $(a(s), b(s)) \in B \times A$.

Therefore $\xi_{X_{11}}(a, b)(V) \subseteq U(A, B)$, which proves that $\xi_{X_{11}} \circ (a, b)$ is continuous. $\square$

Theorem 7.2 has a converse of the following form, whose proof we leave to the reader.

THEOREM 7.3. *Let X be a boolean space, $X_1 \supseteq X_{11}$ closed subsets of X, and $a, b : X_1 \to X_{11}$ functions such that $a = b =$ identity on X_{11} and the composition $\xi_{X_{11}} \circ (a, b) : X_1 \to \mathcal{P}_{1,2}(X_{11})$ is continuous. Let L denote the ordered group of those locally constant functions $p : X \to (1/2)\mathbb{Z}$ such that $p(X_{11}) \subseteq \mathbb{Z}$ and*

$p = \frac{1}{2}(pa + pb)$ on X_1. Then L is a Riesz group with order-unit 1, and this order-unit has index 2 provided $X_1 \neq X$. $\square$

Suppose that we try to modify the method of Theorem 4.3 to realize some $(L, 1)$ as in Theorem 7.2 as K_0 of a locally matricial algebra over a field F. This would involve building an algebra R of locally constant functions $r : X \to M_2(F)$ such that $r(s)$ is a scalar matrix for $s \in X_{11}$ and a diagonal matrix for $s \in X_{12}$. To ensure that $K_0(R)$ satisfies the relations corresponding to the equations $p(s) = \frac{1}{2}pa(s) + \frac{1}{2}pb(s)$ for $p \in L$ and $s \in X_{12}$, we need the value of any $r \in R$ at any $s \in X_{12}$ to be a diagonal matrix whose diagonal entries are the scalars $ra(s)$ and $rb(s)$. This involves a choice: either $r(s) = \begin{pmatrix} ra(s) & 0 \\ 0 & rb(s) \end{pmatrix}$ or $r(s) = \begin{pmatrix} rb(s) & 0 \\ 0 & ra(s) \end{pmatrix}$. To make these choices consistently for all $r \in R$ requires that the functions a and b should each be continuous. In case a and b can be chosen to be continuous, we can construct an algebra R along the line suggested above and obtain $(K_0(R), [R]) \cong (L, 1)$, as follows.

Recall that a *retraction* of a topological space Y onto a subspace Y' is any continuous map $Y \to Y'$ which is the identity on Y'.

LEMMA 7.4. *Let X be a boolean space, $X_1 \supseteq X_{11}$ closed subsets of X, and $a, b : X_1 \to X_{11}$ two retractions. Let $p : X \to (1/2)\mathbb{Z}$ be a locally constant function such that $p(X_{11}) \subseteq \mathbb{Z}$. Then $X = V_1 \cup \cdots \cup V_k$ for some pairwise disjoint clopen subsets V_j such that*

(a) *p is constant on each V_j.*

(b) *For each $j = 1, \ldots, k$, there exist $j', j'' \in \{1, \ldots, k\}$ such that $a(V_j \cap X_1) \subseteq V_{j'}$ and $b(V_j \cap X_1) \subseteq V_{j''}$.*

PROOF. Since X is compact and p is locally constant, p takes only finitely many distinct values, say $n(1)$ values. Then $X = W_{11} \cup \cdots \cup W_{1,n(1)}$ for some nonempty pairwise disjoint clopen sets W_{1i} on which p is constant.

Since a is continuous, pa is a locally constant function on X_1, say with $n(2)$ distinct values. Hence, $X_1 = A_1 \cup \cdots \cup A_{n(2)}$ for some nonempty pairwise disjoint relatively clopen sets A_i on which pa is constant. Since X_1 is compact, each A_i is compact and thus closed in X. Consequently, there exist pairwise disjoint clopen subsets $W_{2i} \subseteq X$ such that $X = W_{21} \cup \cdots \cup W_{2,n(2)}$ and $W_{2i} \cap X_1 = A_i$ for all i. Similarly, $X = W_{31} \cup \cdots \cup W_{3,n(3)}$ for some pairwise disjoint clopen subsets W_{3i} such that pb is constant on $W_{3i} \cap X_1$ for each i and pb takes different values on different $W_{3i} \cap X_1$'s.

Now let $J = \{j : \{1,2,3\} \to \mathbb{N} \mid j(i) \leq n(i) \text{ for } i = 1,2,3\}$, and set

$$V_j = W_{1,j(1)} \cap W_{2,j(2)} \cap W_{3,j(3)}$$

for $j \in J$. These V_j are pairwise disjoint clopen subsets of X whose union equals X. Since $V_j \subseteq W_{1,j(1)}$, the function p is constant on V_j. Similarly, the functions pa and pb are constant on $V_j \cap X_1$. Since p is constant on $a(V_j \cap X_1)$, we must have $a(V_j \cap X_1) \subseteq W_{1,j'(1)}$ for some $j'(1)$. Note that since $a(V_j \cap X_1) \subseteq X_{11}$, the functions a and b act as the identity on $a(V_j \cap X_1)$, and so pa and pb are constant on this set. Hence, $a(V_j \cap X_1) \subseteq W_{2,j'(2)} \cap W_{3,j'(3)}$ for some $j'(2)$, $j'(3)$. Therefore we have constructed $j' \in J$ such that $a(V_j \cap X_1) \subseteq V_{j'}$. Likewise, $b(V_j \cap X_1) \subseteq V_{j''}$ for some $j'' \in J$. $\square$

THEOREM 7.5. *Let X be a boolean space, $X_1 \supseteq X_{11}$ closed subsets of X, and $a, b : X_1 \to X_{11}$ two retractions. Let L denote the ordered group of those locally constant functions $p : X \to (1/2)\mathbb{Z}$ such that $p(X_{11}) \subseteq \mathbb{Z}$ and $p = \frac{1}{2}(pa + pb)$ on X_1.*

Given any field F, there exists a locally matricial F-algebra R such that $(K_0(R), [R]) \cong (L, 1)$.

PROOF. Let I_2 denote the 2×2 identity matrix, and let $\sigma : F \cdot I_2 \to F$ be the unique F-algebra isomorphism. Set $X_{12} = X_1 \setminus X_{11}$; then X is the disjoint union of the subsets X_{11}, X_{12}, X_2. Define

$$T_s = \begin{cases} F \cdot I_2 & \text{if } s \in X_{11} \\ \begin{pmatrix} F & 0 \\ 0 & F \end{pmatrix} & \text{if } s \in X_{12} \\ M_2(F) & \text{if } s \in X_2 \end{cases}$$

for all $s \in X$, and set $T(V) = \bigcap\{T_s \mid s \in V\}$ for all $V \subseteq X$. Finally, let R denote the set of those locally constant functions $r : X \to M_2(F)$ such that $r(s) \in T_s$ for all $s \in X$ and $r(s) = \begin{pmatrix} \sigma r a(s) & 0 \\ 0 & \sigma r b(s) \end{pmatrix}$ for all $s \in X_1$. (The latter condition only needs to be verified for $s \in X_{12}$, since it is automatically satisfied for $s \in X_{11}$.) It is clear that R is an F-subalgebra of $M_2(F)^X$.

Let us use the term *special decomposition of* X to denote a list $V_1, \dots, V_k$ of pairwise disjoint clopen subsets of X such that $X = V_1 \cup \cdots \cup V_k$ and condition (b) of Lemma 7.4 is satisfied. Given such a decomposition, set

$$R(V_1, \dots, V_k) = \{r \in R \mid r \text{ is constant on each } V_j\}.$$

We claim that $R(V_1, \dots, V_k)$ is a matricial subalgebra of R, and that any finite subset of R is contained in such a subalgebra.

After renumbering $V_1, \dots, V_k$ if necessary, we may assume that there is an index $m \leq k$ such that those indices j for which $V_j \cap X_{11}$ is empty while $V_j \cap X_{12}$ is nonempty are precisely the indices $j > m$. Thus for $j > m$ there are indices $\alpha(j), \beta(j) \leq m$ such that $a(V_j \cap X_1) \subseteq V_{\alpha(j)}$ and $b(V_j \cap X_1) \subseteq V_{\beta(j)}$. There is a natural embedding $r \mapsto (r_1, \dots, r_k)$ of $R(V_1, \dots, V_k)$ into $T(V_1) \times \cdots \times T(V_k)$, where r_j denotes the value of r on V_j. The image of this embedding consists of those k-tuples $(r_1, \dots, r_k)$ such that $r_j = \begin{pmatrix} \sigma r_{\alpha(j)} & 0 \\ 0 & \sigma r_{\beta(j)} \end{pmatrix}$ for all $j > m$. (There are no restrictions on r_j for $j \leq m$ since for such j either $V_j \cap X_{11} \neq \varnothing$ or $V_j \subseteq X_2$.) Consequently, the natural projection

$$(\dagger) \qquad R(V_1, \dots, V_k) \longrightarrow T(V_1) \times \cdots \times T(V_m)$$

is an isomorphism, and therefore $R(V_1, \dots, V_k)$ is matricial.

Now consider any finite list of elements $c_1, \dots, c_m \in R$. Let $n(i)$ be the number of distinct values of c_i; then $X = W_{i1} \cup \cdots \cup W_{i,n(i)}$ for some pairwise disjoint nonempty clopen subsets W_{it} on which c_i is constant. Since $\begin{pmatrix} \sigma c_i a(s) & 0 \\ 0 & \sigma c_i b(s) \end{pmatrix} = c_i(s)$ is constant for $s \in W_{it} \cap X_1$, we see that $c_i a$ and $c_i b$ are constant on $W_{it} \cap X_1$, whence $a(W_{it} \cap X_1) \subseteq W_{i,t'}$ and $b(W_{it} \cap X_1) \subseteq W_{i,t''}$ for some t', t''. Let

$$J = \{j : \{1, \dots, m\} \to \mathbb{N} \mid j(i) \le n(i) \text{ for all } i\}.$$

The sets $V_j = \bigcap_{i=1}^m W_{i,j(i)}$ for $j \in J$ form a special decomposition of X, and $c_1, \dots, c_m$ all lie in $R(V_j \mid j \in J)$.

This verifies the claim above. In particular, it follows that R is a locally matricial F-algebra.

Evaluation at any $s \in X$ defines a homomorphism $R \to M_2(F)$, from which we obtain a pseudo-rank function N_s on R such that $N_s(r) = \frac{1}{2} \operatorname{rank} r(s)$ for all $r \in R$. Let ϕ_s be the corresponding state on $(K_0(R), [R])$, and observe that $\phi_s([eR^n]) = \frac{1}{2} \operatorname{rank} e$ for any idempotent matrix $e \in M_n(R)$. In particular, ϕ_s is a morphism from $(K_0(R), [R])$ to $((1/2)\mathbb{Z}, 1)$.

The maps ϕ_s induce a morphism $\phi : (K_0(R), [R]) \to ((1/2)\mathbb{Z}^X, 1)$. Consider any element $e \in R$. Since e is locally constant on X, so is the function $\phi([eR])$. For $s \in X_{11}$, we have $e(s) \in T_s = F \cdot I_2$, whence $\operatorname{rank} e(s) \in \{0, 2\}$ and so $\phi([eR])(s) \in \{0, 1\}$. Note also that $\phi([eR])(s) = \frac{1}{2}\phi([eR])(a(s)) + \frac{1}{2}\phi([eR])(b(s))$ because $a(s) = b(s) = s$. For $s \in X_{12}$, we have $e(s) = \begin{pmatrix} \sigma e a(s) & 0 \\ 0 & \sigma e b(s) \end{pmatrix}$ and so

$$\operatorname{rank} e(s) = \operatorname{rank} \sigma e a(s) + \operatorname{rank} \sigma e b(s) = \tfrac{1}{2} \operatorname{rank} e a(s) + \tfrac{1}{2} \operatorname{rank} e b(s).$$

Thus $\phi([eR])(s) = \frac{1}{2}\phi([eR])(a(s)) + \frac{1}{2}\phi([eR])(b(s))$ in this case also. This shows that $\phi([eR]) \in L$. Since $K_0(R)$ is generated by the classes $[eR]$ for $e \in R$, it follows that $\phi(K_0(R)) \subseteq L$.

Thus we may now view ϕ as a morphism from $(K_0(R), [R])$ to $(L, 1)$.

Any $c \in \ker \phi$ can be expressed in the form $c = [eR^n] - [fR^n]$ where e and f are idempotent matrices in some $M_n(R)$. There exists a special decomposition $X = V_1 \cup \cdots \cup V_k$ such that e and f are constant on each V_j. From $\phi(c) = 0$ it follows that $\operatorname{rank} e(s) = \operatorname{rank} f(s)$ for all $s \in X$, and hence $e(s)T(V_j)^n \cong f(s)T(V_j)^n$ whenever $s \in V_j$. Since the matricial algebra $Q = R(V_1, \dots, V_k)$ is a direct product of some of the $T(V_j)$'s, we conclude that $eQ^n \cong fQ^n$, whence $eR^n \cong fR^n$ and so $c = 0$. Therefore ϕ is injective.

It remains to show that $\phi(K_0(R)^+) = L^+$. Since L^+ is generated as a monoid by its unit interval $[0, 1]$, it suffices to show that any $p \in L$ satisfying $0 \le p \le 1$ is in $\phi(K_0(R)^+)$. By Lemma 7.4, there is a special decomposition $X = V_1 \cup \cdots \cup V_k$ such that p is constant on each V_j. As above, we may assume that there is an index $m \le k$ such that those indices j for which $V_j \cap X_{11}$ is empty while $V_j \cap X_{12}$ is nonempty are precisely the indices $j > m$.

Let p_j denote the value of p on V_j. If $j < m$ and $V_j \cap X_{11}$ is nonempty, then $p_j \in \{0, 1\}$, and we choose $e_j \in F \cdot I_2 = T(V_j)$ with $\operatorname{rank} e_j = 2p_j$. If

$j \leq m$ and $V_j \cap X_{11}$ is empty, then $V_j \subseteq X_2$ and $p_j \in \{0, \frac{1}{2}, 1\}$; we choose $e_j \in M_2(F) = T(V_j)$ with $\operatorname{rank} e_j = 2p_j$. In view of the isomorphism (†), there exists $e \in R(V_1, \dots, V_k)$ such that $e(s) = e_j$ for $j = 1, \dots, m$ and $s \in V_j$. Hence, $\phi([eR]) = p$ on $V_1 \cup \cdots \cup V_m$. Given $j > m$, there exists $s \in V_j \cap X_{12}$, and $a(s) \in V_\alpha$ and $b(s) \in V_\beta$ for some $\alpha, \beta \leq m$. Hence,

$$\phi([eR])(s) = \tfrac{1}{2}\phi([eR])(a(s)) + \tfrac{1}{2}\phi([eR])(b(s)) = \tfrac{1}{2}pa(s) + \tfrac{1}{2}pb(s) = p(s).$$

Since $\phi([eR])$ and p are constant on V_j, we conclude that they agree there. Thus $\phi([eR]) = p$, which completes the proof that $\phi(K_0(R)^+) = L^+$.

Therefore ϕ is an isomorphism of $(K_0(R), [R])$ onto $(L, 1)$. □

We conclude with an example showing that the functions a and b in Theorem 7.2 cannot always be chosen to be continuous. This example is based on the existence of boolean spaces which do not admit "continuous symmetric selection functions", as follows. We are indebted to George Bergman for discovering such spaces, and for permission to include the following particular case here.

PROPOSITION 7.6. [Bergman] *There exists a boolean space Y which does not admit any continuous function $f : Y \times Y \to Y$ such that $f(y, z) = f(z, y) \in \{y, z\}$ for all $y, z \in Y$.*

PROOF. Let Y be the Stone-Čech compactification of an infinite discrete space Y_0. The closure in Y of any subset of Y_0 is homeomorphic to its Stone-Čech compactification; in particular, the closure in Y of any infinite subset of Y_0 is uncountable.

Suppose there exists a continuous function $f : Y \times Y \to Y$ as described. Define a relation $\leq$ on Y by decreeing that $y \leq z$ if and only if $f(y, z) = y$. It follows from the assumptions on f that this relation is reflexive and antisymmetric, and that any two elements of Y are comparable. Even though $\leq$ is not necessarily transitive, we shall treat it in the manner of a partial ordering. Observe that if $a, b, c \in Y$ are any three elements such that either $a \leq b, c$ or $a \geq b, c$, then $\leq$ is transitive on the set $\{a, b, c\}$.

We next build an infinite subset of Y_0 on which $\leq$ is transitive. To start, choose any $y_0 \in Y_0$. Since y_0 is comparable with all elements of Y_0, it must be either $<$ infinitely many or $>$ infinitely many. Hence, there is an infinite set $Y_1 \subseteq Y_0 \setminus \{y_0\}$ such that either $y_0 < y$ for all $y \in Y_1$ or $y_0 > y$ for all $y \in Y_1$. Similarly, if we choose $y_1 \in Y_1$ then there is an infinite set $Y_2 \subseteq Y_1 \setminus \{y_1\}$ such that either $y_1 < y$ for all $y \in Y_2$ or $y_1 > y$ for all $y \in Y_2$. Continuing by induction, we obtain an infinite sequence of distinct elements $y_0, y_1, \dots \in Y_0$ such that for each i, either $y_i < y_j$ for all $j > i$ or $y_i > y_j$ for all $j > i$. It follows from the remark above that $\leq$ is transitive on any 3-element set $\{y_i, y_j, y_k\}$, and therefore $\leq$ is transitive on the set $Z = \{y_0, y_1, y_2, \dots\}$.

Now $(Z, \leq)$ is an infinite totally ordered set. It cannot satisfy both ACC and DCC, and so Z must have an infinite subset N such that $(N, \leq)$ is either order-isomorphic or order-antiisomorphic to the ordinal ω.

Since $N \subseteq Y_0$, its closure $\overline{N}$ is uncountable. Pick two distinct points $z_1, z_2 \in \overline{N} \setminus N$. Either $z_1 < z_2$ or $z_1 > z_2$, say $z_1 < z_2$. Thus $f(z_1, z_2) = z_1$. Choose disjoint open sets $U_1, U_2 \subseteq Y$ such that each $z_i \in U_i$. Since f is continuous, there exists an open set $V \subseteq Y \times Y$ such that $(z_1, z_2) \in V \subseteq f^{-1}(U_1)$. There is no loss of generality in assuming that V is a basic open set of the form $V_1 \times V_2$ where the V_i are open subsets of Y. After replacing the V_i by $V_i \cap U_i$, we may also assume that each $V_i \subseteq U_i$. Given any points $v_i \in V_i$, note that $f(v_1, v_2) \in U_1$ and so $f(v_1, v_2) \notin V_2$, whence $f(v_1, v_2) = v_1$. Therefore $v_1 < v_2$ for all $v_1 \in V_1$ and $v_2 \in V_2$.

Finally, note that since each V_i meets $\overline{N} \setminus N$, the sets $V_i \cap N$ must be infinite. But this is impossible, since neither ω nor its order-dual contains infinite sets W_1, W_2 such that $w_1 < w_2$ for all $w_i \in W_i$. This contradiction establishes the proposition. □

EXAMPLE 7.7. *There exist a boolean space X, closed subsets $X_1 \supseteq X_{11}$, and functions $a, b : X_1 \to X_{11}$ such that:*

(a) $a = b =$ identity *on X_{11} and the composition $\xi_{X_{11}} \circ (a, b) : X_1 \to \mathcal{P}_{1,2}(X_{11})$ is continuous.*

(b) *There do not exist any continuous functions $a', b' : X_1 \to X_{11}$ such that $\xi_{X_{11}} \circ (a', b') = \xi_{X_{11}} \circ (a, b)$.*

(c) *$X \setminus X_1$ is dense in X.*

PROOF. Let Y be as in Proposition 7.6, and set $X_1 = \mathcal{P}_{1,2}(Y)$. Let $\overline{\mathbb{N}} = \mathbb{N} \cup \{\infty\}$ be the one-point compactification of $\mathbb{N}$, and set $X = X_1 \times \overline{\mathbb{N}}$. Then X is a boolean space, $X_1 \times \{\infty\}$ is a closed subspace homeomorphic to X_1, and $X \setminus (X_1 \times \{\infty\})$ is dense in X. We identify X_1 with $X_1 \times \{\infty\}$.

Let X_{11} be the collection of one-point subsets of Y. Then X_{11} is closed in X_1, and there is a homeomorphism $\sigma : Y \to X_{11}$ given by the rule $\sigma(y) = \{y\}$. Choose functions $a_0, b_0 : X_1 \to Y$ such that $x = \{a_0(x), b_0(x)\}$ for all $x \in X_1$. Then $a = \sigma a_0$ and $b = \sigma b_0$ are functions from X_1 to X_{11} such that $a = b =$ identity on X_{11} and $x = a(x) \cup b(x)$ for all $x \in X_1$. Moreover, $\xi_{X_{11}} \circ (a, b)$ coincides with the map from X_1 to $\mathcal{P}_{1,2}(X_{11})$ induced by σ, whence $\xi_{X_{11}} \circ (a, b)$ is a homeomorphism.

It remains to prove part (b). Suppose, to the contrary, that $\xi_{X_{11}} \circ (a, b) = \xi_{X_{11}} \circ (a', b')$ for some continuous functions $a', b' : X_1 \to X_{11}$. It follows that $x = a(x) \cup b(x) = a'(x) \cup b'(x)$ for all $x \in X_1$, and so $\sigma^{-1} a'(x) \in x$ for all $x \in X_1$. But then $f = \sigma^{-1} a' \xi_Y : Y \times Y \to Y$ is a continuous map such that $f(y, z) = f(z, y) \in \{y, z\}$ for all $y, z \in Y$, contradicting Proposition 7.6. Therefore part (b) is proved. □

If the data from Example 7.7 are used in Theorem 7.3, we obtain a Riesz group L with an order-unit 1 of index 2, to which Theorem 7.5 cannot be applied. Therefore we leave as an open problem whether there exists a regular ring R of index 2 such that $(K_0(R), [R]) \cong (L, 1)$.

References

1. W. D. Burgess and J.-M. Goursaud, *K_0 d'un anneau dont les localisés centraux sont simples artiniens*, Canad. Math. Bull. **25** (1982), 344-347.
2. W. D. Burgess and W. Stephenson, *Pierce sheaves of non-commutative rings*, Communic. in Algebra **4** (1976), 51-75.
3. A. B. Carson, *Representation of regular rings of finite index*, J. Algebra **39** (1976), 512-526.
4. ———, *Homogeneous direct summands of regular rings of finite index*, Communic. in Algebra **10** (1982), 1361-1368.
5. K. R. Goodearl, *Von Neumann Regular Rings*, Pitman, London, 1979; *Second Ed.*, Krieger, Melbourne, FL, 1991.
6. ———, *Metrically complete regular rings*, Trans. Amer. Math. Soc. **272** (1982), 275-310.
7. ———, *Partially Ordered Abelian Groups with Interpolation*, Surveys and Monographs 20, Amer. Math. Soc., Providence, 1986.
8. K. R. Goodearl and D. E. Handelman, *Homogenization of regular rings of bounded index*, Pacific J. Math. **84** (1979), 63-78.
9. K. R. Goodearl, D. E. Handelman, and J. W. Lawrence, *Affine representations of Grothendieck groups and applications to Rickart C^*-algebras and $\aleph_0$-continuous regular rings*, Memoirs Amer. Math. Soc. No. 234 (1980).
10. J. Hannah, *Homogenization of regular rings of bounded index, II*, Pacific J. Math. **94** (1981), 107-112.
11. I. Kaplansky, *Topological representations of algebras. II*, Trans. Amer. Math. Soc. **68** (1950), 62-75.
12. M. Kutami, *Projective modules over regular rings of bounded index*, Math. J. Okayama Univ. **30** (1988), 53-62.
13. E. Pardo, *On a density condition for K_0^+ of von Neumann regular rings*, Communic. in Algebra **22** (1994), 707-719.
14. R. S. Pierce, *Modules over commutative regular rings*, Memoirs Amer. Math. Soc. No. 70 (1967).

Department of Mathematics, University of California, Santa Barbara, CA 93106

E-mail address: goodearl@math.ucsb.edu

Contemporary Mathematics
Volume **171**, 1994

Torsion in Quotients of the Multiplicative Group of a Number Field

DARREN HOLLEY AND ROGER WIEGAND

March 2, 1994

Dedicated to the memory of Richard S. Pierce, who was Wiegand's advisor, mathematical colleague and friend.

ABSTRACT. Let E and F be subfields of the algebraic number field K. We show that the torsion subgroup of K^*/E^*F^* is finite. The analogous result for three subfields is false.

Let K/k be a finite separable extension, and let E and F be intermediate fields. We are interested in the structure of the group $\Gamma = K^*/E^*F^*$, where L^* denotes the multiplicative group of the field L. We allow the possiblity that $E = F$, but we always assume that neither E nor F is equal to K. It is known that Γ is always non-trivial. In fact Γ has infinite torsion-free rank unless k is an algebraic extension of a finite field. (See [**W, (1.2)**], [**GW, (5.3)**] and [**GW, (1.12)**] for proofs of these facts.) The torsion subgroup of Γ, however, is rather small. We show that Γ has a bounded subgroup A of exponent dividing the degree $[L : k]$, where L/k is the Galois closure of K/k, such that Γ/A embeds in a direct product of finitely many copies of L^*. If k has characteristic 0, the group A is in fact finite. It follows immediately that the torsion subgroup of Γ is finite if K is an algebraic number field. (More recently, [**CGW**], the subgroup A has been shown to be trivial in all characteristics.)

Both properties (infinite rank and finiteness of the torsion) can fail in the case of *three* intermediate fields. For example, let E_1, E_2 and E_3 be the three proper intermediate fields for the extension $K/\mathbf{Q}$, where $K = \mathbf{Q}(\sqrt{2}, i)$. In this

1991 *Mathematics Subject Classification.* Primary 11R32, 12F10; Secondary 13A15.
Key words and phrases. Multiplicative group, Picard group, algebraic number field.
The research of the second named author was partially supported by a grant from the National Science Foundation.
This paper is in final form and will not be submitted elsewhere.

case $K^*/E_1^*E_2^*E_3^*$ is an infinite elementary abelian 2-group. (See [**GW, (1.2), (1.5)**].)

Our approach is to view the group $\Gamma = K^*/E^*F^*$ as the Picard group of a certain one-dimensional affine domain R over k. When we tensor this domain with the Galois closure L/k, the affine curve corresponding to R breaks up into a union of lines, and a straightforward matrix argument shows that the Picard group of $L \otimes_k R$ is isomorphic to a direct product of copies of L. The kernel A of the map $\mathrm{Pic}(R) \to \mathrm{Pic}(L \otimes_k R)$ is a bounded torsion group whose exponent divides the degree $[L:k]$. The main result of [**W**] then implies that A is finite if $[L:K]$ is prime to the characteristic of k.

Let L/k be a finite Galois extension with Galois group G, and let $F \subseteq K$ be intermediate fields. We want to examine carefully what happens to the inclusion $F \to K$ when we change rings by applying $L \otimes_k (\ \ )$. Let $r = [F:k], s = [K:F], m = rs = [K:k]$.

Write $F = k(u) \cong k[X]/(f)$, where $f \in k[X]$ is the minimal polynomial for u. Then $L \otimes_k F \cong L[X]/(f) \cong L^r$. The isomorphism $L \otimes_k F \to L^r$ takes $1 \otimes u$ to $(u_1, \dots, u_r)$, where the u_i are the roots of f in L. In order to describe this isomorphism canonically, we choose right coset representatives $\theta_1, \dots, \theta_r$ for $\mathrm{Gal}(L/F)$ in G. Then the isomorphism $L \otimes_k F \cong L^r$ takes $\gamma \otimes \alpha$ to the r-tuple $(\gamma\alpha^{\theta_i})$. Thus, the "change of rings" map $F \to L^r$ takes α to the r-tuple (α^{θ_i}).

Next, let $\phi_1, \dots, \phi_s$ be right coset representatives for $\mathrm{Gal}(L/K)$ in $\mathrm{Gal}(L/F)$. Then the m elements $\phi_j\theta_i$ are right coset representatives for $\mathrm{Gal}(L/K)$ in G, and we have the isomorphism $L \otimes_k K \cong L^m$ taking $\gamma \otimes \beta$ to the m-tuple $(\gamma\beta^{\phi_j\theta_i})$. The change of rings map $K \to L^m$ takes β to the m-tuple $(\beta^{\phi_j\theta_i})$.

When we make the identifications $L \otimes F \cong L^r$ and $L \otimes K \cong L^m$, the map $L^r \to L^m$ induced by the inclusion $F \to K$ takes the r-tuple $(\gamma\alpha^{\theta_i})$ to the m-tuple $(\gamma\alpha^{\phi_j\theta_i})$. Since $\alpha^{\phi_j} = \alpha$, this mapping is nothing more than the direct product of r copies of the diagonal embedding $L \to L^s$.

We have the following commutative diagram, in which the vertical arrows are change of rings and the horizontal arrows are the obvious ones:

$$\begin{array}{ccc} L^r & \xrightarrow{\text{diag}} & L^m \\ \uparrow & & \uparrow \\ F & \xrightarrow{\text{incl}} & K \end{array}$$

In the following theorem there would be no loss of generality in taking $K = L$, but the discussion above seems clearer when they are given different names.

THEOREM. *Let L/k be a finite Galois extension, and let E, F and K be intermediate fields with $K \supseteq E, F$. Let $m = [K:k]$ and $n = [L:k]$. There is an exact sequence*

$$0 \to A \to K^*/E^*F^* \to L^{*t},$$

where $t < m$ and A is a bounded torsion group with exponent dividing n.

PROOF. Choose a primitive element v for K/E, and let $R = \{f \in E[X] | f(v) \in F\}$. Letting π be the map from $E[X]$ onto K taking X to v, we obtain a representation of R as a pullback:

$$\begin{array}{ccc} R & \longrightarrow & E[X] \\ \downarrow & & \downarrow{\scriptstyle \pi} \\ F & \longrightarrow & K \end{array}$$

It follows from the Mayer-Vietoris exact sequence [**B, Chap. IX, (5.3)**] that the group K^*/E^*F^* is isomorphic to the Picard group of R.

When we tensor this diagram with L over k, we obtain another pullback diagram

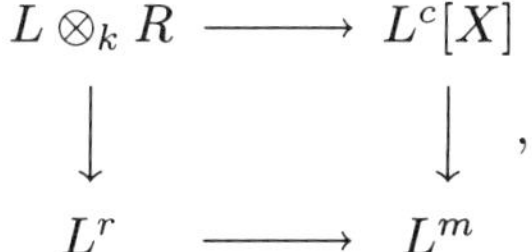

where $c = [E : k]$ and $r = [F : k]$. Moreover, the bottom horizontal arrow is the product of r copies of the diagonal embedding $L \to L^{m/r}$, and the action of the right-hand vertical arrow on L^c is the product of c copies of the diagonal embedding $L \to L^{m/c}$.

Let's consider the kernel A of the natural map $\mathrm{Pic}(R) \to \mathrm{Pic}(L \otimes_k R)$. If I is an invertible ideal of R that represents an element of A, then $L \otimes_k I \cong L \otimes_k R$ (as $L \otimes_k R$-modules, hence as R-modules). Now $L \otimes_k I \cong \oplus^n I$ and $L \otimes_k R \cong \oplus^n R$ as R-modules. Taking n^{th} exterior powers, we conclude that $I^n \cong R$. This shows that A is a torsion group with exponent dividing n.

Finally, we will show that $\mathrm{Pic}(L \otimes_k R) \cong L^{*t}$. The Mayer-Vietoris sequence implies that $\mathrm{Pic}(L \otimes_k R)$ is isomorphic to L^{*m} modulo the join of the images of L^{*c} and L^{*r}. Since these two groups are mapped "diagonally" into L^{*m}, the following lemma tells us exactly what we need to know:

LEMMA. *Let G be an additively written abelian group, and let A and B be matrices of 0's and 1's, defining maps $G^c \to G^m$ and $G^r \to G^m$ respectively. (Thus A is $m \times c$ and B is $m \times r$.) Assume that each row of A has exactly one non-zero entry, and that each row of B has exactly one non-zero entry. Then the cokernel of the $m \times (c + r)$ matrix $[A|B]$ is isomorphic to G^t, where $t = m - \mathrm{rank}([A|B])$.*

PROOF. We describe a procedure for reducing $[A|B]$ to an equivalent matrix with an identity matrix in the top left corner and 0's elsewhere. (Technically, we are letting $\mathrm{Aut}(G^m)$ act on the left and $\mathrm{Aut}(G^{c+r})$ act on the right, but the only autmorphisms of G we need are 1 and -1.)

We may harmlessly assume that every column of A contains at least one 1 (since a column of 0's does not affect the cokernel). We begin by cleaning up A while keeping in mind what effect our row operations have on B. Rearrange the rows of A so that A has block-diagonal form, with each block a column of 1's. Use the top entry of each block to wipe out the other entries in that block, and now move all the non-zero rows of A to the top. The matrix $[A|B]$ has now been transformed to the following form:

$$\begin{bmatrix} I_c & C \\ 0 & D \end{bmatrix},$$

where D is what we call a *balanced* matrix, that is, each of its non-zero rows contains exactly two non-zero entries, namely, a 1 and a -1. Next we do column operations to make $C = 0$.

Finally we work on the $(n-c) \times r$ matrix D. If $D = 0$ we are done. If not, we can assume, by permuting rows and columns if necessary, that the top left entry of D is non-zero. Moreover, after multiplying some of the rows of D by -1 if necessary, we can turn every non-zero entry of the first column of D into a 1. Now use the top left entry of D to wipe out all the other entries in the first column, and then to wipe out all the other entries in the first row of D. The rest of D (that is, the lower-right $(n-c-1) \times (r-1)$ block) is still balanced, and we have brought our original matrix to the following form:

$$\begin{bmatrix} I_{c+1} & 0 \\ 0 & E \end{bmatrix},$$

where E is balanced. By continuing this procedure, we eventually get $[A|B]$ into the desired form.

COROLLARY. *Let L be a field containing only finitely many roots of unity (e.g., any field finitely generated over the prime field). Let L/k be Galois (and if* $\operatorname{char}(k) = p$ *assume $[L : k]$ is prime to p). Let E, F and K be intermediate fields, with $K \supseteq E, F$. The the torsion subgroup of K^*/E^*F^* is finite.*

PROOF. The torsion subgroup of L^{*t} is finite, and the group A (as in the Theorem) is finite by [**W, (0.1)**].

It has recently been shown [**CGW**] that the Corollary is still valid without the parenthetical assumption.

REFERENCES

[B] H. Bass, *Algebraic K-Theory*, W. A. Benjamin, New York, 1968.

[CGW] J.-L. Colliot-Thélène, R. Guralnick and R. Wiegand, *The multiplicative group of a field modulo the product of subfields*, preprint.

[GW] R. Guralnick and R. Wiegand, *Galois groups and the multiplicative structure of field extensions*, Trans. Amer. Math. Soc. **331** (1992), 563–588.

[W] R. Wiegand, *Torsion in Picard groups of affine rings*, Proceedings of the 1992 Joint Summer Research Conference on Commutative Algebra: Syzygies, Multiplicities and Birational Algebra, Amer. Math. Soc., Providence, RI (to appear).

OMAHA NORTH HIGH SCHOOL, 4410 N. 36TH STREET, OMAHA, NE 68111-2217

E-mail address: dholley@cwis.unomaha.edu

DEPARTMENT OF MATHEMATICS & STATISTICS, UNIVERSITY OF NEBRASKA, LINCOLN, NE 68588-0323

E-mail address: rwiegand@unl.edu

Contemporary Mathematics
Volume **171**, 1994

On p^α-Injective Abelian Groups

PATRICK KEEF

ABSTRACT. In the category of p-local abelian groups, the injectives for the functor $p^\alpha \operatorname{Ext}$ are discussed. Of particular interest is a long exact sequence involving the maps $p^\alpha \operatorname{Ext}^k(A/B, G) \longrightarrow p^\alpha \operatorname{Ext}^k(A, G)$ when B is a subgroup of $p^\alpha A$.

Dedicated to the memory of Dick Pierce.

1. Introduction

By the term "group" we will mean a p-local abelian group, where p is a prime fixed for the duration. The restriction to local groups is not, strictly speaking, necessary, but it is natural, since we are only concerned with behavior at p. Capital roman letters will signify groups and α, β, γ and λ will signify ordinals. By $p^\infty G$ we mean the maximal divisible subgroup of G.

A short exact sequence $0 \longrightarrow G \longrightarrow X \longrightarrow A \longrightarrow 0$ is said to be *p^α-pure* if it represents an element of $p^\alpha \operatorname{Ext}(A, G)$ (*p^∞-purity* is defined similarly). Of course, G is *p^α-injective* if $p^\alpha \operatorname{Ext}(—, G) \equiv 0$, and A is *p^α-projective* if $p^\alpha \operatorname{Ext}(A, —) \equiv 0$. Purity in the usual sense is equivalent to p^ω-purity. An excellent exposition of the main elements of the theory of p^α-purity is available in [**2**] and we will assume the results contained there. One of the most useful facts is that if H is p^α-projective, then $\operatorname{Ext}(H, G)$ is p^α-injective and $\operatorname{Tor}(H, G)$ is p^α-projective.

Quite a bit is known about the p^α-injectives and projectives. A p^α-injective is always cotorsion. In fact, the cotorsion groups are precisely the p^∞-injectives ([**7**], Prop. 4.1.ii). A p^α-projective is the direct sum of a free group and a torsion p^α-projective. In fact, the p^∞-projectives are the direct sums of a free group and an arbitrary torsion group ([**7**], Prop. 4.1.i). Since any group has both

1991 *Mathematics Subject Classification.* Primary 20K35, 20K40.
Final version.

p^α-injective and p^α-projective resolutions, we are free to talk about the higher derived functors $p^\alpha \operatorname{Ext}^k$ for $k < \omega$.

For $n < \omega$, the $p^{\omega+n}$-injectives and projectives are well-known. By [**7**], Th. 6.8 and Th. 5.1.iii, the cotorsion group G is $p^{\omega+n}$-injective iff $p^{\omega+n}G$ is divisible (and hence a summand). By [**7**], Cor. 6.5, the group A is $p^{\omega+n}$-projective if there is $P \subseteq A[p^n]$ such that A/P is a direct sum of cyclics. The general properties of the p^α-projectives for an arbitrary α have been more thoroughly studied than those of the p^α-injectives. The purpose of this note is to address this inequality.

One of the most crucial tools in the study of p^α-purity is the following result, which, in various forms, is due to several authors (see [**1**],56.1):

THEOREM 1. *If K is a subgroup of $p^\alpha G$, then there is an exact sequence,*

$$\begin{aligned} 0 \longrightarrow \operatorname{Hom}(A,K) \longrightarrow \operatorname{Hom}(A,G) \longrightarrow \operatorname{Hom}(A,G/K) \\ \longrightarrow \operatorname{Ext}(A,K) \longrightarrow p^\alpha \operatorname{Ext}(A,G) \longrightarrow p^\alpha \operatorname{Ext}(A,G/K) \longrightarrow 0. \end{aligned}$$

It is logical to ask if this useful result can be dualized to subgroups B of $p^\alpha A$. In Theorem 2 we provide such a dual. The result is a long exact sequence connecting the natural homomorphisms $p^\alpha \operatorname{Ext}^k(A/B,G) \longrightarrow p^\alpha \operatorname{Ext}^k(A,G)$ for all $k < \omega$. To establish this sequence we analyze the standard construction of a p^α-injective resolution of G.

If λ is a limit, the λ-*topology* on G is defined utilizing $\{p^\beta G\}_{\beta<\lambda}$ as a neighborhood base of 0. It follows that $G/p^\lambda G$ embeds in its completion $L_\lambda G = \varprojlim G/p^\beta G$ (where the inverse limit is constructed using the natural surjections $G/p^\beta G \longrightarrow G/p^\gamma G$ for $\gamma < \beta < \lambda$), and we let $E_\lambda G = L_\lambda G/(G/p^\lambda G)$. One of the most beautiful features of the theory of cotorsion groups is the observation that if G is reduced, then there is a natural split exact sequence,

$$0 \longrightarrow \operatorname{Ext}(\mathbf{Z}_{p^\infty}, p^\lambda G) \longrightarrow p^\lambda \operatorname{Ext}(\mathbf{Z}_{p^\infty}, G) \longrightarrow \operatorname{Hom}(\mathbf{Z}_{p^\infty}, E_\lambda G) \longrightarrow 0.$$

This means that the Ulm subgroups (and using [**1**],57.2, the Ulm factors) of the cotorsion hull of G have a tractable structure.

For any α there is defined a "generalized Prüfer group" H_α, which is a totally projective group of length α (and hence p^α-projective). Just as the cotorsion hull of a group G is $p^\infty G \oplus \operatorname{Ext}(\mathbf{Z}_{p^\infty}, G)$, when constructing a p^α-pure injective resolution for G, the group $D_0 \oplus \operatorname{Ext}(H_\alpha, G)$ appears, where D_0 is the divisible hull of $p^\alpha G$. So a reduced group is p^α-injective iff it is a summand of $\operatorname{Ext}(H_\alpha, G)$. It would be nice if the description of the Ulm subgroups of $\operatorname{Ext}(\mathbf{Z}_{p^\infty}, G)$ generalized to a description of the Ulm subgroups of $\operatorname{Ext}(H, G)$ for any totally projective group H. This is accomplished in Corollary 5 which implies that when λ is a limit, A is torsion and $A/p^\lambda A$ is a p^λ-projective, then there is a natural split exact sequence

$$0 \longrightarrow \operatorname{Ext}(p^\lambda A, p^\lambda G) \longrightarrow p^\lambda \operatorname{Ext}(A,G) \longrightarrow \operatorname{Hom}(p^\lambda A, E_\lambda G) \longrightarrow 0.$$

So the Ulm subgroups and factors of p^α injectives of the form $\operatorname{Ext}(H_\alpha, G)$ can be described. It is worth noting that when $\lambda = \omega$ and $A/p^\alpha A$ is a direct sum of cyclics, this sequence is mentioned in [**1**] on page 246, where it is said that nothing much is known about it. In addition, for an arbitrary λ this sequence essentially appears in [**9**], Th. 1.11, where is it proven that it is pure.

We next turn to proving various properties of p^α-injectives which are dual to known results for p^α-projectives. For example, we prove that if G is p^α-injective, where $\alpha = \beta + \gamma$, then $p^\beta G$ is p^γ-injective and $G/p^\beta G$ is p^α-injective. We also construct examples of p^α-injectives which are *proper* in the sense that they fail to be p^β-injectives for all $\beta < \alpha$.

One nice characterization of cotorsion groups was given in [**5**], where it was proved that G is cotorsion iff $E_\lambda G$ is reduced for every limit ordinal λ. This is sharpened for p^α-injectives. It is shown that if $\alpha = \lambda + \gamma$ and G is a p^α-injective, then $p^\gamma E_\lambda G = 0$.

Finally, we define a group to be $p^{<\infty}$-injective if it is p^α-injective for some ordinal α. We show that the $p^{<\infty}$-injectives form the smallest class of groups closed under certain elementary operations. This is almost a dual to the description of the class of so-called *subprojectives* mentioned in [**7**].

2. A Long Exact Sequence

We begin this section with a somewhat detailed consideration of the usual method of constructing a p^α-injective resolution for a given group. Recall that there is a sequence $0 \to \mathbf{Z} \to T_\alpha \to H_\alpha \to 0$, where $\mathbf{Z} = p^\alpha T_\alpha$. Given a group G, we then have an exact sequence

$$0 \longrightarrow p^\alpha G \longrightarrow G \xrightarrow{\delta_G} \operatorname{Ext}(H_\alpha, G) \xrightarrow{\mu} \operatorname{Ext}(T_\alpha, G) \longrightarrow 0.$$

Let D_0 be the divisible hull of $p^\alpha G$ and $E_0 = D_0/p^\alpha G$. Note that the inclusion $p^\alpha G \to D_0$ extends to a map $f : G \to D_0$. Then $(f, \delta_G) : G \to D_0 \oplus \operatorname{Ext}(H_a, G)$ will be a p^α-pure injection which we will assume is an inclusion. This determines a p^α-pure sequence,

$$0 \longrightarrow G \longrightarrow D_0 \oplus \operatorname{Ext}(H_\alpha, G) \xrightarrow{\mu'} E_0 \oplus \operatorname{Ext}(T_\alpha, G) \longrightarrow 0.$$

The only thing to justify is the explicit description of the last term. Note that $p^\alpha G \subseteq D_0$, since $\delta_G(p^\alpha G) = 0$. Therefore, $(D_0 \oplus \operatorname{Ext}(H_\alpha, G))/p^\alpha G \cong E_0 \oplus \operatorname{Ext}(H_\alpha, G)$. Note that $G/p^\alpha G$ embeds in this last group, and since δ_G determines an injection $G/p^\alpha G \to \operatorname{Ext}(H_\alpha, G)$, we have $(G/p^\alpha G) \cap E_0 = 0$. Therefore, E_0 embeds in $(E_0 \oplus \operatorname{Ext}(H_\alpha, G))/(G/p^\alpha G)$, and since it is divisible, it is a summand. The cokernel of this embedding is

$$(\operatorname{Ext}(H_\alpha, G) \oplus E_0)/(G/p^\alpha G + E_0) \cong \operatorname{Ext}(H_\alpha, G)/\delta_G(G) \cong \operatorname{Ext}(T_\alpha, G).$$

Observe that the restriction of μ' to D_0 is simply the natural homomorphism onto E_0, though the restriction to $\operatorname{Ext}(H_\alpha, G)$ may not simply be μ since there may be some "bleeding" onto E_0.

Continuing this construction, let $Q_1 = p^\alpha \operatorname{Ext}(T_\alpha, G)$, D_1 be the divisible hull of Q_1 and E_1 be the quotient D_1/Q_1. We then have a p^α-pure sequence,

$$0 \longrightarrow E_0 \oplus \operatorname{Ext}(T_\alpha, G) \longrightarrow E_0 \oplus D_1 \oplus \operatorname{Ext}(H_\alpha, \operatorname{Ext}(T_\alpha, G)) \longrightarrow \\ E_1 \oplus \operatorname{Ext}(T_\alpha, \operatorname{Ext}(T_\alpha, G)) \longrightarrow 0.$$

Note first that the maps on the divisible summands are the natural ones. In addition, by results of [**6**],

$$\operatorname{Ext}(H_\alpha, \operatorname{Ext}(T_\alpha, G)) \cong \operatorname{Ext}(\operatorname{Tor}(H_\alpha, T_\alpha), G)$$

and

$$\operatorname{Ext}(T_\alpha, \operatorname{Ext}(T_\alpha, G)) \cong \operatorname{Ext}(\operatorname{Tor}(T_\alpha, T_\alpha), G).$$

This leads us to the following set of definitions: Let $T_\alpha^0 = \mathbf{Z}_{p^\infty}$, $T_\alpha^1 = T_\alpha$ and for $k > 1$, let $T_\alpha^k = \operatorname{Tor}(T_\alpha, T_\alpha, \ldots, T_\alpha)$ (k-copies). Note that for $k > 1$ these definitions really only depend upon the torsion subgroup of T_α. For $k \geq 0$ set $J_k = \operatorname{Ext}(\operatorname{Tor}(H_\alpha, T_\alpha^k), G)$. Also, let $Q_0 = p^\alpha G$ and for $k \geq 1$, let $Q_k = p^\alpha \operatorname{Ext}(T_\alpha^k, G)$. Let D_k be the divisible hull of Q_k and $E_k = D_k/Q_k$. Finally, if $I_0 = D_0 \oplus J_0$, and for $k \geq 1$, $I_k = E_{k-1} \oplus D_k \oplus J_k$, then we have a p^α-pure injective resolution $I_0 \longrightarrow I_1 \longrightarrow I_2 \longrightarrow I_3 \ldots$ of G which we will denote by $\mathbf{I}$. Notice that $p^\alpha J_k = 0$, so that $p^\alpha I_k = E_{k-1} \oplus D_k$. Further, on the divisible parts of $\mathbf{I}$ the coboundary homorphisms are determined by the natural surjections $D_k \longrightarrow E_k$.

If A is another group and B is a subgroup of $p^\alpha A$, then for every k there is a left exact sequence

$$0 \longrightarrow \operatorname{Hom}(A/B, I_k) \longrightarrow \operatorname{Hom}(A, I_k) \longrightarrow \operatorname{Hom}(B, I_k).$$

We claim that the image of the last map can be identified with $\operatorname{Hom}(B, E_{k-1} \oplus D_k)$. To see this note that a homomorphism $f : B \longrightarrow I_k$ is in the image of this map iff it can be extended to a map $g : A \longrightarrow I_k$. If $f(B) \subseteq E_{k-1} \oplus D_k$ this can clearly be done. Conversely, if such an extension exists, then considering the composition $A \longrightarrow I_k \longrightarrow J_k$, and using the fact that $p^\alpha J_k = 0$, we can conclude that $f(B) \subseteq g(p^\alpha G) \subseteq E_{k-1} \oplus D_k$.

If $\mathbf{D}$ is the complex $D_0 \longrightarrow E_0 \oplus D_1 \longrightarrow E_1 \oplus D_2 \longrightarrow \ldots$, then we have shown there is an exact sequence of complexes

$$0 \longrightarrow \operatorname{Hom}(A/B, \mathbf{I}) \longrightarrow \operatorname{Hom}(A, \mathbf{I}) \longrightarrow \operatorname{Hom}(B, \mathbf{D}) \longrightarrow 0.$$

It follows that their cohomology groups are connected by a long exact sequence. Note that the cohomology groups of the first complex are $p^\alpha \operatorname{Ext}^k(A/B, G)$, and similarly the cohomology groups of the second are $p^\alpha \operatorname{Ext}^k(A, G)$. Finally, the cohomology groups of the last complex are easily determined. Since the maps in $\mathbf{D}$ are constructed from the injective resolutions $D_k \longrightarrow E_k$ of Q_k, it follows that the kth cohomology group of this complex is isomorphic to $\operatorname{Ext}(B, Q_{k-1}) \oplus \operatorname{Hom}(B, Q_k)$. If we put this all into one statement, we get the following:

THEOREM 2. *Suppose B is a subgroup of $p^\alpha A$. If $Q_0 = p^\alpha G$ and for $k \geq 1$, $Q_k = p^\alpha \operatorname{Ext}(T_\alpha^k, G)$, then there is an exact sequence*

$$\begin{aligned} 0 \longrightarrow \operatorname{Hom}(A/B, G) &\longrightarrow \operatorname{Hom}(A, G) \longrightarrow \operatorname{Hom}(B, p^\alpha G) \longrightarrow \\ p^\alpha \operatorname{Ext}(A/B, G) &\longrightarrow p^\alpha \operatorname{Ext}(A, G) \longrightarrow \\ &\operatorname{Ext}(B, p^\alpha G) \oplus \operatorname{Hom}(B, p^\alpha \operatorname{Ext}(T_\alpha, G)) \longrightarrow \dots \\ \dots \longrightarrow p^\alpha \operatorname{Ext}^k(A/B, G) &\longrightarrow p^\alpha \operatorname{Ext}^k(A, G) \longrightarrow \\ &\operatorname{Ext}(B, Q_{k-1}) \oplus \operatorname{Hom}(B, Q_k) \longrightarrow \dots \end{aligned}$$

We separate out one important special case of the last result in the following.

COROLLARY 3. *If $A/p^\alpha A$ is p^α-projective, then for $k \geq 1$ there are isomorphisms*

$$p^\alpha \operatorname{Ext}^k(A, G) \cong \operatorname{Ext}(p^\alpha A, Q_{k-1}) \oplus \operatorname{Hom}(p^\alpha A, Q_k).$$

We are primarily concerned with the case where A is a p-group. To this end, we prove the following result:

LEMMA 4. *If λ is a limit, then the torsion subgroups of $p^\lambda \operatorname{Ext}(T_\lambda, G)$ and $E_\lambda G = (L_\lambda G)/(G/p^\lambda G)$ are isomorphic.*

PROOF. Let Y be the torsion subgroup $E_\lambda G$ and suppose that $X \subseteq L_\lambda G$ satisfies $X/(G/p^\lambda G) = Y$. Similarly, let Z be the torsion subgroup of $\operatorname{Ext}(T_\lambda, G)$ and suppose that $W \subseteq \operatorname{Ext}(H_\lambda, G)$ satisfies $W/(G/p^\lambda G) = Z$. Since $\operatorname{Ext}(H_\lambda, G) \cong \operatorname{Ext}(\oplus_{\beta<\lambda} H_\beta, G) \cong \prod_{\beta<\lambda} \operatorname{Ext}(H_\beta, G)$, and $\operatorname{Ext}(H_\beta, G)$ is discrete in the λ-topology, it follows that $\operatorname{Ext}(H_\lambda, G)$ is complete in the λ-topology. Since $G/p^\lambda G$ embeds as an isotype subgroup in $\operatorname{Ext}(H_\lambda, G)$, it follows that we can extend this to an embedding of $L_\lambda G$, which, in turn, determines an embedding of $E_\lambda G$ into $\operatorname{Ext}(T_\lambda, G)$. It follows that these embeddings take Y into Z and hence X into W. Identify these embeddings with inclusions. We need to show $Y = p^\lambda Z$. Note that if $y \in Y$, then $y = [x]$ for some $x \in X$. It follows that for every $\beta < \lambda$ there is an $x_\beta \in p^\beta \operatorname{Ext}(H_\lambda, G)$ and $g_\beta \in G/p^\lambda G$ such that $x = x_\beta + g_\beta$. Therefore $y = [x_\beta]$ has height at least β, and since this is true for all $\beta < \lambda$, $y \in p^\lambda Z$. Conversely, if $\beta < \lambda$, then since Z is torsion, we have that $0 \longrightarrow p^\beta G/p^\lambda G \longrightarrow p^\beta W \longrightarrow p^\beta Z \longrightarrow 0$ is short exact (this follows easily by induction on [**2**], Th. 91). So if $z \in p^\lambda Z$, then there exist $w_\beta \in p^\beta W$ such that $z = [w_\beta]$. Note that $w_0 - w_\beta$ is a neat λ-Cauchy net in $G/p^\lambda G$, converging to w_0. Therefore, $w_0 \in L_\lambda G$, so that $z = [w_0] \in E_\lambda G$, but since z has finite order, $z \in Y$, as required.

It is worth observing that if $\alpha = \lambda + n$, where λ is a limit and $n < \omega$, then $T_\alpha = T_\lambda$ (see [**2**], Th. 69). Therefore, the torsion subgroup of $p^\alpha \operatorname{Ext}(T_\alpha, G) = p^n p^\lambda \operatorname{Ext}(T_\lambda, G)$ is isomorphic to $p^n E_\lambda G$.

COROLLARY 5. *If $\alpha = \lambda + n$, where λ is a limit and $n < \omega$, A is torsion and $A/p^\alpha A$ is p^α-projective, then*

$$p^\alpha \operatorname{Ext}(A, G) \cong \operatorname{Ext}(p^\alpha A, p^\alpha G) \oplus \operatorname{Hom}(p^\alpha A, p^n E_\lambda G).$$

PROOF. The condition that A is torsion allows us to replace $p^\alpha \operatorname{Ext}(T_\alpha, G)$ with $p^n E_\lambda G$.

The last result, together with [**1**], 57.2, makes it possible to describe all of the Ulm factors of $\operatorname{Ext}(H, G)$, whenever H is totally projective. We now turn to a related situation. In [**3**] the p-group A was defined to be p^α-*extending* if for every G, every homomorphism $p^\alpha A \longrightarrow p^\alpha G$ extends to a homomorphism $A \longrightarrow G$.

THEOREM 6. *The p-group A is p^α-extending iff for every G the map*

$$p^\alpha \operatorname{Ext}(A/p^\alpha A, G) \longrightarrow p^\alpha \operatorname{Ext}(A, G)$$

is injective. If this is the case, and $\alpha = \lambda + n$, where λ is a limit and $n < \omega$, then the sequence

$$0 \longrightarrow p^\alpha \operatorname{Ext}(A/p^\alpha A, G) \longrightarrow p^\alpha \operatorname{Ext}(A, G) \longrightarrow \\ \operatorname{Ext}(p^\alpha A, p^\alpha G) \oplus \operatorname{Hom}(p^\alpha A, p^n E_\lambda G) \longrightarrow 0$$

is splitting exact.

PROOF. Notice that A is p^α-extending iff the map

$$\operatorname{Hom}(A, G) \longrightarrow \operatorname{Hom}(p^\alpha A, p^\alpha G)$$

is surjective for all G. This clearly implies the first statement. For any A call the above sequence $S(A)$. We need to show that if A is p^α-extending, then $S(A)$ is splitting exact. As in [**3**], Th. 3, let B be a p-group such that $p^\alpha B \cong p^\alpha A$ and $B/p^\alpha B$ is p^α-projective. It follows as in [**3**], Th. 3, that $A \oplus B/p^\alpha B \cong A/p^\alpha A \oplus B$. It is readily checked that all terms of $S(B/p^\alpha B)$ are 0, so if we interpret the direct sum of short exact sequences to be their term-by-term direct sum, then $S(A) \cong S(A) \oplus S(B/p^\alpha B) \cong S(A \oplus B/p^\alpha B) \cong S(A/p^\alpha A \oplus B) \cong S(A/p^\alpha A) \oplus S(B)$. Finally, it can easily be seen that $S(A/p^\alpha A)$ and $S(B)$ split (in fact, they each have only two non-zero, isomorphic terms).

In light of Theorem 2, it might be interesting to ask about the behavior of the natural maps $p^\alpha \operatorname{Ext}^k(A, G) \longrightarrow p^\alpha \operatorname{Ext}^k(A, G/K)$ when K is a subgroup of $p^\alpha G$ and $k \geq 2$.

THEOREM 7. *If $k \geq 2$ and K is a subgroup of $p^\alpha G$, then the natural map*

$$p^\alpha \operatorname{Ext}^k(A, G) \longrightarrow p^\alpha \operatorname{Ext}^k(A, G/K)$$

is an isomorphism.

PROOF. We compare the p^α-injective resolutions **I** and **I**' for G and G/K constructed above. Since H_α is p^α-projective, it follows easily from Theorem 1 that $\operatorname{Ext}(H_\alpha, G)$ is isomorphic to $\operatorname{Ext}(H_\alpha, G/K)$. Therefore,

$$\operatorname{Ext}(T_\alpha, G) \cong \operatorname{Ext}(H_\alpha, G)/(G/p^\alpha G) \cong \\ \operatorname{Ext}(H_\alpha, G/K)/(\{G/K\}/\{p^\alpha G/K)\}) \cong \operatorname{Ext}(T_\alpha, G/K)$$

It follows that the only difference between **I** and **I**' is that that the former contains D_0, the divisible hull of $p^\alpha G$, in I_0, and $E_0 = D_0/p^\alpha G$ in I_1, whereas the latter contains D_0', the divisible hull of $p^\alpha G/K$, in I_0', and $E_0' = D_0'/(p^\alpha G/K)$ in I_1. For $k \geq 2$ they are the same and so their cohomology groups are isomorphic. However, this is just what we are trying to prove.

An examination of the resolutions **I** and **I**' in this last proof give another way to understand the exact sequence in Theorem 1. If D_0'' is the divisible hull of K, $E_0'' = D_0''/K$ and **D** is the injective resolution $D_0'' \to E_0'' \to 0 \to \dots$, of K, then there is an exact sequence of complexes:

$$0 \longrightarrow \operatorname{Hom}(A, \mathbf{D}) \longrightarrow \operatorname{Hom}(A, \mathbf{I}) \longrightarrow \operatorname{Hom}(A, \mathbf{I}') \longrightarrow 0$$

and looking at the corresponding long exact sequence for the cohomology groups gives both Theorem 1 and Theorem 7.

3. General Properties

The next two results are parallel to [**9**], Cor. 2.4

THEOREM 8. *Assume $\alpha = \beta + \gamma$ and G is p^α-injective. Then $p^\beta G$ is p^γ-injective.*

PROOF. If X is any group, then there is a group A such that $p^\beta A \cong X$ and $A/p^\beta A$ is p^β-projective (in fact, this last quotient can be assumed to be totally projective). It follows from Corollary 3 that $\operatorname{Ext}(X, p^\beta G) \cong \operatorname{Ext}(p^\beta A, p^\beta G)$ is isomorphic to a summand of $p^\beta \operatorname{Ext}(A, G)$. Therefore, $p^\gamma \operatorname{Ext}(X, p^\beta G) \subseteq p^\gamma p^\beta \operatorname{Ext}(A, G) = p^\alpha \operatorname{Ext}(A, G) = 0$.

THEOREM 9. *Assume $\alpha = \beta + \gamma$ and G is p^α-injective. Then $G/p^\beta G$ is p^α-injective.*

PROOF. We may clearly assume that G is reduced. Since G is cotorsion, $p^\alpha G \cong p^\alpha \operatorname{Ext}(\mathbf{Z}_{p^\infty}, G) = 0$. It follows that there is a diagram with p^α-pure rows:

$$\begin{array}{ccccc}
 & & & & \operatorname{Ext}(H_\alpha, p^\beta G) \\
 & & & & \downarrow{\scriptstyle g} \\
0 & \longrightarrow & G & \xrightarrow{\delta_G} & \operatorname{Ext}(H_\alpha, G) \\
 & & {\scriptstyle f}\downarrow & & \downarrow \\
0 & \longrightarrow & G/p^\beta G & \xrightarrow{\delta_{G/p^\beta G}} & \operatorname{Ext}(H_\alpha, G/p^\beta G) \\
 & & \downarrow & & \downarrow \\
 & & 0 & & 0
\end{array}$$

Since G is p^α-injective, there is a splitting map $\phi : \operatorname{Ext}(H_\alpha, G) \longrightarrow G$ for δ_G. Note that the image of g is contained in $p^\beta \operatorname{Ext}(H_\alpha, G)$, so $f \circ \phi \circ g$ is the zero map. This implies that $f \circ \phi$ determines a homomorphism $\phi' : \operatorname{Ext}(H_\alpha, G/p^\beta G) \to G/p^\beta G$ which will be a splitting map for $\delta_{G/p^\beta G}$. Therefore, $G/p^\beta G$ is p^α-injective.

It is worth noting that for p^α-projectives there is a partial converse to these results which states that if $G/p^\beta G$ is p^β-projective and $p^\beta G$ is p^γ-projective and $\alpha = \beta + \gamma$, then G is p^α-projective. The obvious dual of this for p^β-injectivity is false. If G is a cotorsion group of length $\omega 2$, then both $G/p^\omega G$ and $p^\omega G$ are algebraically compact, and hence p^ω-injective. By [**7**], Theorems 5.1 and 6.8, there are such groups which fail to be $p^{\omega 2}$-injective. In fact, in a later work we will show how to construct such groups which are not p^α-injective for any ordinal α. On the other hand, if G is p^α-injective, then $G/p^{\omega+n}G$ is actually $p^{\omega+n}$-injective whenever $n < \omega$.

The following is a dual to [**9**], Prop. 3.1

LEMMA 10. *If* $(*) : 0 \longrightarrow X \longrightarrow G \longrightarrow Y \longrightarrow 0$ *is* $p^{\alpha+1}$*-pure,* G *is* p^α*-injective and* X *is cotorsion, then* $(*)$ *splits.*

PROOF. By the p^α-injectivity of G and the p^α-purity of $(*)$ we have an exact sequence

$$\dots \longrightarrow \operatorname{Hom}(Y, G) \longrightarrow \operatorname{Hom}(Y, Y) \xrightarrow{\nu} p^\alpha \operatorname{Ext}(Y, X) \longrightarrow 0$$

But since the image of ν is actually contained in $p^{\alpha+1} \operatorname{Ext}(Y, X)$, it follows that $p^\alpha \operatorname{Ext}(Y, X)$ is divisible. But

$$\operatorname{Ext}(Y, X) \cong \operatorname{Ext}(Y, \operatorname{Ext}(\mathbf{Z}_{p^\infty}, X)) \cong \operatorname{Ext}(\operatorname{Tor}(Y, \mathbf{Z}_{p^\infty}), X),$$

and since $\operatorname{Tor}(Y, \mathbf{Z}_{p^\infty})$ is a torsion group, $\operatorname{Ext}(Y, X)$ must be reduced. Therefore, $p^\alpha \operatorname{Ext}(Y, X) = 0$. Since $(*) \in p^\alpha \operatorname{Ext}(Y, X)$, it must split.

The following generalizes [**1**], 53.5.

LEMMA 11. *Suppose* $(*) : 0 \longrightarrow X \longrightarrow Y \xrightarrow{\pi} Z \longrightarrow 0$ *is* p^α*-pure and* K *is the image of the connecting homomorphism* $\operatorname{Hom}(X, W) \longrightarrow \operatorname{Ext}(Z, W)$*. Then*

$$0 \longrightarrow \operatorname{Ext}(Z, W)/K \xrightarrow{\operatorname{Ext}(\pi, W)} \operatorname{Ext}(Y, W) \longrightarrow \operatorname{Ext}(X, W) \longrightarrow 0$$

is p^α*-pure.*

PROOF. Consider the homomorphism

$$\delta_{\operatorname{Ext}(Z,W)} : \operatorname{Ext}(Z, W) \longrightarrow \operatorname{Ext}(H_\alpha, \operatorname{Ext}(Z, W)).$$

By the p^α-purity of $(*)$, $K \subseteq p^\alpha \operatorname{Ext}(Z, W)$, and so by Theorem 1,

$$\operatorname{Ext}(H_\alpha, \operatorname{Ext}(Z, W)) \cong \operatorname{Ext}(H_\alpha, \operatorname{Ext}(Z, W)/K).$$

We can therefore view $\delta_{\operatorname{Ext}(Z,W)}$ as inducing $\delta_{\operatorname{Ext}(Z,W)/K}$ on $\operatorname{Ext}(Z, W)/K$.

By [**2**], Th. 76 there is also a natural homomorphism $\partial_Z : \operatorname{Tor}(H_\alpha, Z) \longrightarrow Z$, an isomorphism $\mu : \operatorname{Ext}(\operatorname{Tor}(H_\alpha, Z), W) \longrightarrow \operatorname{Ext}(H_\alpha, \operatorname{Ext}(Z, W))$ and $\mu \circ \operatorname{Ext}(\partial_Z, W) = \delta_{\operatorname{Ext}(Z,W)}$. Since $(*)$ is p^α-pure, by [**2**], Lem. 85(a) there is a homomorphism $\nu : \operatorname{Tor}(H_\alpha, Z) \longrightarrow Y$ such that $\pi \circ \nu = \partial_Z$. Consider $\nu' = \mu \circ \operatorname{Ext}(\nu, W) : \operatorname{Ext}(Y, W) \longrightarrow \operatorname{Ext}(H_\alpha, \operatorname{Ext}(Z, W))$. Since $\nu' \circ \operatorname{Ext}(\pi, W) = \mu \circ \operatorname{Ext}(\nu, W) \circ \operatorname{Ext}(\pi, W) = \mu \circ \operatorname{Ext}(\pi \circ \nu, W) = \mu \circ \operatorname{Ext}(\partial_Z, W) = \delta_{\operatorname{Ext}(Z,W)}$, the result follows from [**2**], Lem. 85(b).

A similar argument shows that if L is the image of $\mathrm{Hom}(W,Z) \to \mathrm{Ext}(W,X)$, then

$$0 \to \mathrm{Ext}(W,X)/L \to \mathrm{Ext}(W,Y) \to \mathrm{Ext}(W,Z) \to 0$$

is p^α-pure.

LEMMA 12. *If $\mathrm{Ext}(p^\alpha A, p^\alpha G) \neq 0$, then $p^\alpha \mathrm{Ext}(A,G) \neq 0$.*

PROOF. Suppose $0 \to p^\alpha G \to X \to p^\alpha A \to 0$ represents a non-zero element of $\mathrm{Ext}(p^\alpha A, p^\alpha G)$. We can complete the following diagram with exact rows:

$$\begin{array}{ccccccccc}
0 & \longrightarrow & p^\alpha G & \longrightarrow & X & \longrightarrow & p^\alpha A & \longrightarrow & 0 \\
 & & \| & & \downarrow & & \downarrow & & \\
0 & \longrightarrow & p^\alpha G & \longrightarrow & Y & \longrightarrow & A & \longrightarrow & 0 \\
 & & \downarrow & & \downarrow & & \| & & \\
0 & \longrightarrow & G & \longrightarrow & Z & \longrightarrow & A & \longrightarrow & 0
\end{array}$$

By the surjectivity of $\mathrm{Ext}(A, p^\alpha G) \to \mathrm{Ext}(p^\alpha A, p^\alpha G)$, the second row exists (though it is not unique). By Theorem 1 the bottom row is p^α-pure. Identify all of the vertical maps with inclusions. We claim that $X = p^\alpha Z$. This will follow if we can show that $X/p^\alpha G = p^\alpha(Z/p^\alpha G)$. However, by considering the last two rows of our diagram, we have $Z/p^\alpha G \cong G/p^\alpha G \oplus Y/p^\alpha G$, so that $p^\alpha(Z/p^\alpha G) = p^\alpha(Y/p^\alpha G) \cong p^\alpha A \cong X/p^\alpha G$. We now need only observe that if the bottom row split, then so would the top sequence, since it consists of functorial subgroups.

It is perhaps worth observing that the argument in this lemma could be utilized to conclude that in Theorem 2 the image of the map

$$\mathrm{Ext}(A/B, G) \to \mathrm{Ext}(B, p^\alpha G) \oplus \mathrm{Hom}(B, Q_1)$$

contains the first summand.

The following is a dual to [**8**], Th. 3.4.

THEOREM 13. *Suppose that A is torsion, $p^\alpha A \neq 0$ and G is cotorsion. If $\mathrm{Ext}(A,G)$ is p^α-injective, then G is p^α-injective.*

PROOF. There is no loss of generality in assuming that G is reduced. Since A is torsion, $\mathrm{Ext}(A,G)$ is reduced. Therefore, since $\mathrm{Ext}(A,G)$ is p^α-injective, we must have $p^\alpha \mathrm{Ext}(A,G) = 0$. By Lemma 12, this implies that $\mathrm{Ext}(p^\alpha A, p^\alpha G) = 0$. However, since G is reduced, we can conclude that $p^\alpha G = 0$. Let K be a p^α-high subgroup of A (i.e., a subgroup maximal with respect to intersecting $p^\alpha A$ trivially). If $D = A/K$, then D is a non-zero divisible group, and by [**2**], Th. 92, $0 \to K \to A \to D \to 0$ is $p^{\alpha+1}$-pure. Consider

$$\to \mathrm{Ext}(D,G) \to \mathrm{Ext}(A,G) \to \mathrm{Ext}(K,G) \to 0.$$

Clearly $\mathrm{Ext}(D,G) \cong \mathrm{Ext}(\oplus \mathbf{Z}_{p^\infty}, G) \cong \prod \mathrm{Ext}(\mathbf{Z}_{p^\infty}, G) \cong \prod G$, and since the image of the first arrow is contained in $p^\alpha \prod G = 0$, it follows that this is a short exact sequence. In fact, by Lemma 11, it is $p^{\alpha+1}$-pure, and since $\mathrm{Ext}(A, G)$ is p^α-injective, by Lemma 10 it must split. Therefore, G is p^α-injective.

A p^α-injective is called *proper* if it is not p^β-injective for any $\beta < \alpha$. If $B \cong \oplus_{n<\omega} \mathbf{Z}_{p^n}$, then by a result of [**4**], $\mathrm{Ext}(\mathbf{Z}_{p^\infty}, B)$ is not p^α-injective for any α. Therefore,

$$\mathrm{Ext}(H_\alpha, \mathrm{Ext}(\mathbf{Z}_{p^\infty}, B)) \cong \mathrm{Ext}(\mathrm{Tor}(H_\alpha, \mathbf{Z}_{p^\infty}), B) = \mathrm{Ext}(H_\alpha, B)$$

is p^α-injective, and by the last result, it is not p^β-injective for any $\beta < \alpha$. This construction is dual to [**7**], Prop. 6.10.

It was proven in [**5**] that the group G is cotorsion iff $E_\lambda G$ is reduced for every limit ordinal λ. Since the p^α-injective groups are cotorsion, it makes sense that there is some relationship between them and this condition.

THEOREM 14. *Suppose λ is a limit, $\alpha = \lambda + \gamma$ and G is p^α-injective. Then $p^\gamma E_\lambda G = 0$.*

PROOF. There is a p^λ-pure sequence

$$0 \to G/p^\lambda G \to \mathrm{Ext}(H_\lambda, G) \to \mathrm{Ext}(T_\lambda, G) \to 0.$$

As in the proof of Lemma 4, there is an embedding of $L_\lambda G$ in $\mathrm{Ext}(H_\lambda G)$ which determines an embedding of $E_\lambda G$ in $\mathrm{Ext}(T_\lambda, G)$, and $E_\lambda G \subseteq p^\lambda \mathrm{Ext}(T_\lambda, G)$. Therefore, $p^\gamma E_\lambda G \subseteq p^\alpha \mathrm{Ext}(T_\lambda, G) = 0$, proving the result.

In fact, this last result can be viewed as a new way to see that every p^α-injective is cotorsion. It might be tempting to conjecture that the converse of this result is valid, i.e., if $p^\gamma E_\lambda G = 0$ whenever $\alpha = \lambda + \gamma$, then G is p^α-injective. However, this is false. If $p^\omega G$ and $G/p^\omega G$ are algebraically compact, then $E_\lambda G = 0$ for all λ. However, such a group may fail to be $p^{\omega 2}$-injective.

Using terminology from [**10**], A and B are *$\mathcal{A}/\mathcal{B}$-isomorphic* if there is a homomorphism $A \to B$ whose kernel and cokernel are bounded. The terminology comes from the notion of a quotient category, where $\mathcal{A}$ denotes the category of abelian groups and $\mathcal{B}$ denotes the subcategory consisting of the bounded groups. Of course, this property is symmetric. We will say G is $p^{<\infty}$-injective if it is p^α-injective for some α. The following relates these two notions.

THEOREM 15. *The reduced $p^{<\infty}$-injective groups form the smallest non-empty class* **C** *closed under $\mathcal{A}/\mathcal{B}$-isomorphisms, summands and direct products.*

PROOF. Observe first that the reduced $p^{<\infty}$-injectives are clearly closed under summands. If $\{G_i\}_{i\in I}$ is a collection such that G_i is p^{α_i}-injective for each $i \in I$, then $\prod_{i\in I} G_i$ is p^α-injective, where $\alpha = \sup\{\alpha_i : i \in I\}$. Next suppose G is a reduced p^α-injective and G is $\mathcal{A}/\mathcal{B}$-isomorphic to G'. So there is an exact sequence $0 \to B \to G \to G' \to B' \to 0$ where $p^j B = p^k B' = 0$. If α is finite, G is bounded, so that G' is bounded, so assume α is infinite. If $L = G/B$, then clearly L is reduced. For any X, considering the exact sequence

$$\mathrm{Ext}(X, B) \to \mathrm{Ext}(X, G) \to \mathrm{Ext}(X, L) \to 0,$$

since $p^j \operatorname{Ext}(X, B) = 0$, it follows that L is $p^{\alpha+j}$-injective. Now there is a short exact sequence $0 \to L \to G' \to B' \to 0$ and it follows that G' is reduced. For any X, considering the exact sequence

$$\operatorname{Hom}(X, B') \to \operatorname{Ext}(X, L) \to \operatorname{Ext}(X, G') \to \operatorname{Ext}(X, B') \to 0,$$

since $p^k \operatorname{Hom}(X, B') = p^k \operatorname{Ext}(X, B') = 0$, it follows that G' is $p^{\alpha+j+k}$-injective. Therefore, every group in $\mathbf{C}$ is $p^{<\infty}$-injective.

Conversely, we now want to show that every reduced $p^{<\infty}$-injective is in $\mathbf{C}$. If G is a reduced p^α-injective, then the map $\delta_G : G \to \operatorname{Ext}(H_\alpha, G)$ is a splitting monomorphism. So it is sufficient to prove that $\operatorname{Ext}(H_\alpha, G)$ is in $\mathbf{C}$. Our conditions imply that any bounded group is in $\mathbf{C}$. Since for finite α, $\operatorname{Ext}(H_\alpha, G)$ is bounded, it is in $\mathbf{C}$. Now, if α is a limit ordinal, then $\operatorname{Ext}(H_\alpha, G) \cong \operatorname{Ext}(\oplus_{\beta<\alpha} H_\beta, G) \cong \prod_{\beta<\alpha} \operatorname{Ext}(H_\beta, G)$, is, by induction, in $\mathbf{C}$. If $\alpha = \beta + 1$, then there is an exact sequence $0 \to \mathbf{Z}_p \to H_\alpha \to H_\beta \to 0$ which gives an exact sequence

$$\operatorname{Hom}(\mathbf{Z}_p, G) \to \operatorname{Ext}(H_\beta, G) \to \operatorname{Ext}(H_\alpha, G) \to \operatorname{Ext}(\mathbf{Z}_p, G) \to 0.$$

Therefore, $\operatorname{Ext}(H_\beta, G)$ and $\operatorname{Ext}(H_\alpha, G)$ are $\mathcal{A}/\mathcal{B}$-isomorphic. Since, by induction, $\operatorname{Ext}(H_\beta, G)$ is in $\mathbf{C}$, $\operatorname{Ext}(H_\alpha, G)$ is in $\mathbf{C}$, too.

This is close to being a dual to [**7**], Prop. 6.6. If we define a $p^{<\infty}$-projective to be a group which is p^α-projective for some α, then the following is true:

THEOREM 16. *The torsion $p^{<\infty}$-projective groups form the smallest non-empty class closed under $\mathcal{A}/\mathcal{B}$-isomorphisms, summands and direct sums.*

In [**7**], however, the focus is on the groups isomorphic to subgroups of $p^{<\infty}$-projectives, which are called the subprojectives. It is tempting to dualize this to consider the class of groups which are epimorphic images of $p^{<\infty}$-injectives. However, this is not a worthwhile notion, since every cotorsion group is the epimorphic image of an algebraically compact, and hence p^ω-injective, group.

REFERENCES

1. L. Fuchs, *Infinite Abelian Groups, Vols. I and II*, Academic Press, New York and London, 1970 and 1973.
2. P. Griffith, *Infinte Abelian Groups*, Chicago Lectures in Mathematics, Chicago and London, 1970.
3. P. Keef, *On generalizations of purity in primary abelian groups*, J. of Algebra (to appear).
4. ———, *On p^α-torsion injective abelian groups*, submitted.
5. R. Mines, *Torsion and cotorsion completions*, Studies on Abelian Groups, Dunod, Paris, 1968, pp. 301-303.
6. R. Nunke, *Modules of extensions over Dedekind rings*, Illinois J. Math. **3** (1959), 41-55.
7. ———, *Purity and subfunctors of the identity*, Topics in Abelian Groups, Scott, Foresman and Company, Chicago, 1963, pp. 121-171.
8. ———, *On the structure of Tor*, Proceedings of the Colloquium on Abelian Groups, Publ. House of the Hung. Acad. of Sci., Budapest, 1964, pp. 115-124.
9. ———, *Homology and direct sums of countable abelian groups*, Math. Z. **22** (1967), 182-212.

10. E. A. Walker, *Quotient categories and quasi-isomorphisms of abelian groups*, Proceedings of the Colloquium on Abelian Groups, Publ. House of the Hung. Acad. of Sci., Budapest, 1964, pp. 147-162..

Department of Mathematics, Whitman College, Walla Walla, Washington, 99362

E-mail address: keef@whitman.edu

Contemporary Mathematics
Volume **171**, 1994

Abelian groups with contractions I

FRANZ–VIKTOR KUHLMANN

20. 3. 1994

ABSTRACT. On the value group of nonarchimedean exponential fields with respect to their natural valuation, the exponential induces a map called contraction. A corresponding theory of abelian groups with contractions is axiomatized and, by a detailed study of their algebraic properties, shown to be model complete, complete, decidable and to admit quantifier elimination. Finally, other concepts of contractions are discussed.

1. Introduction

In the last years, the model theory of the reals with exponentiation has gone through a remarkable progress by the works of A. Wilkie and other authors (cf. [W1], [W2], [D–M–M]). This has increased the interest in the structure of ordered fields with an exponential map, in particular if they are (at least in part) models of the above mentioned theory. In [KS] and [K–KS], the structure of nonarchimedean exponential fields and in particular, of their value groups with respect to the natural valuation, has been studied. It has turned out that an exponential induces a map χ, called *contraction*, on the value group. It is essentially given as follows: if v is the natural valuation and x a field element with $vx < 0$, then $\chi vx = v \log x$, where log is the inverse of the exponential. The contraction has some amazing properties: although "contracting" archimedean classes, it is still surjective. For the discussion how to obtain contractions from exponentials, see chapter 3 of [K–KS]. In the present paper, we will take over the axiom system that was derived there to study the model theoretic properties of divisible ordered abelian groups with centripetal contractions. The property "centripetal" corresponds to a growth axiom satisfied by the usual exponential.

1991 *Mathematics Subject Classification.* Primary 03C60, 06F20; Secondary 12L12.
Key words and phrases. Abelian groups, ordered abelian groups, exponential fields.
This paper is in final form and no version of it will be submitted for publication elsewhere.

In section 2, we will give some preliminary facts about ordered abelian groups and their natural valuations.

At the beginning of section 3, the axioms for contractions on ordered abelian groups will be listed and discussed. Afterwards, we study precontraction groups (on which the contraction is not assumed to be surjective) and their extensions. By means of embedding lemmas, the model theory of divisible centripetal contraction groups will be shown to be model complete, complete, decidable and to admit quantifier elimination (Theorems 3.23, 3.26, 3.27 and 3.24).

In section 4, we will finally discuss some other possible concepts for contractions on (not necessarily ordered) abelian groups.

In a forthcoming paper [K2], we will study the terms built up with contractions, obtaining a description of the definable sets. This description together with the quantifier elimination result of the present paper will show that the theory of divisible centripetal contraction groups is weakly o-minimal.

At this point, I would like to thank my wife, Salma Kuhlmann, for her support and many invaluable discussions. Further, I am endebted to Angus MacIntyre for bringing my attention to the contractions on value groups of exponential fields.

2. Preliminaries

In this paper, we will assume $\mathbb{N}$, the set of natural numbers, to contain 0. Further, whenever we will talk of ordered groups, we will mean ordered abelian groups.

Let G be an ordered group. We write $G^{<0} = \{g \in G \mid g < 0\}$ and $G^{>0} = \{g \in G \mid g > 0\}$. We set $\mathrm{sign}(0) = 0$ and for $a \in G$, we set $\mathrm{sign}(a) = 1$ if $a > 0$, and $\mathrm{sign}(a) = -1$ if $a < 0$. Further, we set $|a| := \max\{a, -a\} = \mathrm{sign}(a) \cdot a$. Two elements $a, b \in G$ are called *archimedean equivalent* if there is some $n \in \mathbb{N}$ such that $n|a| \geq |b|$ and $n|b| \geq |a|$. Let va denote the equivalence class of a. The set of equivalence classes is ordered as follows: $va < vb$ if and only if $|a| > |b|$ and $va \neq vb$. We write $\infty := v0$; this is the maximal element in the ordered set of equivalence classes. The map $a \mapsto va$ is a group valuation, that is, it satisfies

(V0) $\forall x : \; vx = \infty \Leftrightarrow x = 0\,,$

(V1) $\forall x, y : \; v(x - y) \geq \min\{vx, vy\}\,.$

From these rules, we may deduce

(V2) $\forall x : \; vx = v(-x)\,,$

(V3) $v(\sum_{1 \leq i \leq n} x_i) = \min_{1 \leq i \leq n} vx_i$ if all nonzero x_i have different values,

(V4) $\forall x, y : \; v(x - y) > \min\{vx, vy\} \Rightarrow vx = vy\,.$

The ordered set $vG := \{vg \mid 0 \neq g \in G\}$ will be called the *value set* of G.

The valuation that we have defined above, is canonically associated to every ordered group and is called the *natural valuation*. The order type of vG is called the *rank* of the ordered group G. Hence, G is archimedean if and only if its rank is 1, that is, vG consists of just one element. Note that for every $a \in G$ and every $n \in \mathbb{Z} \setminus \{0\}$, the element $na \in G$ is archimedean equivalent to a and so, the natural valuation satisfies the axiom scheme

(NV1) $\forall x : v(nx) = vx \qquad (0 \neq n \in \mathbb{Z})$.

Further, the natural valuation is compatible with the ordering, in the following sense:

(NV2) $\forall x, y \in G^{<0} : (vx < vy \Rightarrow x < y) \wedge (x < y \Rightarrow vx \leq vy)$.

Let us also note:

(NV3) $\operatorname{sign}(\sum_{1\leq i\leq n} x_i) = \operatorname{sign}(x_m)$ if $vx_m < vx_i$ for all $i \neq m$,

(NV4) $\forall x, y : v(x - y) > vx \Rightarrow \operatorname{sign}(x) = \operatorname{sign}(y)$.

Further, (NV3) may be generalized to

(NV5) $x_m < x'_m \Rightarrow \sum_{1\leq i\leq n} x_i < \sum_{1\leq i\leq n} x'_i$ if $vx_m < vx_i$ and $vx'_m < vx'_i$ for all $i \neq m$.

In the sequel, v will always denote the natural valuation, if not stated otherwise.

Given an ordered Hahn sum $\coprod_{i\in I} A_i$ or Hahn product $\mathbf{H}_{i\in I} A_i$ over archimedean ordered groups A_i, then the ordered index set I is order isomorphic to the ordered set of archimedean classes, and via this isomorphism, the natural valuation is (equivalent to) the minimum support valuation, that is, for $a = (a_i)_{i\in I}$ we may set

$$va = \min\{i \in I \mid a_i \neq 0\} .$$

Let $(S, <)$ be a totally ordered set. If $S_1, S_2 \subset S$ and $a \in S$, we will write $a < S_2$ if $a < b$ for all $b \in S_2$, and further, we will write $S_1 < S_2$ if $a < S_2$ for all $a \in S_1$. Similarly, we use the relations $>$, $\leq$ and $\geq$. A pair (S_1, S_2) of two convex subsets of S satisfying $S_1 \cup S_2 = S$ will be called a *quasicut in* S if $S_1 \leq S_2$, and it will be called a *cut* if it even satisfies $S_1 < S_2$. We allow S_1 or S_2 to be empty, with the convention that $\emptyset < G$ and $G < \emptyset$. We will say that $a \in S$ *realizes the (quasi)cut* (S_1, S_2) if $S_1 \leq a \leq S_2$. If the ordered set T contains S and if $b \in T$, then $(\{a \in S \mid a \leq b\}, \{a \in S \mid a > b\})$ will be called the *cut induced by* b *in* S. Note that a (quasi)cut in a densely ordered set can be realized by at most one element of that set.

If (G_1, G_2) is a quasicut in G and $g \in G$, then

$$(G_1, G_2) - g \;:=\; (\{a - g \mid a \in G_1\}\,,\, \{a - g \mid a \in G_2\})$$

is again a quasicut in G, called a *shift* of (G_1, G_2). If (G_1, G_2) is a cut, then also $(G_1, G_2) - g$ is a cut. If (G_1, G_2) is the cut induced by b in G, then $(G_1, G_2) - g$ is the cut induced by $b - g$ in G. Note that two elements $a \neq a'$ may determine the same shift $(G_1, G_2) - a = (G_1, G_2) - a'$.

To every quasicut (G_1, G_2) in G, we may associate a quasicut in vG in the following way. If G_2 contains an element ≤ 0, then we set $S_1 = v(G_1)$ and $S_2 = v(G^{<0} \cap G_2)$ and find that (S_1, S_2) is a quasicut in vG by virtue of (NV2). If on the other hand, G_2 is contained in $G^{>0}$, then $-G_1$ contains an element ≤ 0, and we set $S_1 = v(-G_2) = v(G_2)$ and $S_2 = v(G^{<0} \cap -G_1) = v(G^{>0} \cap G_1)$; again, (S_1, S_2) is a quasicut in vG. If (S_1, S_2) is a cut, then (G_1, G_2) will be called a *v-cut*. From this definition, we may deduce the following criterion: (G_1, G_2) is a v-cut if and only if there is no pair (g_1, g_2) of elements $g_1 \in G_1$ and $g_2 \in G_2$ such that $vg_1 = vg_2$ and $\mathrm{sign}(g_1) = \mathrm{sign}(g_2)$. Using this criterion together with (NV2), we see that every element b in an extension of $(G, <)$ with $vb \notin vG \cup \{\infty\}$ will induce a v-cut in G. There is also a converse, stated as part b) of the following lemma:

LEMMA 2.1. *Let G be divisible and (G_1, G_2) be a cut in G.*

a) Among all shifts of (G_1, G_2), there is at most one v-cut. If (G_1, G_2) is realized by some element $a \in G$, then $(G_1, G_2) - a$ is this unique v-cut, realized by 0.

b) Assume (G_1, G_2) to be a v-cut which is not realized in G. If $(H, <)$ is an extension of $(G, <)$ and $b \in H$ realizes (G_1, G_2), then $vb \notin vG \cup \{\infty\}$.

Proof: a): Suppose that $(G_1, G_2) - g$ and $(G_1, G_2) - g'$ are v-cuts; we have to show that they are equal. Replacing G_i by $G_i - g' = \{h - g' \mid h \in G_i\}$, $i = 1, 2$, and g by $g - g'$, we may assume from the start that $g' = 0$. Further, let us assume that $-g \in G_1$; for $-g \in G_2$, the proof is symmetrical.

Suppose that $(G_1 - g) \cap G_2 \neq \emptyset$ and choose $a \in G_1$ such that $a - g \in G_2$. If $a \geq -g$, it follows that $2a \in G_2$ since $2a \geq a - g$. If $a \leq -g$, it follows that $-2g \in G_2$ since $-2g \geq a - g$. In view of $a, -g \in G_1$, both is impossible by our criterion for v-cuts.

Now suppose that $G_1 \cap (G_2 - g) \neq \emptyset$ and choose $b \in G_2$ such that $b - g \in G_1$. Then $b - 2g = (b - g) - g \in G_1 - g$. Applying our criterion to (G_1, G_2), we find that also $b/2 \in G_2$ and thus, $b/2 - g \in G_2 - g$. Applying our criterion to $(G_1, G_2) - g$, we obtain $b - 2g = 2(b/2 - g) \in G_2 - g$, contradicting $b - 2g \in G_1 - g$.

We have now shown that (G_1, G_2) and $(G_1, G_2) - g$ are equal. As to the last assertion of a), we leave it to the reader to show that $(G_1, G_2) - a$ is a v-cut realized by 0. The uniqueness then follows from what we have already proved.

b): Assume that b realizes (G_1, G_2) and that there is some $a \in G$ such that $va = vb$, that is, a and b are archimedean equivalent. Since G is assumed to be divisible, this implies the existence of some $q \in \mathbb{Q} \setminus \{0\}$ such that b lies between a and $qa \in G$. Since a and qa are archimedean equivalent, we have $va = vqa$ and our above criterion shows that (G_1, G_2) cannot be a v-cut. □

Note that the proof shows that instead of requiring G to be divisible, it suffices to suppose that G is p-divisible for some prime p (which need not be equal to 2). Without this divisibility condition, the lemma is not even true for densely ordered groups. Indeed, in the lexicographic product $\mathbb{Z} \amalg \mathbb{Q}$, both

$$(\{(x,y) \mid x \leq 0\}, \{(x,y) \mid x \geq 1\})$$

and

$$(\{(x,y) \mid x \leq -1\}, \{(x,y) \mid x \geq 0\}) = (\{(x,y) \mid x \leq 0\}, \{(x,y) \mid x \geq 1\}) - (1,0)$$

are v-cuts.

The cut (G_1, G_2) will be called a *shifted v-cut* if there exists some $g \in G$ such that $(G_1, G_2) - g$ is a v-cut. We have:

COROLLARY 2.2. *Let the situation be as in the previous lemma. Assume that (G_1, G_2) is not realized in G but is realized by $b \in H$, where $(H, <)$ is an extension of $(G, <)$. Then (G_1, G_2) is a shifted v-cut if and only if there is some $g \in G$ such that $v(b - g) \notin vG \cup \{\infty\}$. In this case, $(G_1, G_2) - g$ is the unique v-cut among all shifts of (G_1, G_2).*

Proof: Assume that $(G_1, G_2) - g$ is a v-cut. Since b realizes (G_1, G_2), the element $b - g$ realizes $(G_1, G_2) - g$. Hence by part b) of Lemma 2.1, $v(b - g) \notin vG \cup \{\infty\}$. Conversely, if for some $g \in G$ we have $v(b - g) \notin vG \cup \{\infty\}$, then by our remark preceding Lemma 2.1, $(G_1, G_2) - g$ is a v-cut. The uniqueness assertion follows directly from Lemma 2.1. □

The model theoretical facts that we will use in this paper, can be found in most of the books on general model theory, e.g. [C–K] or [P]. Also, they can be found in [K1]. In particular, let us mention the following facts (cf. Korollar 2.19 of [P], or [K1]):

LEMMA 2.3. *Let $\mathcal{S}'$ be a substructure of the structure $\mathcal{S}$. If $\mathcal{S}'$ is existentially closed in $\mathcal{S}$, then $\mathcal{S}$ is embeddable over $\mathcal{S}'$ in every $|\mathcal{S}|^+$-saturated elementary extension of $\mathcal{S}'$. Conversely, if $\mathcal{S}$ is embeddable over $\mathcal{S}'$ in some elementary extension of $\mathcal{S}'$, then $\mathcal{S}'$ is existentially closed in $\mathcal{S}$.*

Note that $|\mathcal{S}|^+$ denotes the successor cardinal of the cardinality of (the underlying set of) $\mathcal{S}$.

We will need the following well known facts about ordered groups. Their proofs can be found in [K1].

LEMMA 2.4. *Let $(G, <) \subset (H, <)$ be an extension of ordered abelian groups and $b \in H \setminus G$. Assume that G be divisible. If $b' \notin G$ is also an element of some extension of $(G, <)$, inducing in G the same cut as b, then the assignment $b \mapsto b'$ defines an order preserving isomorphism from $G + \mathbb{Z}b$ onto $G + \mathbb{Z}b'$. On the other hand, in every $|G|^+$-saturated extension of $(G, <)$ there is an element $b' \notin G$ inducing in G the same cut as b.*

(The assertions of this lemma can easily be used to show the model completeness of the elementary class of nontrivial divisible ordered abelian groups.)

LEMMA 2.5. *Let $(G,<) \subset (H,<)$ and $(G,<) \subset (H',<)$ be extensions of ordered abelian groups. Suppose that b_i and b'_i, $i \in I$, are elements of H and H' respectively, such that*

1) all vb_i, $i \in I$, are different and not contained in vG,

2) $sign(b_i) = sign(b'_i)$ for all $i \in I$, and the assignment $vb_i \mapsto vb'_i$ establishes an order isomorphism

$$\tau : vG \cup \{vb_i \mid i \in I\} \to vG \cup \{vb'_i \mid i \in I\} .$$

Then the assignment $b_i \mapsto b'_i$ defines order and valuation preserving isomorphisms

$$G + \sum_{i\in I} \mathbb{Z}b_i \longrightarrow G + \sum_{i\in I} \mathbb{Z}b'_i$$

$$G + \sum_{i\in I} \mathbb{Q}b_i \longrightarrow G + \sum_{i\in I} \mathbb{Q}b'_i$$

both inducing τ. (Here, the groups are endowed with the restrictions of the orders of the divisible hulls of H and H' which are uniquely determined by those of H and H'.) Note that $vG \cup \{vb_i \mid i \in I\}$ is the value set of both groups on the left hand side, and $vG \cup \{vb'_i \mid i \in I\}$ is the value set of both groups on the right hand side.

Alternatively, assume 1) and the following condition:

2') the assignment $b_i \mapsto b'_i$ establishes an isomorphism

$$\iota : G \cup \{b_i \mid i \in I\} \to G \cup \{b'_i \mid i \in I\}$$

as ordered sets.

Then ι extends linearly to order and valuation preserving isomorphisms of the above groups.

Note that by virtue of $G \cap \{b_i \mid i \in I\} = \emptyset$, the ordering on $G \cup \{b_i \mid i \in I\}$ is already determined by the cuts induced in G by the elements b_i and by the ordering on the set $\{b_i \mid i \in I\}$.

LEMMA 2.6. *Let $(H,<)$ be generated by n elements over $(G,<)$. Then $vH \backslash vG$ consists of at most n values.*

3. Ordered abelian groups with contractions

Let us now introduce contraction maps. We will work in the language $\mathcal{L}_{cg} = \{+,-,0,<,\chi\}$ where $<$ is a binary relation symbol, $+$ is a binary function symbol, and $-$ and χ are unary function symbols. If $(G,+,-,0,<,\chi)$ is an $\mathcal{L}_{cg}$-structure, then it will be called a *precontraction group* if it satisfies

(OAG) $(G,+,-,0,<)$ is an ordered abelian group,

(C0) $\forall x: \chi x = 0 \Leftrightarrow x = 0$,

(C≤) χ preserves $\leq$,

(C−) $\forall x: \chi(-x) = -\chi x$,

(CA) if x is archimedean equivalent to y and $\text{sign}(x) = \text{sign}(y)$, then $\chi x = \chi y$.

If these axioms hold, then χ will be called a *precontraction*. If in addition,

(CS) χ is surjective,

then χ will be called a *contraction* and the group is a *contraction group*. Axioms (CA) and (CS) together show that every archimedean ordered contraction group must be trivial. Further, $(G, +, -, 0, <, \chi)$ will be called *centripetal*, if it satisfies

(CP) $\forall x \in G \setminus \{0\}: |x| > |\chi x|$,

and it will be called *centrifugal*, if it satisfies

(CF) $\forall x \in G \setminus \{0\}: |x| < |\chi x|$.

In the sequel, we will write (G, χ) instead of $(G, +, -, 0, <, \chi)$.

Axiom (CA) may be expressed by the following recursive axiom scheme:

$$\forall x, y: x \geq y > 0 \wedge ny \geq x \implies \chi x = \chi y \qquad (n \in \mathbb{N}).$$

Observe that axioms (C0), (C−) and (CA) together imply

(CZ) $\forall x \in G: \chi(zx) = \text{sign}(z) \cdot \chi x \qquad (z \in \mathbb{Z})$.

Observe further that by axiom (C≤), $x \leq y \leq z$ and $\chi x = \chi z$ implies $\chi y = \chi z$. This shows that in the presence of axiom (C≤), the axiom scheme (CA) may be replaced by the single axiom

(CA′) $\forall x \in G: \chi(2x) = \chi x$.

This and all other axioms are immediately seen to be elementary in the language $\mathcal{L}_{cg}$. All axioms are universal, except for the surjectivity axiom (CS). Since properties described by universal axioms are inherited by substructures, we have:

LEMMA 3.1. *Every substructure S of a precontraction group (G, χ) is again a precontraction group. If (G, χ) is centripetal (resp. centrifugal), then so is S.*

We will need a further axiom scheme which is not universal. Namely, $(G, +, 0)$ is *divisible* if it satisfies

(D) $\forall x \, \exists y: ny = x \qquad (0 \neq n \in \mathbb{N})$.

We will consider the theory of divisible centripetal contraction groups. Some results also hold for the theory of divisible centrifugal contraction groups. We will frequently need that the groups are nontrivial, that is, they satisfy the axiom $\exists x: x \neq 0$. Before we continue, let us put together some technical preliminaries. Using the natural valuation v, we may express axiom (CA) in the following way:

(CV1) $\forall x, y: vx = vy \wedge \text{sign}(x) = \text{sign}(y) \implies \chi x = \chi y$

(but note that v is neither a symbol in our language $\mathcal{L}_{cg}$ nor definable in the theory of divisible centripetal or centrifugal contraction groups). Sometimes, we will only be interested in equality up to the sign; instead of writing $|a| = |b|$, we will then write $a = \pm b$. Then (CV1) reads as follows:

(CV2) $\forall x, y:\ vx = vy \Longrightarrow \chi x = \pm \chi y\,.$

From (V3), (NV3) and (V4), (NV4) we way infer:

(CV3) $\chi(\sum_{1\leq i\leq n} x_i) = \chi x_m$ if $vx_m < vx_i$ for all $i \neq m$,

(CV4) $\forall x, y:\ v(x-y) > vx \Rightarrow \chi x = \chi y\,.$

From (CV2) and (NV2) together with (C$\leq$), one may deduce

(CV5) $\forall x, y:\ vx \leq vy \Longrightarrow |\chi x| \geq |\chi y|\,.$

The following lemma collects some generalities about precontractions.

LEMMA 3.2. *Let (G,χ) be a precontraction group. Then the following assertions hold:*

a) sign(a) = sign(χa) for every $a \in G$, hence $\chi G^{<0} \subset G^{<0}$ and $\chi G^{>0} \subset G^{>0}$. Moreover, $\chi G^{<0} = -\chi G^{>0}$.

b) χ is centripetal if and only if $v(\chi a) > va$ for all $a \in G \setminus \{0\}$. Similarly, χ is centrifugal if and only if $v(\chi a) < va$ for all $a \in G \setminus \{0\}$.

c) Every nontrivial centripetal precontraction group is densely ordered. The same is true for every nontrivial centrifugal contraction group.

Proof: a): If $a < 0$ then by (C$\leq$) and (C0), $\chi a < \chi 0 = 0$. Similarly, $a > 0$ is shown to imply $\chi a > 0$. The last assertion follows directly from (C$-$).

b): Assume that χ is centripetal. Then for every $a \in G\setminus\{0\}$, we have $|\chi a| < |a|$ and thus, $v(\chi a) \geq va$. But for every $b \in G$ with $va = vb$ it follows from (CA) and (CP) that $|\chi a| = |\chi b| < |b|$. Hence, $v(\chi a) = va$ is impossible. The proof for centrifugal precontractions is similar.

c): A centripetal precontraction group cannot have a least positive element 1 since by (C0), (C$\leq$) and (CP), $0 < \chi g < g$ for every $g \in G^{>0}$. Now assume 1 to be the least positive element in a centrifugal contraction group (G,χ). Then there is some $a \in G$ such that $\chi a = 1$. But then by (C0), (C$\leq$) and (CF), $0 < a < 1$, a contradiction. □

LEMMA 3.3. *Assume $(G,\chi) \subset (H,\chi)$ to be an extension of precontraction groups. Let $b \in H$ such that $vb \notin vG \cup \{\infty\}$ and $\chi b = a \in G$. Then $(G + \mathbb{Z}b, \chi)$ is a precontraction group with $\chi(G+\mathbb{Z}b) = \chi G \cup \{a, -a\} \subset G$. Moreover, the extension of χ from (G,χ) to $G + \mathbb{Z}b$ is uniquely determined by the assignment $\chi b = a$.*

Proof: Since $vb \notin vG$, (V3) shows that every element in $G + \mathbb{Z}b$ has value either in $vG \cup \{\infty\}$ or equal to vb. Let $d = g + zb \in G + \mathbb{Z}b$. Then $\chi d = \chi zb = \text{sign}(z)\cdot a$ if $vd = vb$, and $\chi d = \chi g$ if $vd = vg \in vG \cup \{\infty\}$. Hence, the extension of χ from G to $G + \mathbb{Z}b$ is uniquely determined by $\chi b = a$. It also proves $\chi(G+\mathbb{Z}b) = \chi G \cup \{a, -a\} \subset G$. Consequently, $G+\mathbb{Z}b$ is closed under χ and thus a precontraction group by virtue of Lemma 3.1. □

LEMMA 3.4. *Assume $(G,\chi) \subset (H,\chi)$ to be an extension of precontraction groups. Let $a \in G \setminus \chi G$ and $b \in H$ such that $a = \chi b$. Then $vb \notin vG \cup \{\infty\}$, and $(G+\mathbb{Z}b,\chi)$ is a precontraction group satisfying $\chi(G+\mathbb{Z}b) \subset G$. Moreover, if $(G,\chi) \subset (H',\chi')$ is a second extension of precontraction groups and if $b' \in H'$ such that $a = \chi' b'$, then the assignment $b \mapsto b'$ induces an isomorphism $G+\mathbb{Z}b \to G+\mathbb{Z}b'$ of precontraction groups over G.*

Proof: If there would exist some $c \in G$ such that $vb = vc$ then in view of (CV2), $a = \chi b = \pm\chi c \in \chi G$. Hence, $a \notin \chi G$ implies $vb \notin vG \cup \{\infty\}$, and from Lemma 3.3 it follows that $(G+\mathbb{Z}b,\chi)$ is a precontraction group with $\chi(G+\mathbb{Z}b) \subset G$.

We show that the ordering on $vG \cup \{vb\}$ is uniquely determined by (G,χ) and the assignment $\chi b = a$. Assume w.l.o.g. that $a \in G^{<0}$. Then $b \in H^{<0}$ by part a) of Lemma 3.2. Let $c \in G^{<0}$. Since $vc \neq vb$, (NV2) shows that $vc < vb$ if and only if $c < b$ and $vc > vb$ if and only if $c > b$. Then by virtue of (C≤), $vc < vb$ if and only if $\chi c < a$ and $vc > vb$ if and only if $\chi c > a$.

Now let also b' be as in the hypothesis. By what we have proved already, $vb \notin vG \cup \{\infty\}$ and $vb' \notin vG \cup \{\infty\}$, and the assignment $vb \mapsto vb'$ induces an order isomorphism over vG of the two sets $vG \cup \{vb\}$ and $vG \cup \{vb'\}$ (which are endowed with the restrictions of the ordering of vH resp. vH'). Further, b and b' have the same sign since $\chi b = \chi' b'$. From Lemma 2.5 it follows that the groups $G+\mathbb{Z}b$ and $G+\mathbb{Z}b'$, endowed with the restrictions of the ordering of H resp. H', are isomorphic over G by sending b to b'. By the uniqueness statement of Lemma 3.3, this isomorphism also preserves the precontraction. □

There are very simple examples for precontraction groups. The map "sign" is itself a precontraction on $\mathbb{Z}$, but it is neither centrifugal nor centripetal. It satisfies $\forall x \in G: |x| \geq |\chi x|$, but this does not remain true in extensions of $\mathbb{Z}$ where 1 is not the least positive element. $\mathbb{Z}$ does not admit a centrifugal or centripetal precontraction since it is archimedean. Indeed, it follows from part b) of Lemma 3.2 that a nontrivial centrifugal or centripetal precontraction group must have infinite rank.

Consider the Hahn sums $\coprod_{\mathbb{N}} \mathbb{Z}$ and $\coprod_{-\mathbb{N}} \mathbb{Z}$ where $\mathbb{Z}$ stands for the ordered group $(\mathbb{Z},+,-,0,<)$, and $\mathbb{N}$ resp. $-\mathbb{N}$ stands for the positive resp. negative integers with their usual ordering. On the first Hahn sum, we may define a precontraction in the following way: if $(z_i)_{i\in\mathbb{N}} \in \coprod_{\mathbb{N}} \mathbb{Z}$ and if i_0 is the minimal index such that $z_{i_0} \neq 0$, then we set

$$\chi\,(z_i)_{i\in\mathbb{N}} = \operatorname{sign}(z_{i_0}) \cdot e_{i_0+1}$$

where e_i denotes the element of $\coprod_{\mathbb{N}} \mathbb{Z}$ which has a 1 at the index i and zeros everywhere else. The resulting centripetal precontraction group will be denoted by $\mathcal{P}_{cp}$. Analogously, if $(z_i)_{i\in -\mathbb{N}} \in \coprod_{-\mathbb{N}} \mathbb{Z}$ and if i_0 is the maximal index such that $z_{i_0} \neq 0$, then we set

$$\chi\,(z_i)_{i\in\mathbb{N}} = \operatorname{sign}(z_{i_0}) \cdot e_{i_0-1}$$

where $e_i \in \coprod_{-\mathbb{N}} \mathbb{Z}$ is defined as above. The resulting centrifugal precontraction group will be denoted by $\mathcal{P}_{cf}$. These examples are as representative as they can be:

LEMMA 3.5. *$\mathcal{P}_{cp}$ is the prime structure of the elementary class of nontrivial centripetal precontraction groups. Analogously, $\mathcal{P}_{cf}$ is the prime structure of the elementary class of nontrivial centrifugal precontraction groups.*

Proof: We will show that every substructure S of a centripetal precontraction group generated by one element $a \neq 0$ is isomorphic to $\mathcal{P}_{cp}$. Indeed, we infer from part b) of Lemma 3.2 that $v(\chi^n a)$, $n \in \mathbb{N}$, is a strictly increasing sequence in the value set vS. Let S_0 be the subgroup of S which is generated by all the elements $\chi^n a$. Every element s of S_0 is then a finite sum $\sum_{i=0}^{\infty} z_i\chi^i a$ with coefficients $z_i \in \mathbb{Z}$, only finitely many of them nonzero. In view of (V3) and (NV1), we find $vs = v(\chi^{i_0}a)$ if $i_0 = \min\{i \in \mathbb{N} \mid z_i \neq 0\}$. Then it follows from (CV2) that $\chi s = \pm\chi^{i_0+1}a$. This shows that S_0 is closed under χ and hence, $S_0 = S$.

Observe that a and $-a$ generate the same substructure. Hence, we may assume $a > 0$, which yields that $\chi^n a > 0$ for all $n \in \mathbb{N}$. Now set $h(\chi^n a) = e_n$ for $n \in \mathbb{N}$. By Lemma 2.5, this assignment defines an order and valuation preserving isomorphism from S onto $\coprod_{\mathbb{N}} \mathbb{Z}$.

The proof for the centrifugal case is similar. In this case, the sequence $v(\chi^n a)$, $n \in \mathbb{N}$, is strictly decreasing, and we have to set $i_0 = \max\{i \in \mathbb{N} \mid z_i \neq 0\}$. □

An extension $(G, v) \subset (H, v)$ of valued groups is called *rank preserving* if $vH = vG$ (precisely speaking, if the induced embedding of vG in vH is onto). For example, the divisible hull $\tilde{G}$ of G is a rank preserving extension of (G, v) since every element in $\tilde{G}$ is archimedean equivalent to some element of G. But if $va = vb$, then every precontraction on H will satisfy $\chi a = \pm\chi b$ by virtue of (CV2). On the other hand, if χ is a precontraction on G, then it may be extended to H by setting $\chi a = \operatorname{sign}(a) \cdot \chi|b|$. This yields $\chi H = \chi G$, showing that H is closed under χ and thus, that (H, χ) is a precontraction group. Hence, we have:

LEMMA 3.6. *Let (G, χ) be a precontraction group. Then for every extension $(H, <)$ of $(G, <)$ which is rank preserving with respect to the natural valuation, χ extends in a unique way to a precontraction χ on H (under preservation of the properties "centripetal" and "centrifugal"), and we have $\chi H = \chi G$. More precisely, if $(G, \chi) \subset (H', \chi')$ is an extension of precontraction groups such that $(H, <) \subset (H', <)$, then the restriction of χ' to H coincides with χ. In particular, these assertions hold for $H = G$.*

If the rank preserving extension $G \subset H$ is nontrivial, then $\chi H = \chi G \neq H$. Hence:

COROLLARY 3.7. *Let $(G,\chi) \subset (H,\chi)$ be a proper rank preserving extension of precontraction groups. Then (H,χ) cannot be a contraction group. In particular, this holds for the divisible hull of a non-divisible precontraction group.*

To prove even more, we introduce a new map which is associated to χ. We define $\rho_\chi : vG \to G^{<0}$ as follows: if $\alpha = va \in vG$, $a \in G^{<0}$, then $\rho_\chi \alpha = \chi a$. This is well defined since by (CV1), the definition does not depend upon the choice of the negative element a of value α. We have $\rho_\chi(vG) = \chi G^{<0}$: if $b = \chi a \in \chi G$, then $b = \rho_\chi va$. Hence, ρ_χ is surjective if and only if χ is. Moreover, ρ_χ preserves $\leq$: on the one hand, χ preserves $\leq$ by (C$\leq$); on the other hand, $va < va'$ for $a, a' \in G^{<0}$ implies $a < a'$.

THEOREM 3.8. *Let $(G,\chi) \subset (H,\chi)$ be a nontrivial extension of precontraction groups. If (H,χ) is a contraction group, then $vH \setminus vG$ is infinite and thus, $(G,v) \subset (H,v)$ is not rank preserving.*

Proof: Since (H,χ) is a contraction group, we have $\chi H = H$. By assumption, $G \subset H$ is a nontrivial extension, hence there is an element $b \in \chi H \setminus \chi G$. Since ρ_χ is surjective, there is some $\beta \in vH \setminus vG$ such that $\rho_\chi \beta = b$. The fact that $vH \setminus vG$ is nonempty implies that $H \setminus G$ is infinite. Thus, the same is true for $\chi H \setminus \chi G$, in view of $\chi H = H$ and $\chi G \subset G$. Again by the surjectivity of ρ_χ, it follows that $vH \setminus vG$ is infinite. □

We will now show that every precontraction group is embeddable in a divisible contraction group.

LEMMA 3.9. *Every (centripetal resp. centrifugal) precontraction group (G,χ) is embeddable in a divisible (centripetal resp. centrifugal) contraction group (H,χ).*

Proof: (H,χ) will be the union over a chain of precontraction groups (G_n,χ), $n \in \mathbb{N}$. Let G_1 be the Hahn product $\mathbf{H}_{vG}\,\mathbb{R}$. Then there is an order and valuation preserving embedding of G in G_1, and the extension $(G,v) \subset (G_1,v)$ is rank preserving. Hence, there is a unique extension of χ from G to G_1.

Having constructed (G_n,χ), let us show how to obtain (G_{n+1},χ). We set

$$\Gamma := vG_n \cup (G_n^{<0} \setminus \chi G_n^{<0}) \,.$$

To define an ordering on Γ, we extend the orderings which already exist on vG_n and on G_n. Hence, we only have to give the order relation between two elements $\beta \in vG_n$ and $a \in G_n^{<0} \setminus \chi G_n^{<0}$. We let $a < \beta$ if $a < \rho_\chi \beta$; otherwise, we let $a > \beta$. Now, we let G_{n+1} be the Hahn product $\mathbf{H}_\Gamma\,\mathbb{R}$. Then there is a natural order and valuation preserving embedding of G_n in G_{n+1} which is induced by the embedding $vG_n \subset \Gamma$. We identify G_n with its image in G_{n+1}. It remains to define the extension of χ. Let $b \in G_{n+1}^{<0}$. If $vb = vb' \in vG_n$ for some $b' \in G_n^{<0}$ then necessarily, in view of (CV1), $\chi b = \chi b'$. If $vb = a \in G_n^{<0} \setminus \chi G_n^{<0}$, then we set $\chi b = a \in G_n \subset G_{n+1}$. For $b \in G_{n+1}^{>0}$, we set $\chi b = -\chi(-b)$. In this way, every element of G_n becomes an element of the range of χ in G_{n+1}. Consequently, χ will be surjective on the union of the G_n. Since all other axioms are universal, we see that this union is a contraction group. □

Let us now consider the question whether there are "closures" of precontraction groups in contraction groups. Let $(G, \chi) \subset (G', \chi)$ be an extension of precontraction groups. We will call (G', χ) a *contraction hull* of (G, χ) if it is a contraction group and has the following universal property:

(CH) if $(G, \chi) \subset (H, \chi)$ is any extension of precontraction groups and (H, χ) is a contraction group, then there is an embedding of (G', χ) in (H, χ) over (G, χ).

Similarly, we will call (G', χ) a *divisible contraction hull* of (G, χ) if it is a divisible contraction group and has the following universal property:

(CHD) if $(G, \chi) \subset (H, \chi)$ is any extension of precontraction groups and (H, χ) is a divisible contraction group, then there is an embedding of (G', χ) in (H, χ) over (G, χ).

Note that if (G', χ) is a contraction hull of a divisible precontraction group (G, χ) but is not itself divisible, then its divisible hull is *not* a divisible contraction hull of (G, χ) (cf. Theorem 3.8).

LEMMA 3.10. *For every precontraction group (G, χ) there exists a contraction hull and a divisible contraction hull. Such a hull (G', χ) may be chosen such that ρ_χ induces an order preserving bijection $vG' \setminus vG \to G'^{<0} \setminus \chi G$. Moreover, we can assume that for every $b \in G'$ there is some $n \in \mathbb{N}$ such that $\chi^n b \in G$.*

Proof: By Lemma 3.9, we may embed (G, χ) in some divisible contraction group (H, χ). We will construct a subgroup of H as follows. For the construction of the contraction hull, we set $G_0 = G$. For the construction of the divisible contraction hull, we let G_0 be the divisible hull of G and endow it with the unique extension of the precontraction χ (cf. Lemma 3.6). Note that $\chi G_0 = \chi G \subset G$. Assume that G_ν for some ordinal ν is already constructed such that ρ_χ induces a bijection $vG_\nu \setminus vG \to \chi G_\nu^{<0} \setminus \chi G$ (for $\nu = 0$, this bijection is empty). If χ is not surjective on G_ν and hence there is some $a \in G_\nu^{<0} \setminus \chi G_\nu$, then we choose $b \in H$ such that $\chi b = a$, and we set $G_{\nu+1} := G_\nu + \mathbb{Z} b$ (resp. $G_{\nu+1} := G_\nu + \mathbb{Q} b$ if we are constructing the divisible contraction hull). From Lemma 3.3 we know that $G_{\nu+1}$ together with the restriction of χ is a precontraction group having $vG_\nu \cup \{vb\}$ as its value set and satisfying $\chi G_{\nu+1} = \chi G_\nu \cup \{\pm a\} \subset G_\nu$. The map ρ_χ sends vb to $\chi b = a$ which is the only element in $\chi G_{\nu+1}^{<0} \setminus \chi G_\nu = \{a\}$. Hence, ρ_χ induces a bijection $vG_{\nu+1} \setminus vG \to \chi G_{\nu+1}^{<0} \setminus \chi G$. Since $\chi G_{\nu+1} \subset G_\nu$, $G_{\nu+1}$ inherits the property that for every element b there is some $n \in \mathbb{N}$ such that $\chi^n b \in G$.

If λ is a limit ordinal and if we have constructed G_ν for all $\nu < \lambda$, then we let G_λ be the union over all G_ν. Then G_λ is again a precontraction group (the theory of precontraction groups being universal). Still, ρ_χ induces a bijection

$vG_\lambda \setminus vG \to \chi G_\lambda^{<0} \setminus \chi G$, and for every $b \in G_\lambda$ there is some $n \in \mathbb{N}$ such that $\chi^n b \in G$. Since at every step we are constructing a nontrivial extension but remain in H, this process is bounded by the successor cardinal κ^+ of the cardinality κ of H. Hence, we will arrive at some G_μ for an ordinal $\mu < \kappa^+$ where χ is surjective. That is, (G_μ, χ) is a contraction group (resp. a divisible contraction group). We set $G' := G_\mu$. Since now $G' = \chi G'$ holds, we have that ρ_χ induces a bijection $vG' \setminus vG \to G'^{<0} \setminus \chi G$ (which is order preserving since ρ_χ preserves $\leq$).

Now let $(G, \chi) \subset (H', \chi')$ be any extension of precontraction groups and assume that (H', χ') is a contraction group (resp. a divisible contraction group). Assume that we have already embedded (G_ν, χ) in (H', χ') over (G, χ); we may identify it with its image in (H', χ'). Now we have to show how this embedding extends to $(G_{\nu+1}, \chi)$. Let $a \in G_\nu$ and $b \in H$ be chosen as above. Choose $b' \in H'$ such that $\chi' b' = a$. Then Lemma 3.4 shows that the precontraction groups $(G_{\nu+1}, \chi)$ and $(G_\nu + \mathbb{Z}b', \chi')$ (resp. $(G_\nu + \mathbb{Q}b', \chi')$, using also Lemma 3.6) are isomorphic over (G_ν, χ). For a limit ordinal λ, the embeddings of the (G_ν, χ), $\nu < \lambda$, extend canonically to an embedding of (G_λ, χ). It follows that (G', χ) is embeddable in (H', χ') over (G, χ). $\square$

Note that the universal property of the contraction hull is somewhat weak: the embedding is not necessarily unique. Indeed, we have seen in the proof above that there is some arbitrariness in the construction of the isomorphism, namely, the choice of $b' \in H'$ satisfying $\chi' b' = a$ is not canonical: we can take any $b'' \in H'$ with $vb'' = vb'$ instead. So we are not able to deduce the uniqueness of the contraction hull from the usual standard argument of category theory. However, uniqueness is not needed for our further results.

Applying the last lemma to the prime structures $\mathcal{P}_{cp}$ and $\mathcal{P}_{cf}$, we obtain:

COROLLARY 3.11. *The elementary classes of nontrivial centripetal (resp. centrifugal) contraction groups and of nontrivial divisible centripetal (resp. centrifugal) contraction groups have prime models.*

Further, the contraction hulls have a good model theoretic property:

LEMMA 3.12. *Let $(G, \chi) \subset (G'', \chi)$ be an extension of precontraction groups and assume that (G, χ) is a contraction group (resp. a divisible contraction group). If (G, χ) is existentially closed in (G'', χ), then it is existentially closed in every contraction hull (resp. divisible contraction hull) (G', χ) of (G'', χ).*

Proof: If (G, χ) is existentially closed in (G'', χ), then by Lemma 2.3, (G'', χ) is embeddable in every $|G''|^+$-saturated elementary extension $(G, \chi)^*$ of (G, χ). We identify (G'', χ) with its image in $(G, \chi)^*$. As an elementary extension, $(G, \chi)^*$ is a (divisible) contraction group like (G, χ). It follows from the universal property of (divisible) contraction hulls that every (divisible) contraction hull (G', χ) of (G'', χ) is embeddable in $(G, \chi)^*$ over (G'', χ). Again by Lemma 2.3, it now follows that (G, χ) is existentially closed in (G', χ). $\square$

Let $(G,\chi) \subset (H,\chi)$ be an extension of precontraction groups. For a given $b \in H$, we will consider the cut

$$(1) \qquad (\{g \mid G \ni g \leq b\}, \{g \mid G \ni g > b\})$$

induced by b in G and its image under χ:

$$(2) \qquad (\{\chi g \mid G \ni g \leq b\}, \{\chi g \mid G \ni g > b\}) .$$

By virtue of (C$\leq$),

$$(3) \qquad \{\chi g \mid G \ni g \leq b\} \leq \chi b \leq \{\chi g \mid G \ni g > b\} .$$

Now assume (G,χ) to be a contraction group. Then $\chi G = G$ and (2) is a quasicut in G which is realized by χb. If the sets of (2) have a nonempty intersection, then it consists of just one element which must be χb, implying that $\chi b \in G$. If the intersection of the two sets is empty, then (2) is a cut. In particular, this is the case if $\chi b \notin G$ and then, (2) is the cut induced by χb in G. If (2) is a cut which is not realized by any element of G, then $\chi b \notin G$ and the cut (1) determines uniquely the cut induced by χb in G. But if there is an element $a \in G$ which realizes (2), then the cut (1) may not determine whether $\chi b \in G$. Indeed, we then only know that there is no element of G between a and χb. This happens if $a = \chi b$, but it may also happen that H contains an element d such that $vd > vG$ and $\chi b = a + d$. Let us discuss the situation in general:

LEMMA 3.13. *Let $(G,<) \subset (H,<)$ be an extension of ordered groups. Assume that two elements $b_1 \in H$ and $b_2 \in G$ realize the same quasicut in $(G,<)$. Then there is no element of G properly between b_1 and b_2 ; hence, there is no element of G properly between $b_1 - b_2$ and 0. We have two cases:*

a) If $b_1 - b_2$ is archimedean equivalent to some element of $G\setminus\{0\}$ (i.e., $v(b_1 - b_2) \in vG$), then $v(b_1 - b_2)$ is the maximal element in vG and $(G,<)$ is discretely ordered.

b) If $b_1 - b_2$ is not archimedean equivalent to some element of $G \setminus \{0\}$, then it is archimedean smaller than all nonzero elements of G, that is, $v(b_1 - b_2) > vG$. It follows that $vb_1 = vb_2$ and sign$(b_1) =$ sign(b_2).

In particular, if $(G,<)$ is dense and vG is cofinal in vH then necessarily, $b_1 = b_2$.

Proof: Assume that there is no element of G properly between $b_1 - b_2$ and 0. If there were $a \in G \setminus \{0\}$ such that $va > v(b_1 - b_2)$ then a would be archimedean smaller than $b_1 - b_2$ and thus, a or $-a$ would lie properly between $b_1 - b_2$ and 0, contrary to our assumption.

Suppose that $b_1 - b_2$ is archimedean equivalent to $a \in G \setminus \{0\}$, that is, $va = v(b_1 - b_2)$. Then by what we have shown, it follows that va is the maximal element of vG. Consequently, the convex subgroup $\{x \in G \mid vx \geq va\}$ is isomorphic to $\{x \in G \mid vx \geq va\}/\{x \in G \mid vx > va\}$ and hence archimedean. It is an ordered subgroup of the archimedean group $\{x \in H \mid vx \geq va\}/\{x \in H \mid vx > va\}$ which contains the nonzero image of $b_1 - b_2$. By our assumption on $b_1 - b_2$ it follows that there is no element of $\{x \in G \mid vx \geq va\}$ properly between this image and zero. Since both groups are archimedean, it follows that $\{x \in G \mid vx \geq va\}$ must be discretely ordered, and the same is consequently true for G.

Now suppose that $b_1 - b_2$ is not archimedean equivalent to any element of $G \setminus \{0\}$. Then by our initial argument, $v(b_1 - b_2) > vG$. Since $b_2 \in G$ by assumption, this yields $v(b_1 - b_2) > vb_2$, showing that $vb_1 = vb_2$ and $\mathrm{sign}(b_1) = \mathrm{sign}(b_2)$ by virtue of (V4) and (NV4).

If $(G, <)$ is dense then it is not discretely ordered and the first case is impossible. If vG is cofinal in vH then the second case is impossible unless $v(b_1 - b_2) = \infty$, that is, $b_1 = b_2$. □

We will now apply this lemma to the quasicut (2).

LEMMA 3.14. *Let $(G, \chi) \subset (H, \chi)$ be an extension of centripetal or centrifugal precontraction groups and assume that (G, χ) is a contraction group. Let $0 \neq b \in H$.*

a) Suppose that $a \in G \setminus \{0\}$ realizes (2). Then $v\chi b = va \in vG$ and thus, $\chi^2 b = \pm \chi a \in G$.

b) Suppose that vG is cofinal in vH and $a \in G \setminus \{0\}$ realizes (2). Then $\chi b = a$.

c) If (2) is not realized by any element of G, then $\chi b \notin G$ and the cut induced by χb in G is equal to (2) and thus uniquely determined by the cut (1). The same holds if 0 realizes (2).

Proof: Since we assume (G, χ) to be a centripetal or centrifugal contraction group, it is dense by part c) of Lemma 3.2, showing that part a) of the foregoing lemma cannot apply to (G, χ). Further, $0 \neq b \in H$ implies $\chi b \neq 0$. Hence, part a) and part b) of our present lemma are direct consequences of the foregoing lemma. The first assertion of c) was already discussed above.

Now assume that 0 realizes (2). Then the foregoing lemma shows that $v\chi b = v(\chi b - 0) > vG$. Since $\chi b \neq 0$, this yields $\chi b \notin G$. Consequently, (2) is the cut induced by χb in $(G, <)$. □

LEMMA 3.15. *Let $(G, \chi) \subset (H, \chi)$ and $(G, \chi) \subset (H', \chi)$ be two extensions of centripetal or centrifugal contraction groups. Further, let $b \in H$ be such that $\chi^n b \notin G$ for all $n \in \mathbb{N}$. Suppose that some $b' \in H' \setminus \{0\}$ induces in G the same cut as b. Then also $\chi^n b' \notin G$ for all $n \in \mathbb{N}$, and both $(G + \sum_{i \in \mathbb{N}} \mathbb{Z}\chi^i b, \chi)$ and $(G + \sum_{i \in \mathbb{N}} \mathbb{Z}\chi^i b', \chi)$ are precontraction groups. Moreover, there is an isomorphism of these two groups over (G, χ), sending $\chi^n b$ to $\chi^n b'$ for all $n \in \mathbb{N}$.*

Proof: We show that the sets $G \cup \{\chi^n b \mid n \in \mathbb{N}\}$ and $G \cup \{\chi^n b' \mid n \in \mathbb{N}\}$, endowed with the restrictions of the orders of H resp. H', are order isomorphic over G by sending $\chi^n b$ to $\chi^n b'$ for every $n \in \mathbb{N}$, and that $v\chi^n b \notin vG$ for every $n \in \mathbb{N}$. By assumption, b and $b' \neq 0$ induce the same cut in G. Since $\chi^2 b \notin G$, we have $vb \notin vG$, and part a) of Lemma 3.14 shows that $b' \notin G$. We proceed by induction on $n \geq 0$: Suppose that we have already shown that $v\chi^n b \notin vG$ and $\chi^n b' \notin G$ and that

$$\mathrm{Cut}_n(b) := (\{\chi^n g \mid G \ni g \leq b\}, \{\chi^n g \mid G \ni g > b\})$$

is the cut induced by both $\chi^n b$ and $\chi^n b'$ in G. Since we know that $\chi^2(\chi^n b) = \chi^{n+2} b \notin G$, it follows that $v\chi^{n+1} b \notin vG$, and part a) of Lemma 3.14 shows that $\mathrm{Cut}_{n+1}(b)$ is not realized by any element of $G \setminus \{0\}$. We have $\chi^n b, \chi^n b' \neq 0$ since $\chi^n b, \chi^n b' \notin G$. Using part c) of Lemma 3.14, we may deduce that $\chi^{n+1} b' \notin G$ and that $\mathrm{Cut}_{n+1}(b)$ is the cut induced by both $\chi^{n+1} b$ and $\chi^{n+1} b'$ in G. This completes the induction step. It follows that for all $n \in \mathbb{N}$, the order relation between $\chi^n b$ and a given element $g \in G$ is the same as between $\chi^n b'$ and g. Furthermore, the ordering on the sets $\{\chi^n b \mid n \in \mathbb{N}\}$ and $\{\chi^n b' \mid n \in \mathbb{N}\}$ is given by our condition that both (H, χ) and (H', χ) be centripetal or centrifugal contraction groups. Hence, $\chi^n b \mapsto \chi^n b'$ induces an order isomorphism between the ordered sets $G \cup \{\chi^n b \mid n \in \mathbb{N}\}$ and $G \cup \{\chi^n b' \mid n \in \mathbb{N}\}$. By Lemma 2.5 it now follows that this isomorphism extends linearly to an order and valuation preserving isomorphism σ between the groups $G_b := G + \sum_{i \in \mathbb{N}} \mathbb{Z}\chi^i b$ and $G_{b'} := G + \sum_{i \in \mathbb{N}} \mathbb{Z}\chi^i b'$.

On the other hand, (V3) shows that every element in G_b has value either in $vG \cup \{\infty\}$ or equal to $v\chi^n b$ for some $n \in \mathbb{N}$. Consequently, the group G_b is closed under χ and thus, (G_b, χ) is a precontraction group. The order isomorphism between $G \cup \{\chi^n b \mid n \in \mathbb{N}\}$ and $G \cup \{\chi^n b' \mid n \in \mathbb{N}\}$ induces an order isomorphism between $vG \cup \{v\chi^n b \mid n \in \mathbb{N}\}$ and $vG \cup \{v\chi^n b' \mid n \in \mathbb{N}\}$. Hence, for every element $d = g + \sum_{i \in \mathbb{N}} z_n \chi^i b$ in G_b, we have $vd = v\chi^n b$ if and only if $v\sigma d = v(g + \sum_{i \in \mathbb{N}} z_n \chi^i b') = v\chi^n b'$ and $vd = vg$ if and only if $v\sigma d = vg$. We obtain that $\chi d = \mathrm{sign}(z_n)\chi^{n+1} b$ if and only if $\chi\sigma d = \mathrm{sign}(z_n)\chi^{n+1} b' = \sigma\mathrm{sign}(z_n)\chi^{n+1} b$ and $\chi d = \chi g$ if and only if $\chi d = \chi g = \sigma\chi g$. This proves that σ also preserves the precontraction. □

COROLLARY 3.16. *Let $(G, \chi) \subset (H, \chi)$ be an extension of centripetal or centrifugal precontraction groups and assume that (G, χ) is a nontrivial divisible contraction group. Let $b \in H$ be such that $\chi^n b \notin G$ for all $n \in \mathbb{N}$. Then (G, χ) is existentially closed in the precontraction group $(G_b, \chi) := (G + \sum_{i \in \mathbb{N}} \mathbb{Z}\chi^i b, \chi)$.*

Proof: Let $(G, \chi)^*$ be a $|G|^+$-saturated elementary extension of (G, χ). Since $(G, \chi)^*$ is divisible like (G, χ), Lemma 2.4 shows that there is some nonzero element b' of $(G, \chi)^*$ which induces the same cut in G as b. Hence, the previous lemma gives an embedding of (G_b, χ) in $(G, \chi)^*$ over (G, χ). Now it follows from Lemma 2.3 that (G, χ) is existentially closed in (G_b, χ). □

COROLLARY 3.17. *Let $(G, \chi) \subset (H, \chi)$ be an extension of nontrivial divisible centripetal contraction groups. Then there exists a divisible centripetal contraction group $(G', \chi) \subset (H, \chi)$ such that vG' is cofinal in vH and (G, χ) is existentially closed in (G', χ).*

Proof: If vG is not cofinal in vH then choose $b \in H \setminus \{0\}$ such that $vb > vG$. Since (H, χ) is centripetal, we have $v\chi^n b > vG$ for all $n \in \mathbb{N}$. By the foregoing lemma, (G, χ) is existentially closed in the precontraction group (G_b, χ). By Lemma 3.12, (G, χ) is also existentially closed in every divisible contraction hull of that group. By its universal property, such a divisible contraction hull can be chosen in (H, χ). If the value set of this new group is still not cofinal in vH, we may proceed by (possibly transfinite) induction (bounded by the cardinality of vH) to construct (G', χ) such that vG' is cofinal in vH. □

For centripetal contraction groups, we will now generalize the principle that we have used in the proof of Lemma 3.15. The elements b, b' appearing in that lemma seem to be "transcendental" in some sense. For the centripetal case, we will give a corresponding definition.

In the sequel, all precontraction groups are assumed to be centripetal.

Let $(G, \chi) \subset (H, \chi)$ be an extension of precontraction groups, G divisible, and (H', χ) a substructure of (H, χ) generated over G by one element b. Then as an abelian group, H' is not necessarily generated over G by only the element b. We form a sequence of generators for H' as an abelian group over G in the following way. If $G + \mathbb{Z}b$ is a rank preserving extension of G, then by Lemma 3.6, we have $\chi(G + \mathbb{Z}b) \subset G$. In this case, we find $H' = G + \mathbb{Z}b$, and our sequence only consists of the element $b_1 := b$; we set $g_1 := 0$ for later use. In this case, g_1, b_1 and vb_1 are uniquely determined by b. If on the other hand, the extension is not rank preserving, then there is an element g_1 in the divisible group G such that $v(b - g_1) \notin vG \cup \{\infty\}$. We set $b_1 := b - g_1$ and find that $G + \mathbb{Z}b = G + \mathbb{Z}b_1$. Note that vb_1 is uniquely determined by b, by virtue of Lemma 2.6. In view of (V3), every element of $G + \mathbb{Z}b_1$ has value either in $vG \cup \{\infty\}$ or equal to vb_1. Hence by (CV1), $\chi(G + \mathbb{Z}b_1) \subset G + \mathbb{Z}\chi b_1 \subset G + \mathbb{Z}b_1 + \mathbb{Z}\chi b_1$. Having constructed b_n and the group $G + \sum_{i=1}^n \mathbb{Z}b_i$ such that $vb_1, \dots, vb_n$ is a strictly increasing sequence of values not in $vG \cup \{\infty\}$ and uniquely determined by b, and such that $\chi(G + \sum_{i=1}^n \mathbb{Z}b_i) \subset G + \sum_{i=1}^n \mathbb{Z}b_i + \mathbb{Z}\chi b_n$, we proceed as follows.

Since χ is centripetal, we have $v\chi b_n > vb_n \geq \{vb_1, \dots, vb_n\}$. By virtue of (V3), $\chi b_n \in G + \sum_{i=1}^n \mathbb{Z}b_i$ will thus imply $\chi b_n \in G$. If this is the case, then $G + \sum_{i=1}^n \mathbb{Z}b_i$ is closed under χ, and we let our sequence end with b_n.

If $G + \mathbb{Z}\chi b_n$ is a nontrivial rank preserving extension of G then we set $b_{n+1} := \chi b_n$ and $g_{n+1} := 0$. Then $vb_{n+1} = v\chi b_n > vb_n$. Note that in this case again, g_{n+1}, b_{n+1} and vb_{n+1} are uniquely determined by χb_n. This in turn is, up to the sign, uniquely determined by vb_n. Hence, vb_{n+1} is uniquely determined by vb_n and thus by b. The nonzero elements $b_1, \dots, b_n$ have different values $\notin vG = v(G + \mathbb{Z}b_{n+1})$ which shows that every element d of $G + \sum_{i=1}^{n+1} \mathbb{Z}b_i$ has value either in $vG \cup \{\infty\}$ or equal to vb_i for some $i \leq n$. Hence by virtue of (CV3), $\chi(G + \sum_{i=1}^{n+1} \mathbb{Z}b_i) = \chi(G + \sum_{i=1}^n \mathbb{Z}b_i) \subset G + \sum_{i=1}^{n+1} \mathbb{Z}b_i$. Consequently, the group $G + \sum_{i=1}^{n+1} \mathbb{Z}b_i$ is closed under χ, and we let our sequence end with b_{n+1}.

If the extension is not rank preserving, then there is an element $g_{n+1} \in G$ such that $v(\chi b_n - g_{n+1}) \notin vG \cup \{\infty\}$ and we set $b_{n+1} := \chi b_n - g_{n+1}$. We find that $vb_{n+1} = v(\chi b_n - g_{n+1}) \geq v\chi b_n > vb_n$ since otherwise, $v(\chi b_n - g_{n+1}) = vg_{n+1} \in vG \cup \{\infty\}$ by (V3). (Note that it is precisely the $\geq$-sign in the foregoing inequality that makes it impossible to construct analogous sequences with strictly decreasing values in centrifugal precontraction groups.) Again by Lemma 2.6, vb_{n+1} is uniquely determined by χb_n ; as before, it follows that vb_{n+1} is uniquely determined by b. All elements $b_1, \ldots, b_{n+1}$ have different values $\notin vG \cup \{\infty\}$, so every element d of $G + \sum_{i=1}^{n+1} \mathbb{Z}b_i$ has value either in $vG \cup \{\infty\}$ or equal to vb_i for some $i \leq n+1$. If $vd = vb_{n+1}$ then $\chi d = \pm\chi b_{n+1}$ by (CV2). Otherwise, $\chi d \in \chi(G + \sum_{i=1}^{n} \mathbb{Z}b_i) \subset G + \sum_{i=1}^{n} \mathbb{Z}b_i + \mathbb{Z}\chi b_n = G + \sum_{i=1}^{n+1} \mathbb{Z}b_i$. Hence, $\chi(G + \sum_{i=1}^{n+1} \mathbb{Z}b_i) \subset G + \sum_{i=1}^{n+1} \mathbb{Z}b_i + \mathbb{Z}\chi b_{n+1}$. Since $vb_{n+1} \notin vG \cup \{\infty\}$, the induction step will again be repeated.

The sequence of elements b_i constructed in this way may be finite or infinite. If it is finite, the element b will be called *χ-algebraic over* (G, χ), and otherwise, b will be called *χ-transcendental over* (G, χ). These notions are welldefined, although the *shift elements* g_i and the sequence members b_i may not be uniquely determined. Indeed, in our construction, all values vb_i were uniquely determined by b, which also yields that all sequences constructed from b must have the same length. Every such sequence will be called a *characteristic sequence* of b over (G, χ). If $b_m \neq 0$ and $vb_m > vG$ for some m, then the sequence is infinite and $vb_i > vG$ holds for all $i \geq m$; the sequence beginning with b_m will then be called a *supersequence*. If vG is cofinal in vH, then there are no supersequences in H over (G, χ).

Note that "χ-algebraic" does not mean "algebraic" in the model theoretic sense. We will show in [K2] that only the elements of the divisible hull $\tilde{G}$ are algebraic over a precontraction group (G, χ).

By construction, the abelian group generated over G by all the elements of a characteristic sequence, is closed under χ and thus a precontraction group. On the other hand, all elements b_i are contained in H' which was the substructure of (H, χ) generated by b. We find that the elements of a characteristic sequence form a set of generators of H' as an abelian group over G. By construction, it is even a minimal set of generators.

For arbitrary b_n in a characteristic sequence, all inclusions in the chain $G \subset G + \mathbb{Z}b_n \subset G + \mathbb{Z}b_n + \mathbb{Z}b_{n-1} \subset \ldots \subset G + \sum_{i=1}^{n} \mathbb{Z}b_i$ are proper. Indeed, from our construction it follows that $vb_{j-1} < \{vb_j, \ldots, vb_n\}$ and there is thus no element of value vb_{j-1} in the group $G + \sum_{i=j}^{n} \mathbb{Z}b_i$, for every $j \leq n$. Since $\chi(G + \sum_{i=j}^{n} \mathbb{Z}b_i) \subset G + \sum_{i=j+1}^{n} \mathbb{Z}b_i + \mathbb{Z}\chi b_n$, we also see that

$$(4) \qquad b_j \notin \chi(G + \sum_{i=j}^{n} \mathbb{Z}b_i) \quad \text{for every } j \leq n.$$

Note that our assertions also hold for $\mathbb{Q}$ in the place of $\mathbb{Z}$. If $b_1, \ldots, b_n$ is the characteristic sequence of some χ-algebraic element b over (G,χ), then all groups in this chain are precontraction groups since for every $i \le n$, the sequence $b_i, \dots, b_n$ is a characteristic sequence of b_i. The construction carried through in the proof of Lemma 3.10 may be applied to $(G + \mathbb{Z}b_n, \chi)$ in the place of (G,χ). By virtue of (4), we may take the first adjoined elements to be $b_{n-1}, \dots, b_1$ successively. In this way, we obtain a contraction hull of $(G + \mathbb{Z}b_n, \chi)$ in (H,χ) which contains all elements of the given characteristic sequence of b. Hence, it also contains the precontraction group generated by b over (G,χ). If H is divisible, then we may replace $\mathbb{Z}$ by $\mathbb{Q}$ to obtain a divisible contraction hull. We have proved:

LEMMA 3.18. *Let $(G,\chi) \subset (H,\chi)$ be an extension of centripetal precontraction groups and assume that G is divisible and (H,χ) is a contraction group (resp. a divisible contraction group). Let $b \in H$ be χ-algebraic over (G,χ) with characteristic sequence $b_1, \dots, b_n$. Then there exists a contraction hull (resp. a divisible contraction hull) of $(G + \mathbb{Z}b_n, \chi)$ in (H,χ) which contains the precontraction group generated by b over (G,χ).*

In the sequel, let us assume (G,χ) to be a divisible centripetal contraction group. Let $b \notin G$. We will now examine the question whether the cuts induced in G by the elements of characteristic sequences of b are already determined by the cut

$$\mathrm{Cut}(b) := (\{g \mid G \ni g < b\}, \{g \mid G \ni g > b\})$$

induced by b in G. To this end, we need the notion of a v-cut that we have introduced in section 2. Bearing in mind the proof of Lemma 3.15 where we defined an isomorphism by an assignment $\chi^i b \mapsto \chi^i b'$, we will treat the following problem. Assume that $(G,\chi) \subset (H',\chi)$ is a second extension of centripetal precontraction groups and that $b' \in H'$ also induces the cut $\mathrm{Cut}(b)$ in G. Is it then possible to construct characteristic sequences of b and b' of equal length and having the same shift elements g_i for every i?

Let us fix a characteristic sequence $b_1, \dots$ of b over (G,χ). For every b_i in this sequence, we have

$$\mathrm{Cut}(b_i) = (\{g \mid G \ni g < b_i\}, \{g \mid G \ni g > b_i\})$$

since by definition, $b_i \notin G$.

Assume that b_m, $m \ge 1$, is an element of our characteristic sequence and that we have already shown $\mathrm{Cut}(b_{m-1})$ to depend only on $\mathrm{Cut}(b)$ (and not on the chosen characteristic sequence or the element $b \notin G$ inducing $\mathrm{Cut}(b)$). For $m = 1$, this assumption is trivially true. If $m > 1$, then assume in addition that we have already constructed the first elements $b'_1, \dots b'_{m-1}$ of a characteristic sequence of b', satisfying $b'_1 = b' - g_1$ and $b'_i = \chi b_{i-1} - g_i$ as well as $\mathrm{Cut}(b_i) =$

$\mathrm{Cut}(b'_i)$ for all $1 < i < m$. For the case $m = 1$, we set $C_1 := \mathrm{Cut}(b) = \mathrm{Cut}(b')$. For $m > 1$, we set

$$C_m := (\{\chi g \mid G \ni g < b_{m-1}\}, \{\chi g \mid G \ni g > b_{m-1}\}) . \tag{5}$$

Note that a priori, we only know that (5) is a quasicut. But since b_m is an element of our characteristic sequence, $b_m + g_m \notin G$. As $\chi b_{m-1} = b_m + g_m$ realizes (5), this shows that (5) is a cut. Since by induction hypothesis, $\mathrm{Cut}(b_{m-1}) = \mathrm{Cut}(b'_{m-1})$ depends only on $\mathrm{Cut}(b)$, also C_m depends only on $\mathrm{Cut}(b)$, and by virtue of (C$\leq$), both χb_{m-1} and $\chi b'_{m-1}$ realize C_m. In the sequel, let us treat the cases $m = 1$ and $m > 1$ simultaneously by taking the undefined expression "χb_0" as "b".

Assume that C_m is not realized in G. There are two possibilities for C_m. If C_m is not a shifted v-cut, then by Corollary 2.2, there is no $g \in G$ such that $v(\chi b_{m-1} - g) \notin vG \cup \{\infty\}$ or $v(\chi b'_{m-1} - g) \notin vG \cup \{\infty\}$. Then necessarily, $g_m = 0$ and $b_m = \chi b_{m-1}$ as well as $b'_m = \chi b'_{m-1}$ by our construction of characteristic sequences, implying also that $\mathrm{Cut}(b_m) = C_m = \mathrm{Cut}(b'_m)$. Moreover, b_m and b'_m are the last elements of the respective characteristic sequences.

If on the other hand, C_m is a shifted v-cut, then by Corollary 2.2, there is some $g \in G$ such that $v(\chi b_{m-1} - g) \notin vG \cup \{\infty\}$. Then by our construction, $vb_m = v(\chi b_{m-1} - g_m) \notin vG \cup \{\infty\}$ which by Corollary 2.2 yields that $C_m - g_m$ is the unique v-cut among all shifts of C_m. By part b) of Lemma 2.1, we find that $v(\chi b'_{m-1} - g_m) \notin vG \cup \{\infty\}$, so we may take $b'_m := \chi b'_{m-1} - g_m$. But also any other elements $\tilde{b}_m$, $\tilde{b}'_m$ continuing the characteristic sequences would have to satisfy $v\tilde{b}_m$, $v\tilde{b}'_m \notin vG \cup \{\infty\}$ and would thus induce v-cuts, which are then equal to $C_m - g_m$ because of its uniqueness. Consequently, $\mathrm{Cut}(b_m) = C_m - g_m = \mathrm{Cut}(b'_m)$ only depends on $\mathrm{Cut}(b)$.

Now assume that C_m *is* realized by some $a \in G$. Then in view of Lemma 3.13, of which only part b) can apply here since G is dense, our construction of characteristic sequences yields $vb_m > vG$, and b_m is the beginning of a supersequence. Then b is χ-transcendental and for all $i \geq m$, we have $vb_i > vG$ and all cuts $\mathrm{Cut}_i(b)$ are the same and equal to one of the two possible cuts in G which are realized by 0 (these are $(G^{<0}, \{0\} \cup G^{>0})$ and $(G^{<0} \cup \{0\}, G^{>0})$). But the following pathological case may appear: we cannot exclude the possibility that $\chi b'_{m-1} = a$. In this case, the characteristic sequence of b' would end with b'_{m-1} and b' would be χ-algebraic. Then certainly, the assignment $b \mapsto b'$ would not extend to an isomorphism of precontraction groups over (G, χ). On the other hand, if our characteristic sequence of b does not eventually run into a supersequence, then for none of its elements b_i, the cut $\mathrm{Cut}(b_i)$ is realized in G. In this case, our procedure of constructing b'_i may be accomplished through the full length of the characteristic sequence of b, showing that the characteristic sequences of b' are at least as long as those of b. If symmetrically, we know that also the characteristic sequences of b' do not run into supersequences, then we may conclude that the characteristic sequences of b and b' are of equal length.

If the characteristic sequences of $b \in H$ over (G, χ) are infinite but do not run into supersequences, then b will be called *bounded χ-transcendental over* (G,χ). If vG is cofinal in vH, then there are no supersequences in H over (G,χ), and every χ-transcendental element is already bounded χ-transcendental.

Assume that $b \in H$ is bounded χ-transcendental over (G,χ) and that b' is as above, satisfying $\mathrm{Cut}(b) = \mathrm{Cut}(b')$. Then our construction yields an infinite characteristic sequence $b'_1, \dots$ of b' with the same shift elements g_i and such that the cuts $\mathrm{Cut}(b_i) = \mathrm{Cut}(b'_i)$ are not realized in G, for all i. In particular, also b' is bounded χ-transcendental over (G,χ). Moreover, we know that all b_i and b'_i have values not contained in $vG \cup \{\infty\}$. Let us further note that every isomorphism ι between the ordered groups $(G+\sum_{i=1}^{\infty} \mathbb{Z}b_i\,, <)$ and $(G+\sum_{i=1}^{\infty} \mathbb{Z}b'_i\,, <)$ sending b_i to b'_i, will automatically respect the precontraction. Indeed, since every element in the first group has value either in $vG \cup \{\infty\}$ or equal to some vb_i, we only have to check whether $\iota\chi b_i = \chi b'_i$. But this is immediate: $\iota\chi b_i = \iota(b_{i+1}+g_{i+1}) = b'_{i+1} + g_{i+1} = \chi b'_i$.

Now the proofs of the following lemma and corollary are analogous to those of Lemma 3.15 and Corollary 3.16.

LEMMA 3.19. *Let $(G,\chi) \subset (H,\chi)$ and $(G,\chi) \subset (H',\chi)$ be two extensions of centripetal contraction groups and assume that G is divisible. Further, let $b \in H$ be bounded χ-transcendental over (G,χ). Suppose that some $b' \in H'$ induces in G the same cut as b. Then also b' is bounded χ-transcendental over (G,χ), and there is an isomorphism of the precontraction groups generated by b resp. b' over (G,χ), sending b to b'.*

COROLLARY 3.20. *Let $(G,\chi) \subset (H,\chi)$ be an extension of centripetal precontraction groups and assume that (G,χ) is a nontrivial divisible contraction group. If (G_b,χ) is the precontraction group generated over G by one bounded χ-transcendental element $b \in H$, then (G,χ) is existentially closed in (G_b,χ).*

Next, we treat the case of extensions generated by one χ-algebraic element.

LEMMA 3.21. *Let $(G,\chi) \subset (H,\chi)$ be an extension of centripetal precontraction groups and assume that (G,χ) is a nontrivial divisible contraction group. If (G_b,χ) is the precontraction group generated over G by one χ-algebraic element $b \in H$, then (G,χ) is existentially closed in (G_b,χ).*

Proof: We may assume $b \notin G$ since otherwise, the assertion is trivial. Pick a characteristic sequence $b_1, \dots, b_n$ of b over (G,χ). Then the group $G + \mathbb{Z}b_n$ is a precontraction group, generated over (G,χ) by b_n. According to Lemma 3.18, there is a divisible contraction hull (H_0,χ) of $G + \mathbb{Z}b_n$ in (H,χ) which contains the precontraction group (G_b,χ) generated by b over (G,χ). If we are able to show that (G,χ) is existentially closed in $(G + \mathbb{Z}b_n,\chi)$, then it will follow from Lemma 3.12 that (G,χ) is also existentially closed in (H_0,χ) and thus also in (G_b,χ). This shows: w.l.o.g. we may assume from the start that $n = 1$ and $b_1 - b \notin G$. Then the cut induced by b in G is not realized in G, since otherwise,

b_1 would be the beginning of a supersequence and b could not be χ-algebraic over (G,χ). We also know that $\chi b \in G$. Since (G,χ) is a contraction group, we may pick some $c \in G$ such that $\chi c = \chi b$. Let us assume that $c < b$; for $c > b$ the proof is analogous. Consider the following set of assertions:

(6) $\{\text{“}g < x \wedge \chi g = \chi x\text{”} \mid g \in G \wedge c \leq g < b\} \cup \{\text{“}g > x\text{”} \mid g \in G \wedge g > b\}$.

This set is finitely satisfiable in (G,χ). For this, we only have to show that for every finite subset $\mathcal{F}$ of (6), there is some $x \in G$ for which all assertions of $\mathcal{F}$ are true. We write $\mathcal{F} = \mathcal{F}_1 \cup \mathcal{F}_2$ with $\mathcal{F}_1$ a subset of the first and $\mathcal{F}_2$ a subset of the second set in (6). Let g_1 be the maximal element appearing in the assertions of $\mathcal{F}_1$. Then there is some element $g' \in G$ such that $g_1 < g' < b$ since otherwise, g_1 would realize the cut induced by b in G. Since $c \leq g_1 < g' < b$ and $\chi c = \chi b$, it follows from (C$\leq$) that $\chi g = \chi b = \chi g'$ for all g appearing in the assertions of $\mathcal{F}_1$, and for these, we also have $g \leq g_1 < g'$. On the other hand, for every $g > b$ we have $g > g'$. This shows that $x = g'$ satisfies all assertions in $\mathcal{F}$. Thus we have shown that the set (6) is finitely satisfiable in (G,χ). Hence, it is satisfiable in every $|G|^+$-saturated elementary extension $(G,\chi)^*$ of (G,χ). That is, there is some element b' in $(G,\chi)^*$ which induces in G the same cut as b and which satisfies $\chi b' = \chi b$. This yields that the assignment $b \mapsto b'$ defines an order preserving isomorphism from $G + \mathbb{Z}b$ onto $G + \mathbb{Z}b'$ over G. We have to show that this isomorphism is also an isomorphism of contraction groups. If $G \subset G + \mathbb{Z}b$ is not rank preserving, then by our choice of b (to be the first element of some characteristic sequence of b) we know that $vb \notin vG \cup \{\infty\}$, and our assertion follows from the uniqueness assertion of Lemma 3.3. If $G \subset G + \mathbb{Z}b$ is rank preserving, then the uniqueness assertion of Lemma 3.6 shows that every order preserving isomorphism from $G + \mathbb{Z}b$ onto $G + \mathbb{Z}b'$ over G will automatically be an isomorphism of contraction groups.

We have shown that (G_b,χ) embeds over (G,χ) in every $|G|^+$-saturated elementary extension of (G,χ). Hence, (G,χ) is existentially closed in (G_b,χ), as contended. □

Now we are able to prove our main lemma.

LEMMA 3.22. *Let $(G,\chi) \subset (H,\chi)$ be an extension of centripetal precontraction groups and assume that (G,χ) is a nontrivial divisible contraction group. Then (G,χ) is existentially closed in (H,χ).*

Proof: In view of Lemma 3.9, we may assume w.l.o.g. that (H,χ) is a divisible contraction group. By Corollary 3.17, we know that there exists a divisible contraction group (G',χ) in (H,χ) such that vG' is cofinal in vH and that (G,χ) is existentially closed in (G',χ). Now, it remains to show that (G',χ) is existentially closed in (H,χ). Hence, we may assume from the start that vG is cofinal in vH.

It suffices to show that (G,χ) is existentially closed in every substructure of (H,χ) which is finitely generated over G. Again in view of Lemma 3.9, we may thus assume w.l.o.g. that (H,χ) is a divisible contraction hull of a precontraction group which is finitely generated over (G,χ). Picking one of the generators out of a minimal set of, say, m generators, we may consider the substructure (G_1,χ) of (H,χ) generated by this element. We know that (G_1,χ) is a precontraction group. If we are able to show that (G,χ) is existentially closed in (G_1,χ), then by Lemma 3.10 and Lemma 3.12 there is a divisible contraction hull (G_1',χ) of (G_1,χ) in (H,χ) such that (G,χ) is existentially closed in (G_1',χ). Now (H,χ) is a divisible contraction hull of a precontraction group which is generated over (G_1',χ) by $m-1$ generators. This shows that we may proceed by induction on the number of generators. That is, we only have to show that (G,χ) is existentially closed in every substructure of (H,χ) which is generated over G by one element b. If b is χ-algebraic over (G,χ), then this is the assertion of Lemma 3.21. If b is χ-transcendental over (G,χ), then it is bounded χ-transcendental since vG is cofinal in vH, and an application of Lemma 3.20 now completes our proof. □

By Robinson's Test, the foregoing lemma implies:

THEOREM 3.23. *The theory of nontrivial divisible centripetal contraction groups is model complete.*

Now assume that (G,χ) is a common substructure of the two divisible centripetal contraction groups (H,χ) and (H',χ). By Lemma 3.1, (G,χ) is a centripetal precontraction group. By Lemma 3.10, there is a divisible contraction hull (G',χ) of (G,χ) in (H,χ) which embeds in (H',χ) over (G,χ). Let us identify (G',χ) with its image in (H',χ). By Lemma 3.5, the theory of centripetal contraction groups admits $\mathcal{P}_{cp}$ as its prime structure. Hence if G is the trivial group, then we may replace (G,χ) by $\mathcal{P}_{cp}$ to obtain that (G',χ) is nontrivial. Now (G',χ) is a nontrivial divisible centripetal contraction group, so from the model completenes stated in the preceding theorem we may infer that (H,χ) and (H',χ) are equivalent over (G',χ) and hence also over (G,χ). We have thus shown that the theory of divisible centripetal contraction groups is substructure complete, that is,

THEOREM 3.24. *The theory of nontrivial divisible centripetal contraction groups admits elimination of quantifiers.*

Combining this result with Lemma 3.9, we obtain:

THEOREM 3.25. *The theory of nontrivial divisible centripetal contraction groups is the model completion of the theory of centripetal precontraction groups.*

Since every two centripetal contraction groups contain the trivial group as a common substructure, substructure completeness yields

THEOREM 3.26. *The theory of nontrivial divisible centripetal contraction groups is complete.*

Since the axiom system {(OAG), (C0), (CS), (C$\leq$), (C$-$), (CA$'$), (CP), (D), $\exists x : x \neq 0$} is recursive, this theorem implies

THEOREM 3.27. *The theory of nontrivial divisible centripetal contraction groups is decidable.*

4. Other concepts of contraction groups

Although we have constantly used the natural valuation in the last sections, we have not put a symbol for it into the language of contraction groups. So the question arises whether we obtain similar results if we do. This would mean that we add a symbol for a binary relation which is interpreted by "$vx < vy$" or "$vx \leq vy$". Then certainly, to obtain completeness, there have to be axioms telling us about the relation between the valuation and the contraction. Note that the natural valuation of an ordered group is not elementarily axiomatizable. It is elementarily axiomatizable that a valuation be compatible with the ordering, but then there will always be models whose valuation is a proper coarsening of the natural valuation. The most direct way to fix the relation between valuation and contraction is to say that the valuation is the *valuation associated to* χ:

(VC) $\forall x, y \in G : \; vx = vy \;\Leftrightarrow\; \chi x = \pm \chi y \, .$

This valuation is always a coarsening of the natural valuation; it coincides with the natural valuation for instance on the prime structure and prime models that we have constructed. Since it is definable in the theory of precontraction groups, the definition just being axiom (VC), all model theoretic results remain true for this expansion.

It is also possible to axiomatize elementarily that for every nonzero element a the set $\{vb \mid \chi b = a\}$ contains precisely n values, where $n > 0$ is a fixed natural number or ∞. It is rather likely that the model theoretic results also carry over to these expansions.

Let us now consider valued abelian groups which are not ordered. In this case, an adequate axiom system for contraction groups might be (OAG), (C0), (CS) together with

(CV$'$) $\forall x, y \in G : \; vx = vy \;\Rightarrow\; \chi x = \chi y \, ,$

and in view of part b) of Lemma 3.2, we would express the property "centripetal" by the axiom $\forall x \in G : \; x \neq 0 \;\Rightarrow\; v\chi x > vx$, and the property "centrifugal" by the axiom $\forall x \in G : \; x \neq 0 \;\Rightarrow\; v\chi x < vx$.

Open Problem 1: Determine the algebraic and model theoretic properties of this theory.

Note that the contraction transports (a coarsening of) the ordering of the value set into the group. But this ordering will not be compatible with the addition.

We do not know whether this is a sound situation. If so, then there arises a further question:

Open Problem 2: Do there exist o-minimal or weakly o-minimal expansions or models of this theory?

Finally, on abelian groups which are not ordered or valued, there may still be classes of elements which are appropriate to be contracted. For example, if the group is an R-module for some ring R, we may consider the axiom scheme

(CR) $\forall x \in G : rx \neq 0 \Rightarrow \chi rx = \chi x \qquad (r \in R)$,

that is, for every element a in the group, the set χRa is a singleton. Combined again with the surjectivity, this describes rather "big" modules.

Open Problem 3: Study (CR) or similar axioms in combination with the surjectivity (CS).

References

[C–K] Chang, C. C. — Keisler, H. J.: Model Theory, North Holland, Amsterdam – London (1973)

[D–M–M] van den Dries, L. – Macintyre, A. – Marker, D.: The elementary theory of restricted analytic functions with exponentiation, *preprint*

[K1] Kuhlmann, F.-V.: Valuation theory of fields and modules, *to appear in the* "Algebra, Logic and Applications" *series*, eds. A. MacIntyre and R. Göbel, Gordon & Breach, New York

[K2] Kuhlmann, F.-V.: Abelian groups with contractions II: weak o-minimality, *preprint*

[KS] Kuhlmann, S.: On the structure of nonarchimedean exponential fields I, *submitted*

[K–KS] Kuhlmann, F.-V. – Kuhlmann, S.: On the structure of nonarchimedean exponential fields II, *to appear in Comm. Algebra*

[P] Prestel, A.: Einführung in die mathematische Logik und Modelltheorie, vieweg studium, Braunschweig (1986)

[W1] Wilkie, A. J.: Model completeness results for expansions of the real field I: restricted Pfaffian functions, *to appear in Amer. J. Math.*

[W2] Wilkie, A. J.: Model completeness results for expansions of the real field II: the exponential function, *to appear in Amer. J. Math.*

Mathematisches Institut der Universität Heidelberg, Im Neuenheimer Feld 288, D-69120 Heidelberg, Germany

E-mail address: fvk@harmless.mathi.uni-heidelberg.de

Contemporary Mathematics
Volume 171, 1994

Typesets and Cotypesets of Finite-Rank Torsion-Free Abelian Groups

By Reiff S. Lafleur

Throughout this paper we shall use "group" for "finite-rank torsion-free abelian group". Every group G gives rise to two important sets of types:

$$\text{typeset}(G) = \{\text{type}(x): 0 \neq x \in G\},$$

$$\text{cotypeset}(G) = \{\text{type}(X): X \text{ is a rank-1 factor of } G\}.$$

In 1961 Beaumont and Pierce [2] posed the problem of characterizing those sets of types (necessarily countable) which arise as typesets of rank-2 groups. They showed that for a countable set of types T, if T = typeset(G) for some G of rank-2, then there exists a type τ_0 such that $\tau_0 = \sigma\wedge\tau$ for all $\sigma \neq \tau$ in T, where $\tau_0 \in T$ if T is finite. However, in 1965 Dubois [3] constructed a countable set of types T such that $\sigma\wedge\tau = \text{type}(\mathbb{Z})$ for $\sigma \neq \tau$ in T, while T is not the typeset of a rank-2 group. Further results were obtained by Koehler [5] in 1964, Dubois [3] in 1965-1966, and Ito [4] in 1975.

In 1978 Schultz [6] introduced the term cotypeset for the set of types of all rank-1 factors of a group. In 1982 Vinsonhaler and Wickless [7] characterized cotypesets of rank-2 groups. They showed that for a countable set of types T, T = cotypeset(G) for some G of rank-2 if and only if there exists a type τ_0 such that $\tau_0 = \sigma\vee\tau$ for all $\sigma \neq \tau$ in T, with τ_0 required to be in T if T is finite. The following year Vinsonhaler and Wickless [8] gave a characterization of cotypesets of rank-n groups for any n. However, typesets of rank-2 groups have not been characterized. The best results presently known about typesets of rank-2 groups appear in Arnold and Vinsonhaler [1]. Little has been done with typesets of groups whose ranks are greater

---

Subject Classification: Primary 20K15, Secondary 06A15, 06A12
The paper is in final form and no version of it will be submitted elsewhere.
It contains results from my thesis at the University of Connecticut.

than two.

This paper gives some new alternative characterizations of cotypesets of rank-n groups. Using these characterizations, for any $n \geq 3$ we give an example of a set of types which is the cotypeset of a rank-n group but is not the cotypeset of any group of rank higher than n. Also, we give necessary conditions on typesets of rank-n groups. Using these and the Warfield dual (see Warfield [9]) we can give for any $n \geq 3$ an example of a set of types which is the typeset of a rank-n group but is not the typeset of any group of higher rank.

An expanded version of these results will appear as my dissertation at the University of Connecticut. I would like to thank my major advisor Professor William Wickless for his encouragement and assistance while these results were developed. I would like to thank my associate advisors Professors Charles Vinsonhaler and Eugene Spiegel for discussions on these topics which clarified several points. I would like to thank Professor James Schmerl of the University of Connecticut for his extensive help with the proof of the critical Theorem 2. Finally, I would like to thank Professor J. B. Nation of the University of Hawaii for providing a key step for the proof of Theorem 2.

We will assume the standard facts on types and the typeset and cotypeset of a group. See, for example, Arnold and Vinsonhaler [1] and Vinsonhaler and Wickless [8]. We begin with some

DEFINITIONS.

Let T be a set of types. Define $\wedge T = T \cup \{$all infs of finite subsets of T$\}$ and $\vee T = T \cup \{$all sups of finite subsets of T$\}$. We refer to these as the "inf closure" and "sup closure" of T respectively. (All sups and infs considered in this paper are of nonempty collections and are computed in the set of all types.) Define $S^n = \{$nonzero subspaces of $\mathbb{Q}^n\}$, and $S_1^n = \{$1-dimensional subspaces of $\mathbb{Q}^n\}$.

THEOREM 1.

Let T be a finite or countable set of types. Then the following are equivalent:

(1) T is the cotypeset of a rank-n group.

(2) There is a surjective map $\psi: S_1^n \longrightarrow T$ such that $Z \leq \sum_{i=1}^k Z_i$ implies $\psi(Z) \leq \sup\{\psi(Z_i): 1 \leq i \leq k\}$ for all $Z, Z_1, \ldots Z_k \in S_1^n$.

(3) There is a map $\varphi: S^n \longrightarrow \vee T$ such that $\varphi(S_1^n) = T$ and for all V and $W \in S^n$, $\varphi(V+W) = \varphi(V) \vee \varphi(W)$.

(4) There are maps $\varphi: S^n \longrightarrow \vee T$ and $\theta: \vee T \longrightarrow S^n$ such that $\varphi(S_1^n) = T$ and for all $V \in S^n$ and $\tau \in \vee T$, $V \leq \theta(\tau)$ iff $\varphi(V) \leq \tau$.

REMARK.

Both S^n and $\vee T$ are upper semilattices. The second part of condition **3** states that φ is an upper semilattice homomorphism. This implies that φ is monotone. We will see below that the monotonicity of both φ and θ follows from condition **4** as well. In order to help the reader distinguish between elements of S^n and elements of S_1^n, we consistently use Z's to denote elements of S_1^n.

PROOF.

1⇔2 Vinsonhaler and Wickless [8]

3⇒2. Set $\psi = \varphi | S_1^n$ and note that $Z \le \Sigma_{i=1}^k Z_i$ implies $\psi(Z) = \varphi(Z) \le \varphi(\Sigma_{i=1}^k Z_i) = \sup\{\varphi(Z_i):1 \le i \le k\} = \sup\{\psi(Z_i):1 \le i \le k\}$, while $\varphi(S_1^n) = T$ implies that ψ is surjective.

2⇒3. Let $k = \dim(W)$ and choose $Z_1, Z_2, \ldots, Z_k$ 1-dimensional with $\Sigma_{i=1}^k Z_i = W$. We refer to $\{Z_i\}$ as a basis for W. Define $\varphi(W) = \sup\{\psi(Z_i):1 \le i \le k\}$. As a result of condition 2, $\varphi(W)$ is independent of the choice of the Z_i and equals $\sup\{\psi(Z):Z \in S_1^n, Z \le W\}$. For V, $W \in S^n$, by expanding a basis for $V \cap W$ to bases for V and W respectively, we see that $\varphi(V+W) = \varphi(V)+\varphi(W)$. Finally, $Z \in S_1^n$ implies $\varphi(Z) = \psi(Z)$, so $\varphi(S_1^n) = \psi(S_1^n) = T$.

3⇒4. We need to define θ. Choose $\tau \in \vee T$ and consider $F = \{V \in S^n : \varphi(V) \le \tau\}$. There is $\sigma \in T$ such that $\sigma \le \tau$. Since $\varphi(S_1^n) = T$, there is $Z \in S_1^n$ such that $\varphi(Z) = \sigma$. Then $Z \in F$, so $F \ne \phi$ and $\cup F = \cup\{W : W \in F\}$ is nonempty and nonzero. We claim $\cup F$ is a subspace of $\mathbb{Q}^n$. It is clearly closed under scalar multiplication. Suppose v_1 and v_2 are in $\cup F$. Then $v_1 \in V_1$ and $v_2 \in V_2$ with $V_1, V_2 \in F$. So $v_1+v_2 \in V_1+V_2$ and $\varphi(V_1+V_2) = \varphi(V_1) \vee \varphi(V_2) \le \tau$. Thus, $v_1+v_2 \in \cup F$. Hence $\cup F \in S^n$ and we may put $\theta(\tau) = \cup F$. Clearly $\varphi(V) \le \tau$ implies $V \le \theta(\tau)$. For the converse, first suppose $Z \in S_1^n$. Then $Z \le \theta(\tau)$ implies $Z \cap \theta(\tau) \ne 0$. So $Z \cap V \ne 0$ for some V with $\varphi(V) \le \tau$. Since Z is 1-dimensional, $Z \le V$. Thus $\varphi(Z) \le \varphi(V) \le \tau$. Now consider $V \in S^n$ with $V \le \theta(\tau)$. Choose a basis $\{Z_i:1 \le i \le k\}$ for V. Then $Z_i \le V \le \theta(\tau)$ implies $\varphi(Z_i) \le \tau$ for every i. Thus $\varphi(V) = \varphi(\Sigma_{i=1}^k Z_i) = \sup\{\varphi(Z_i):1 \le i \le k\} \le \tau$.

4⇒3. We claim that φ is the desired map. We need only prove the second part of 3. We first prove that φ is monotone. Take $V \le W$ in S^n. The second part of 4 applied to $\varphi(W) \le \varphi(W)$ yields $W \le \theta\varphi(W)$. So $V \le \theta\varphi(W)$. Hence $\varphi(V) \le \varphi(W)$, again by the second part of 4. Thus φ is monotone. Now $V \le V+W$ implies $\varphi(V) \le \varphi(V+W)$. Similarly, $\varphi(W) \le \varphi(V+W)$. So $\varphi(V) \vee \varphi(W) \le \varphi(V+W)$. Next suppose τ in $\vee T$ satisfies $\varphi(V) \le \tau$ and $\varphi(W) \le \tau$. Then the second part of 4 yields $V \le \theta(\tau)$ and $W \le \theta(\tau)$. So $V+W \le \theta(\tau)$. Thus $\varphi(V+W) \le \tau$, again by the second part of 4. Taking $\tau = \varphi(V) \vee \varphi(W)$ (which is in $\vee T$ since $\vee T$ is closed under finite sups), we see that $\varphi(V+W) \le \varphi(V) \vee \varphi(W)$. So $\varphi(V+W) = \varphi(V) \vee \varphi(W)$ as desired.

PROPERTIES OF φ AND θ.

We have seen above that whenever φ and θ satisfy the conditions of the theorem, φ is monotone and $W \le \theta\varphi(W)$ for all $W \in S^n$. A dual proof shows that θ is monotone and $\varphi\theta(\tau) \le \tau$ for all $\tau \in \vee T$. We now prove that, in addition, for all $\tau \in \vee T$, $\tau \le \varphi\theta(\tau)$, so that $\varphi\theta(\tau) = \tau$. Let $\tau \in \vee T$ and choose $\tau_1,\ldots,\tau_k \in T$ such that $\tau = \sup\{\tau_i : 1 \le i \le k\}$. Since $\varphi(S^n_1) = T$, there are $Z_1,\ldots,Z_k \in T$ such that $\varphi(Z_i) = \tau_i$ for each i. Then $\varphi(Z_i) \le \tau$ implies $Z_i \le \theta(\tau)$. So $\tau_i = \varphi(Z_i) \le \varphi\theta(\tau)$ for every i. Thus $\tau = \sup\{\tau_i : 1 \le i \le k\} \le \varphi\theta(\tau)$, as desired. Next, suppose $\theta(\sigma) \le \theta(\tau)$ for some $\sigma, \tau \in \vee T$. Then $\sigma = \varphi\theta(\sigma) \le \varphi\theta(\tau) = \tau$. Hence, $\theta(\sigma) = \theta(\tau)$ implies $\sigma = \tau$ and θ is injective. Also, $\sigma < \tau$ implies $\theta(\sigma) < \theta(\tau)$. Thus a necessary condition for T to be the cotypeset of rank-n group is that there is an embedding of $\vee T$ into S^n. (For convenience, we use "embedding" as a synonym for "injective mapping", even though this may not be appropriate in the context of partially ordered sets.) For future reference, we collect these consequences of the conditions of Theorem 1 (suppressing leading universal quantifiers):

(A) $V \le W \Rightarrow \varphi(V) \le \varphi(W)$.

(B) $\sigma \le \tau \Leftrightarrow \theta(\sigma) \le \theta(\tau)$.

(C) $\varphi(V) \le \tau$ iff $V \le \theta(\tau)$.

(D) $W \le \theta\varphi(W)$.

(E) $\tau = \varphi\theta(\tau)$.

(F) $\sigma < \tau \Rightarrow \theta(\sigma) < \theta(\tau)$.

DEFINITIONS.

If C is a finite chain of a poset, then we define the length of C to be the number of vertices of C. We define the height of an element σ of a poset to be the sup of the lengths of the chains which have σ as the top element. We also define the coheight of an element σ of a poset to be the sup of the lengths of the chains which have σ as the bottom element. We define the height of a poset P to be the sup of the lengths of the chains of P. We denote these by lgth(C), ht(σ), cht(σ), and ht(P), respectively. With these definitions, $ht(S^n) = n$, and for all $V \in S^n$, ht(V) equals the dimension of V, while cht(V) is one more than the dimension of a complementary subspace of V.

NECESSARY CONDITIONS ON T.

In light of the preceding results, we are now able to obtain some concrete necessary conditions for T to be the cotypeset of a rank-n group. We already know that, for such a T, there exist maps φ and θ satisfying the conditions of Theorem 1.

We know that θ is injective and $ht(S^n) = n$, so it follows that $ht(\vee T) \leq n$. It is clear that $\vee T$ is closed under finite sups. In fact, we will show that $\vee T$ is closed under arbitrary sups. (We remind the reader that all sups and infs considered in this paper are of nonempty collections and are computed in the set of all types. Thus, we are asserting that for any nonempty subset S of $\vee T$, the sup of S in the set of all types exists, and, furthermore, this sup is contained in $\vee T$.) In fact, we claim that for every $S \subseteq \vee T$, every $k \leq n$, and every choice of $\sigma_1,\sigma_2,\ldots,\sigma_k \in S$ such that $\sigma_1 < \sigma_1\vee\sigma_2 < \ldots < \sigma_1\vee\sigma_2\vee\ldots\vee\sigma_k$, there is an m between k and n and there are $\sigma_{k+1},\ldots,\sigma_m$ such that $\sigma_1\vee\sigma_2\vee\ldots\vee\sigma_m = \sup S$. For any nonempty $S \subseteq \vee T$, taking $k = 1$ and any $\sigma_1 \in S$ in the above claim yields that $\sup S$ exists and is in $\vee T$ since $\vee T$ is closed under finite sups. We now prove the claim. Notice that if $A \subseteq S$ and $\sigma = \sup A$, then σ is an upper bound for S implies $\sigma = \sup S$. (Since we are not assuming $\sup S$ exists, by $\sigma = \sup S$ we mean σ is a least upper bound for S.) It follows that if $\sigma_1,\sigma_2,\ldots,\sigma_k \in S$ with $\tau_k = \sigma_1\vee\sigma_2\vee\ldots\vee\sigma_k \neq \sup S$, then there is $\sigma_{k+1} \in S$ such that $\tau_{k+1} = \tau_k\vee\sigma_{k+1} > \tau_k$. (Take $A = \{\sigma_1,\sigma_2,\ldots,\sigma_k\}$.) Assume $\sigma_1,\sigma_2,\ldots,\sigma_k \in S$ satisfy $\sigma_1 < \sigma_1\vee\sigma_2 < \ldots < \sigma_1\vee\sigma_2\vee\ldots\vee\sigma_k$. Put $\tau_1 = \sigma_1$, $\tau_2 = \sigma_1\vee\sigma_2,\ldots,\tau_k = \sigma_1\vee\sigma_2\vee\ldots\vee\sigma_k$. If $\tau_k = \sup S$, we are done. Otherwise, there is $\sigma_{k+1} \in S$ with $\tau_{k+1} = \tau_k\vee\sigma_{k+1} > \tau_k$. If $\tau_{k+1} = \sup S$, we are done. If we continue this process, we must obtain $\tau_m = \sup S$ for some $m \leq n$, since $ht(\vee T) \leq n$, and $\tau_i \in \vee T$ for each i. This completes the proof of the claim. As observed above, it follows that $\vee T$ is closed under arbitrary sups. Also, by setting $S = \vee T$ we see that if $\tau_s = \sup\vee T$, then for all $k \leq n$ and $\sigma_1,\sigma_2,\ldots,\sigma_k \in \vee T$, with $\sigma_1 < \sigma_1\vee\sigma_2 < \ldots < \sigma_1\vee\sigma_2\vee\ldots\vee\sigma_k$, there is an m between k and n and there are $\sigma_{k+1},\ldots\sigma_m$ such that $\sigma_1\vee\sigma_2\vee\ldots\vee\sigma_m = \tau_s$.

For the next condition we need a definition. If $\tau \in \vee T\backslash T$, then we call $\{\tau_i{:}i\in I\} \subseteq \vee T$ a **separating set for** τ if $\tau_i < \tau$ for all i and whenever $\sigma \in T$ is less than τ there is an $i \in I$ such that $\sigma \leq \tau_i$. So $\{\tau_i\}$ separates τ from the elements of T below it. We will show that there are no finite separating sets for any of the types in $\vee T\backslash T$. For suppose that $\{\tau_1,\tau_2,\ldots,\tau_k\}$ is a separating set for $\tau \in \vee T\backslash T$. Then, by property F above, $\tau_i < \tau$ implies $\theta(\tau_i) < \theta(\tau)$ for each i. Thus the $\theta(\tau_i)$'s are proper subspaces of $\theta(\tau)$ with $\cup\theta(\tau_i) \leq \theta(\tau)$. Now let Z be a 1-dimensional subspace of $\theta(\tau)$. Then $Z \leq \theta(\tau)$ implies $\varphi(Z) \leq \tau$ by property C above. Since $\varphi(Z) \in T$ and $\{\tau_1,\tau_2,\ldots,\tau_k\}$ is a separating set for τ, there is an i between 1 and k such that $\varphi(Z) \leq \tau_i \leq \tau$. Then $Z \leq \theta(\tau_i) \leq \cup\theta(\tau_i)$, again by property C. Since every one dimensional subspace of $\theta(\tau)$ is contained in $\cup\theta(\tau_i)$, it follows that $\theta(\tau) \leq \cup\theta(\tau_i)$. Thus $\cup\theta(\tau_i) = \theta(\tau)$, that is $\theta(\tau)$ is the union of finitely many proper subspaces, which is impossible. Therefore, there are no finite separating sets.

We next claim that if τ_1 and τ_2 are elements of $\vee T$ which satisfy $ht(\tau_1)+ht(\tau_2) > n$, then there is σ in $\vee T$ such that $\sigma \leq \tau_1$ and $\sigma \leq \tau_2$. For $ht(\tau_1)+ht(\tau_2) > n$ implies $ht[\theta(\tau_1)]+ht[\theta(\tau_2)] > n$ because θ is an embedding. Since

$\theta(\tau_1)$ and $\theta(\tau_2)$ are subspaces of $\mathbb{Q}^n$, it follows that $\theta(\tau_1)\cap\theta(\tau_2) \neq 0$. Then $\theta(\tau_1)\cap\theta(\tau_2) \leq \theta(\tau_1)$ implies $\varphi[\theta(\tau_1)\cap\theta(\tau_2)] \leq \varphi\theta(\tau_1) = \tau_1$. Similarly, $\varphi[\theta(\tau_1)\cap\theta(\tau_2)] \leq \tau_2$. So $\sigma = \varphi[\theta(\tau_1)\cap\theta(\tau_2)]$ is the desired type.

Now if $\sigma_1 < \sigma_2 < \ldots < \sigma_k$ is a chain in S^n, we say the chain is below τ_1 and τ_2 if $\sigma_k \leq \tau_1$ and $\sigma_k \leq \tau_2$. We would like to get an upper bound for the lengths of chains that are below τ_1 and τ_2. We first note that, since θ is an embedding, we must have $\mathrm{ht}(\tau) \leq \dim(\theta(\tau)) \leq n+1 - \mathrm{cht}(\tau)$ for any $\tau \in S^n$. In addition, if τ_1 and τ_2 are incomparable types, then $\theta(\tau_1)$ and $\theta(\tau_2)$ are also incomparable, so that $\dim[\theta(\tau_1)\cap\theta(\tau_2)] < \min\{\dim[\theta(\tau_1)],\dim[\theta(\tau_2)]\} \leq 1+\min\{n-\mathrm{cht}(\tau_1),n-\mathrm{cht}(\tau_2)\}$. Letting $m = \min\{n-\mathrm{cht}(\tau_1),n-\mathrm{cht}(\tau_2)\}$ we then have that $\dim[\theta(\tau_1)\cap\theta(\tau_2)] \leq m$. Now any chain below τ_1 and τ_2 must be taken by θ to a chain below $\theta(\tau_1)$ and $\theta(\tau_2)$. Therefore its length cannot be greater than m. Thus m is the desired upper bound. We note that this does not limit the number of incomparable types below τ_1 and τ_2 unless m is 0 or 1.

We now collect all of our derived necessary conditions for T to be the cotypeset of a rank-n group:

(a) There is an embedding of $\vee T$ into S^n.

(b) $\mathrm{ht}(\vee T) \leq n$.

(c) $\vee T$ is closed under arbitrary sups.

(d) For every $S \subseteq \vee T$, every $k \leq n$, and every choice of $\sigma_1,\sigma_2,\ldots,\sigma_k \in S$ such that $\sigma_1 < \sigma_1\vee\sigma_2 < \ldots < \sigma_1\vee\sigma_2\vee\ldots\vee\sigma_k$, there is an m between k and n and there are $\sigma_{k+1},\ldots,\sigma_m \in S$ such that $\sigma_1\vee\sigma_2\vee\ldots\vee\sigma_m = \sup S$.

(e) In particular, if $\tau_s = \sup T$, then for all $k \leq n$ and $\sigma_1,\sigma_2,\ldots,\sigma_k \in T$, with $\sigma_1 < \sigma_1\vee\sigma_2 < \ldots < \sigma_1\vee\sigma_2\vee\ldots\vee\sigma_k$, there is an m between k and n and there are $\sigma_{k+1},\ldots\sigma_m \in T$ such that $\sigma_1\vee\sigma_2\vee\ldots\vee\sigma_m = \tau_s$.

(f) There are no finite separating sets for any of the types in $\vee T\setminus T$.

(g) If τ_1 and τ_2 in $\vee T$ satisfy $\mathrm{ht}(\tau_1)+\mathrm{ht}(\tau_2) > n$, then there is σ in $\vee T$ such that $\sigma \leq \tau_1$ and $\sigma \leq \tau_2$.

(h) If τ_1 and τ_2 in $\vee T$ are incomparable, then there can be no chains below both of them whose length is greater than $\min\{n-\mathrm{cht}(\tau_1),n-\mathrm{cht}(\tau_2)\}$.

Later in this paper, we will show that all of these conditions together are not sufficient to prove that T is a cotypeset, for any rank bigger than or equal to 3.

QUESTION.

Theorem 1 states that a set of types T is the cotypeset of a rank-n group iff there is an upper semilattice homomorphism φ from S^n to $\vee T$ with $\varphi(S_1^n) = T$. In view of this fact, the following question arises: For $k > n$, does there exist an upper semilattice homomorphism δ from S^k to S^n with $\delta(S_1^k) = S_1^n$. If such a map exists, then any cotypeset of a rank-n group would also be the cotypeset of a rank-k group. For, by Theorem 1, φ would exist as above. Then $\varphi\circ\delta$ would map S^k to $\vee T$ with $\varphi\circ\delta(S_1^k) = T$, as required. For $n = 1$ or 2, and for any $k > n$ the answer is yes, such a map does exist. The trivial proof will be left to the reader. The answer for $k > n \geq 3$ is given in the following theorem whose proof will be deferred until later:

THEOREM 2.

For $k > n \geq 3$ there does not exist an upper semilattice homomorphism δ from S^k to S^n with $\delta(S_1^k) = S_1^n$.

THEOREM 3.

For any $n \geq 3$, there is a set of types T_n which is the cotypeset of a rank-n group, but is not the cotypeset of any higher rank group.

PROOF.

It suffices to find a set of types T_n for which there exists an upper semilattice isomorphism $\varphi_n:S^n \rightarrow \vee T_n$ with $\varphi_n(S_1^n) = T_n$. (For then T_n would be the cotypeset of a rank-n group. However, if T_n were the cotypeset of a rank-k group with $k > n$, then there would be $\varphi:S^k \rightarrow \vee T_n$ with $\varphi(S_1^k) = T_n$. But then $\varphi_n^{-1}\circ\varphi:S^k \rightarrow S^n$ would be an upper semilattice homomorphism satisfying $\varphi_n^{-1}\circ\varphi(S_1^k) = S_1^n$, which is impossible by Theorem 2.) We first define φ_n. List the subspaces of $\mathbb{Q}^n$ of dimension $n - 1$ as X_1, X_2, X_3, etc. (We call these co-dimension 1 subspaces.) Partition the set of prime numbers as an infinite union of infinite sets, $\Pi = \cup P_i$. For $V \in S^n$, $\varphi_n(V)$ will be the type with the following characteristic: $\varphi_n(V)(p) = 0$ if $p \in P_i$ and $V \leq X_i$ and $\varphi_n(V)(p) = 1$ if $p \in P_i$ and $V \nleq X_i$. Finally, set $T_n = \varphi_n(S_1^n)$. It is easy to see that $\varphi_n(V+W) = \varphi_n(V)\vee\varphi_n(W)$. It follows that $\varphi_n(S^n) = \vee T_n$. Also φ_n is injective since V properly contained in W implies there is X_i which contains V but not W. So $\varphi_n:S^n \rightarrow \vee T_n$ is an upper semilattice isomorphism, and T_n is the desired set of types.

We now turn to typesets. The proof of the following theorem is dual to the proofs of the corresponding parts of Theorem 1.

THEOREM 1*.

Let T^* be a finite or countable set of types. Then the following are equivalent:

(2^*) There is a surjective map $\psi^*:S_1^n \rightarrow T^*$ such that $Z \leq \sum_{i=1}^k Z_i$ implies $\psi^*(Z) \geq \inf\{\psi^*(Z_i):1 \leq i \leq k\}$ for all $Z, Z_1, \ldots Z_k \in S_1^n$.

(3^*) There is a map $\varphi^*:S^n \rightarrow \wedge T^*$ such that $\varphi^*(S_1^n) = T^*$ and for all V and

$W \in S^n$, $\varphi^*(V+W) = \varphi^*(V)\wedge\varphi^*(W)$.

(4*) There are maps $\varphi^*: S^n \longrightarrow \wedge T^*$ and $\theta^*: \wedge T^* \longrightarrow S^n$ such that $\varphi^*(S_1^n) = T^*$ and for all $V \in S^n$ and $\tau \in \wedge T^*$, $V \le \theta^*(\tau)$ iff $\varphi^*(V) \ge \tau$.

REMARK.

The maps φ^* and θ^* are semilattice anti-homorphisms. These conditions are necessary for typesets since the map $\varphi^*: S^n \longrightarrow \wedge T^*$ defined by $\varphi^*(V)$ = inner type($G \cap V$) is easily seen to satisfy condition 3*. Furthermore, dualizing previous results gives the following necessary conditions for φ^*, θ^* and T^* when T^* is the typeset of a rank-n group:

PROPERTIES OF φ^* AND θ^*.

(A*) $V \le W \Rightarrow \varphi^*(V) \ge \varphi^*(W)$.
(B*) $\sigma \ge \tau \Leftrightarrow \theta^*(\sigma) \le \theta^*(\tau)$.
(C*) $\varphi^*(V) \ge \tau$ iff $V \le \theta^*(\tau)$.
(D*) $W \le \theta^*\varphi^*(W)$.
(E*) $\tau = \varphi^*\theta^*(\tau)$.
(F*) $\sigma > \tau \Rightarrow \theta^*(\sigma) < \theta^*(\tau)$.

NECESSARY CONDITIONS ON T^*.

(a*) There is an anti-embedding of $\wedge T^*$ into S^n.
(b*) $ht(\wedge T^*) \le n$
(c*) $\wedge T^*$ is closed under arbitrary infs.
(d*) For every $S \in \wedge T^*$, every $k \le n$, and every choice of $\sigma_1,\sigma_2,\ldots,\sigma_k \in S$ such that $\sigma_1 > \sigma_1\wedge\sigma_2 > \ldots > \sigma_1\wedge\sigma_2\wedge\ldots\wedge\sigma_k$, there is an m between k and n and there are $\sigma_{k+1},\ldots,\sigma_m \in S$ such that $\sigma_1\wedge\sigma_2\wedge\ldots\wedge\sigma_m = \inf S$.

(e*) In particular, if $\tau_i = \inf T^*$, then for all $k \le n$ and $\sigma_1,\sigma_2,\ldots,\sigma_k \in T^*$, with $\sigma_1 > \sigma_1\wedge\sigma_2 > \ldots > \sigma_1\wedge\sigma_2\wedge\ldots\wedge\sigma_k$, there is an m between k and n and there are $\sigma_{k+1},\ldots\sigma_m \in T^*$ such that $\sigma_1\wedge\sigma_2\wedge\ldots\wedge\sigma_m = \tau_i$.

(f*) There are no finite co-separating sets for any of the types in $\wedge T^* \backslash T^*$.

(g*) If τ_1 and τ_2 in $\wedge T^*$ satisfy $cht(\tau_1)+cht(\tau_2) > n$, then there is σ in $\wedge T^*$ such that $\sigma \ge \tau_1$ and $\sigma \ge \tau_2$.

(h*) If τ_1 and τ_2 in $\wedge T^*$ are incomparable, then there can be no chains above both of them whose length is greater than $\min\{n-ht(\tau_1), n-ht(\tau_2)\}$.

REMARK.

It is easy to see that conditions (a^*) through (h^*) are not sufficient to prove that T^* is a typeset even for rank-2. In this case they reduce to the requirement that there is a type τ_0 such that $\sigma\wedge\tau = \tau_0$ for all $\sigma \neq \tau$ in T^*, which must be in T^* if T^* is finite. It is well known that this is not sufficient for rank-2 typesets. We can also point out at this time that the dual conditions (a) through (h) for cotypesets reduce to the requirement that there is a type τ_0 such that $\sigma\vee\tau = \tau_0$ for all $\sigma \neq \tau$ in T, which must be in T if T is finite. This is sufficient for rank-2 cotypesets. However, the example T_3 that we defined above satisfies (a) through (h) for all $n > 3$, but is not a cotypeset for any rank greater than 3. Conditions (a) through (h) are also not sufficient to prove that T is a rank-3 cotypeset, since $T = T_3 \cup \{[<1,1,1,...>]\}$ fails condition 3 of Theorem 1, but satisfies (a) through (h) for rank-3.

THEOREM 3*.

For any $n \geq 3$, there is a set of types T_n^* which is the typeset of a rank-n group, but is not the typeset of any higher rank group.

PROOF.

To define the necessary set of types T_n^*, interchange 0's and 1's in the definition of T_n. Then $\wedge T_n^*$ looks like $\vee T_n$ turned upside down, so there is a semilattice anti-isomorphism $\varphi_n^*: S^n \longrightarrow \wedge T_n^*$ with $\varphi_n^*(S_1^n) = T_n^*$. So T_n^* cannot be the typeset of a group of rank-k higher than n. (For then by Theorem 1^* there would exist an anti-homomorphism $\varphi^*: S^k \longrightarrow \wedge T_n^*$ satisfying $\varphi^*(S_1^k) = T_n^*$. But then $\varphi_n^{*-1} \circ \varphi^*: S^k \longrightarrow S^n$ would be a semilattice homomorphism satisfying $\varphi_n^{*-1} \circ \varphi^*(S_1^k) = S_1^n$, contradicting Theorem 2.) We use the Warfield dual to see that T_n^* is the typeset of a group of rank-n. Take X to be a rank-1 group with type [<1,1,1,...>]. The functor Hom(_,X) is an exact, contravariant, rank preserving duality. (See Warfield [9].) Let G be a rank-n group which has T_n as cotypeset. Let G^* be Hom(G,X). We claim that typeset(G^*) = T_n^*. Suppose A is a pure rank-1 subgroup of G^*. Then we have a short exact sequence

$$0 \longrightarrow A \longrightarrow G^* \longrightarrow B \longrightarrow 0.$$

Applying Hom(_,X) yields

$$0 \longrightarrow B^* \longrightarrow G^{**} \longrightarrow A^* \longrightarrow 0.$$

Now, $G \cong G^{**}$ and $\mathrm{rank}(A^*) = \mathrm{rank}(A)$, so $\mathrm{type}(A^*) \in \mathrm{cotypeset}(G^{**}) = \mathrm{cotypeset}(G) = T_n$. Since $\mathrm{type}(A^*)$ is just $\mathrm{type}(A)$ with the 0's and 1's interchanged, then $\mathrm{type}(A) \in T_n^*$. So we have shown that $\mathrm{typeset}(G^*) \subseteq T_n^*$. The proof of the other containment is similar.

The rest of the paper will be devoted to the proof of Theorem 2, which we restate for the benefit of the reader.

THEOREM 2.

For $k > n \geq 3$ there does not exist an upper semilattice homomorphism δ from S^k to S^n with $\delta(S_1^k) = S_1^n$.

PRELIMINARIES.

We will need some facts about ($\mathbb{Q}$-)vector spaces. It is well known that a vector space cannot be the union of finitely many proper subspaces. Using this one can show that if V is n-dimensional, $V_1,V_2,\ldots,V_k$ are proper subspaces of V with dimensions $n_1,n_2,\ldots,n_k$ respectively and $m \leq n\text{-}\max(n_1,n_2,\ldots,n_k)$, then there is an m-dimensional subspace W of V which is disjoint from $V_1,V_2,\ldots,V_k$. The next result we will call

LEMMA 1.

In the semilattice S^3, let Z_1, Z_2, Z_3 and Z_4 be four 1-dimensional subspaces such that any three are independent. Let $L_0 = \{Z_1,Z_2,Z_3,Z_4\}$. For each m, let $P_{m+1} = \{$planes generated by pairs of lines from $L_m\}$ and $L_{m+1} = L_m \cup \{$lines which are intersections of pairs of planes of $P_{m+1}\}$. Then every line is contained in L_m for some m.

PROOF OF LEMMA 1.

Let $\{\mathbf{i},\mathbf{j},\mathbf{k}\}$ be the standard basis for $\mathbb{Q}^3$. Given Z_1 through Z_4 as above, there is a linear isomorphism of $\mathbb{Q}^3$ such that $\mathbb{Q}\mathbf{i}$, $\mathbb{Q}\mathbf{j}$, $\mathbb{Q}\mathbf{k}$, and $\mathbb{Q}(\mathbf{i}+\mathbf{j}+\mathbf{k})$ are mapped to Z_1, Z_2, Z_3, and Z_4 respectively. Since linear isomorphisms preserve sums of pairs of lines and intersections of pairs of planes, we may assume $Z_1 = \mathbb{Q}\mathbf{i}$, $Z_2 = \mathbb{Q}\mathbf{j}$, $Z_3 = \mathbb{Q}\mathbf{k}$ and $Z_4 = \mathbb{Q}(\mathbf{i}+\mathbf{j}+\mathbf{k})$. It suffices to prove that for all non-zero (a,b,c) in $\mathbb{Z}^3$ with $\gcd(a,b,c) = 1$, the line determined by $a\mathbf{i}+b\mathbf{j}+c\mathbf{k}$ is contained in some L_m. We first notice that $\mathbb{Q}(\mathbf{i}+\mathbf{j}) = [\mathbb{Q}\mathbf{i}+\mathbb{Q}\mathbf{j}]\cap[\mathbb{Q}(\mathbf{i}+\mathbf{j}+\mathbf{k})+\mathbb{Q}\mathbf{k}] \in L_1$. Similarly, $\mathbb{Q}(\mathbf{i}+\mathbf{k})$ and $\mathbb{Q}(\mathbf{j}+\mathbf{k})$ are in L_1. Then $\mathbb{Q}(\mathbf{i}-\mathbf{j}) = [\mathbb{Q}\mathbf{i}+\mathbb{Q}\mathbf{j}]\cap[\mathbb{Q}(\mathbf{i}+\mathbf{k})+\mathbb{Q}(\mathbf{j}+\mathbf{k})] \in L_2$. Similarly, $\mathbb{Q}(\mathbf{i}-\mathbf{k})$ and $\mathbb{Q}(\mathbf{j}-\mathbf{k})$ are in L_2. Next $\mathbb{Q}(\mathbf{i}+\mathbf{j}-\mathbf{k}) = [\mathbb{Q}(\mathbf{i}+\mathbf{j})+\mathbb{Q}\mathbf{k}]\cap[\mathbb{Q}\mathbf{i}+\mathbb{Q}(\mathbf{j}-\mathbf{k})] \in L_3$. Similarly, $\mathbb{Q}(\mathbf{i}-\mathbf{j}+\mathbf{k})$ and $\mathbb{Q}(\mathbf{i}-\mathbf{j}-\mathbf{k}) = \mathbb{Q}(-\mathbf{i}+\mathbf{j}+\mathbf{k})$ are in L_3. We define the content of (a,b,c) to be $|a|+|b|+|c|$. The proof is by induction on content. All triples of content 1 appear in L_0. Now suppose all triples whose content is less than n have appeared and let (a,b,c) have content n. We will express $\mathbb{Q}(ai+bj+ck)$ as the intersection of sums of lines that have already appeared. We may assume that there are at least two non-zero components, and in fact that $a > 0$ and $b \neq 0$. If $c \neq 0$, then we have $\mathbb{Q}(ai+bj+ck) = [\mathbb{Q}(ai+bj)+\mathbb{Q}k] \cap [\mathbb{Q}i+\mathbb{Q}(bj+ck)]$. If $c = 0$ and $b > 0$, then $\mathbb{Q}(ai+bj) = [\mathbb{Q}i+\mathbb{Q}j] \cap [\mathbb{Q}(i+j-k)+\mathbb{Q}((a-1)i+(b-1)j+k]$. If $c = 0$ and $b < 0$ then $\mathbb{Q}(ai+bj) = [\mathbb{Q}i+\mathbb{Q}j] \cap [\mathbb{Q}(i-j-k)+\mathbb{Q}((a-1)i+(b+1)j+k]$. This completes the proof of Lemma 1.

PROOF OF THEOREM 2.

We suppose that such a map δ exists and make some simple observations. First of all, δ is monotone. Secondly, we must have $\delta(\mathbb{Q}^k) = \mathbb{Q}^n$. Next, suppose $Z_1, Z_2, \ldots, Z_l$ is a basis for $U \leq \mathbb{Q}^k$. Then $\delta(U) = \delta(\Sigma_{i=1}^k Z_i) = \Sigma_{i=1}^k \delta(Z_i)$. So $\delta(Z_1), \delta(Z_2), \ldots, \delta(Z_l)$ is a spanning set for $\delta(U)$. It follows that $\dim[\delta(U)] \leq \dim[U]$. Now if V is a complimentary summand of U, then $U+V = \mathbb{Q}^k$, and $\delta(U)+\delta(V) = \mathbb{Q}^n$. So $n-\dim[\delta(U)] \leq \dim[\delta(V)] \leq \dim[V] = k-\dim[U]$. Thus $\dim[U]-(k-n) \leq \dim[\delta(U)] \leq \dim[U]$. So under δ subspaces cannot go up in dimension, nor can they go down by any more than k-n. For the rest of the proof, when we say that U "drops", we mean that $\dim[\delta(U)] < \dim[U]$.

CASE 1: $k = 4$ and $n = 3$.

If $\dim[U] = 2$, then $\dim[\delta(U)] = 1$ or 2. Suppose U and V were distinct 2-dimensional subspaces with $\dim[\delta(U)] = \dim[\delta(V)] = 1$. We consider the following diagrams of subspaces:

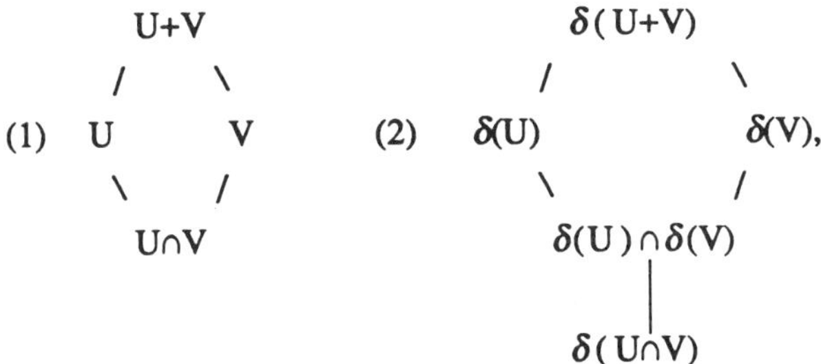

where $\delta(U \cap V)$ does not appear unless $U \cap V \neq 0$. These subspaces satisfy the following equations which we will use without mention in the rest of the proof: $\dim[U+V]+\dim[U \cap V] = \dim[U]+\dim[V]$, and $\dim[\delta(U+V)]+\dim[\delta(U) \cap \delta(V)] = \dim[\delta(U)]+\dim[\delta(V)]$. If $U \cap V \neq 0$ then $\dim[U \cap V] = 1 \Rightarrow \dim[\delta(U \cap V)] = 1 \Rightarrow \delta(U) = \delta(U) \cap \delta(V) = \delta(V) = \delta(U+V)$. So $\dim[\delta(U+V)] = 1$. But $\dim[U+V] = 3$ and $k-n = 4-3 = 1$. So we must have $U \cap V = 0$. Then $\dim[U \cap V] = 0 \Rightarrow \dim[U+V] = 4 \Rightarrow \dim[\delta(U+V)] = 3$. But this is impossible since $\dim[\delta(U)] = \dim[\delta(V)] = 1$. So we conclude that there can be at most one 2-dimensional subspace that drops. We will assume that there is exactly one such subspace, the case where there are none is similar. Call the unique such subspace U_1. Choose 2-dimensional subspaces U_2, U_3, and U_4 such that U_1 through U_4 are pairwise disjoint. Then $\dim[\delta(U)] = \dim[\delta(V)] = 2$, so $\delta(U) \cap \delta(V) \neq 0$. Choose Z a 1-dimensional subspace of $\delta(U) \cap \delta(V)$. Choose a 1-dimensional subspace Z' of $\mathbb{Q}^k$ such that $\delta(Z') = Z$. Then Z' is disjoint from U_2 or U_3; by renaming we may assume $Z' \cap U_2 \neq 0$. Then $\dim[Z'+U_2] = 3$, and $\delta(Z'+U_2) = \delta(Z')+\delta(U_2) = \delta(U_2)$. If we let $V_2 = Z'+U_2$, then we have shown that U_2 is contained in a 3-dimensional subspace V_2 that drops to a 2-dimensional subspace under δ. Similarly, using U_3 and U_4, we can show that U_3 is contained in a subspace V_3 which is 3-dimensional and goes to a 2-dimensional subspace under δ. Now consider diagrams (1) and (2) with $U = V_2$ and $V = V_3$. Then $\dim[V_2] = \dim[V_3] = 3$ and $\dim[V_2+V_3] = 4$, so $\dim[V_2 \cap V_3] = 2$. Also $\dim[\delta(V_2)] = \dim[\delta(V_3)] = 2$ and $\dim[\delta(V_2+V_3)] = 3$, so $\dim[\delta(V_2) \cap \delta(V_3)] = 1$. This implies that $\dim[\delta(V_2 \cap V_3)] = 1$ and, thus, that $V_2 \cap V_3 = U_1$. But then $\mathbb{Q}^4 = U_1+U_2 \leq V_2$. This contradiction finishes Case 1.

<u>CASE 2:</u> $k > n = 3$.

Assume δ is an upper semilattice homomorphism from S^k to S^3 with $\delta(S_1^k) = S_1^3$. For $W \leq \mathbb{Q}^k$, let S^W denote the set of nonzero subspaces of W, and let S_1^W denote the 1-dimensional subspaces of W. The idea is to find a 4-dimensional subspace W of $\mathbb{Q}^k$ such that $\delta(S_1^W) = S_1^3$. Since $\delta | S^W$ automatically preserves sums, and since S^W is just a copy of S_1^4, the existence of $\delta | S^W$ would contradict Case 1. Let Z_1, Z_2 and Z_3 be independent 1-dimensional subspaces of $\mathbb{Q}^3$. Let Z_1^*, Z_2^* and Z_3^* be 1-dimensional subspaces of $\mathbb{Q}^k$ such that $\delta(Z_i^*) = Z_i$ for $i = 1,2,3$. Let $P = Z_1^*+Z_2^*+Z_3^*$. Let Z_4^* be a 1-dimensional subspace of P such that any three of Z_1^* through Z_4^* are independent, and let $Z_4 = \delta(Z_4^*)$. Then any three of Z_1 through Z_4 are independent. We claim that the 1-dimensional subspaces of P are mapped onto S_1^3 by δ. Suppose that U and V are distinct planes contained in P, and consider diagrams (1) and (2). Now $\dim[U] = \dim[V] = 2 \Rightarrow \dim[U \cap V] = 1 \Rightarrow \dim[\delta(U \cap V)] = 1$. Also since P does not drop, neither can U or V so $\dim[\delta(U)] = \dim[\delta(V)] = 2$. Thus, $\dim[\delta(U) \cap \delta(V)] = 1$. Since $\delta(U \cap V)$ is always contained in $\delta(U) \cap \delta(V)$, they must be equal. So δ preserves intersections of distinct planes contained in P, as well as sums of distinct lines contained in P. Thus, by Lemma 1, we have a set of generators of the semilattice of subspaces of P being mapped onto a set of generators of the semilattice of subspaces of $\mathbb{Q}^3$, while δ preserves the two operations in the generating process. Therefore the subspaces of P must be mapped onto the subspaces of $\mathbb{Q}^3$. In particular, we must have

$\delta(S_1^P) = S_1^3$. Now take any 4-dimensional extension W of P. Then $S_1^3 = \delta(S_1^P) \le \delta(S_1^W) \le S_1^3$. As observed above, this is impossible by Case 1.

<u>CASE 3:</u> $k > n > 3$.

The proof of the general case will be by induction on n. The case n = 3 has already been completed. We will show that if there is an upper semilattice homomorphism δ mapping $\mathbb{Q}^k$ to $\mathbb{Q}^n$ with $\delta(S_1^k) = S_1^n$, then there is $W \le \mathbb{Q}^k$ and $X < \mathbb{Q}^n$ with $\delta(W) = X$ and $\dim[X] = n-1$ such that $\delta(S_1^W) = S_1^X$. Since $\dim[X] = n-1$, the existence of $\delta|S^W$ would contradict the induction hypothesis. We first claim that there there is $W \le \mathbb{Q}^k$ and $X < \mathbb{Q}^n$ such that $\delta(W) = X$ and $\dim[X] < \dim[W]$. Otherwise nothing whose dimension is less than n drops. But then choose U and V disjoint subspaces of $\mathbb{Q}^k$ with dimensions n-1 and 2 respectively. Then $\dim[\delta(U)] = n-1$ and $\dim[\delta(V)] = 2$. These cannot be disjoint so there is a 1-dimensional subspace Z contained in $\delta(U) \cap \delta(V)$. Choose Z' such that $\delta(Z') = Z$. Now Z' cannot meet both U and V. If $U \cap Z' = 0$ then take $W = U+Z'$, while if $V \cap Z' = 0$ then take $W = V+Z'$. In either case let $X = \delta(W)$. This proves the claim. So $B = \{X<\mathbb{Q}^n : \exists W \le \mathbb{Q}^k \text{ such that } \delta(W) = X \text{ and } \dim[X] < \dim[W]\}$ is nonempty. Choose $X \in B$ of maximal dimension. We claim that $\dim[X] = n-1$. If not, choose a 1-dimensional subspace Z which is disjoint from X, and Z' with $\delta(Z') = Z$. Then $\delta(W+Z') = X+Z$ and W+Z' drops onto X+Z. Since $X+Z < \mathbb{Q}^n$, it follows that $X+Z \in B$, contradicting the maximality of the dimension of X. So $\dim[X] = n-1$. Now $C = \{W<\mathbb{Q}^k : \delta(W) = X \text{ and } \dim[X] < \dim[W]\}$ is nonempty. Choose $W \in C$ of maximal dimension. Then we claim that for every 1-dimensional $Z \le X$ there is a 1-dimensional $Z' \le W$ such that $\delta(Z') = Z$. We know that there is a 1-dimensional $Z' \le \mathbb{Q}^k$ such that $\delta(Z') = Z$. Then $\delta(W+Z') = \delta(W)+\delta(Z') = X+Z = X$. By maximality of W, $W = W+Z'$ and $Z' \le W$. So $\delta(S_1^W) = S_1^X$. Since $\dim[X] = n-1$, the existence of $\delta|S^W : S^W \to S^X$ is impossible by the induction hypothesis. This completes the proof of Theorem 2.

REFERENCES.

[1] D. Arnold and C. Vinsonhaler
Typesets and Cotypesets of Rank-2 Torsion Free Abelian Groups
Pac. J. Math., 114 (1984), 1-21

[2] R. Beaumont and R. S. Pierce
Torsion-free Groups of Rank Two
Mem. Amer. Math. Soc., 38 (1961)

[3] D. Dubois
Applications of Analytic Number Theory to the Study of
Type sets of Torsionfree Abelian Groups I and II
Publ. Math. Debrecen, 12 (1965), 59-63 and 13 (1966) 1-8

[4] R. Ito
On Type-sets of Torsion-Free Abelian Groups of Rank 2
Proc. Amer. Math. Soc., 48 (1975), 39-42

[5] J. Koehler
The Type Set of a Torsion-Free Group of Finite Rank
Ill. J. Math., 9 (1965), 66-86

[6] P. Schultz
The Typeset and Cotypeset of a Rank 2 Abelian Group
Pac. J. Math., 78 (1978), 503-517

[7] C. Vinsonhaler and W. Wickless
The Cotypeset of a Torsion Free Abelian Group of Rank Two
Proc. Amer. Math. Soc., 84 (1982), 467-473

[8] C. Vinsonhaler and W. Wickless
The Cotypeset of a Torsion-Free Abelian Group of Finite Rank
Jour. of Alg., 83 (1983), 380-386

[9] R. B. Warfield, Jr.
Homomorphisms and Duality for Torsion-free Groups
Math. Z., 107 (1968), 189-200

University of Connecticut
Department of Mathematics, Box U-9
Storrs, CT, 06269-3009
USA

Contemporary Mathematics
Volume **171**, 1994

A Generalization of Butler Groups

A. MADER, O. MUTZBAUER AND K. M. RANGASWAMY

Dedicated to the memory of Richard S. Pierce

1. Introduction

An important characterization by Butler [**7**, Theorem 4] states that a finite rank torsion-free abelian group is a pure subgroup of a completely decomposable group if and only if G satisfies the conditions: (1) The typeset of G is finite, (2) for each type τ, $G(\tau) = G_\tau \oplus G^\sharp(\tau)$, where G_τ is τ-homogeneous completely decomposable, (3) for each type τ, $G^\sharp(\tau)/G^*(\tau)$ is a bounded torsion group. Such groups have been extensively studied under the name of "Butler groups", and possess many interesting properties. Recent attempts to generalize the concept of Butler groups to the infinite rank case lead to the introduction of the so-called B_1– and B_2–groups [**5**].

In this article, we investigate torsion-free abelian groups of arbitrary rank which satisfy the conditions (1), (2), (3) above. We call such groups BR–**groups** (B for Butler, R for regulating subgroup).

In Section 3 we show how some of the classical results of finite rank Butler groups carry over to BR–groups. Sample results are:

(i) BR–groups possess regulating subgroups and regulators with bounded quotient (3.4).
(ii) A local version of Burkhardt's description of regulators is true (3.9, 3.12).
(iii) BR–groups with special typesets have completely decomposable regulating subgroups or are completely decomposable (3.17, 3.18).

The defining properties of BR–groups are exactly what is needed to guarantee the existence of regulating subgroups and regulators. However, the general

---

1991 *Mathematics Subject Classification.* 20K15, 20K20.

This work was done when the first and third authors visited the Mathematische Institut, Universität Würzburg, during the summer 1993. The first author gratefully acknowledges support by the German Academic Exchange Service while the third author expresses his gratitude to the Universität Würzburg for a Visiting Professorship at the Mathematische Institut.
This paper is in final form and no version of it will be submitted for publication elsewhere.

situation is rather complex, and there are more questions than answers at this point. Some of our examples show that rather unexpected and counterintuitive phenomena can occur. It is hoped that the questions raised and the concepts proposed will create some interest in the study of regulating subgroups and regulators of Butler groups, and more generally of BR–groups.

In Section 4 we consider BR–groups G which contain a completely decomposable regulating subgroup. It is shown that then every regulating subgroup of such a group G is completely decomposable. Furthermore, these groups are exactly those which are isomorphic to a subgroup of a completely decomposable group with a bounded quotient. Burkhardt's Theorem on regulators for almost completely decomposable groups extends to BR–groups with a completely decomposable regulating subgroup.

In Section 5 examples are constructed to show that the BR–groups differ in significant aspects from infinite rank Butler groups. Specifically we show that there are B_2–groups which are not BR–groups (5.1) and that there are countable BR–groups which are not B_2–groups, not even finitely Butler (5.2). Our example also shows that a countable group may be both a B_2–group and a BR–group, but still can contain a pure subgroup which is not a BR–group.

2. Preliminaries

All the groups that we consider here are torsion–free abelian groups. For the general notation and terminology we refer to Arnold [**3**] and Fuchs [**8**]. Pure subgroups of finite rank of completely decomposable groups are called **Butler groups**. A torsion–free group (of arbitrary rank) is called **finitely Butler** if every pure subgroup of finite rank in G is a Butler group. A group G is a **B_2–group** if G is the union of a smooth well–ordered ascending chain of pure subgroups

$$0 = G_0 \subset G_1 \subset \cdots \subset G_\alpha \subset G_{\alpha+1} \subset \cdots, \quad \alpha < \lambda,$$

where λ is some ordinal and where, for all $\alpha < \lambda$, $G_{\alpha+1} = G_\alpha + B_\alpha$ for some finite rank Butler group B_α. A countable group G is a B_2–group if and only if it is finitely Butler (see [**5**, 3.4]). An uncountable B_2–group is always finitely Butler, but not conversely as the group $\prod \mathbb{Z}$, a product of copies of $\mathbb{Z}$ shows.

Let G be a torsion–free group. $\mathrm{tp}(x) = \mathrm{tp}^G(x)$ denotes the type of x in G. Also $\mathrm{T}(G) = \{\mathrm{tp}(x) : 0 \neq x \in G\}$ is the **typeset** of G. For any type τ, we have the so–called **type subgroups** $G(\tau) = \{x \in G : \mathrm{tp}(x) \geq \tau\}$, $G^*(\tau) = \langle\{x \in G : \mathrm{tp}(x) > \tau\}\rangle = \sum_{\rho > \tau} G(\rho)$, and $G^\sharp(\tau) = G^*(\tau)_*$, the pure subgroup generated by $G^*(\tau)$. Thus we have $G^*(\tau) \subset G^\sharp(\tau) \subset G(\tau)$. A type $\tau \in \mathrm{T}(G)$ is a **critical type** of G if $G(\tau) \neq G^\sharp(\tau)$. The set of all types $\tau \in \mathrm{T}(G)$ which are critical is denoted by $\mathrm{T}_{cr}(G)$.

Butler [**7**, Theorem 4] proved that a Butler group has the so–called "Butler decompositions" $G(\tau) = G_\tau \oplus G^\sharp(\tau)$ for every $\tau \in \mathrm{T}_{cr}(G)$, where G_τ is a τ–

homogeneous completely decomposable group.

In general, if H is a subgroup of G, then $H \cap G(\tau)$ may or may not equal $H(\tau)$, and $H \cap G^\sharp(\tau)$ may or may not equal $H^\sharp(\tau)$. However, if G/H is bounded we have equality in both cases. Additional important properties are indicated in Lemma 2.1 below and these will be used in the sequel without explicit reference.

LEMMA 2.1. *Let G be any torsion–free group and H a subgroup such that $eG \subset H$ for a positive integer e. Then $H \cap G(\tau) = H(\tau)$ and $H \cap G^\sharp(\tau) = H^\sharp(\tau)$ for any type τ. Furthermore,* $\mathrm{T}(G) = \mathrm{T}(H)$ *and* $\mathrm{T}_{cr}(G) = \mathrm{T}_{cr}(H)$. *The group $G(\tau)/G^\sharp(\tau)$ is τ–homogeneous completely decomposable if and only if $H(\tau)/H^\sharp(\tau)$ is τ–homogeneous completely decomposable. If so, both groups have Butler decompositions $G(\tau) = G_\tau \oplus G^\sharp(\tau)$ and $H(\tau) = H_\tau \oplus H^\sharp(\tau)$.*

PROOF. The first two identities are well–known. Also for $x \in G$, $\mathrm{tp}^G(x) = \mathrm{tp}^H(ex)$ and for $y \in H$, $\mathrm{tp}^H(y) = \mathrm{tp}^G(y)$. Hence $\mathrm{T}(G) = \mathrm{T}(H)$. Using the Dedekind identity, we have an exact sequence

$$0 \to \frac{H(\tau)}{H^\sharp(\tau)} \to \frac{G(\tau)}{G^\sharp(\tau)} \to \frac{H + G(\tau)}{H + G^\sharp(\tau)} \to 0$$

Herein $(H + G(\tau))/(H + G^\sharp(\tau))$ is bounded by e, hence the torsion–free groups on the left are either both zero or both non–zero. Thus $\mathrm{T}_{cr}(G) = \mathrm{T}_{cr}(H)$. Also the quotient $H(\tau)/H^\sharp(\tau)$ is τ–homogeneous if and only if $G(\tau)/G^\sharp(\tau)$ is τ–homogeneous. If $G(\tau)/G^\sharp(\tau)$ is τ–homogeneous completely decomposable then so is $H(\tau)/H^\sharp(\tau)$ by [**8**, 86.6], and conversely, since $G(\tau)/G^\sharp(\tau)$ is also embedded in $H(\tau)/H^\sharp(\tau)$ with bounded index. If these quotients are in fact τ–homogeneous completely decomposable, then the Butler decompositions follow at once from the Baer Lemma. □

We will also have occasion to use the following general and easy fact.

LEMMA 2.2. *Let $K = L \oplus M$ be a direct decomposition of R–modules, R any ring. Then the set of direct complements of M in K is in bijective correspondence with* $\mathrm{Hom}_R(L, M)$, *the correspondence being given by*

$$\phi \in \mathrm{Hom}_R(L, M) \mapsto L(1 + \phi) = \{x(1 + \phi) = x + x\phi : x \in L\}.$$

We will need the following fact about completely decomposable groups. Incidentally, it indicates that considerable complications must be expected if the critical typeset is allowed to be arbitrary.

LEMMA 2.3. *Let $A = \bigoplus_{\rho \in \mathrm{T}_{cr}(A)} A_\rho$ be a decomposition of the completely decomposable group A into homogeneous components. If $A(\tau) = B_\tau \oplus A^\sharp(\tau)$ for all $\tau \in \mathrm{T}_{cr}(A)$ then $B = \sum_{\rho \in \mathrm{T}_{cr}(A)} B_\rho$ is the direct sum of the B_ρ. Furthermore, $A = B$ for each such "regulating subgroup" B if and only if the critical typeset $\mathrm{T}_{cr}(A)$ contains no infinite ascending chains.*

PROOF. (a) We show first that $A = B = \sum_{\rho \in \mathrm{T}_{cr}(A)} B_\rho$ if $B_\rho = A_\rho$ for all but one ρ. Let τ be the exceptional type. Then

$$B_\tau \cap \bigoplus_{\rho \neq \tau} A_\rho \subset B_\tau \cap A(\tau) \cap \bigoplus_{\rho \neq \tau} A_\rho \subset B_\tau \cap A^\sharp(\tau) = 0.$$

Let $U = B_\tau \oplus \bigoplus_{\rho \neq \tau} A_\rho$. To show that $U = A$, we only need that $A_\tau \subset U$. But

$$A_\tau \subset A(\tau) = B_\tau \oplus A^\sharp(\tau) = B_\tau \oplus \bigoplus_{\rho > \tau} A_\rho \subset U.$$

By finite induction it follows that any finite number of summands A_τ can be replaced by any B_τ while preserving the direct decomposition and equality with A. Since independence has finite character, this also shows that any regulating subgroup B is a direct sum of its subgroups B_τ.

(b) We show next that there is a regulating subgroup B properly contained in A if $\mathrm{T}_{cr}(A)$ contains an infinite ascending chain. Let $\tau_1 < \tau_2 < \cdots$ be an ascending chain of critical types of A. Let A_i be rank–one summands of A of type τ_i, and choose monomorphisms $\phi_i : A_i \to A_{i+1}$. Without loss of generality $A = \bigoplus_{i=1}^\infty A_i$. By 2.2 $A(\tau_i) = A_i \oplus A^\sharp(\tau_i) = A_i(1 + \phi_i) \oplus A^\sharp(\tau_i)$. Setting $B_i = A_i(1 + \phi_i)$, the group $B = \bigoplus_{i=1}^\infty B_i$ is a regulating subgroup of A. However, B does not contain A_1, so is a proper subgroup of A.

(c) Suppose now that $\mathrm{T}_{cr}(A)$ does not contain any ascending infinite chain. Let $B = \sum_{\rho \in \mathrm{T}_{cr}(A)} B_\rho$ be a regulating subgroup of A. We prove by induction on depth that $A_\tau \subset B$ for every critical type τ. If τ is maximal, then $A_\tau = A(\tau) = B_\tau \subset B$. Let τ be arbitrary and suppose that $A_\sigma \subset B$ for every type $\sigma > \tau$. By 2.2 applied to $A(\tau) = A_\tau \oplus A^\sharp(\tau) = B_\tau \oplus A^\sharp(\tau)$, there is $\phi \in \mathrm{Hom}(A_\tau, A^\sharp(\tau)) = \mathrm{Hom}(A_\tau, \bigoplus_{\rho > \tau} A(\rho))$ such that $B_\tau = A_\tau(1 + \phi)$. Hence $A_\tau \subset B_\tau + A_\tau \phi \subset B_\tau + \bigoplus_{\rho > \tau} A_\rho \subset B$ by induction hypothesis. □

Finally, $(p_1 p_2 \cdots p_k)^{-1}\mathbb{Z}$ denotes the ring of rational numbers whose denominators (in the reduced form) are products of powers of the primes $p_1, \ldots, p_k$.

3. The Structure of BR–groups

In this section we investigate the structural properties of BR–groups. The existence of regulating subgroups is assured by the very definition of BR–group. These and associated constructs are our main concern. We show that some of the classical results on finite rank almost completely decomposable groups carry over to BR–groups.

We shall begin with the formal definition of a BR–group.

DEFINITION 3.1. *A BR–**group** is a torsion–free group G with the following properties:*

(1) *The typeset $\mathrm{T}(G)$ is finite.*

(2) *For each type* $\tau \in \mathrm{T}(G)$, $G(\tau) = A_\tau \oplus G^\sharp(\tau)$ *for some* τ*-homogeneous completely decomposable group* A_τ.

(3) *For every type* $\tau \in \mathrm{T}(G)$, *the quotient group* $G^\sharp(\tau)/G^*(\tau)$ *is bounded.*

We begin by mentioning some of the subgroups which inherit the BR–property.

LEMMA 3.2. *Let* G *be a* BR*-group. If* $H \subset G$ *such that* $\exp(G/H)$ *is finite, then* H *is a* BR*-group. The type subgroups* $G(\tau)$ *and* $G^\sharp(\tau)$ *are also* BR*-groups.*

PROOF. The first claim follows immediately from Lemma 2.1. The other claims are routine to check. □

As we shall see in Example 5.2, a countable BR–groups may contain a pure subgroup of finite rank which is not a Butler group, in contrast to the behavior of countable Butler groups. However, a feature shared by BR–groups of any rank is the existence and theory of regulating subgroups and regulators. If G is a BR–group, then for each $\tau \in \mathrm{T}_{cr}(G)$, consider the groups A_τ that show up in a decomposition $G(\tau) = A_\tau \oplus G^\sharp(\tau)$. Generalizing Lady [**10**] and Arnold [**2**], we call the subgroup $A = \sum_{\rho \in \mathrm{T}_{cr}(G)} A_\rho$ a **regulating subgroup** of G. The set of all regulating subgroups of G is denoted by $\mathrm{Regg}(G)$. The intersection $\bigcap\{A : A \in \mathrm{Regg}(G)\}$ is called the **regulator** of G and is denoted by $\mathrm{R}(G)$. This concept is due to Burkhardt [**6**]. We also consider the Σ**–regulator** defined as $\mathrm{R}_\Sigma(G) = \sum\{A :\ A \in \mathrm{Regg}(G)\}$.

The definition of regulating subgroup and the description 2.2 of direct complements immediately provide the following presentation of all regulating subgroups in terms of a given one.

LEMMA 3.3. *Let* G *be a* BR*-group, and let* $A = \sum_{\rho \in \mathrm{T}_{cr}(G)} A_\rho$ *be a regulating subgroup. Then the regulating subgroups are exactly the groups of the form* $A = \sum_{\rho \in \mathrm{T}_{cr}(G)} A_\rho(1 + \phi_\rho)$, *where* $\phi_\tau \in \mathrm{Hom}(A_\tau, G^\sharp(\tau))$.

For each $\tau \in \mathrm{T}(G)$, let $e^G_\tau = \exp(G^\sharp(\tau)/G^*(\tau))$, the **exponent** or **bound** of the group $G^\sharp(\tau)/G^*(\tau)$.

We have the following fundamental result.

THEOREM 3.4. *Let* G *be a* BR*-group and* A *a regulating subgroup of* G. *Then the exponent* $\exp(G/A)$ *divides* $\prod_{\rho \in \mathrm{T}(G)} e^G_\rho$ *and also divides* e^d *where* $e = \mathrm{lcm}\{e^G_\rho : \rho \in \mathrm{T}(G)\}$ *and* d *is the depth of* $\mathrm{T}(G)$, *i.e. the largest integer* d *for which there is a chain*

$$\tau_0 < \tau_1 < \cdots < \tau_d, \quad \tau_i \in \mathrm{T}(G).$$

In particular, the regulator exponent $\exp(G/\mathrm{R}(G)) = \mathrm{lcm}\{\exp(G/A) : A \in \mathrm{Regg}(G)\}$ *is finite, and hence* $\mathrm{R}(G)$ *is again a* BR*-group.*

PROOF. The proofs for finite rank Butler groups can be found in [**11**, 2.1 and 2.2] and carry over verbatim to BR–groups. □

The Σ–regulator of a BR–group can be easily described.

PROPOSITION 3.5. *Let G be any* BR*–group. Then* $\mathrm{R}_\Sigma(G) = \sum_{\rho\in\mathrm{T}_{cr}(G)} G(\rho)$, *a fully invariant subgroup of G.*

PROOF. The inclusion of the Σ–regulator in the sum $\sum_{\rho\in\mathrm{T}_{cr}(G)} G(\rho)$ is obvious. To show the other inclusion, let τ be a critical type of G. Then $G(\tau) = A_\tau \oplus G^\sharp(\tau)$ where A_τ is a non–zero τ–homogeneous completely decomposable group. By definition, the Σ–regulator contains A_τ and by 3.3 also $A_\tau(1+\phi)$ where $\phi \in \mathrm{Hom}(A_\tau, G^\sharp(\tau))$. Hence the Σ–regulator contains $A_\tau\phi$ for every $\phi \in \mathrm{Hom}(A_\tau, G^\sharp(\tau))$. However, these images $A_\tau\phi$ generate $G^\sharp(\tau)$, and hence $G(\tau) = A_\tau \oplus G^\sharp(\tau) \subset \mathrm{R}_\Sigma(G)$. □

The following example, due to C. Vinsonhaler (oral communication), shows that some critical types may not be "truly critical", even in a finite rank Butler group, in contrast to the situation for almost completely decomposable groups. This suggests that a distinction between critical types may be useful.

EXAMPLE 3.6. *There is a finite rank Butler group G with a regulating subgroup $A = \sum_{\rho\in\mathrm{T}_{cr}(G)} A_\rho$ such that $G/\sum_{\rho\in S} A_\rho$ is a bounded group for some proper subset S of $\mathrm{T}_{cr}(G)$.*

PROOF. Let $p_1, p_2, \ldots$ be a listing of all prime numbers. Set $A_{ij} = (p_ip_j)^{-\infty}\mathbb{Z}$, and

$$A = A_{12} \oplus A_{13} \oplus A_{45} \oplus A_{46} \subset \mathbb{Q} \oplus \mathbb{Q} \oplus \mathbb{Q} \oplus \mathbb{Q}.$$

The desired group is

$$G = \frac{A}{D}, \quad \text{where} \quad D = \langle (1,1,1,1) \rangle.$$

If $^{-} : A \to A/D = G$ denotes the natural epimorphism, then $\overline{(1,1,0,0)} = \overline{(0,0,-1,-1)}$ has type $\tau = \mathrm{tp}(A_{14})$. The type τ is critical since $G(\tau) = \langle \overline{(1,1,0,0)} \rangle$ and $G^\sharp(\tau) = 0$. The group $B = \langle \overline{(1,1,0,0)} \rangle + \overline{A_{12}} + \overline{A_{13}} + \overline{A_{45}} + \overline{A_{46}}$ is a regulating subgroup of G and $G/(\overline{A_{12}} + \overline{A_{13}} + \overline{A_{45}} + \overline{A_{46}}) = 0$ is bounded although $\langle \overline{(1,1,0,0)} \rangle$ has been dropped. □

The regulator is much more elusive than the Σ–regulator. Our goal is to come as close as possible to the description of Burkhardt of the regulator of an almost completely decomposable group (of finite rank). We begin with some "local" properties which work the same as for almost completely decomposable groups.

LEMMA 3.7. *Let G be a* BR*–group.*

(1) $\mathrm{T}(G^\sharp(\tau)) \subset \mathrm{T}(G(\tau)) = \{\rho \in \mathrm{T}(G) : \rho \geq \tau\}$, $\mathrm{T}_{cr}(G^\sharp(\tau)) = \{\rho \in \mathrm{T}_{cr}(G) : \rho > \tau\}, \mathrm{T}_{cr}(G(\tau)) = \{\rho \in \mathrm{T}_{cr}(G) : \rho \geq \tau\}$.
(2) *If* $A = \sum_{\rho\in\mathrm{T}_{cr}(G)} A_\rho \in \mathrm{Regg}(G)$, *then* $\sum_{\rho\geq\tau} A_\rho \in \mathrm{Regg}(G(\tau))$ *and* $A = \sum_{\rho>\tau} A_\rho \in \mathrm{Regg}(G^\sharp(\tau))$.
(3) *Let* $B \in \mathrm{Regg}(G^\sharp(\tau))$ *and let* $G(\tau) = B_\tau \oplus G^\sharp(\tau)$. *Then* $C = B_\tau \oplus B \in \mathrm{Regg}(G(\tau))$ *and* $C^\sharp(\tau) = B$.

(4) *Let* $B = \sum_{\rho\in \mathrm{T}_{cr}(G),\, \rho\geq\tau} B_\rho \in \mathrm{Regg}(G(\tau))$. *Then*

$$B^\sharp(\tau) = \sum_{\rho\in \mathrm{T}_{cr}(G),\, \rho>\tau} B_\rho \in \mathrm{Regg}(G^\sharp(\tau))$$

and $B = B_\tau \oplus B^\sharp(\tau)$.

(5) *For* $\tau \in \mathrm{T}_{cr}(G)$,

$$b_\tau^G = \exp\frac{G^\sharp(\tau)}{\mathrm{R}(G^\sharp(\tau))} = \exp\frac{G(\tau)}{\mathrm{R}(G(\tau))}.$$

The positive integers b_τ^G *are called the* **Burkhardt invariants** *of the* BR*–group* G.

(6) *For* $\sigma, \tau \in \mathrm{T}_{cr}(G)$ *and* $\sigma \geq \tau$, $b_\sigma^G = b_\sigma^{G(\tau)}$.

PROOF. These facts are well–known for almost completely decomposable groups and finite rank Butler groups, cf. [**2**, 3.1]. The same routine proofs work for BR–groups. □

The next lemma shows that "globally" the situation for BR–groups as well as for finite rank Butler groups is complicated by the fact that regulating subgroups themselves may be ill–behaved. This is again in sharp contrast with the situation for almost completely decomposable groups. See Example 3.15 (8) in this context.

LEMMA 3.8. *Let* G *be a* BR*–group, and let* $A = \sum_{\rho\in\mathrm{T}_{cr}(G)} A_\rho \in \mathrm{Regg}(G)$. *Then* $A(\tau) = \sum_{\rho\geq\tau} A_\rho$ *and* $A^\sharp(\tau) = \sum_{\rho>\tau} A_\rho$ *are true for all* $\tau \in \mathrm{T}_{cr}(G)$ *if and only if* $A^*(\tau) = A^\sharp(\tau)$ *for all* $\tau \in \mathrm{T}_{cr}(G)$.

PROOF. $\Rightarrow$: Let $\tau \in \mathrm{T}_{cr}(G)$. Then $A(\tau) = A_\tau \oplus A^\sharp(\tau) = A_\tau \oplus \sum_{\rho>\tau} A_\rho \subset A_\tau \oplus A^*(\tau)$. Since $A^\sharp(\tau) \supset A^*(\tau)$, it follows that $A^*(\tau) = A^\sharp(\tau)$.

$\Leftarrow$: We prove the desired identities by induction on depth in the critical typeset. If τ is a maximal type, then $\sum_{\rho\geq\tau} A_\rho = A_\tau = A(\tau)$ and $\sum_{\rho>\tau} A_\rho = 0 = A^\sharp(\tau)$. Now suppose that τ is not maximal in $\mathrm{T}_{cr}(G)$ and assume that $A(\sigma) = \sum_{\rho\geq\sigma} A_\rho$ and $A^\sharp(\sigma) = \sum_{\rho>\sigma} A_\rho$ for every critical type $\sigma > \tau$. Then

$$A(\tau) = A_\tau \oplus A^\sharp(\tau) = A_\tau \oplus A^*(\tau) = A_\tau \oplus \sum_{\rho>\tau} A(\rho) = A_\tau \oplus \sum_{\rho>\tau} A_\rho = \sum_{\rho\geq\tau} A_\rho.$$

Further,

$$A(\tau) = A_\tau \oplus \sum_{\rho>\tau} A_\rho = A_\tau \oplus A^\sharp(\tau)$$

and $A^\sharp(\tau) \supset \sum_{\rho>\tau} A_\rho$, hence $A^\sharp(\tau) = \sum_{\rho>\tau} A_\rho$. □

We now consider regulators. A number of containments are true for BR–groups in general and are easily established.

LEMMA 3.9. *Let* G *be a* BR*–group, and let* $A = \sum_{\rho\in\mathrm{T}_{cr}(G)} A_\rho$ *be a regulating subgroup of* G. *Then the following hold:*

(1) *For each* $\tau \in \mathrm{T}_{cr}(G)$,

$$\mathrm{R}(G^\sharp(\tau)) \subset \mathrm{R}(G(\tau)) \subset \mathrm{R}(G)(\tau) \quad \textit{and} \quad \mathrm{R}(G^\sharp(\tau)) \subset \mathrm{R}(G)^\sharp(\tau).$$

(2)

$$\sum_{\rho \in \mathrm{T}_{cr}(G)} b_\rho^G A_\rho \subset \sum_{\rho \in \mathrm{T}_{cr}(G)} b_\rho^G G(\rho) \subset \sum_{\rho \in \mathrm{T}_{cr}(G)} \mathrm{R}(G(\rho)) \subset \mathrm{R}(G).$$

PROOF. (1) For the arbitrary regulating subgroup $A = \sum_{\rho \in \mathrm{T}_{cr}(G)} A_\rho$ of G, we have that $\sum_{\rho \geq \tau} A_\rho$ is a regulating subgroup of $G(\tau)$. Hence

$$\mathrm{R}(G(\tau)) \subset \sum_{\rho \geq \tau} A_\rho \subset A$$

and thus $\mathrm{R}(G(\tau)) \subset \mathrm{R}(G)$. Since the exponent of $G/\mathrm{R}(G)$ is finite, we obtain that $\mathrm{R}(G(\tau)) \subset G(\tau) \cap \mathrm{R}(G) = \mathrm{R}(G)(\tau)$. Similarly, $\mathrm{R}(G^\sharp(\tau)) \subset \mathrm{R}(G)^\sharp(\tau)$.

Let $B \in \mathrm{Regg}(G(\tau))$. Then $B^\sharp(\tau) \in \mathrm{Regg}(G^\sharp(\tau))$ by 3.7 (4), and hence $\mathrm{R}(G^\sharp(\tau)) \subset B^\sharp(\tau) \subset B$, thus $\mathrm{R}(G^\sharp(\tau)) \subset \mathrm{R}(G(\tau))$ since B was arbitrary.

(2) Using 3.7 (5), and (1), we have $b_\tau^G A_\tau \subset b_\tau^G G(\tau) \subset \mathrm{R}(G(\tau)) \subset \mathrm{R}(G)$. □

In this context we mention a relationship between Burkhardt invariants which is true under special assumptions.

LEMMA 3.10. *Let G be a* BR*-group such that for all $A \in \mathrm{Regg}(G)$ and all $\tau \in \mathrm{T}_{cr}(G)$, $A^*(\tau) = A^\sharp(\tau)$. If $\sigma, \tau \in \mathrm{T}_{cr}(G)$ and $\sigma > \tau$, then b_σ^G divides b_τ^G.*

PROOF. Let C be a regulating subgroup of $G(\sigma)$. Then there is a regulating subgroup B of $G(\tau)$ such that $B \supset C$ and there is a regulating subgroup A of G such that $A \supset B$. By hypothesis and Lemma 3.8, $A(\tau) = B$ and $B(\sigma) = C$. Now

$$b_\tau^G G(\sigma) \subset b_\tau^G G(\tau) \cap G(\sigma) \subset B \cap G(\sigma) = B \cap G(\tau)(\sigma) = B(\sigma) = C.$$

Hence $b_\tau^G G(\sigma) \subset \mathrm{R}(G(\sigma))$ and b_σ^G divides b_τ^G as claimed. □

In the case of almost completely decomposable groups of finite rank, the well–known theorem of Burkhardt on regulators says that there is equality everywhere in the chain (2) of Lemma 3.9. It is unknown whether this is true in the general case of BR–groups. However, a local version of Burkhardt's theorem holds. The following lemma is needed in the proof and again later.

LEMMA 3.11. *Let G be a* BR*-group, $G(\tau) = A_\tau \oplus G^\sharp(\tau)$ and $0 \neq a \in A_\tau$. If $a\phi \in \mathrm{R}(G^\sharp(\tau))$ for every $\phi \in \mathrm{Hom}(A_\tau, G^\sharp(\tau))$, then $a \in b_\tau^G A_\tau$.*

PROOF. Since $b_\tau^G = \exp(G^\sharp(\tau)/\mathrm{R}(G^\sharp(\tau)))$, there exists $z \in G^\sharp(\tau)$ such that the order of $z + \mathrm{R}(G^\sharp(\tau))$ is b_τ^G, i.e. if $ez \in \mathrm{R}(G^\sharp(\tau))$, then b_τ^G divides e. Since $\mathrm{tp}(a) \leq \mathrm{tp}(z)$, there is $\tilde{a} \in A_\tau$ such that $a = e\tilde{a}$ and $\chi^G(\tilde{a}) \leq \chi(z)$ where χ is the characteristic. Also $\langle \tilde{a} \rangle_*$ is a direct summand of A_τ. Hence there is a homomorphism $\phi : A_\tau \to G^\sharp(\tau)$ such that $\tilde{a}\phi = z$. Then $ez = e\tilde{a}\phi = a\phi \in \mathrm{R}(G^\sharp(\tau))$, hence b_τ^G divides e and $a \in b_\tau^G A_\tau$ as desired. □

THEOREM 3.12. *Let G be a* BR*–group and let $A = \sum_{\rho \in \mathrm{T}_{cr}(G)} A_\rho$ be a regulating subgroup of G. Then the following hold:*

(1) *For each $\tau \in \mathrm{T}_{cr}(G)$, $\mathrm{R}(G(\tau)) = b^G_\tau A_\tau \oplus \mathrm{R}(G^\sharp(\tau))$.*

(2) *$\mathrm{T}_{cr}(G) = \mathrm{T}_{cr}(R(G))$ and for each critical type τ, $\mathrm{R}(G)(\tau) = b^G_\tau A_\tau \oplus \mathrm{R}(G)^\sharp(\tau)$. Thus $\sum_{\rho \in \mathrm{T}_{cr}(G)} b^G_\rho A_\rho$ is a regulating subgroup of* $\mathrm{R}(G)$.

(3) *For each critical type τ, $\mathrm{R}(G^\sharp(\tau)) = \mathrm{R}(G)^\sharp(\tau)$, and $\mathrm{R}(G(\tau)) = \mathrm{R}(G)(\tau)$.*

PROOF. (1) It has already been shown that $\mathrm{R}(G(\tau)) \supset b^G_\tau A_\tau \oplus \mathrm{R}(G^\sharp(\tau))$ (3.9). To show containment in the other direction, let $x \in \mathrm{R}(G(\tau)) \subset G(\tau) = A_\tau \oplus G^\sharp(\tau)$. Write $x = a + y$ with $a \in A_\tau$ and $y \in G^\sharp(\tau)$. If B is a regulating subgroup of $G^\sharp(\tau)$, then $A_\tau \oplus B$ is a regulating subgroup of $G(\tau)$, so that $x \in A_\tau \oplus B$. Since $B \subset G^\sharp(\tau)$, it follows from the uniqueness of the representation $x = a + y$ that $y \in B$. Since B was an arbitrary regulating subgroup of $G^\sharp(\tau)$, we conclude that $y \in \mathrm{R}(G^\sharp(\tau))$. If $\phi \in \mathrm{Hom}(A_\tau, G^\sharp(\tau))$, then $A(1+\phi)$ is another complement of $G^\sharp(\tau)$ in $G(\tau)$, i.e. $G(\tau) = A_\tau(1+\phi) \oplus G^\sharp(\tau)$. Hence we have

$$x = a + y = a'(1+\phi) + y' = a' + (a'\phi + y')$$

for some $a' \in A_\tau$ and $y' \in \mathrm{R}(G^\sharp(\tau))$ as before. Since $a'\phi + y' \in G^\sharp(\tau)$, it follows that $a = a'$ and $y = a\phi + y'$. Hence for every $\phi \in \mathrm{Hom}(A_\tau, G^\sharp(\tau))$, we have $a\phi = y - y' \in \mathrm{R}(G^\sharp(\tau))$. By 3.11 $a \in b^G_\tau A_\tau$ as claimed.

(2) Since $\mathrm{R}(G)(\tau) \supset b^G_\tau A_\tau$, it follows from the decomposition in (1) that

$$\mathrm{R}(G)(\tau) = b^G_\tau A_\tau \oplus [\mathrm{R}(G)(\tau) \cap \mathrm{R}(G^\sharp(\tau))].$$

Further

$$\mathrm{R}(G)(\tau) \cap \mathrm{R}(G^\sharp(\tau)) \subset \mathrm{R}(G)(\tau) \cap \mathrm{R}(G)^\sharp(\tau) = \mathrm{R}(G)^\sharp(\tau) \subset \mathrm{R}(G)(\tau)$$

and $b^G_\tau A_\tau \cap \mathrm{R}(G)^\sharp(\tau) = b^G_\tau A_\tau \cap \mathrm{R}(G) \cap G^\sharp(\tau) = 0$. From

$$\mathrm{R}(G)(\tau) = b^G_\tau A_\tau \oplus [\mathrm{R}(G)(\tau) \cap \mathrm{R}(G^\sharp(\tau))] \subset b^G_\tau A_\tau \oplus \mathrm{R}(G)^\sharp(\tau) \subset \mathrm{R}(G(\tau))$$

it now follows that $b^G_\tau A_\tau \oplus \mathrm{R}(G)^\sharp(\tau) = \mathrm{R}(G(\tau))$, proving (2).

Also we obtain by the last paragraph that $\mathrm{R}(G^\sharp(\tau)) \supset \mathrm{R}(G)(\tau) \cap \mathrm{R}(G^\sharp(\tau)) = \mathrm{R}(G)^\sharp(\tau) \supset \mathrm{R}(G^\sharp(\tau))$, so $\mathrm{R}(G)^\sharp(\tau) = \mathrm{R}(G^\sharp(\tau))$ as claimed. Finally, $\mathrm{R}(G(\tau)) = b^G_\tau A_\tau \oplus \mathrm{R}(G)^\sharp(\tau) = \mathrm{R}(G)(\tau)$ by (1) and (2). $\square$

Recall that for each type $\tau \in \mathrm{T}(G)$, $e^G_\tau = \exp(G^\sharp(\tau)/G^*(\tau))$. Further, for $\tau \in \mathrm{T}_{cr}(G)$, set $s^G_\tau = \exp(G^\sharp(\tau)/\mathrm{R}_\Sigma(G^\sharp(\tau)) = \exp(G(\tau)/\mathrm{R}_\Sigma(G(\tau))$. Like b^G_τ, the values e^G_τ and s^G_τ are invariants of G. These invariants are related as indicated below.

PROPOSITION 3.13. *Let G be a* BR*–group and let τ be a critical type. Then*

(1) *e^G_τ divides s^G_τ and s^G_τ divides b^G_τ.*

(2) *If $b^G_\tau = 1$ for all critical types $\tau \in \mathrm{T}_{cr}(G)$ and $G(\tau) = A_\tau \oplus G^\sharp(\tau)$, then $\mathrm{R}(G) = \sum_{\rho \in \mathrm{T}_{cr}(G)} A_\rho$ is the unique regulating subgroup of G and $G^*(\tau) = G^\sharp(\tau)$ is pure for each critical type τ.*

PROOF. (1)

$$\mathrm{R}(G^\sharp(\tau)) \subset \mathrm{R}_\Sigma(G^\sharp(\tau)) = \sum_{\rho\in \mathrm{T}_{cr}(G),\rho>\tau} G(\rho) \subset G^*(\tau) \subset G^\sharp(\tau)$$

and the claim follows.

(2) This follows immediately from Lemma 3.9 and (1). □

The following theorem generalizes the theorem of Burkhardt on regulators ([**6**, Lemma 1]) and provides a new direct proof even in the case of almost completely decomposable groups of finite rank.

THEOREM 3.14. *Let G be a* BR*-group, and suppose that $A = \sum_{\rho\in \mathrm{T}_{cr}(G)} A_\rho$ is a regulating subgroup of G such that for every critical type $\mu \in \mathrm{T}_{cr}(G)$ the following identities are true:*

$$A^\sharp(\mu) \cap \sum_{\rho\not>\mu} A_\rho = 0 \quad \textit{and} \quad A_\mu \cap \sum_{\rho\not\geq\mu} A_\rho = 0.$$

Then

$$\mathrm{R}(G) = \sum_{\rho\in \mathrm{T}_{cr}(G)} b^G_\rho A_\rho = \sum_{\rho\in \mathrm{T}_{cr}(G)} b^G_\rho G(\rho) = \sum_{\rho\in \mathrm{T}_{cr}(G)} \mathrm{R}(G(\rho)).$$

PROOF. Because of 3.9 we only need to show that

$$\mathrm{R}(G) \subset \mathrm{Bu}(A) = \sum_{\rho\in \mathrm{T}_{cr}(G)} b^G_\rho A_\rho.$$

Let x be a non-zero element of $\mathrm{R}(G)$, write $x = \sum_{\rho\in \mathrm{T}_{cr}(G)} a_\rho$, $a_\rho \in A_\rho$, and let μ be maximal among the critical types ρ with $a_\rho \neq 0$. We will show that $a_\mu \in b^G_\mu A_\mu$. We then have $x - a_\mu \in \mathrm{R}(G)$ and it follows by induction that $x \in \mathrm{Bu}(A)$ as desired.

Let $B \in \mathrm{Regg}(G^\sharp(\mu))$ and $\phi \in \mathrm{Hom}(A_\mu, G^\sharp(\mu))$. Then

$$\sum_{\rho\not\geq\mu} A_\rho + A_\mu(1+\phi) + B \in \mathrm{Regg}(G)$$

and

$$x = \sum_{\rho\not>\mu} a_\rho = \sum_{\rho\not\geq\mu} a'_\rho + a'_\mu(1+\phi) + b, \quad \text{with} \quad a'_\rho \in A_\rho, b \in B.$$

Now

$$\sum_{\rho\not>\mu} a_\rho - \sum_{\rho\not>\mu} a'_\rho = a'_\mu\phi + b \in A \cap G^\sharp(\mu) = A^\sharp(\mu).$$

Since $A^\sharp(\mu) \cap \sum_{\rho\not>\mu} A_\rho = 0$ by hypothesis, we have $a'_\mu\phi = -b \in B$. Using the second hypothesis on A, we have $a_\mu - a'_\mu = \sum_{\rho\not\geq\mu}(a'_\rho - a_\rho) \in A_\mu \cap \sum_{\rho\not\geq\mu} A_\rho = 0$. So $a'_\mu = a_\mu$, and since B was an arbitrary regulating subgroup of $G^\sharp(\mu)$, we conclude that $a_\mu\phi \in \mathrm{R}(G^\sharp(\mu))$ for every $\phi \in \mathrm{Hom}(A_\mu, G^\sharp(\mu))$. By 3.11 we now have $a_\mu \in b^G_\mu A_\mu$ as needed. □

Many questions arise concerning regulating subgroups and regulators of BR–groups. For example, for a regulating subgroup $A = \sum_{\rho\in \mathrm{T}_{cr}(G)} A_\rho$ let us call $\mathrm{Bu}(A) = \sum_{\rho\in \mathrm{T}_{cr}(G)} b^G_\rho A_\rho$ the **Burkhardt transform** of A. It is not clear whether $\mathrm{Bu}(A) = \mathrm{Bu}(A')$ for two distinct regulating subgroups of G. The following example of Mutzbauer [**11**, 2.3] gives some insight.

EXAMPLE 3.15. *Let p be a prime other than $3, 5, 7, 11, 13, 17$, $A = 3^{-\infty}\mathbb{Z}$, $B = (3\cdot 5)^{-\infty}\mathbb{Z}$, $C = (3\cdot 7)^{-\infty}\mathbb{Z}$, $D = 11^{-\infty}\mathbb{Z}$, $X = 13^{-\infty}\mathbb{Z}$, $Y = 17^{-\infty}\mathbb{Z}$, with types $\tau_A, \tau_B, \ldots$ respectively. In the vector space $\mathbb{Q}a \oplus \mathbb{Q}b \oplus \mathbb{Q}c \oplus \mathbb{Q}d$ let*

$$G = (Aa \oplus Bpb \oplus Cpc \oplus Dpd) + X(a + b + d) + Y(c - d).$$

Then the following can be verified:

(1) $\mathrm{T}(G) = \{\tau_A, \tau_B, \tau_C, \tau_D, \tau_X, \tau_Y, \mathrm{tp}(\mathbb{Z})\}$

(2) $G(\tau_A) = Aa + Bpb + Cpc + A(b + c)$, $G^\sharp(\tau_A) = Bpb + Cpc + A(b + c)$, $G^*(\tau_A) = Bpb + Cpc$.

(3) $G(\tau_B) = Bpb$, $G(\tau_C) = Cpc$, $G(\tau_D) = Dpd$, $G(\tau_X) = X(a + b + d)$, $G(\tau_Y) = Y(c - d)$, $G^\sharp(\tau_B) = G^\sharp(\tau_C) = G^\sharp(\tau_D) = G^\sharp(\tau_X) = G^\sharp(\tau_Y) = 0$

(4) $b^G_{\tau_A} = p$, $b^G_{\tau_B} = b^G_{\tau_C} = b^G_{\tau_D} = b^G_{\tau_X} = b^G_{\tau_Y} = 1$.

(5) *G contains p (not necessarily distinct) regulating subgroups*

$$V_i = A(a+i(b+c))+Bpb+Cpc+Dpd+X(a+b+d)+Y(c-d), \quad 0 \le i < p.$$

(6) $\mathrm{R}(G) = \mathrm{Bu}(V_0) = \mathrm{Bu}(V_1) = \ldots = \mathrm{Bu}(V_{p-1}) = V_1 \subset V_0 = G$, $V_1 \subset V_i$ *for all i. In particular, note that* $\mathrm{Bu}(V_1) = V_1$.

(7) $V_1(\tau_A) = A(a + b + c) + Bpd + Cpc$, $V_1^*(\tau_A) = V_1^\sharp(\tau_A) = Bpd + Cpc$, *and* $\mathrm{R}(\mathrm{R}(G)) = \mathrm{R}(V_1) = V_1 = \mathrm{R}(G)$.

(8) *$G = V_0$ does not satisfy the hypotheses of Lemma 3.8 while V_1 does.*

A result of Butler [**7**, Theorem 7] on finite rank Butler groups generalizes to BR-groups.

DEFINITION 3.16. *A partially ordered set $(T, \le)$ is said to have property $(*)$ if each $\tau \in T$ has either at most one cover in T or there are exactly two covers τ_1 and τ_2 and there is no $\sigma \in T$ with $\sigma \ge \tau_1, \tau_2$.*

THEOREM 3.17. *If G is a BR–group whose typeset T has property $(*)$, then every regulating subgroup of G is completely decomposable.*

PROOF. Let $A = \sum_{\rho\in \mathrm{T}_{cr}(G)} A_\rho$ be a regulating subgroup of G. We use induction on $|T|$. If $|T| = 1$ then G is homogeneous completely decomposable and $A = G$. Suppose that $|T| > 1$ and that the claim holds for all BR–groups whose ty property $(*)$ and contains less than $|T|$ elements. Let $\mu = \inf T$, the inner type of G, which belongs to T. Since $|T| > 1$, μ has at least one cover $\tau \in T$.

CASE 1. τ is the only cover of μ. Then $G^*(\mu) = G(\tau) = G^\sharp(\mu)$. Hence $G = G(\mu) = A_\mu \oplus G^\sharp(\mu) = A_\mu \oplus G(\tau)$. The typeset of $G(\tau)$ satisfies the conditions and of the induction hypothesis and contains fewer elements that $\mathrm{T}(G)$. Also,

$\sum_{\rho\geq\tau} A_\rho$ is a regulating subgroup of $G(\tau)$. By induction hypothesis $\sum_{\rho\geq\tau} A_\rho$ is completely decomposable and so is $A = A_\mu \oplus \sum_{\rho\geq\tau} A_\rho$.

CASE 2. μ has exactly two covers τ_1 and τ_2. Then $G^*(\mu) = G(\tau_1) + G(\tau_2) = G(\tau_1) \oplus G(\tau_2)$, the sum being direct since otherwise $G(\tau_1) \cap G(\tau_2)$ contains an element of type $\geq \tau_1, \tau_2$. By induction hypothesis, the regulating subgroups $\sum_{\rho\geq\tau_1} A_\rho$ of $G(\tau_1)$ and $\sum_{\rho\geq\tau_2} A_\rho$ of $G(\tau_2)$ are completely decomposable and hence so is $A = A_\mu \oplus \sum_{\rho\geq\tau_1} A_\rho \oplus \sum_{\rho\geq\tau_2} A_\rho$. □

We end this section by recording the following property of BR–groups which follows by simple induction on the typeset.

THEOREM 3.18. *A* BR–*group with a linearly ordered typeset is completely decomposable.*

4. BR–groups with completely decomposable regulating subgroups

In this section we consider BR–groups with a completely decomposable regulating subgroup. For the sake of brevity we call such groups **BRCD–groups** (where "CD" stands for completely decomposable.) The class of BRCD–groups coincides with the class of torsion–free groups containing a completely decomposable group with finite typeset and bounded quotient. It is crucial that every regulating subgroup of a BRCD–group is completely decomposable. Note that, by Albrecht–Hill [**1**, 4.4], BRCD–groups are always B_2–groups. The infinite rank analogue of Burkhardt's theorem on regulators [**6**, Lemma 1] holds for these groups.

THEOREM 4.1. *Let X be a torsion–free group containing a completely decomposable subgroup A such that $\mathrm{T}_{cr}(A)$ is finite and X/A is bounded. Then X is a* BR–*group and every regulating subgroup of X is completely decomposable.*

PROOF. By Lemma 2.1, X and A have the same typesets and the same critical typesets. Hence condition (1) of the definition of BR–group is satisfied by A. Let $e = \exp(X/A)$. Then

$$eX^\sharp(\tau) \subset A \cap X^\sharp(\tau) = A^\sharp(\tau) = A^*(\tau) \subset X^*(\tau),$$

so the second condition is satisfied also. By the Dedekind identity we have the short exact sequence

$$0 \to A_\tau \cong \frac{A(\tau)}{A^\sharp(\tau)} \to \frac{X(\tau)}{X^\sharp(\tau)} \to \frac{A + X(\tau)}{A + X^\sharp(\tau)} \to 0$$

and $e(A + X(\tau)) \subset A + X^\sharp(\tau)$, so that $X(\tau)/X^\sharp(\tau)$ is τ–homogeneous. Now

$$\frac{X(\tau)}{X^\sharp(\tau)} \cong e\frac{X(\tau)}{X^\sharp(\tau)} \subset \frac{A(\tau) + X^\sharp(\tau)}{X^\sharp(\tau)} \cong \frac{A(\tau)}{A^\sharp(\tau)} \cong A_\tau.$$

By [**8**, 86.6], $X(\tau)/X^\sharp(\tau)$ is completely decomposable. By the Baer Lemma, $X(\tau) = B_\tau \oplus X^\sharp(\tau)$ for some τ–homogeneous completely decomposable group

B_τ. To establish (3), we need to show that $\sum_{\rho\in \mathrm{T}_{cr}(X)} B_\rho$ is in fact a direct sum $\bigoplus_{\rho\in \mathrm{T}_{cr}(X)} B_\rho$. We have

$$B_\tau \cap \bigoplus_{\rho\neq\tau} A_\rho \subset X(\tau) \cap \bigoplus_{\rho\neq\tau} A_\rho = A^\sharp(\tau).$$

Hence

$$B_\tau \cap \bigoplus_{\rho\neq\tau} A_\rho \subset B_\tau \cap A^\sharp(\tau) \subset B_\tau \cap X^\sharp(\tau) = 0.$$

Let $A^\tau = B_\tau \oplus \left(\bigoplus_{\rho\neq\tau} A_\rho\right)$. Recall that $e = \exp(X/A)$. We claim that $e^2X \subset A^\tau$. In fact, $eA_\tau \subset eX(\tau) = e(B_\tau \oplus X^\sharp(\tau)) \subset B_\tau \oplus A^\sharp(\tau) = B_\tau \oplus A^*(\tau) \subset A^\tau$. Now $e^2X \subset eA \subset eA_\tau + \sum_{\rho\neq\tau} A_\rho \subset A^\tau$. Thus we are again in the initial situation and can replace one by one the A_ρ by B_ρ while keeping the sums direct, and obtain $B = \bigoplus_{\rho\in \mathrm{T}_{cr}(X)} B_\rho$ after finitely many steps. □

PROPOSITION 4.2. *Let X be a* BR*–group containing a completely decomposable group with bounded quotient. Then X does not possess distinct nested regulating subgroups.*

PROOF. Suppose that B_1, B_2 are regulating subgroups of X and that $B_1 \subset B_2$. Then B_1 is a regulating subgroup of B_2 and by 2.3, $B_1 = B_2$. □

The following generalization of Burkhardt's theorem [**6**, Lemma 1] is an immediate consequence of 3.14.

THEOREM 4.3. *Let G be a* BRCD*–group, and let $A = \sum_{\rho\in \mathrm{T}_{cr}(G)} A_\rho$ be any regulating subgroup of G. Then*

$$\mathrm{R}(G) = \sum_{\rho\in \mathrm{T}_{cr}(G)} b_\rho^G A_\rho = \sum_{\rho\in \mathrm{T}_{cr}(G)} b_\rho^G G(\rho) = \sum_{\rho\in \mathrm{T}_{cr}(G)} \mathrm{R}(G(\rho)).$$

5. BR–Groups and infinite rank Butler groups

In this section we explore the connection between BR–groups and B_2–groups. Three examples are constructed to indicate that the class of BR–groups is distinct from the known classes of infinite rank Butler groups. Specifically we show that a B_2–group with a finite typeset need not be a BR–group and, conversely, a BR–group need not be a B_2–group, not even a finitely Butler group. It is well–known that countable B_2–groups are closed under pure subgroups. But a countable BR–group may be a B_2–group and still may contain a pure subgroup which is not a BR–group. Our examples also indicate how some of the properties of finite rank Butler groups do not carry over to the infinite rank case. For example, a B_2–group with a linearly ordered typeset need not be completely decomposable, and a BR–group, when localized at a prime, need not be completely decomposable.

EXAMPLE 5.1. *A B_2–group with a finite typeset need not be a* BR*–group.*

PROOF. Let S be a rational group containing 1 with $\chi^S(1) = (1,1,\dots)$. Let τ be the type of S. Let $p_1, p_2, \dots$ be an indexing of all primes. Let $B = \bigoplus_{i\in\mathbb{N}} Sx_i \subset \bigoplus_{i\in\mathbb{N}} \mathbb{Q}x_i$ and

$$A = \langle \mathbb{Z}y \oplus B, p_i^{-2}(x_i + p_i y) : i \in \mathbb{N} \rangle \subset \mathbb{Q}y \oplus \bigoplus_{i\in\mathbb{N}} \mathbb{Q}x_i.$$

Then $\text{tp}(y) = \text{tp}(\mathbb{Z})$ and A has the typeset $\{\text{tp}(\mathbb{Z}, \tau\}$. Since $A^\sharp(\text{tp}(\mathbb{Z})) = B$ and $A = A(\text{tp}(\mathbb{Z}))$ we get $A(\text{tp}(\mathbb{Z}))/A^\sharp(\text{tp}(\mathbb{Z})) = A/B \cong S$ which is not of type $\text{tp}(\mathbb{Z})$. So A is not a BR–group. But A is a countable B_2–group since it is the union of the increasing chain of pure subgroups

$$\langle y \rangle_* \subset \langle y, x_1 \rangle_* \subset \langle y, x_1, x_2 \rangle_* \subset \cdots$$

where for each n, $\langle y, x_1, \dots, x_n \rangle_* = \mathbb{Z}y + \sum_{i=1}^n Sx_i + \sum_{i=1}^n \mathbb{Z}p_i^{-2}(x_i + p_i y)$ is a finite rank Butler group. $\square$

REMARK. Since a completely decomposable group always has a Butler decomposition, the group A of Example 5.1 is not completely decomposable. This shows that a B_2–group with a linearly ordered typeset need not be completely decomposable.

EXAMPLE 5.2. *A* BR*–group need not be a* B_2*–group, not even be finitely Butler.*

Let

$$V = \bigoplus_{i=1}^{\infty} \mathbb{Q}a_i \oplus \bigoplus_{i=1}^{\infty} \mathbb{Q}b_i.$$

Let p, q, r, s be different primes, and let

$$A = p^{-\infty}\mathbb{Z}, B = q^{-\infty}\mathbb{Z}, C = r^{-\infty}\mathbb{Z},$$

with corresponding types $\tau_A = \text{tp}(A)$, $\tau_B = \text{tp}(B)$, $\tau_C = \text{tp}(C)$. In addition let $\tau_\mathbb{Z}$ denote the type of $\mathbb{Z}$. Next, let π and ρ be irrational s–adic units such that π/ρ is irrational. Write

$$\pi = \sum_{i=0}^{\infty} \pi_i s^i, \quad 0 \le \pi_i < s, \quad \pi_0 \ne 0,$$

and similarly

$$\rho = \sum_{i=0}^{\infty} \rho_i s^i, \quad 0 \le \rho_i < s, \quad \rho_0 \ne 0.$$

For brevity we set

$$\pi^{(j)} = \sum_{i=0}^{j-1} \pi_i s^i, \quad \textit{and} \quad \rho^{(j)} = \sum_{i=0}^{j-1} \rho_i s^i.$$

Let

$$c_i = s^{-1}(\rho^{(i)} a_1 + \pi^{(i)} b_1) + a_{i+1} + b_{i+1}$$

and

$$F_a = \bigoplus_{i=1}^{\infty} Aa_i, \quad F_b = \bigoplus_{i=1}^{\infty} Bb_i, \quad F_c = \bigoplus_{i=1}^{\infty} Cc_i.$$

The example is

$$X = F_a + F_b + F_c \subset V.$$

The group

$$P = \left\langle p^{-\infty}\mathbb{Z}a_1, q^{-\infty}\mathbb{Z}b_1, s^{-j}(\rho^{(j)}a_1 + \pi^{(j)}b_1) : j \in \mathbb{N} \right\rangle$$

will be shown to be a pure subgroup of X which is not a Butler group. More precisely, we claim that

(1) F_a, F_b, F_c are all pure in X.
(2) X is a BR–group with typeset $\mathrm{T}(X) = \{\tau_A, \tau_B, \tau_C, \tau_{\mathbb{Z}}\}$. Moreover, $X(\tau_A) = F_a$, $X(\tau_B) = F_b$, $X(\tau_C) = F_c$.
(3) P is a rank–2 pure subgroup of X, P is not a Butler group, and hence X is not a B_2–group.

PROOF. (0.1) We will use repeatedly the following representation of an arbitrary element of X in terms of the $\mathbb{Q}$–basis $\{a_1, a_2, \dots, b_1, b_2, \dots\}$ of V.

$$\begin{aligned} x &= \sum_{i=1}^{\infty} \alpha_i a_i + \sum_{i=1}^{\infty} \beta_i b_i + \sum_{i=1}^{\infty} \gamma_i c_i \\ &= (\alpha_1 + \sum_{i\geq 1} \gamma_i s^{-i}\rho^{(i)})a_1 + \sum_{i\geq 2} (\alpha_i + \gamma_{i-1})a_i \\ &\qquad + (\beta_1 + \sum_{i\geq 1} \gamma_i s^{-i}\pi^{(i)})b_1 + \sum_{i\geq 2} (\beta_i + \gamma_{i-1})b_i. \end{aligned}$$

Hereby $\alpha_i \in p^{-\infty}\mathbb{Z}$, $\beta_i \in q^{-\infty}\mathbb{Z}$, $\gamma_i \in r^{-\infty}\mathbb{Z}$.

(0.2) The representation of arbitrary element of F_c in terms of the $\mathbb{Q}$–basis $\{a_1, a_2, \dots, b_1, b_2, \dots\}$ reads as follows.

$$\sum_{i=1}^{\infty} \tilde{\gamma}_i c_i = \left(\sum_{i\geq 1} \tilde{\gamma}_i s^{-i}\rho^{(i)}\right) a_1 + \sum_{i\geq 2} \tilde{\gamma}_{i-1} a_i + \left(\sum_{i\geq 1} \tilde{\gamma}_i s^{-i}\pi^{(i)}\right) b_1 + \sum_{i\geq 2} \tilde{\gamma}_{i-1} b_i,$$

where $\tilde{\gamma}_i \in r^{-\infty}\mathbb{Z}$.

(0.3) The following formula is an immediate consequence of the definition of the elements c_i.

$$sc_i = c_{i-1} + \rho_{i-1}a_1 + \pi_{i-1}b_1 + (sa_{i+1} - a_i) + (sb_{i+1} - b_i).$$

(0.4) It is readily seen that the quotient group $X/(F_a \oplus F_b)$ is the direct sum of a divisible s–primary group and a divisible r–primary group.

(1.1) We now show that F_a is pure in X. Let $x \in X$ and suppose that $nx \in F_a$. Using the representation of (0.1) it follows that

(1) $\beta_i = -\gamma_{i-1} \in q^{-\infty}\mathbb{Z} \cap r^{-\infty}\mathbb{Z} = \mathbb{Z}$ for $i \geq 2$,

(2) $\beta_1 = -\sum_{i=1}^{k} \gamma_i s^{-i}\pi^{(i)} \in q^{-\infty}\mathbb{Z} \cap r^{-\infty}\mathbb{Z} = \mathbb{Z}$,

(3) $n(\alpha_1 + \sum_{i=1}^{k} \gamma_i s^{-i}\rho^{(i)}) \in p^{-\infty}\mathbb{Z}$.

Hence

$$x = (\alpha_1 + \sum_{i=1}^{k} \gamma_i s^{-i}\rho^{(i)})a_1 + \sum_{i\geq 2}(\alpha_i + \gamma_{i-1})a_i$$

whereby $\alpha_i + \gamma_{i-1} \in p^{-\infty}\mathbb{Z} + \mathbb{Z} = p^{-\infty}\mathbb{Z}$. Thus $x \in F_a$ (and F_a is pure) if and only if it can be shown that $\sum_{i=1}^{k} \gamma_i s^{-i}\rho^{(i)}) \in p^{-\infty}\mathbb{Z} \cap r^{-\infty}\mathbb{Z} = \mathbb{Z}$.

Since π is an s–adic unit, there is an s–adic integer σ such that $\pi\sigma = \rho$. Hence there exist integers m_i such that $\pi^{(i)}\sigma^{(k)} = \rho^{(i)} + m_i s^i$ for $i \leq k$. We now obtain

$$\sum_{i=1}^{k} \gamma_i s^{-i}\rho^{(i)} = \sum_{i=1}^{k} \gamma_i s^{-i}(\pi^{(i)}\sigma^{(k)} - m_i s^i) = \sigma^{(k)} \sum_{i=1}^{k} \gamma_i s^{-i}\pi^{(i)} - \sum_{i=1}^{k} \gamma_i m_i \in \mathbb{Z}.$$

This proves purity of F_a.

By symmetry, F_b is also pure in X. The purity of F_c is established by a similar procedure of comparing coefficients using the representations (0.1) and (0.2).

By straightforward calculations the typeset of X is shown to be $\mathrm{T}(X) = \{\tau_A, \tau_B, \tau_C, \tau_{\mathbb{Z}}\}$. Since the types τ_A, τ_B, τ_C are maximal, $X^\sharp(\tau_A) = X^\sharp(\tau_B) = X^\sharp(\tau_C) = 0$. Moreover, $X(\tau_A)$, $X(\tau_B)$, $X(\tau_C)$ are homogeneous completely decomposable being respectively equal to F_a, F_b, F_c. Further $X = X^\sharp(\tau_{\mathbb{Z}})$. Thus X is a BR–group.

Next we show that X is not finitely Butler and hence not a B_2–group. The group P is the promised non–Butler pure subgroup. Recall that $P = \langle P_1, y_i : i \in \mathbb{N}\rangle$ where $P_1 = p^{-\infty}\mathbb{Z}a_1 \oplus q^{-\infty}\mathbb{Z}b_1$ and $y_i = s^{-i}(\rho^{(i)}a_1 + \pi^{(i)}b_1)$. By simple calculations we find

(i): $sy_{i+1} = y_i + \rho_i a_1 + \pi_i b_1$,

(ii): $c_i = y_i + a_{i+1} + b_{i+1}$.

Then the purity of P is established by observing that P *is pure in* X *if and only if* P *is* s*–pure in* X *and* P_1 *is* t*–pure in* X *for all primes* t *different from* s. The t–purity of P_1 in X for all primes t other than s is a matter of simple calculation of heights.

The group P, a so–called Pontryagin group, is somewhat similar to Example 5 of [**8**, Section 88]. It satisfies $P^*(\tau_{\mathbb{Z}}) = P_1 = p^{-\infty}\mathbb{Z}a_1 \oplus q^{-\infty}\mathbb{Z}b_1$, and $P^\sharp(\tau_{\mathbb{Z}}) = P$. Now P is not Butler, since P/P_1 is not finite, being a non–zero divisible s–group. □

REMARK.

(1) The Example 5.2 shows that a pure subgroup of a BR–group need not be again a BR–group. The group X of Example 5.2 is a "B_0–group" in the sense that $X^\sharp(\tau) = X^*(\tau)$ for all types τ.

(2) The group X also serves as an example of a BR–group which does not become completely decomposable when localized at a prime, namely at the prime s. This is because P_s is a pure subgroup of X_s and P_s is indecomposable, as the irrational ratio π/ρ of irrational s-adic units ρ and π in the definition of P remains irrational under localization at s.

Bican–Salce [**5**, 3.4] proved that the countable rank pure subgroups of completely decomposable groups are always B_2–groups. Since the finite rank pure subgroups of completely decomposable groups are Butler, they are clearly BR–groups. But the same thing does not hold for countable rank pure subgroups, as the next example shows.

EXAMPLE 5.3. *A pure subgroup of a countable completely decomposable group with a finite typeset need not be a* BR*–group.*

PROOF. We utilize an example of Shiflet [**12**] who used it for different purposes.

(a) Let $2 < p_1 < p_2 < \cdots$ be a listing of all prime numbers in increasing order. Let

$$G = \left\langle \mathbb{Z}a_0 \oplus \bigoplus_{i=1}^{\infty} 2^{-\infty}\mathbb{Z}a_i, \frac{a_0 + a_i}{p_i} : i = 1, 2, \ldots \right\rangle \subset \bigoplus_{i=0}^{\infty} \mathbb{Q}a_i.$$

Properties of this groups are the following:

(1) $\mathrm{T}(G) = \{\mathrm{tp}(2^{-\infty}\mathbb{Z}), \mathrm{tp}(\mathbb{Z})\}$.

(2) G is not a BR–group since there is no decomposition $G(\mathrm{tp}(\mathbb{Z})) = A \oplus G^{\sharp}(\mathrm{tp}(\mathbb{Z}))$ with A homogeneous completely decomposable (hence free).

Property (2) follows if one notes that $B = G^{\sharp}(\mathrm{tp}(\mathbb{Z})) = \langle a_i : i \geq 1\rangle$ and that the rank–1 group G/B cannot be cyclic, since a_0 is divisible modulo B by p_i, for all i.

We will show next that G can be imbedded as a pure subgroup in a countable completely decomposable group.

(b) Let S be the subgroup of the additive group of rational numbers $\mathbb{Q}$ which is generated by all fractions $1/p_i$, $i = 1, 2, \ldots$. We consider the following completely decomposable group

$$C = Sb \oplus \bigoplus_{i=0}^{\infty} 2^{-\infty}\mathbb{Z}b_i \subset \mathbb{Q}b \oplus \bigoplus_{i=0}^{\infty} \mathbb{Q}b_i.$$

The linear transformation

$$\phi : \bigoplus_{i=0}^{\infty} \mathbb{Q}a_i \longrightarrow \mathbb{Q}b \oplus \bigoplus_{i=0}^{\infty} \mathbb{Q}b_i \quad \text{given by} \quad a_0 \mapsto b - b_0, a_i \mapsto b_0 + p_i b_i, i \geq 1,$$

obviously is an embedding. We have

$$G\phi = \left\langle \mathbb{Z}(b-b_0) \oplus \bigoplus_{i=1}^{\infty} 2^{-\infty}\mathbb{Z}(b_0+p_ib_i), \frac{1}{p_i}(b+p_ib_i) \right\rangle$$

and

$$2^{\omega}(G\phi) = \bigoplus_{i=1}^{\infty} 2^{-\infty}\mathbb{Z}(b_0+p_ib_i).$$

(c) We claim that $G\phi$ is pure in C. One needs only to show that $2^{\omega}(G\phi)$ is pure in $2^{\omega}C$ and $G\phi/2^{\omega}(G\phi)$ is pure in $C/2^{\omega}G$. We leave it to the reader to do the necessary calculations. □

REMARK. Note that the group G of Example 5.3 is a B_2–group, being isomorphic to a pure subgroup of a completely decomposable group. Thus it is another example of a B_2–group which is not a BR–group.

6. Open questions

We summarize the questions which have been casually mentioned along the way.

(1) Let G be a BR–group. Call a critical type τ "truly critical" if for every regulating subgroup $A = \sum_{\rho\in \mathrm{T}_{cr}(G)} A_\rho$, the quotient $G/\sum_{\rho\neq\tau} A_\rho$ is unbounded. Call a critical type "superfluous" if $G/\sum_{\rho\neq\sigma} A_\rho$ is bounded for every A. Investigate "truly critical" and "superfluous types". Develop a theory of regulating subgroups and regulators based on types which are not superfluous rather than all critical types.

(2) Which of the containments in 3.9 (2) are in fact equalities? For which classes of groups?

(3) Establish precise relationships between the invariants e_τ^G, s_τ^G, b_τ^G.

(4) Prove or disprove that the Burkhardt transforms of distinct regulating subgroups are equal.

(5) If the formation of regulators is iterated, does the chain of iterated regulators become constant?

REFERENCES

1. U. Albrecht and P. Hill, *Butler groups of infinite rank and Axiom* 3, Czech. Math. J. **37** (1987), 293 – 309.
2. D. M. Arnold, *Pure subgroups of finite rank completely decomposable groups*, in Lect. Notes Math. **874**, Springer Verlag (1981), 1 – 31.
3. D. Arnold, Finite Rank Torsion Free Abelian Groups and Rings, Lect. Notes Math. **931**, Springer Verlag (1982).
4. L. Bican, *Splitting in abelian groups*, Czech. Math. J. **28**(103) (1978), 356 – 364.
5. L. Bican and L. Salce, *Butler groups of infinite rank*, in Abelian Group Theory (Honolulu), Lect. Notes in Math. **1006**, Springer Verlag (1983), 171 – 189.
6. R. Burkhardt, *On a special class of almost completely decomposable torsion free abelian groups* I, in Abelian Groups and Modules, C. I. S. M. **287**, Springer Verlag (1984), 141 – 150.
7. M. C. R. Butler, *A class of torsion-free abelian groups of finite rank*, Proc. Lond. Math. Soc. (3) **15** (1965), 680 – 698.

8. L. Fuchs, *Infinite Abelian Groups* I,II, Academic Press (1970, 1973).
9. K.-J. Krapf and O. Mutzbauer, *Classification of almost completely decomposable groups*, in Abelian Groups and Modules, C. I. S. M. **287**, Springer Verlag (1984), 151 – 161.
10. E. L. Lady, *Almost completely decomposable torsion free abelian groups*, Proc. Amer. Math. Soc. **45** (1974), 41 – 47.
11. O. Mutzbauer, *Regulating subgroups of Butler groups*, Abelian Groups: Proc. 1991 Curaçao Conf., Lect. Notes Pure Appl. Math., **146**, Marcel Dekker (1993), 209 – 217.
12. A. Shiflet, *Almost completely decomposable groups*, Dissertation, Vanderbilt University, 1976.

ADOLF MADER, 2565 THE MALL, DEPARTMENT OF MATHEMATICS, UNIVERSITY OF HAWAII AT MANOA, HONOLULU, HAWAII 96822, U. S. A.
E-mail address: adolf@math.hawaii.edu

OTTO MUTZBAUER, MATHEMATISCHES INSTITUT, UNIVERSITÄT WÜRZBURG, AM HUBLAND, 97074 WÜRZBURG, GERMANY
E-mail address: mutzbauer@vax.rz.uni-wuerzburg.d400.de

K. M. RANGASWAMY, DEPARTMENT OF MATHEMATICS, UNIVERSITY OF COLORADO, COLORADO SPRINGS, CO 80933-1750, U. S. A.
E-mail address: kmranga@colospgs.bitnet

Contemporary Mathematics
Volume **171**, 1994

Endomorphisms Over Incomplete Discrete Valuation Domains

WARREN MAY

Dedicated to the memory of Dick Pierce, colleague and friend

ABSTRACT. Let R be a discrete valuation domain with completion $\hat{R}$. The notion of $\hat{R}$-hull of a reduced R-module is defined. For a certain class of mixed R-modules, it is shown that isomorphisms of endomorphism algebras are induced by isomorphisms of $\hat{R}$-hulls.

1. Introduction

We wish to consider left R-modules where R denotes a discrete valuation domain with prime element p and completion $\hat{R}$. All modules will be understood to be R-modules unless otherwise specified. If R is incomplete, we know that it is hopeless to try to classify all torsion-free modules F with endomorphism algebras isomorphic to R, say, since developments based on work of Shelah show that there are arbitrarily large such F (see [**1**]). In the complete case, however, the situation for torsion-free modules is vastly different. Precisely, if F is a torsion-free module over $\hat{R}$, and if $\mathrm{End}_{\hat{R}}F \cong \mathrm{End}_{\hat{R}}F'$, then F' is isomorphic either to F, or possibly to $\oplus_I K (K = R(p^\infty))$, if F is a complete module ([**5,7,8,13**]).

Similiar questions for mixed modules have been largely unexplored. However, recent techniques used in [**9**] for endomorphism algebras of certain classes of mixed modules over complete discrete valuation domains lead to conclusions as decisive as those in the torsion-free case. In this paper, we wish to show that these techniques allow one to exhibit a connection between $\hat{R}$-modules and endomorphisms of certain mixed modules over R. In brief, with proper assumptions on M, any R-algebra isomorphism $\mathrm{End}_R M \cong \mathrm{End}_R N$ must be induced by an isomorphism of $\hat{R}$-modules closely related to M and N. This contrasts strongly with the situation for torsion-free modules if R is incomplete.

1991 *Mathematics Subject Classification.* Primary 20K30, 20K21, 16S50. Secondary 13F30.
This paper is in final form, and no version of it will be submitted for publication elsewhere.

To describe the results, we need a definition. Let M be a reduced module. Then M is naturally embedded in the cotorsion hull $M^\bullet = \operatorname{Ext}^1_R(K, M)$, which is an $\hat{R}$-module. We define the *$\hat{R}$-hull* of M, denoted $\hat{R}M$, to be the $\hat{R}$-submodule of $M^\bullet$ which is generated by M. Then $\hat{R}M$ is a reduced $\hat{R}$-module containing M as a generating subset, and the assignment of $\hat{R}M$ to M is a functor from the category of reduced R-modules to the category of reduced $\hat{R}$-modules. In [**10**], it is shown that under appropriate hypotheses on a reduced mixed module M, there exist many nonisomorphic modules with endomorphism algebras isomorphic to that of M. One observes from the construction of these modules that they are all contained in the $\hat{R}$-hull $\hat{R}M$. We want to show that this is not accidental. Namely, for certain mixed modules M, the modules N with $\operatorname{End}_R M \cong \operatorname{End}_R N$ can be regarded as lying in $\hat{R}M$. In fact, we shall show more; although M and N may not be isomorphic, the isomorphism of their endomorphism algebras is induced by an isomorphism of their $\hat{R}$-hulls. Note that there is a natural embedding of $\operatorname{End}_R M$ in $\operatorname{End}_{\hat{R}} \hat{R}M$.

THEOREM. *Let M be a reduced mixed R-module which is divisible modulo torsion, and let λ denote the length of the torsion submodule. Assume that $p^\lambda \hat{R}M$ has finite rank over $\hat{R}$, and that every full $\hat{R}$-submodule of $\hat{R}M$ contains a nice $\hat{R}$-submodule B with $\hat{R}M/B$ totally projective. Then every R-algebra isomorphism $\operatorname{End}_R M \cong \operatorname{End}_R N$ is induced by an isomorphism $\hat{R}M \cong \hat{R}N$.*

COROLLARY 1. *Let M be as above. Then every automorphism of $\operatorname{End}_R M$ is induced by an inner automorphism of $\operatorname{End}_{\hat{R}} \hat{R}M$.*

In the corollaries to follow, we assume that M is a reduced mixed R-module which is divisible modulo torsion of length λ. The first two give special cases of modules which meet the hypotheses of the theorem.

COROLLARY 2. *Assume that M has finite torsion-free rank and totally projective torsion submodule. Then every isomorphism $\operatorname{End}_R M \cong \operatorname{End}_R N$ is induced by an isomorphism $\hat{R}M \cong \hat{R}N$.*

COROLLARY 3. *Assume that M is a Warfield module and that $p^\lambda M$ has finite rank. Then every isomorphism $\operatorname{End}_R M \cong \operatorname{End}_R N$ is induced by an isomorphism $\hat{R}M \cong \hat{R}N$. If N is also a Warfield module, then $M \cong N$.*

Finally, we will show that the isomorphism theorem proved in [**11**] for rank one Warfield modules can be retrieved as

COROLLARY 4. *Assume that M is Warfield module of torsion-free rank one. If N is a module of torsion-free rank one, then every isomorphism $\operatorname{End}_R M \cong \operatorname{End}_R N$ is induced by an isomorphism $M \cong N$.*

2. Preliminaries

We shall use [**3**] as a reference for definitions and well-known properties. In particular, if L is a module and σ is an ordinal, see §37 for definitions of $p^\sigma L$ and L^σ. The length of L, denoted $\ell(L)$, is the smallest σ such that $p^{\sigma+1} L = p^\sigma L$.

First, we collect some straightforward facts about the $\hat{R}$-hull $\hat{R}M$ of a reduced module M.

LEMMA 1. *Let M and $M_i (i \in I)$, be reduced R-modules.*

(1) *$\hat{R}M$ is isomorphic to $\hat{R} \otimes_R M$ modulo its maximal divisible submodule.*
(2) *$\hat{R}M$ is characterized as a reduced $\hat{R}$-module containing M as an R-module, generated by it as an $\hat{R}$-module, and such that the quotient is torsion-free.*
(3) *The torsion-free rank of $\hat{R}M$ as an $\hat{R}$-module does not exceed that of M as an R-module.*
(4) *The torsion submodule of $\hat{R}M$ is that of M.*
(5) *M is an isotype submodule of $\hat{R}M$.*
(6) *$\hat{R}(\oplus_I M_i) \cong \oplus_I \hat{R}M_i$.*
(7) *For σ an ordinal, $p^\sigma \hat{R}M \supseteq \hat{R}p^\sigma M$. Equality holds if σ is finite, or if M has torsion-free rank one, or if M is a Warfield module.*

PROOF. (1) and (2) follow from the exact sequence $0 \to R \otimes M \to \hat{R} \otimes M \to (\hat{R}/R) \otimes M \to 0$ and the fact that a reduced module containing M such that the quotient is torsion-free and divisible may be embedded in the cotorsion-hull $M^\bullet$. (3) is obvious, and (4) and (5) follow from (2). The fact that cotorsion hull commutes with finite direct sums implies (6). The first part of (7) is an easy induction. If M has torsion-free rank one and $y \in \hat{R}M$ is torsion-free, then $y = ux$ for some unit $u \in \hat{R}$ and $x \in M$. The statement for rank one follows from this, and the claim for Warfield modules follows from (6) after adding an appropriate torsion module to M such that the result is a direct sum of modules of torsion-free rank one. ∎

We will need an additional concept for a reduced module M. Regarding M as embedded in $M^\bullet$, there is a unique maximal ($\mathrm{End}_{\hat{R}} M^\bullet$)-module contained in M; it is the sum of all such modules contained in M. We denote this submodule by $C(M)$. Since $M^\bullet/M$ is torsion-free, it is not hard to see that $M/C(M)$ is torsion-free. Suppose that $x \in M^\bullet$, $rx \in C(M)$ for some $r \in R$, and $\alpha \in \mathrm{End}_{\hat{R}} M^\bullet$. Then $r\alpha(x) = \alpha(rx) \in C(M)$ implies $\alpha(x) \in M$, hence $x \in C(M)$. We conclude that $C(M)$ contains the torsion of M and is isotype in M.

Finally, we state without proof some results from [**9**] which we shall need.

PROPOSITION. *Let M be a reduced module with torsion submodule T such that M/T is divisible. If $\Phi : End_R M \to End_R N$ is an R-algebra isomorphism, then we may regard M and N as embedded in $T^\bullet$ as pure submodules containing T. Moreover, when the endomorphisms of M and N are extended to $T^\bullet$, we may regard Φ as the identity map. We have the further identification, $Hom_R(M, C(M)) = Hom_R(N, C(N))$.*

LEMMA 2. *Let T be an unbounded torsion module such that $T \subseteq \tilde{T}$, where $\tilde{T}$ is a totally projective module of length $< \ell(T) + \omega$. Let σ be the smallest ordinal such that T^σ is bounded. If $\tau < \sigma$, then the torsion-free rank of $(T^\bullet)^\tau$ is at least*

$2^{\aleph_0}$. *The torsion-free rank of* $(T^\bullet)^\sigma$ *is at least* $2^{\aleph_0}$, *except in the case that* σ *is a limit ordinal of cofinality greater than* ω, *in which event* $(T^\bullet)^\sigma$ *is torsion.*

3. Proof of the Theorem

The Theorem will be proved after a sequence of rather technical lemmas. In the first several of these, completeness does not play a role. To fix notation, in this section T will always denote a reduced torsion module of length λ. If L is a module and $x \in L$, the p-height of x in L will be denoted by $|x|$ or $|x|_L$.

LEMMA 3. *Let* $T_0 \subseteq \tilde{T}$, $\tilde{T}$ *totally projective, and* $x \in T_0^\bullet$ *torsion-free. There exists* T_1 *with* $T_0 \subseteq T_1 \subseteq \tilde{T}$ *such that if* $\phi^\bullet : T_0^\bullet \to T_1^\bullet$ *extends the inclusion* $T_0 \subseteq T_1$, *and if we put* $x_1 = \phi^\bullet(x)$, *then* $x_1 \in T_1^\bullet$ *is torsion-free, and* $\langle x_1\rangle_*/\langle x_1\rangle$ *is totally projective of length* $< \ell(\tilde{T}) + \omega$. *(*$\langle\rangle_*$ *denotes purification in* $T_1^\bullet$*.)*

PROOF. First suppose that if we take $T_1 = \tilde{T}$, then x_1 is torsion-free. Then $\langle x_1\rangle_*/\langle x_1\rangle$ is a countably generated extension of $\tilde{T}$, hence totally projective. For the length condition, since $\langle x_1\rangle$ is nice in $\langle x_1\rangle_*$ and $|x_1| < \ell(\tilde{T}) + \omega$, we need only be concerned with $|y|$ such that $y \notin \langle x_1\rangle$, $py = p^k u x_1$ (u unit of $R, k \geq 1$), and $|p^{k-1}x_1| > \ell(\tilde{T})$. Then $y = t + p^{k-1}ux_1$ for some $t \in \tilde{T}, t \neq 0$. Therefore, $|y| \leq \ell(\tilde{T})$, and we are done in this case.

Thus we may assume that $\langle x\rangle$ intersects the kernel of $T_0^\bullet \to \tilde{T}^\bullet$ nontrivially. The inclusion $T_0 \to \tilde{T}$ induces the exact sequence $0 \to Hom_R(K, \tilde{T}/T_0) \to T_0^\bullet \to \tilde{T}^\bullet$. Since x may be replaced by any root of a nonzero multiple, we may assume that $\hat{R}x$ is a pure submodule of $Hom_R(K, \tilde{T}/T_0)$. We may select S between T_0 and $\tilde{T}$ such that S/T_0 is divisible and $Hom_R(K, S/T_0) = \hat{R}x$. Take T_1 between T_0 and $\tilde{T}$ such that $(\tilde{T}/T_0) = (S/T_0) \oplus (T_1/T_0)$. Then $Hom_R(K, \tilde{T}/T_0) = \hat{R}x \oplus Hom_R(K, T_1/T_0)$, and $Hom_R(K, T_1/T_0)$ is the kernel of $\phi^\bullet$. Therefore, x_1 is torsion-free, as desired. Since $\tilde{T}/T_1 \cong S/T_0$ is divisible, we have the exact sequence $0 \to Hom_R(K, S/T_0) \to T_1^\bullet \to \tilde{T}^\bullet \to (\tilde{T}/T_1)^\bullet = 0$, showing that $(T_1^\bullet/\hat{R}x_1) \cong \tilde{T}^\bullet$. But then $\langle x_1\rangle_*/\langle x_1\rangle$ is isomorphic to the torsion submodule of $T_1^\bullet/\hat{R}x_1$, which is isomorphic to $\tilde{T}$, thus totally projective of the proper length. ∎

LEMMA 4. *Let* L *be a reduced module with torsion submodule* T *such that* L/T *is divisible. Suppose that* $B = \langle x\rangle \oplus B'(x \neq 0)$, *is a torsion-free nice submodule of* L *with* L/B *totally projective of length* $< \lambda + \omega$. *Then there exists an ordinal* μ *with* $\mu + \omega \leq \ell(T^\bullet)$ *such that* $|p^k x + B'|_{L/B'} \leq \mu + k (k \geq 0)$.

PROOF. Let T' be the torsion submodule of L/B'. L/B' is reduced since L/B and B/B' are. We have $T \subseteq T' \subseteq L/B$ under natural embeddings, thus $\lambda \leq \ell(T') < \lambda + \omega$. First suppose that $\ell(T^\bullet) > \lambda$. Then $\ell(T^\bullet) = \lambda + \omega$. L/B' is reduced with torsion T', thus there exists m such that $|p^k x + B'| \leq \ell(T') + m + k (k \geq 0)$. Therefore, we may take $\mu = \lambda + m$.

Now suppose that $\ell(T^\bullet) = \lambda$. Let σ be the smallest ordinal such that T^σ is bounded. Lemma 2 applies to tell us that σ is a limit ordinal of cofinality $> \omega$.

But σ is also smallest such that $(T')^\sigma$ is bounded, hence $\ell((T')^\bullet) = \ell(T')$. Put $\mu = \sup_k\{|p^k x + B'|\}$. Since L/B' may be embedded as an isotype submodule of $(T')^\bullet$, the cofinality implies that $\mu+\omega < \ell(T') < \lambda+\omega$. Thus $\mu+\omega \leq \lambda = \ell(T^\bullet)$. ■

The next lemma on endomorphisms of $T^\bullet$ is crucial.

LEMMA 5. *Let L be a pure submodule of $T^\bullet$ containing T. Suppose that $B = \langle x\rangle \oplus B'(x \neq 0)$, is a torsion-free nice submodule of L with L/B totally projective of length λ. Then there exists an ordinal μ with $\mu + \omega \leq \ell(T^\bullet)$ such that:*

(1) *given $z \in p^\mu T^\bullet$, there exists $\alpha \in End_R T^\bullet$ with $\alpha(x) = z$ and $\alpha(B') = 0$;*
(2) *for α as in (1), if $y \in T^\bullet$ and $|p^k\alpha(y)| < \mu$ for all $k \geq 0$, then there exists $\beta \in End_R T^\bullet$ with $\beta(y) = \alpha(y)$ and $\beta(B) = 0$.*

PROOF. (1) Take μ as in Lemma 4 and let $z \in p^\mu T^\bullet$. Then $|p^k x + B'| \leq |p^k z|$ for all k, thus the map $B \to T^\bullet$ sending x to z and B' to 0 does not decrease heights in $T^\bullet$. Consequently, it extends to a map $L \to T^\bullet$, and then to $\alpha \in \mathrm{End}_R T^\bullet$.

(2) Since $|p^k\alpha(y)| < \mu$, and since $\alpha : T^\bullet \to T^\bullet$ with $\alpha(B) \subseteq p^\mu T^\bullet$, we have $|p^k\alpha(y)| = |p^k\alpha(y) + p^\mu T^\bullet| \geq |p^k y + B|_{T^\bullet/B}$. Put $\tilde{L} = \langle L, y\rangle_*$, the purification in $T^\bullet$. Then $\tilde{L}/L$ is torsion-free since L is pure in $T^\bullet$, thus the torsion submodule of $\tilde{L}/B$ is L/B. Moreover, $\tilde{L}/B$ is of torsion-free rank one. The map $p^k y + B \mapsto p^k\alpha(y)$ does not decrease heights, thus it extends to a map $\tilde{L}/B \to T^\bullet$. Form the composition $\tilde{L} \to \tilde{L}/B \to T^\bullet$ and let β be the extension to an endomorphism of $T^\bullet$. ■

We now consider whether the submodule $C(L)$ contains torsion-free elements.

LEMMA 6. *Let L be a pure submodule of $T^\bullet$ containing T. Assume that $T \subseteq \tilde{T}$, where $\tilde{T}$ is totally projective of length λ.*

(1) *If $C(L) \neq T$, then there exists an ordinal μ with $\mu + \omega \leq \ell(T^\bullet)$ and $p^\mu T^\bullet \subseteq C(L)$.*

(2) *If L has a nice torsion-free submodule B such that L/B is totally projective of length λ, and if every complete submodule of B has finite rank over $\hat{R}$, then $C(L) = T$.*

PROOF. (1) Choose a torsion-free element $x \in C(L)$. Apply Lemma 3 to T to obtain $T_1 \supseteq T, \phi^\bullet : T^\bullet \to T_1^\bullet$, and $x_1 = \phi^\bullet(x)$. Apply Lemma 4 to $\langle x_1\rangle_*$ and $\langle x_1\rangle$ to obtain μ with $\mu + \omega \leq \ell(T_1^\bullet)$ and $|p^k x_1| \leq \mu + k (k \geq 0)$. Lemma 2 says that the length of $T^\bullet$ is either λ or $\lambda + \omega$, depending on λ. Since $\ell(T_1) = \lambda$ also, we have $\ell(T_1^\bullet) = \ell(T^\bullet)$, therefore $\mu + \omega \leq \ell(T^\bullet)$. If $z \in p^\mu T^\bullet$, then $p^k x_1 \mapsto p^k z$ does not decrease heights, hence extends to $\langle x_1\rangle_* \to T^\bullet$, and then to $T_1^\bullet \to T^\bullet$. By composition with $\phi^\bullet$, we get $\alpha \in \mathrm{End}_R T^\bullet$ with $\alpha(x) = z \in C(L)$.

(2) Suppose $C(L) \neq T$. Then (1) implies that $p^\mu T^\bullet \subseteq C(L)$. Since $p^\mu T^\bullet$ is cotorsion, so is its image in L/B, thus bounded by p^m for some m. Hence,

$p^{\mu+m}T^\bullet \subseteq B$ is reduced torsion-free cotorsion, thus complete. But our hypothesis implies that $p^{\mu+m}T^\bullet$ has finite rank over $\hat{R}$, contrary to Lemma 2. Therefore, $C(L) = T$. ∎

For the final three lemmas, we assume that M is a reduced and nontorsion R-module with torsion submodule T such that M/T is divisible. We further assume that every full $\hat{R}$-submodule of $\hat{R}M$ contains a nice $\hat{R}$-submodule B such that $\hat{R}M/B$ is totally projective. The first lemma draws some simple conclusions from this hypothesis.

LEMMA 7. (1) *Any full $\hat{R}$-submodule of $\hat{R}M$ of the form $B_1 \oplus p^\lambda \hat{R}M$ contains a nice $\hat{R}$-submodule B of the same form, with $\hat{R}M/B$ totally projective of length λ.*

(2) *Any complete torsion-free $\hat{R}$-submodule of $\hat{R}M$ is free of finite rank.*

PROOF. (1) Given $B_1 \oplus p^\lambda \hat{R}M$, choose a nice $\hat{R}$-submodule B' such that $\hat{R}M/B'$ is totally projective. Put $B = B' + p^\lambda \hat{R}M$. Then $B = (B \cap B_1) \oplus p^\lambda \hat{R}M$, and $B/B' = p^\lambda(\hat{R}M/B')$ since B' is nice, therefore B is nice and $\hat{R}M/B$ has length λ.

(2) Suppose that $F \subseteq \hat{R}M$ is complete torsion-free. Choose a free full $\hat{R}$-submodule of $\hat{R}M$, say B. We may assume that B is nice, therefore $\hat{R}M/B$ is reduced torsion. The image of F in this quotient must be bounded, consequently, F is free since B is. The completeness implies that the rank is finite. ∎

Suppose that $W \subseteq \hat{R}M$ is a submodule such that $\hat{R}M/W$ is reduced. Since we have a natural identification $(\hat{R}M)^\bullet = T^\bullet$, and since $Hom_R(K, \hat{R}M/W) = 0$, we get an exact sequence $0 \to W^\bullet \to T^\bullet$. We use this freely in the following to regard $W^\bullet$ as a submodule of $T^\bullet$.

We now introduce a second module N with the same endomorphisms as M.

LEMMA 8. *Let M be as above, and further assume that M and N are pure submodules of $T^\bullet$ containing T such that $End_R M = End_R N$ (when extended to $T^\bullet$). Then $N \subseteq \hat{R}M + (p^\lambda \hat{R}M)^\bullet$.*

PROOF. We may choose a full $\hat{R}$-submodule of $\hat{R}M$ of form $B_1 \oplus p^\lambda \hat{R}M$, with B_1 free. By Lemma 7, we may assume that B is nice, $\hat{R}M/B$ is totally projective of length λ, and every complete submodule of B has finite rank over $\hat{R}$.

We first claim that $C(N) = T$. Lemma 6 applied to $\hat{R}M$ shows that $C(\hat{R}M) = T$, thus $C(M) = T$. For the sake of contradiction, we assume that $C(N) \neq T$. Since M is nontorsion, we may write $B = \langle x \rangle \oplus B'$ for some torsion-free x. Choose μ to satisfy Lemma 5 for $\hat{R}M$ and Lemma 6 for N. Then $p^\mu T^\bullet \subseteq C(N)$. By Lemma 2, we may choose torsion-free $z \in p^\mu T^\bullet$. By Lemma 5, there exists $\alpha \in \mathrm{End}_R T^\bullet$ with $\alpha(x) = z$ and $\alpha(B') = 0$. We shall show that $\alpha(N) \subseteq C(N)$. Let $y \in N$. Then $\alpha(y) \in C(N)$ if $|p^k\alpha(y)| \geq \mu$ for some k; therefore we may assume $|p^k\alpha(y)| < \mu$ for all k. Then Lemma 5 provides $\beta \in \mathrm{End}_R T^\bullet$ with $\beta(y) = \alpha(y)$ and $\beta(B) = 0$. Thus, $\beta(M) \subseteq T$ and the Proposition implies $\beta(N) \subseteq C(N)$. Hence, $\alpha(y) = \beta(y) \in C(N)$ in this case also. But $\alpha(N) \subseteq C(N)$ implies that $\alpha(M) \subseteq C(M) = T$ by the Proposition. This contradicts $\alpha(x) = z$.

Next we claim that $N \subseteq (B^\bullet)_*$. If $y \in T^\bullet \backslash (B^\bullet)_*$, it will suffice to show that $y \notin N$. The exact sequence $0 \to B \to \hat{R}M \to (\hat{R}M/B) \to 0$ induces $0 \to B^\bullet \to T^\bullet \to (\hat{R}M/B)^\bullet \to 0$. Note that $y + B^\bullet$ is torsion-free in $(\hat{R}M/B)^\bullet$. If we put $A = \langle y + B^\bullet \rangle_*$ in $(\hat{R}M/B)^\bullet$, then A has torsion-free rank one and torsion $\hat{R}M/B$ totally projective of length λ. By Lemma 2, we may choose torsion-free $z \in T^\bullet$ such that $|p^k y + B^\bullet| \leq |p^k z|$ for all $k \geq 0$. Thus, there exists a homomorphism $A \to T^\bullet$ sending $y + B^\bullet$ to z. By composing with the natural map to $\hat{R}M/B$ and extending to $T^\bullet$, we obtain $\alpha \in \mathrm{End}_R T^\bullet$ such that $\alpha(y) = z$ and $\alpha(B) = 0$. Hence, $\alpha(M) \subseteq T$ and the Proposition implies that $\alpha(N) \subseteq T$. Therefore, we must have $y \notin N$.

Now let $y \in N$. Then, for some k we have $p^k y = y_1 + v$ $(y_1 \epsilon B_1^\bullet, v \in (p^\lambda \hat{R}M)^\bullet)$, since $B^\bullet = B_1^\bullet \oplus (p^\lambda \hat{R}M)^\bullet$. If $y_1 \notin B_1$, then replacing B_1 by a suitable full $\hat{R}$-submodule B_2 (replace basis elements by suitable nonzero multiples), we may assume that $y_1 \notin (B_2^\bullet)_*$. We may further assume that $B_2 \oplus p^\lambda \hat{R}M$ is nice in $\hat{R}M$ with the quotient totally projective. But then, $N \subseteq (B_2^\bullet \oplus (p^\lambda \hat{R}M)^\bullet)_*$ by what we have already shown. This implies $y_1 \in (B_2^\bullet)_*$, contrary to choice. Thus, we must have $y_1 \in B_1$, and hence $N \subseteq (B_1 + (p^\lambda \hat{R}M)^\bullet)_*$. This last module equals $\hat{R}M + (p^\lambda \hat{R}M)^\bullet$ since $p^\lambda \hat{R}M$ is dense in $(p^\lambda \hat{R}M)^\bullet$. ■

LEMMA 9. *Let M and N be as in the preceding lemma. Then $\hat{R}M + (p^\lambda \hat{R}M)^\bullet = \hat{R}N + (p^\lambda \hat{R}N)^\bullet$.*

PROOF. For convenience, put $M^* = \hat{R}M + (p^\lambda \hat{R}M)^\bullet$ and $N^* = \hat{R}N + (p^\lambda \hat{R}N)^\bullet$. By Lemma 8, $N^* \subseteq M^*$. Note that $\hat{R}M \cap (p^\lambda \hat{R}M)^\bullet = p^\lambda \hat{R}M$ since $p^\lambda \hat{R}M$ is a pure submodule of the complete module $p^\lambda T^\bullet$. Thus, $\hat{R}M/p^\lambda \hat{R}M$ is naturally isomorphic to $M^*/(p^\lambda \hat{R}M)^\bullet$. On the other hand, $N^* \cap (p^\lambda \hat{R}M)^\bullet = (p^\lambda \hat{R}N)^\bullet$, thus $N^*/(p^\lambda \hat{R}N)^\bullet$ is isomorphic to an $\hat{R}$-submodule of the previous quotient. We may choose a free $\hat{R}$-module $B_1 \oplus B_2$ in $\hat{R}M$ such that $B_1 \oplus B_2 \oplus p^\lambda \hat{R}M$ is full in $\hat{R}M$, and such that $(B_2 \oplus (p^\lambda \hat{R}M)^\bullet)/(p^\lambda \hat{R}M)^\bullet$ is a full $\hat{R}$-submodule of the above submodule. By Lemma 7, we may assume that $B_1 \oplus B_2 \oplus p^\lambda \hat{R}M$ is a nice $\hat{R}$-submodule of $\hat{R}M$ with quotient $\tilde{T}$ totally projective of length λ. Note that the decomposition $B_1 \oplus B_2$ may still be assumed to hold since the original B_1 was free. We have $B_1 \oplus B_2 \oplus (p^\lambda \hat{R}M)^\bullet$ a torsion-free $\hat{R}$-submodule of M^* with quotient isomorphic to $\tilde{T}$.

Since $(p^\lambda \hat{R}N)^\bullet$ is a pure complete submodule of $(p^\lambda \hat{R}M)^\bullet$, we may write $(p^\lambda \hat{R}M)^\bullet = (p^\lambda \hat{R}N)^\bullet \oplus E$ for some E. For the sake of contradiction, we assume $N^* \underset{\neq}{\subseteq} M^*$. Then $B_1 \oplus E \neq 0$. Choose nonzero $x \in B_1 \oplus E$ and put $B^* = B_2 \oplus (p^\lambda \hat{R}N)^\bullet$. Then M^*/B^* is reduced, its torsion submodule may be embedded in $\tilde{T}$, the quotient modulo torsion is divisible, and the image of x is a torsion-free element. We may apply Lemma 3 and familiar arguments to obtain $\alpha \in \mathrm{End}_R T^\bullet$ such that $\alpha(x)$ is torsion-free and $\alpha(B^*) = 0$. Thus $\alpha(N) \subseteq T$, and consequently, $\alpha(M) \subseteq T$. But then $\alpha(p^\lambda \hat{R}M) = 0$, and hence we see that $\alpha(M^*) \subseteq T$. This is contrary to $\alpha(x)$ torsion-free, thus we have the desired contradiction. ■

The Theorem now follows from Lemma 9. The Proposition guarantees that we are in the setting of the lemma. The assumption on $p^\lambda \hat{R}M$ implies that it equals $(p^\lambda \hat{R}M)^\bullet$, thus the same is true of $p^\lambda \hat{R}N$ and the conclusion of the lemma becomes $\hat{R}M = \hat{R}N$.

Corollary 2 follows from Lemma 1(3) and the fact that finitely generated $\hat{R}$-submodules are nice. For Corollary 3, we can apply parts (7) and (3) of Lemma 1. If N is also a Warfield module, we note that the Warfield invariants of M (respectively, N), are the same as for $\hat{R}M$ (respectively, $\hat{R}N$), by Lemma 1(6). Since the Ulm invariants are trivially the same, we obtain the isomorphism. In Corollary 4, we have $\hat{R}M = \hat{R}N$. We may choose torsion-free elements $x \in M, y \in N$, and a unit $u \in \hat{R}$ such that $y = ux$. Multiplication by u gives an isomorphism of M with N. Since all endomorphisms commute with this multiplication, we obtain the corollary.

4. Final Remarks

A question raised by the Theorem is the following. Suppose that M varies over a class of modules to which the Theorem applies. Then, up to isomorphism, $\mathrm{End}_{\hat{R}}\hat{R}M$ depends only on the algebra $\mathrm{End}_R M$, and not on M. Thus, the correspondence $\mathrm{End}_R M \mapsto \mathrm{End}_{\hat{R}}\hat{R}M$ induces a well-defined map from certain isomorphism classes of R-algebras to isomorphism classes of $\hat{R}$-algebras. Can this correspondence be given in another manner, say a ring-theoretic one? For example, for modules of torsion-free rank one, $\mathrm{End}_{\hat{R}}\hat{R}M$ can be interpreted as $\hat{R}\mathrm{End}_R M$.

Finally, we comment on the hypothesis that $p^\lambda \hat{R}M$ have finite rank over $\hat{R}$. For simplicity, take R to be the rational p-adic integers for a prime number p. We shall give modules M and N such that the hypothesis of the Theorem holds, except that $p^\lambda \hat{R}M$ will have rank $\aleph_0$, and such that $\hat{R}M \neq \hat{R}N$. Thus the completion $(p^\lambda \hat{R}M)^\bullet$ in Lemma 9 cannot be avoided as things stand.

Take the algebra $A = R$ in Theorem 2.3 in [**4**], and construct X as in that proof, with $\lambda = \aleph_0$. Since the $\hat{R}$ functor commutes with purification, in that proof we see that $\hat{R}X = \oplus_{i<\lambda} e_i \hat{R}$. We may choose another basic submodule $\oplus_{i<\lambda} e_i' R$ such that $\oplus e_i' \hat{R} \neq \oplus e_i \hat{R}$. Construct X_1 and X_2 using these two different basic submodules of the completion. Then $\hat{X}_1 = \hat{X}_2$, $\mathrm{End}_R X_1 = \mathrm{End}_R X_2$, but $\hat{R}X_1 \neq \hat{R}X_2$. Apply the functor G from [**4**], putting $M = G(X_1)$ and $N = G(X_2)$. Then $\mathrm{End}_R M = \mathrm{End}_R N$ (Theorem 5.1 in [**4**]), and $\hat{R}M$ has torsion-free rank $\aleph_0$ and torsion a direct sum of cyclic modules. The nice $\hat{R}$-submodule condition of the Theorem is met by Lemma 11 in [**9**]. Moreover, the functors $\hat{R}$ and G commute, thus $p^\lambda \hat{R}M = \hat{R}X_1$ has rank $\aleph_0$. But, $p^\lambda \hat{R}N = \hat{R}X_2 \neq \hat{R}X_1$, hence $\hat{R}M \neq \hat{R}N$, as desired.

References

1. M. Dugas and R. Göbel, *Every cotorsion-free algebra is an endomorphism algebra*, Math. Z. **181** (1982), 451–470.
2. S. Files, *Mixed modules over incomplete discrete valuation rings*, Comm. Algebra (to appear).

3. L. Fuchs, *Infinite Abelian Groups, Vol. I, Vol. II*, Academic Press, New York, 1970, 1973.
4. R. Göbel and W. May, *The construction of mixed modules from torsion-free modules*, Arch. Math. **48** (1987), 476–490.
5. B. Goldsmith, *Endomorphism rings of torsion-free modules over a complete discrete valuation ring*, J. London Math. Soc. (2) **18** (1978), 464–471.
6. R. Hunter, F. Richman, and E. Walker, *Warfield Modules. In: Abelian Group Theory, Proceedings Las Cruces Conference 1976*, Lecture Notes in Mathematics, Vol. 616 (D. Arnold, R. Hunter, and E. Walker, eds.), Springer–Verlag, Berlin–Heidelberg–New York, 1977, pp. 87–123.
7. I. Kaplansky, *Infinite Abelian Groups*, University of Michigan Press, Ann Arbor, 1969.
8. W. Liebert, *Endomorphism rings of reduced torsion-free modules over complete discrete valuation rings*, Trans. Amer. Math. Soc. **169** (1972), 347–363.
9. W. May, *Isomorphism of endomorphism algebras over complete discrete valuation rings*, Math. Z. **204** (1990), 485–499.
10. ______, *Endomorphism algebras of not necessarily cotorsion-free modules*, Contemp. Math., Vol. 130, 1992, pp. 257–264.
11. W. May and E. Toubassi, *Endomorphisms of rank one mixed modules over discrete valuation rings*, Pac. J. Math. **108** (1983), 155–163.
12. S. Shelah, *Infinite abelian groups, Whitehead problem and some constructions*, Israel J. Math. **18** (1974), 243–256.
13. K. Wolfson, *Isomorphisms of the endomorphism rings of torsion-free modules*, Proc. Amer. Math. Soc. **13** (1962), 712–714.

Department of Mathematics, University of Arizona, Tucson, Arizona 85721

E-mail address: may@math.arizona.edu

Contemporary Mathematics
Volume **171**, 1994

$Bext^2(G,T)$ CAN BE NONTRIVIAL, EVEN ASSUMING GCH

MENACHEM MAGIDOR AND SAHARON SHELAH

ABSTRACT. Using the consistency of some large cardinals we produce a model of Set Theory in which the generalized continuum hypothesis holds and for some torsion-free abelian group G of cardinality $\aleph_{\omega+1}$ and for some torsion group T

$$Bext^2(G,T) \neq 0.$$

Hence G.C.H. is not sufficient for getting the results of [10].

1. INTRODUCTION

All groups in this paper are abelian groups. For basic terminology about abelian groups in general we refer the reader to [9]. For terminology concerning Butler groups see [2, 1, 3, 10, 8]. It is commonly agreed that the three major questions concerning the infinite rank Butler groups are:

(1) Are B_1-groups necessarily B_2-groups?
(2) Does $Bext^2(G,T) = 0$ hold for all torsion-free groups G and torsion groups T?
(3) Which pure subgroups of B_2-groups are again B_2-groups? In particular: is a balanced subgroup of a B_2-group a B_2-group?

In [2] it is shown that the answer to all these questions is "Yes" for countable groups G. In the series of papers [1, 4, 3] it was shown that under the continuum hypothesis the answer is "Yes" to all three questions for groups G of cardinality $\leq \aleph_\omega$. In [5] it is shown that the answer to question 2 is "No" if the continuum hypothesis fails. In a more recent paper [10] it is shown that in the constructible universe, L the answer is "Yes" to all three questions for *arbitrary* groups G. Actually [10] used only the generalized continuum hypothesis and that the combinatorial principle $\square_\kappa$ holds for every singular cardinal κ whose cofinality is $\aleph_0$. Is the use made in [10] of the additional combinatorial principle really needed or does the affirmitive answer to our three questions follow simply from G.C.H.? Let us mention that a key tool used in [3, 10] was the representation of an arbitrary torsion-free group as the union of a chain of subgroups which are countable unions of balanced subgroups. In [7] it is shown that such a representation is equivalent to a weak version of $\square_\kappa$.

1991 *Mathematics Subject Classification.* 20K20, 20K40;Secondary 03E55, 03E35.
Key words and phrases. Butler Groups, Balanced Extensions, Consistency Proofs.
The research of the second named author was supported by the Basic Research Foundation of the Israeli Academy of Science (Publ. No. 514).
This paper is in final form and no version of it will be submitted for publication elsewhere.

In this paper we show that at least for getting an affirmitive answer to questions 2 and 3, one needs some extra set theoretic assumptions in addition to G.C.H. We do it by producing a model of Set Theory, satisfying G.C.H., in which for some torsion-free G of cardinality $\aleph_{\omega+1}$ and some torsion T, $Bext^2(G,T) \neq 0$. Also in the same model there will be a balanced subgroup of a completely decomposable group which is not a B_2-group. Hence the answer to question 3 in this model is "No". The construction of the model requires the consistency of some large cardinals, which can not be avoided since getting a model in which $\Box_\kappa$ fails for some singular κ requires assumptions stronger than the consistency of Set Theory. Let us stress that the status of question 1 is not known and it is possible (though unlikely) that the implication "every B_1-group is a B_2-group" is a theorem of Set Theory.

Since this paper is aimed at a mixed audience of set theorists and abelian group theorists it is divided into two sections with very different prerequisites. In the next section we describe the construction of the model of Set Theory with certain properties to be listed below. In the following section we shall describe how to use the listed properties to get a group G which will be the counterexample to $Bext^2(G,T) = 0$. A reader who is not familiar with standard set theoretical techniques, like forcing, can skip the set theoretic section and simply assume the properties of the model listed below. We do assume some basic Set Theory at the level introduced by [6].

We now describe the properties of the model which will be used in the construction of the counterexample to questions 2 and 3. The model will naturally satisfy G.C.H. Hence by standard cardinal arithmetic $\aleph_\omega^{\aleph_0} = \aleph_{\omega+1}$. Therefore we can enumerate all the ω-sequences from $\aleph_\omega$ in a sequence of order type $\aleph_{\omega+1}$. Let $\langle f_\alpha | \alpha < \aleph_{\omega+1}\rangle$ be this enumeration. Let F_α be the range of f_α. The important property of the model is the following:

For some stationary subset S of $\aleph_{\omega+1}$ such that every point of S has cofinality $\aleph_1$, and for some choice of a cofinal set C_β in β of order type ω_1, for every $\beta \in S$ and for some fixed countable ordinal δ we have:

(1)
$$\bigcup_{\alpha \in D} F_\alpha$$
has order type δ for every $D \subseteq C_\beta$ which is cofinal subset of C_β and for every $\beta \in S$. In particular for $D = C_\beta$
$$E_\beta = \bigcup_{\alpha \in C_\beta} F_\alpha$$
has order type δ.

(2) *If $\beta \neq \gamma$ both in S, then $E_\beta \cap E_\gamma$ has order type less than δ.*

(3) *δ is an indecomposable ordinal, namely δ can not be represented as a finite sum of smaller ordinals. Or equivalently, δ is not the finite union of sets of ordinals of order type less than δ.*

Denote the conjunction of all the properties above by (*). The main theorem of Section 1 is

Theorem 1. *Assume the consistency of a supercompact cardinal. Then there is a model of Set Theory in which (*) holds. The model also satisfies the Generalized Continuum Hypothesis.*

The construction of the model is very close to the construction in [11]. The main tool that will be used to get in Section 3 an example of a group G satisfying $Bext^2(G,T) \neq 0$ is the notion of $\aleph_0$-prebalancedness (see [8]). We are rephrasing the original definition in a form which is clearly equivalent to the original definition.

Definition 1. *Let G be a pure subgroup of the group H. G is said to be $\aleph_0$-prebalanced in H if for every element $h \in H - G$ there are countably many elements $g_0, g_1, \ldots$ of G such that for every element g of G the type (in H) of $h - g$ is bounded by the the union of finitely many types of the form $lh - g_i$ for some natural number l. More explicitly for some $n, l \in \omega$*

$$\mathbf{t}(h-g) \leq \mathbf{t}(lh-g_0) \cup \ldots \cup \mathbf{t}(lh-g_n).$$

Also the group G is said to admit an $\aleph_0$-prebalanced chain if G can be represented as a continuous increasing union of pure $\aleph_0$-prebalanced subgroups where at the successor stages the factors are of rank 1.

We shall use the following fundamental result of Fuchs ([8]):

Theorem 2. *A torsion-free group G admits an $\aleph_0$-prebalanced chain if and only if in its balanced projective resolution*

$$0 \to B \to C \to G \to 0$$

(where C is completely decomposable) B is a B_2-group. Moreover, if CH holds, then this condition is equivalent to $Bext^2(G,T) = 0$ for all torsion groups T.

The main result of Section 3 will be

Theorem 3. *If (*) holds, then there is a torsion-free group G of cardinality $\aleph_{\omega+1}$ which does not admit an $\aleph_0$-prebalanced chain.*

Using theorem 2 we get

Corollary 4. *If (*) holds, then there is a group G of cardinality $\aleph_{\omega+1}$ such that $Bext^2(G,T) \neq 0$ for some torsion group T.*

By using the balanced projective resolution of G we also get

Corollary 5. *If (*) holds, then there is a balanced subgroup of a completely decomposable group of cardinality $\aleph_{\omega+1}$ which is not a B_2-group.*

2. The Consistency of (*)

In this section we shall prove Theorem 1. We assume familiarity with some basic large cardinals notions like supercompact cardinals and some basic forcing techniques. We start from a ground model V having a supercompact cardinal κ. We can assume without loss of generality that V satisfies G.C.H. We let $\mu = \kappa^{+\omega}$ and $\lambda = \mu^+ = \kappa^{+\omega+1}$. In our final model μ will be $\aleph_\omega$ and λ will be $\aleph_{\omega+1}$. It follows from the results of Menas in [12] that there is a normal ultrafilter U on $P_\kappa(\lambda)$ such that for some set $A \in U$ the map $P \to \sup(P)$ on A is one-to-one. (Recall that $P_\kappa(\lambda)$ is the set of all subsets of λ of cardinality less than κ). Fix such U and A. Also fix an enumeration $\langle g_\alpha \mid \alpha < \lambda\rangle$ of all the ω-sequences in μ. Standard facts about normal ultrafilters on $P_\kappa(\lambda)$ imply that the set of all $P \in P_\kappa(\lambda)$ satisfying the following properties is in U:

(1) The order type of $P \cap \mu$ is a singular cardinal of cofinality ω such that the order type of P is its successor.
(2) For $\alpha \in \lambda$ the range of g_α is a subset of $P \cap \mu$ if and only if $\alpha \in P$.

Hence we can assume without loss of generality that every $P \in A$ satisfies all the above properties. Again standard arguments show that the set $T = \{\sup(P) \mid P \in A\}$ is a stationary subset of λ. For $\alpha \in T$, let P_α be the unique $P \in A$ such that $\sup(P) = \alpha$. Note that for $P \in A$ and $Q \subseteq P$ we have that if Q is cofinal in $\sup(P)$, then the order type of $Q^* = \cup\{range(g_\alpha) \mid \alpha \in Q\}$ is the same as the order type of $P \cap \mu$. This holds since otherwise Q^* has cardinality smaller than δ =the order type of $P \cap \mu$. Hence, by our G.C.H. assumption, we have less than δ α's such that the range of g_α is in Q^*, hence less than the order type of P, which is a regular cardinal. Therefore Q must be bounded in P.

For $\alpha \in T$ the map $\alpha \to$ the order type of $P_\alpha \cap \mu$ maps T into κ. Hence it is fixed on some subset S which is still stationary in λ. Let δ be the fixed value of this map on S. Note that for $\alpha \in S$ the order type of P_α is δ^+.

Claim 6. *Let α and β be two different members of S. Then $P_\alpha \cap P_\beta \cap \mu$ has order type less than δ.*

Proof. Let $X = P_\alpha \cap P_\beta \cap \mu$. Note that if g is an ω-sequence from X, then $g = g_\rho$ for some $\rho \in P_\alpha \cap P_\beta$. If X has order type δ, then (using the fact that δ is a singular cardinal of cofinality ω) we have δ^+ ω-sequences from X, so that $P_\alpha \cap P_\beta$ must have order type which is at least δ^+. Since the order type of both P_α and P_β is δ^+, P_α and P_β must have the same sup. This is a contradiction. □

The model which will witness (*) will be obtained from V by collapsing δ to be countable, followed by the collpasing all the cardinals between δ^{++} and κ to have cardinality δ^{++}. Denote the resulting model by V_1. Note since V satisfies G.C.H. then the resulting model satisfies G.C.H. Also δ is of course countable, δ^+ is $\aleph_1$, μ is $\aleph_\omega$ and λ is $\aleph_{\omega+1}$. Since the cardinality of the forcing notion is $\kappa < \lambda$, S is still a stationary subset of λ. Note that now we have for every $\alpha \in S$ that the cofinality of α is $\aleph_1$. In order to verify (*) in the resulting model we fix an enumeration $\langle f_\gamma \mid \gamma < \lambda\rangle$ of all the ω-sequences from $\aleph_\omega = \mu$. And as in the previous section let F_γ be the range of f_γ. (Note that in V_1 there are new ω-sequences so that the enumeration $\langle g_\gamma \mid \gamma < \lambda\rangle$ we had in V enumerates only a subset of the set of all ω-sequences). For $\gamma < \lambda$ let $\eta(\gamma)$ be the unique η such that $g_\gamma = f_\eta$. Without loss of generality (by reducing S to a subset which is still stationary in λ) we can assume that for $\alpha \in S$ if $\gamma < \alpha$, then $\eta(\gamma) < \alpha$. We can also assume without loss of generality that for $\alpha \in S$, $Q_\alpha = \{\eta(\gamma) \mid \gamma \in P_\alpha\}$ is cofinal in α. This follows since the set $\{\alpha \in S \mid Q_\alpha$ is bounded in $\alpha\}$ is not stationary. So for each $\alpha \in S$ pick C_α which is cofinal in Q_α and has order type $\aleph_1 = \delta^+$. We claim that S, δ and $\langle C_\alpha \mid \alpha \in S\rangle$ are witnesses to the truth of (*) in V_1. As in the introduction we put

$$E_\alpha = \bigcup_{\gamma \in C_\alpha} F_\gamma.$$

Since we clearly have G.C.H. in V_1, since S is stationary and since δ is an indecomposable ordinal (it is a cardinal in V!), we are left with verifying the following claim:

Claim 7. *In V_1*

> **A:** *For $\alpha \neq \beta \in S$ $E_\alpha \cap E_\beta$ has order type less than δ.*
> **B:** *If $D \subseteq C_\alpha$ is cofinal in α, then $\cup\{F_\gamma \mid \gamma \in D\}$ has order type δ.*

Proof. Clause A follows immediately from the fact that for $\alpha \in S$, $E_\alpha \subseteq P_\alpha \cap \mu$, hence $E_\alpha \cap E_\beta \subseteq P_\alpha \cap P_\beta \cap \mu$ and the last set has order type less than δ if $\alpha \neq \beta$.

For proving B note that if $D \subseteq C_\alpha$ is cofinal in α, then the set $F = \{\gamma \mid \eta(\gamma) \in D\}$ is a subset of P_α of cardinality $\aleph_1 = \delta^+$. Our forcing is an iteration of two forcing notions where the first is of cardinality (in V) δ and the second is δ^{++} closed, hence it introduces no new sets of ordinals of order type δ^+. So F contains a subset $Q \in V$ of cardinality δ^+. Q must be cofinal in P_α since P_α has order type δ^+, so by a previous remark $\cup\{range(g_\gamma) \mid \gamma \in Q\}$ has order type δ. But this last set is clearly a subset of $\cup\{F_\rho \mid \rho \in D\}$, so this set clearly has order type at least δ. It can not have order type greater than δ since it is a subset of $P_\alpha \cap \mu$. □

3. A group which does not admit an $\aleph_0$-prebalanced chain

In this section we prove Theorem 3. So we assume (*). Fix the enumeration $\langle f_\alpha \mid \alpha < \aleph_{\omega+1}\rangle$ of the ω-sequences from $\aleph_\omega$. Let F_α be the range of f_α. Also fix the stationary subset S of $\aleph_{\omega+1}$, the countable ordinal δ and for $\beta \in S$ a set C_β cofinal in β, which witness the truth of (*). As in the statement of (*) (for $\beta \in S$) let

$$E_\beta = \bigcup_{\alpha \in C_\beta} F_\alpha.$$

We know that the order type of E_β is δ. Since $\delta \times \omega$ is countable we can assign to every pair $\mu < \delta, n < \omega$ a unique prime number p^n_μ.

We are ready to define the group G that will not admit a chain of $\aleph_0$-prebalanced subgroups. For each $\alpha < \aleph_{\omega+1}$ and $\beta \in S$ fix distinct symbols x_α and y_β. The group G is a subgroup of

$$\sum_{\alpha < \aleph_{\omega+1}} \oplus \boldsymbol{Q} x_\alpha \oplus \sum_{\beta \in S} \oplus \boldsymbol{Q} y_\beta.$$

G is generated by x_α for $\alpha < \aleph_{\omega+1}$, by y_β for $\beta \in S$ and by $\frac{1}{p^n_\mu}(y_\beta - x_\alpha)$ provided α is in C_β and the $f_\alpha(n)$ is the μ-th member of E_β. For $\delta < \aleph_{\omega+1}$ let G_δ be the subgroup of G generated by x_α,y_γ and $\frac{1}{p^n_\mu}(y_\gamma - x_\alpha)$ where α and γ are less than δ.

The sequence $\langle G_\delta \mid \delta < \aleph_{\omega+1}\rangle$ is a filtration of G into a continuous chain of smaller cardinality. If G allows an $\aleph_0$-prebalanced chain, then by standard arguments, the set of $\delta < \aleph_{\omega+1}$ such that G_δ appears in the $\aleph_0$-prebalanced chain contains a closed unbounded subset of $\aleph_{\omega+1}$. This will imply, since S is stationary in $\aleph_{\omega+1}$, that for some $\beta \in S$, G_β is $\aleph_0$ prebalanced in G. The fact that we get a contradiction and that G does not allow an $\aleph_0$-prebalanced chain follows from:

Claim 8. *For $\beta \in S$, G_β is not an $\aleph_0$-prebalanced subgroup of G.*

Proof. Assume that for some fixed $\beta \in S$, G_β is $\aleph_0$-prebalanced in G. We apply the definition of $\aleph_0$-prebalancedness for y_β and get a sequence of elements $z_n \in G_\beta$ such that for every element z of G_β there are e and l such that

$$\boldsymbol{t}(y_\beta - z) \leq \boldsymbol{t}(ly_\beta - z_0) \cup \ldots \cup \boldsymbol{t}(ly_\beta - z_e).$$

C_β has order type $\aleph_1$ and hence for some fixed e and l we get that the set

$$D = \{\alpha \in C_\beta \mid \boldsymbol{t}(y_\beta - x_\alpha) \leq \boldsymbol{t}(ly_\beta - z_0) \cup \ldots \cup \boldsymbol{t}(ly_\beta - z_e)\} \tag{1}$$

is unbounded in C_β. It means that for $\alpha \in D$ there is a natural number d_α such that if p is a prime number greater than d_α and p divides $y_\beta - x_\alpha$, then p divides $ly_\beta - z_i$ for some $0 \leq i \leq e$. Without loss of generality we can assume that for $\alpha \in D$, d_α is some fixed natural number d. Let $D^* = \cup_{\gamma \in D} F_\gamma$. We know that $D^* \subseteq E_\beta$ and that the order type of D^* is δ. We need the following lemma.

Lemma 9. *Let z be a member of G_β with*

$$z = \sum_{i=1}^{k} r_i x_{\alpha_i} + \sum_{j=1}^{g} s_j y_{\beta_j},$$

where $r_i, s_j \in \boldsymbol{Q}$ and $\alpha_i, \beta_j < \beta$ for $1 \leq i \leq k, 1 \leq j \leq g$. Assume also that $ly_\beta - z$ is divisible (in G) by p_μ^n where $p_\mu^n > l$. Then either for some $1 \leq j \leq g$, the μ-th member of E_β is the same as the μ-th member of E_{β_j} or for some $1 \leq i \leq k$, the μ-th member of E_β is in F_{α_i}.

Proof. By assumption $ly_\beta - z$ is divisible by $p = p_\mu^n$ in G. Hence

$$ly_\beta - z = p(\sum_{m=1}^{f} r_m x_{\gamma_m} + \sum_{t=1}^{u} s_t y_{\eta_t} + \sum_{q=1}^{v} \frac{w_q}{p_q}(y_{\nu_q} - x_{\xi_q})). \tag{2}$$

where the r_m's , the s_t's and the w_q's are integers.

Let us define a (bipartite) graph P, whose nodes are all the symbols (x's and y's) appearing in equation 2, where y_ρ is connected by an edge to x_ζ iff for some $1 \leq q \leq v$, $\rho = \nu_q$, $\zeta = \xi_q$ and $p_q = p$. Let W be the connected component of y_β in P and let $a \in \boldsymbol{Q}$ be the sum of all the coefficients in the right side of equation 2 of symbols in W. a is easily seen to be a member of $p\boldsymbol{Q}_p$, where $\boldsymbol{Q}_p$ is the ring of rationals whose denominators are prime to p.This is true because the only summands on the right side of 2, that can possibly add to a a rational number which is not in $p\boldsymbol{Q}_p$, is of the form $\frac{w_q}{p_q}(y_{\nu_q} - x_{\xi_q})$ where $p_q = p$. But in this case y_{ν_q} and x_{ξ_q} are connected by an edge of P, so they are both in W or both outside of W. In both cases the contribution of this summand to a is 0.

We use the fact that the sum of the coefficients of symbols in W must be the same for the left side and the right side of 2. Of course $y_\beta \in W$ and its coefficient in equation 2 is l which is not in $p\boldsymbol{Q}_p$, so there must be a symbol in W appearing in the representation of z, so that either $x_{\alpha_i} \in W$ for some $1 \le i \le k$, or $y_{\beta_j} \in W$ for some $1 \le j \le g$. Our lemma will be verified if we prove

Claim 10. (1) *If $y_\eta \in W$, then the μ-th member of E_η is the same as the μ-th member of E_β.*
(2) *If $x_\gamma \in W$, then $f_\gamma(n)$ is the μ-th member of E_β.*

Proof. The proof is by induction on the length of the path in P leading from y_β to the symbol y_η and x_γ respectively. If this length is 0, we are in the case where the symbol is $y_\eta = y_\beta$, and the claim is obvious. For the induction step, in the first case we are given y_η. Let x_γ be the element preceding y_η in the path leading from y_β to y_η. By the induction assumption $f_\gamma(n)$ is the μ-th member of E_β. x_γ and y_β are connected by an edge of P, so that $\frac{1}{p_\mu^n}(y_\eta - x_\gamma)$ is one of the generators of G. Hence $\gamma \in C_\eta$ and $f_\gamma(n)$ is the μ-th member of E_η, and the claim is verified in this case. The other case (the x_γ case) is argued similary where y_η is now the element in the path preceding x_γ. □

□

For $z \in G_\beta$ let $S(z)$ be the set of all elements γ of E_β such that for some $\mu < \delta$ and $n \in \omega$, γ is the μ-th member of E_β and $ly_\beta - z$ is divisible in G by p_μ^n where $p_\mu^n > l$. It follows from lemma 9 that for $z \in G_\beta$, $S(z)$ is included in a finite union of singletons and of sets of the form $E_\eta \cap E_\beta$ for $\eta < \beta$. So $S(z)$ is a finite union of sets of order type less than δ. δ is an indecomposable ordinal, so for $z \in G_\beta$ the order type of $S(z)$ is less than δ. By definition of D, every element of D^*, except possibly finitely many, is in $\cup_{0 \le i \le e} S(z_i)$. This is because there are only finitely many members of E_β such that if γ is the μ-th member of E_β, then $p_\mu^n \le \max(d,l)$ for some n. So if $\gamma \in D^*$ is not one of these finitely many elements, say γ is the μ-th member of E_β, then $p_\mu^n > \max(d,l)$. Now $\gamma = f_\alpha(n)$ for some $\alpha \in D$ and a natural number n, and hence p_μ^n divides $y_\beta - x_\alpha$, which implies by equation 1 and the definition of d that p_μ^n divides $ly_\beta - z_i$ for some $1 \le i \le e$. We got that D^* is a finite union of sets of order type less that δ, and hence D^* has order type less than δ. We got a contradiction. □

References

1. U. Albrecht and P. Hill, *Butler groups of infinite rank*, Czech. Math. J. **37** (1987), 293–309.
2. L. Bican and L. Salce, *Infinite rank Butler groups*, Lecture Notes In Mathematics, vol. 1006, Springer, 1983.
3. M. Dugas, P. Hill, and K. M. Rangaswamy, *Infinite rank Butler groups II*, Trans. Amer. Math. Soc. **320** (1990), 643–664.
4. M. Dugas and K. M. Rangaswamy, *Infinite rank Butler groups*, Trans. Amer. Math. Soc. **305** (1988), 129–142.
5. M. Dugas and R. Thomé, *The functor Bext under the negation of CH*, Forum Math. **3** (1991), 23–33.
6. P.C. Eklof and A.H.Mekler, *Almost free modules*, North-Holland, 1990.
7. M. Foreman and M. Magidor, *A version of weak □*, to appear.
8. L. Fuchs, *A survey of Butler groups of infinite rank*, this volume.
9. ———, *Infinite abelian groups*, vol. I, Academic Press, New York and London, 1970.

10. L. Fuchs and M. Magidor, *Butler groups of arbitrary cardinality*, Israel Journal of Math. **84** (1993), 239–263.
11. A. Hajnal, I. Juhasz, and S. Shelah, *Splitting strongly almost disjoint families*, Trans. of the A.M.S. **295** (1986), 369–387.
12. T.K.Menas, *A combinatorial property of $P_\kappa\lambda$*, J. of Symbolic Logic **41** (1976), 225–234.

INSTITUE OF MATHEMATICS, HEBREW UNIVERSITY, JERUSALEM ISRAEL 91904
E-mail address: shelah@math.huji.ac.il

Contemporary Mathematics
Volume **171**, 1994

Representations and Duality

R. MINES, C. VINSONHALER AND W.J. WICKLESS

ABSTRACT. We define a duality between the category of representations of a poset over a field and a category of linear-topological representations of the same poset with the opposite ordering. We use this duality to give a characterization of indecomposable cokernels of representation maps of the form

$$0 \to \bigcap_{\nu \in N} A_\nu \to \prod_{\nu \in N} A_\nu$$

where each A_ν is a one-dimensional representation (over any field) of an arbitrary poset and N is any index set.

0. Introduction

Recently, there has been considerable interest by Abelian group theorists in representations of partially ordered sets. This interest stems from early work of M.C.R. Butler [**B**], who established a category equivalence between categories of torsion-free Abelian groups called Butler groups and finite dimensional representations of certain finite partially ordered sets over the field $\mathbf{Q}$ of rational numbers. Butler also described a duality for representations which could be applied to Butler groups via his category equivalence [**B**]. This duality has been applied widely, in particular to classification problems. In [**AV**] for example, a classification, up to isomorphism, of kernels of summation maps,

$$\bigoplus_{i=1}^{n} A_i \to \sum_{i=1}^{n} A_i$$

where A_i are subgroups of the additive rationals, was dualized to cokernels of

$$\bigcap_{i=1}^{n} A_i \to \bigoplus_{i=1}^{n} A_i.$$

1991 *Mathematics Subject Classification.* Primary 16G20.
Key words and phrases. representation of posets, duality, invariants.
This paper is in final form, and no version of it will be submitted for publication elsewhere.

Generally speaking it is easier to work with representations than it is to work with torsion-free Abelian groups. The awkwardness of "quasi" concepts is exchanged for the familiarity of vector space arguments. Furthermore, representation results are more general—any theorem in the representation setting can be translated to one on Butler groups. Adhering to this viewpoint, Rangaswamy and Vinsonhaler [**RV**] extended the results of [**AV**] to a classification of indecomposable kernels of representation maps of the form

$$\bigoplus_{\nu\in N} A_\nu \to \sum_{\nu\in N} A_\nu$$

where each A_ν is a one-dimensional representation (over any field) of an arbitrary poset I, and N is any index set. Such kernels have an obvious "dual," namely, the cokernels of diagonal embeddings

$$\bigcap_{\nu\in N} A_\nu \to \prod_{\nu\in N} A_\nu .$$

Nevertheless, existing dualities do not apply in this setting since they require finite dimension. Out purpose here is to overcome this problem and obtain a duality that is applicable (Theorem 2.1). Our duality is then applied to the results of [**RV**] (Theorem 4.3).

1. Background

Throughout we let I denote a partially ordered set, and K denote a field. A **K-representation over** I, or simply a K-representation is a K-vector space V and a functor from I to the category of subspaces of V. We denote a K-representation by a pair $\mathbf{V} = (V, V(\cdot))$. In this notation we understand that V is a vector space, and that for each i in I we have $V(i)$ is a subspace of V. Since $V(\cdot)$ is a functor it follows that if $i \le j$ in I, then $V(i) \subseteq V(j)$. A **map** $f\colon \mathbf{V} \to \mathbf{W}$ of K-representations is a vector space homomorphism $f\colon V \to W$ that induces a natural transformation from the functor $V(\cdot)$ to the functor $W(\cdot)$. This is just a fancy way of saying $f(V(i)) \subseteq W(i)$ for each i in I. We denote the category of all K-representations and K-representation maps by $\mathbf{Rep}(K, I)$. The category $\mathbf{Rep}(K, I)$ is preabelian, that is an additive category with kernels and cokernels.

Let $f\colon \mathbf{V} \to \mathbf{W}$ be a map in $\mathbf{Rep}(K, I)$. Then f is an **embedding** if f is one-to-one and if $f(V(i)) = f(V) \cap W(i)$ for each i in I. If f is onto and $f(V(i)) = W(i)$ for each i, we say f is an **epimorphism.** It is clear that f is an embedding (resp. epimorphism) if and only if f is a kernel (resp. cokernel) in $\mathbf{Rep}(K, I)$.

Let $\mathbf{V}$ be a K-representation. Then each subspace U of V determines a K-representation, $\mathbf{U}$, by setting $U(i) = U \cap V(i)$. The inclusion map $U \hookrightarrow V$ induces an embedding $\mathbf{U} \hookrightarrow \mathbf{V}$. The K-representation $\mathbf{U}$ is the **subrepresentation determined by** U. Similarly if $f\colon V \to W$ is an onto linear transformation, then we can define a K-representation $\mathbf{W} = f(\mathbf{V})$ by setting $W(i) = f(V(i))$. Then f induces an epimorphism from $\mathbf{V}$ to $\mathbf{W}$.

The diagram

$$\cdots \longrightarrow \mathbf{V}_{n-1} \xrightarrow{f_{n-1}} \mathbf{V}_n \xrightarrow{f_n} \mathbf{V}_{n+1} \longrightarrow \cdots$$

is a **sequence** if $f_n \circ f_{n-1} = 0$ for all n. The same sequence is exact if for each i in I the sequence

$$V_{n-1}(i) \xrightarrow{f_{n-1}} V_n(i) \xrightarrow{f_n} V_{n+1}(i)$$

and the sequence

$$V_{n-1} \xrightarrow{f_{n-1}} V_n \xrightarrow{f_n} V_{n+1}$$

are exact sequences in the category of vector spaces over K. A **short exact sequence** is an exact sequence of the form $0 \longrightarrow \mathbf{U} \longrightarrow \mathbf{V} \longrightarrow \mathbf{W} \longrightarrow 0$.

Notice that if $\mathbf{V} \xrightarrow{f} \mathbf{W} \longrightarrow 0$ is exact, then f is an epimorphism. On the other hand, it is possible for the sequence $0 \longrightarrow \mathbf{U} \xrightarrow{f} \mathbf{V}$ to be exact without f being an embedding: let $U = K$ with $U(i) = 0$ for all i in I, and $V = V(i) = K$ for all i in I. Nevertheless we do have the following result.

THEOREM 1.1. *If*

$$0 \longrightarrow \mathbf{U} \xrightarrow{f} \mathbf{V} \xrightarrow{g} \mathbf{W} \longrightarrow 0$$

is a short exact sequence in $\mathbf{Rep}(K, I)$*, then f is an embedding and g is an epimorphism.*

PROOF. The only problem is in showing $f(U(i)) = f(U) \cap V(i)$, but this follows from the fact that the sequence

$$0 \longrightarrow U(i) \xrightarrow{f} V(i) \xrightarrow{g} W(i) \longrightarrow 0$$

is a short exact sequence in the category of vector spaces over K. □

At this point we need the notion of a topological K-representation. A **topological K-representation** $\mathbf{V} = (V, V(\cdot))$ is a K-representation in which the vector space V has a linear Hausdorff topology, and the subspaces $V(i)$ are closed subspaces. We will let $\mathbf{T\text{-}Rep}(K, I)$ denote the category of all topological K-representations over the poset I. A map in $\mathbf{T\text{-}Rep}(K, I)$ is a $\mathbf{Rep}(K, I)$ map that is also continuous. One can see that a finite-dimensional topological representation has the discrete topology. In particular, the field K always has the discrete topology.

The representation $\mathbf{K} = (K, K(i))$ in $\mathbf{Rep}(K, I)$ (or $\mathbf{T\text{-}Rep}(K, I)$) that satisfies $K(i) = K$ for all i in I is the **cotrivial representation** [**ARV**]. Here, as always, K has the discrete topology.

THEOREM 1.2. *The cotrivial representation* $\mathbf{K}$ *is injective with respect to embeddings in both* $\mathbf{Rep}(K, I)$ *and* $\mathbf{T\text{-}Rep}(K, I)$*. That is, if* $0 \longrightarrow \mathbf{U} \xrightarrow{f} \mathbf{V}$ *is an embedding and* $g\colon \mathbf{U} \to \mathbf{K}$ *is a map, then there exists a map* $\bar{g}\colon \mathbf{V} \to \mathbf{K}$ *such that* $\bar{g} \circ f = g$.

PROOF. We can restrict ourselves to $\mathbf{T\text{-}Rep}(K, I)$. Let $0 \longrightarrow \mathbf{U} \xrightarrow{f} \mathbf{V}$ be an embedding and let $g\colon \mathbf{U} \to \mathbf{K}$ be a map in $\mathbf{T\text{-}Rep}(K, I)$. Since $g\colon U \to K$ is continuous, there is an open subspace N of V satisfying $\ker g = N \cap U$. We can extend g to a map $g'\colon U + N \to K$ with $\ker g' = N$. Now extend g' to a vector space map $g\colon V \to K$. Since $\ker \bar{g} \supseteq N$ it follows that $\bar{g}$ is continuous. Since $\mathbf{K}$ has the property that $K(i) = K$ for all i in I, is clear that $\bar{g}$ is a $\mathbf{T\text{-}Rep}(K, I)$ map. □

2. Dual representations

For I a poset, let I^{op} be the poset with the opposite ordering. We define a contravariant functor, $*$, from **T-Rep**(K, I) to **T-Rep**(K, I^{op}) as follows. If $\mathbf{V} = (V, V(i))$ is an object in **T-Rep**(K, I), we set $\mathbf{V}^* = (V^*, V^*(i))$, where V^* is the set of all continuous linear functionals from V to K and $V^*(i) = V(i)^\circ$, the set of all continuous linear functionals that annihilate $V(i)$. The underlying space V^* is given the compact-open topology. Because each $V(i)^\circ$ is the intersection of kernels of continuous maps into the discrete space K, we see that each $V(i)^\circ$ is closed in V^*. It is not hard to check that $(V^*, V^*(i))$ is an object in **T-Rep**(K, I^{op}). If f is a map in **T-Rep**(K, I), then f^* is the usual dual map. Note that if $f\colon V \to W$ is continuous, then so is $f^*\colon W^* \to V^*$, where W^* and V^* are endowed with the compact-open topology. Thus f^* is a map in **T-Rep**(K, I^{op}).

We now are ready for the main theorem of this section. If each object of **Rep**(K, I) is given the discrete topology, then we can regard **Rep**(K, I) as a full subcategory of **T-Rep**(K, I),

THEOREM 2.1. *The functor $*$ restricted to the full subcategory* **Rep**(K, I), *provides a duality on* **Rep**(K, I); *that is, $(* \circ *)$ restricted to* **Rep**(K, I) *is naturally equivalent to the identity functor on* **Rep**(K, I).

PROOF. The previous discussion shows that $*$ is a contravariant functor. It remains to show that $(* \circ *)$ is equivalent to the identity on **Rep**(K, I). The arguments here are standard (see [**MO**], for example). We include them for the sake of completeness.

Since V^* is given the compact-open topology, for each v in V, the evaluation map e_v is a continuous map from V^* into K (see [**K**], p. 223). Hence, the function $v \to e_v$ is a vector space monomorphism from V into $V^{**} = (V^*)^*$. (Recall that $(V^*)^*$ denotes all continuous maps from V^* into K and K has the discrete topology.) To complete the proof that V is naturally isomorphic to V^{**}, we will show that each continuous map from V^* into K is evaluation at some v in V.

For V_0 a subspace of V let V_0° be the annihilator of V_0 and let $0 \in V^*$ be the zero map. Since both V and K are discrete, it is easy to check that in the compact-open topology for V^* a neighborhood base for 0 is

$$\{V_0^\circ : V_0 \text{ is a finite dimensional subspace of } V\}.$$

Let $0 \neq \vartheta \in V^{**}$. Choose f in V^* such that $\vartheta(f) = 1$. Since ϑ is continuous and K is discrete there exists a finite dimensional subspace $V_0 \subseteq V$ with $\vartheta(f + V_0^\circ) = 1$. Thus $\vartheta(V_0^\circ) = 0$.

Let $V = V_0 \oplus V_1$ where V_1 is any complementing subspace, and let π be the projection of V onto V_0 with kernel V_1. It is easy to check that there is a vector space decomposition $V^* = V_0^\circ \oplus V_1^\circ$. It is also easy to check that the map $\pi^*\colon V_0^* \to V_1^\circ$, given by $\pi^*(f) = f\pi$, is a vector space isomorphism.

Returning to our map ϑ, note that $\vartheta\pi^* \in (V_0^*)^*$. Since V_0 is finite dimensional, the space V_0^* is endowed with the discrete topology. Thus $(V_0^*)^*$ is just the ordinary vector space double dual. By the classical duality for finite dimensional vector spaces $\vartheta\pi^* = e_v$ for some v in V_0.

Let $i\colon V_0 \to V$ be inclusion. We claim that $\vartheta = e_{iv}$. To verify this claim let f be an arbitrary element of V^*. Write $f = f_0 + f_1$ where $f_0 \in V_0^\circ$ and $f_1 \in V_1^\circ$. Choose g in V_0^* with $\pi^*(g) = f_1$. It follows that $\vartheta(f) = \vartheta(f_0 + f_1) = \vartheta(f_1) = \vartheta(\pi^*(g)) = e_v(g) = g(v)$. On the other hand $e_{iv}(f) = e_{iv}(f_0 + f_1) = f_0(iv) + f_1(iv) = f_1(iv) = g\pi(iv) = g(v)$. Since f was an arbitrary element of V^*, we have $\vartheta = e_{iv}$.

To see that the compact-open topology on V^{**} coincides with the discrete topology let e_0 be evaluation at 0 in V. We will show that $\{e_0\}$ is an open set in V^{**}. Since $V^{**} = (V^*)^*$ has the compact-open topology and $\{0\}$ is open in K, it is enough to produce a compact set $C \subseteq V^*$ such that $\{e_0\} = C^\circ$. Let $\{v_j : j \in J\}$ be a basis for V. For each fixed i in J, define f_i in V^* to be the K-linear map such that $f_i(v_i) = 1$ and $f_i(v_j) = 0$ for $j \neq i$. Let $C = \{f_j : j \in J\} \cup \{0\}$. Then C is a compact subset of V^* because any open set in V^* containing 0 must contain all but a finite number of the f_j's (refer to the third paragraph of the proof). Let $\delta \in C^\circ$. We've already shown that $\delta = e_v$ for some v in V. We have $\delta(f_j) = f_j(v) = 0$ for all j in J. It follows that $v = 0$ and we're done. □

THEOREM 2.2. *Let*

$$0 \longrightarrow \mathbf{U} \longrightarrow \mathbf{V} \longrightarrow \mathbf{W} \longrightarrow 0$$

be a short exact sequence in $\mathbf{T}$-$\mathbf{Rep}(K, I)$. *Then*

$$0 \longrightarrow \mathbf{W}^* \longrightarrow \mathbf{V}^* \longrightarrow \mathbf{U}^* \longrightarrow 0$$

is short exact.

PROOF. This follows because the cotrivial representation $\mathbf{K}$ is injective with respect to embeddings, by Theorem 1.2. □

3. Types

A **type** is a representation $\mathbf{T} = (T, T(i))$ in $\mathbf{Rep}(K, I)$ with $T = K$. The cotrivial representation is an example of a type. If $\mathbf{V}$ is a representation in $\mathbf{T}$-$\mathbf{Rep}(K, I)$ and if $\mathbf{T}$ is a type, define $\mathbf{V}(\mathbf{T})$ to be the subrepresentation of $\mathbf{V}$ determined by the subspace $V(\mathbf{T}) = \bigcap\{V(i) : T(i) \neq 0\}$. Note that $V(\mathbf{T})$ is a closed subspace of V, so that this does define a subrepresentation. Dually define $\mathbf{V}[\mathbf{T}] = \bigcap\{\ker f : f \in \operatorname{Hom}(\mathbf{V}, \mathbf{T})\}$. Since $\mathbf{V}[\mathbf{T}]$ is an intersection of kernels it follows that the underlying vector space $V[\mathbf{T}]$ is a closed subspace of V and $\mathbf{V}[\mathbf{T}]$ is the subrepresentation determined by $V[\mathbf{T}]$.

LEMMA 3.1. *Let* $\mathbf{V}$ *be in* $\mathbf{T}$-$\mathbf{Rep}(K, I)$ *and let* $\mathbf{T}$ *be a type. Then, as vector spaces,*

(a) $V(\mathbf{T}) = \sum\{f(T) : f \in \operatorname{Hom}(\mathbf{T}, \mathbf{V})\}$
(b) $V[\mathbf{T}] = \sum\{V(i) : T(i) = 0\}$, *if* V *has the discrete topology.*

PROOF. (a): Suppose $0 \neq f \in \operatorname{Hom}(\mathbf{T}, \mathbf{V})$ and $T(i) \neq 0$. Then $f(T) \subseteq V(i)$. This shows $W = \sum\{f(T) : f \in \operatorname{Hom}(\mathbf{T}, \mathbf{V})\} \subseteq V(\mathbf{T})$. Conversely, any vector space map $T \to W$ is a continuous representation map from $\mathbf{T}$ into $\mathbf{V}$. It follows that $V(\mathbf{T}) \subseteq W$.

(b): Suppose $T(i) = 0$ and $f \in \operatorname{Hom}(\mathbf{V}, \mathbf{T})$. Since f is a representation map, $f(V(i)) = 0$. This shows $W = \sum\{V(i) : T(i) = 0\} \subseteq V[\mathbf{T}]$. Conversely, a vector space map from V/W to T defines a continuous representation map from $\mathbf{V}$ to $\mathbf{T}$, since V is discrete. It follows that $V[\mathbf{T}] \subseteq W$. □

LEMMA 3.2. *Let* $\mathbf{V}$ *be an object in* $\mathbf{T}\text{-}\mathbf{Rep}(K, I)$ *having the discrete topology. Then for each set of types* $\boldsymbol{\mathcal{M}}$ *in* $\mathbf{Rep}(K, I)$ *we have.*

(a) $\left(\bigcap_{\mathbf{T}\in\boldsymbol{\mathcal{M}}} V[\mathbf{T}]\right)^\circ = \sum_{\mathbf{T}\in\boldsymbol{\mathcal{M}}} V^*[\mathbf{T}^*]$

(b) $\left(\sum_{\mathbf{T}\in\boldsymbol{\mathcal{M}}} V^*[\mathbf{T}^*]\right)^\circ = \bigcap_{\mathbf{T}\in\boldsymbol{\mathcal{M}}} V[\mathbf{T}]$.

PROOF. Note that, for $\mathbf{T}$ a type in $\mathbf{Rep}(K, I)$, the type $\mathbf{T}^*$ in $\mathbf{Rep}(K, I^{\mathrm{op}})$ is simply the type defined by $T^*(i) = 0$ if and only if $T(i) = K$. The following equations represent easy vector space computations using Lemma 3.1 and the definitions

$$\begin{aligned}\sum_{\mathbf{T}\in\boldsymbol{\mathcal{M}}} V^*(\mathbf{T}^*) &= \sum_{\mathbf{T}\in\boldsymbol{\mathcal{M}}}\left(\bigcap_{T^*(i)\neq 0} V^*(i)\right) = \sum_{\mathbf{T}\in\boldsymbol{\mathcal{M}}}\left(\bigcap_{T(i)=0} V(i)^\circ\right)\\ &= \sum_{\mathbf{T}\in\boldsymbol{\mathcal{M}}}\left(\sum_{T(i)=0} V(i)\right)^\circ = \left(\bigcap_{\mathbf{T}\in\boldsymbol{\mathcal{M}}}\left(\sum_{T(i)=0} V(i)\right)\right)^\circ\\ &= \left(\bigcap_{\mathbf{T}\in\boldsymbol{\mathcal{M}}} V[\mathbf{T}]\right)^\circ\end{aligned}$$

and

$$\begin{aligned}\bigcap_{\mathbf{T}\in\boldsymbol{\mathcal{M}}} V[\mathbf{T}] &= \bigcap_{\mathbf{T}\in\boldsymbol{\mathcal{M}}}\left(\sum_{T(i)=0} V(i)\right) = \bigcap_{\mathbf{T}\in\boldsymbol{\mathcal{M}}}\left(\sum_{T^*(i)\neq 0} (V^*(i))^\circ\right)\\ &= \bigcap_{\mathbf{T}\in\boldsymbol{\mathcal{M}}}\left(\bigcap_{T^*(i)\neq 0} V^*(i)\right)^\circ = \left(\sum_{\mathbf{T}\in\boldsymbol{\mathcal{M}}}\left(\bigcap_{T^*(i)\neq 0} V^*(i)\right)\right)^\circ\\ &= \left(\sum_{\mathbf{T}\in\boldsymbol{\mathcal{M}}} V^*(\mathbf{T}^*)\right)^\circ. \quad \square\end{aligned}$$

LEMMA 3.3. *Let* $\mathbf{V}$ *be an element of* $\mathbf{T}\text{-}\mathbf{Rep}(K, I)$ *having the discrete topology. Then as vector spaces.*

(a) $V^*(\mathbf{T}^*) = V[\mathbf{T}]^\circ \cong (V/V[\mathbf{T}])^*$

(b) $V[\mathbf{T}] = V^*(\mathbf{T}^*)^\circ \cong (V^*/V^*(\mathbf{T}^*))^\circ$.

PROOF. (a): To show that $V^*(\mathbf{T}^*) = V[\mathbf{T}]^\circ$, we use Lemma 3.1 and observe:

$$V[\mathbf{T}]^\circ = \left(\sum_{T(i)=0} V(i)\right)^\circ = \bigcap_{T(i)=0} V(i)^\circ = \bigcap_{T^*(i)\neq 0} V^*(i) = V^*(\mathbf{T}^*).$$

The isomorphism $V[\mathbf{T}]^\circ \cong (V/V[\mathbf{T}])^*$ is standard.

(b): This has a similar proof, using $V^{**} \cong V$. $\square$

THEOREM 3.4. *If* $\mathbf{V}$ *is an object in* $\mathbf{Rep}(K, I)$, *then the dual of the exact sequence*

$$0 \longrightarrow \mathbf{V}(\mathbf{T}) \longrightarrow \mathbf{V} \longrightarrow \mathbf{V}/\mathbf{V}(\mathbf{T}) \longrightarrow 0$$

is

$$0 \longrightarrow \mathbf{V}^*[\mathbf{T}^*] \longrightarrow \mathbf{V}^* \longrightarrow \mathbf{V}^*/\mathbf{V}^*[\mathbf{T}^*] \longrightarrow 0$$

PROOF. By Lemma 3.3, $\mathbf{V}^*(\mathbf{T}^*) \cong (\mathbf{V}/\mathbf{V}[\mathbf{T}])^*$ as vector spaces. We need to show that, under the isomorphism, their topological and representation structures coincide. If $j \in I^{\mathrm{op}}$, then using Lemma 3.1 and the definitions:

$$\begin{aligned} V^*(\mathbf{T}^*)(j) &= V^*(\mathbf{T}^*) \cap V^*(j) \\ &= \left(\bigcap_{\mathbf{T}^*(i) \neq 0} V^*(i) \right) \cap V^*(j) \\ &= \left(\bigcap_{\mathbf{T}(i) = 0} V(i)^\circ \right) \cap V(j)^\circ \\ &= \left(\sum_{\mathbf{T}(i)=0} V(i) + V(j) \right)^\circ \\ &= (V[\mathbf{T}] + V(j))^\circ \\ &\cong (V/V[\mathbf{T}])^* (j). \end{aligned}$$

Thus, the representation structures are identical. The topology on $\mathbf{V}(\mathbf{T}^*)$ is the subspace topology inherited from $\mathbf{V}^*$. On the other hand, the compact-open topology on $\mathrm{Hom}(V/V[\mathbf{T}], K)$ is easily seen to be the subspace topology if we identify $\mathrm{Hom}(V/V[\mathbf{T}], K)$ with $V[\mathbf{T}]^\circ$ in $\mathrm{Hom}(V, K)$. This establishes that $\mathbf{V}^*(\mathbf{T}^*)$ is the correct first term in the second sequence. The rest of the proof involves routine variations on the arguments given up to here and so is omitted. $\square$

4. The representations $\mathbf{G}[\mathcal{F}]$ and $\mathbf{G}(\mathcal{F})$

Let $\mathcal{F}$ be a family of types. Let $\sup \mathcal{F}$ be the type defined by setting $\sup \mathcal{F} = (K, \sum_{\mathbf{T} \in \mathcal{F}} T(i))$. Then $\sup \mathcal{F}$ is the supremum of the types in $\mathcal{F}$. Note that $\sup \mathcal{F} = \sum_{\mathbf{T} \in \mathcal{F}} \mathbf{T}$. Let $\nabla \colon \bigoplus_{\mathbf{T} \in \mathcal{F}} T \to \sup \mathcal{F}$ denote the codiagonal map. Set $\mathbf{G}(\mathcal{F}) = \ker \nabla$. Dually set $\inf \mathcal{F} = \bigcap \mathcal{F}$ be the infinum of the types in $\mathcal{F}$, and let $\Delta \colon \inf \mathcal{F} \to \prod_{\mathbf{T} \in \mathcal{F}} \mathbf{T}$ denote the diagonal map. Then $\mathbf{G}[\mathcal{F}]$ is $\operatorname{coker} \Delta$.

THEOREM 4.1. *Let* $\mathcal{F}$ *be a family of types in* $\mathbf{Rep}(K, I)$ *and* $\mathcal{F}^*$ *the dual family in* $\mathbf{T}\text{-}\mathbf{Rep}(K, I^{\mathrm{op}})$.

(a) *Then* $\mathbf{G}(\mathcal{F})^* \cong \mathbf{G}[\mathcal{F}^*]$ *in* $\mathbf{T}\text{-}\mathbf{Rep}(K, I^{\mathrm{op}})$.

(b) $\mathbf{G}[\mathcal{F}^*]^* \cong \mathbf{G}(\mathcal{F})$ *in* $\mathbf{Rep}(K, I)$.

PROOF. The exact sequence (in $\mathbf{Rep}(K, I)$)

$$0 \longrightarrow \mathbf{G}(\mathcal{F}) \longrightarrow \bigoplus \mathcal{F} \longrightarrow \sup \mathcal{F} \longrightarrow 0$$

dualizes to

$$0 \longrightarrow \inf \mathcal{F}^* \longrightarrow \prod \mathcal{F}^* \longrightarrow \mathbf{G}[\mathcal{F}^*] \longrightarrow 0$$

since, clearly, the dual of $\bigoplus \mathcal{F}$ is $\prod \mathcal{F}^*$ and the dual of $\sup \mathcal{F}$ is $\inf \mathcal{F}^*$. Thus $\mathbf{G}(\mathcal{F})^* \cong \mathbf{G}[\mathcal{F}^*]$. Because $\prod \mathcal{F}^*$ has the compact-open (product) topology, the dual of $\prod \mathcal{F}^*$ is $\bigoplus \mathcal{F}$. Thus, the dual of the second sequence is the first and $\mathbf{G}[\mathcal{F}^*]^* \cong \mathbf{G}(\mathcal{F})$. □

THEOREM 4.2. [**RV**] *Let $\mathcal{F}_1$ and $\mathcal{F}_2$ be families of types in $\mathbf{Rep}(K, I)$ such that $\mathbf{H}_1 = \mathbf{G}(\mathcal{F}_1)$ and $\mathbf{H}_2 = \mathbf{G}(\mathcal{F}_2)$ are indecomposable. Then $\mathbf{H}_1 \cong \mathbf{H}_2$ if and only if for each set $\mathcal{M}$ of types in $\mathbf{Rep}(K, I)$*

$$\dim\left(\bigcap_{\mathbf{T}\in\mathcal{M}} H_1[\mathbf{T}]\right) = \dim\left(\bigcap_{\mathbf{T}\in\mathcal{M}} H_2[\mathbf{T}]\right)$$

whenever either side is finite.

PROOF. This is a restatement of Theorem 2.9 in [**RV**]. In that theorem, the finiteness condition on dimension is not stated explicitly, but is implicit in the proof. □

THEOREM 4.3. *Let $\mathcal{E}_1$ and $\mathcal{E}_2$ be families of types in $\mathbf{Rep}(K, I^{\mathrm{op}})$ such that $\mathbf{G}_1 = \mathbf{G}[\mathcal{E}_1]$ and $\mathbf{G}_2 = \mathbf{G}[\mathcal{E}_2]$ are indecomposable. Then $\mathbf{G}_1 \cong \mathbf{G}_2$ if and only if for each set $\mathcal{M}^*$ of types in $\mathbf{Rep}(K, I^{\mathrm{op}})$*

$$\dim\left(G_1 \Big/ \sum_{\mathbf{T}^*\in\mathcal{M}^*} G_1(\mathbf{T}^*)\right) = \dim\left(G_2 \Big/ \sum_{\mathbf{T}^*\in\mathcal{M}^*} G_2(\mathbf{T}^*)\right)$$

whenever either side is finite.

PROOF. If $\mathbf{G}_1 \cong \mathbf{G}_2$ the two dimensions are clearly equal. To prove the converse, we impose topologies on $\mathbf{G}_1$ and $\mathbf{G}_2$ and show that they are isomorphic in $\mathbf{T}\text{-}\mathbf{Rep}(K, I^{\mathrm{op}})$. Indeed, note that we may write any type $\mathbf{T}^*$ in $\mathbf{Rep}(K, I^{\mathrm{op}})$ as the dual of a type $\mathbf{T}$ in $\mathbf{Rep}(K, I)$, where $\mathbf{T}(i) = 0$ if and only if $\mathbf{T}^*(i) = K$. Thus, we may take $\mathcal{E}_1 = \mathcal{F}_1^*$ and $\mathcal{E}_2 = \mathcal{F}_2^*$, for $\mathcal{F}_1$ and $\mathcal{F}_2$ families of types in $\mathbf{Rep}(K, I)$. By Theorem 4.1(a), we may impose topologies on $\mathbf{G}_1$ and $\mathbf{G}_2$ by setting $\mathbf{H}_i = \mathbf{G}(\mathcal{F}_i)$ and $\mathbf{G}_i = \mathbf{H}_i$ for $i = 1, 2$. Next note that, by Lemma 3.2(a), for each set of types $\mathcal{M}^*$ in $\mathbf{Rep}(K, I^{\mathrm{op}})$

$$\begin{aligned}\dim\left(G_1/\sum_{\mathbf{T}^*\in\mathcal{M}^*} G_1(\mathbf{T}^*)\right) &= \dim\left(H_1^*/\sum_{\mathbf{T}^*\in\mathcal{M}^*} H_1^*(\mathbf{T}^*)\right)\\ &= \dim\left(H_1^*\Big/\left(\bigcap_{\mathbf{T}\in\mathcal{M}} H_1[\mathbf{T}]\right)^\circ\right)\\ &= \dim \operatorname{Hom}\left(\bigcap_{\mathbf{T}\in\mathcal{M}} H_1[\mathbf{T}], K\right)\end{aligned}$$

If $\dim\bigcap_{\mathbf{T}\in\mathcal{M}} H_1[\mathbf{T}]$ is finite, then we have

$$\dim\bigcap_{\mathbf{T}\in\mathcal{M}} H_1[\mathbf{T}] = \dim\operatorname{Hom}\left(\bigcap_{\mathbf{T}\in\mathcal{M}} H_1[T], K\right) = \dim\left(G_1\Big/\sum_{\mathbf{T}^*\in\mathcal{M}^*} G_1(\mathbf{T}^*)\right).$$

The hypotheses then imply that

$$\dim\bigcap_{\mathbf{T}\in\mathcal{M}} H_1[\mathbf{T}] = \dim\bigcap_{\mathbf{T}\in\mathcal{M}} H_2[\mathbf{T}].$$

By Theorem 4.2, $\mathbf{H}_1 \cong \mathbf{H}_2$. It follows that $\mathbf{G}_1 = \mathbf{H}_1^* \cong \mathbf{H}_2^* = \mathbf{G}_2$ as elements of $\mathbf{T}\text{-}\mathbf{Rep}(K, I^{\mathrm{op}})$ and the proof is complete. □

References

[AV] D. Arnold and C. Vinsonhaler, *Invariants for classes of indecomposable representations of finite posets*, J. Algebra **147** (1992), 245–264.

[ARV] D. Arnold, F. Richman, and C.. Vinsonhaler, *Representations of finite posets and valuated groups*, J. Algebra **155** (1993), 110–126.

[B] M.C.R. Butler, *Torsion-free modules and diagrams of vector spaces*, Proc. London Math. Soc. (3) **18** (1968), 635–652.

[K] J.L. Kelley, *General Topology*, D. Van Nostrand, 1955.

[RW] F. Richman and E. Walker, Ext *in pre-Abelian categories*, Pacific J. Math. (1977), 521–535.

[RV] K.M. Rangaswamy and C. Vinsonhaler, *Butler groups and representations of infinite rank*, J. Algebra (to appear).

[MO] C. Menini and A. Orsatti, *Good dualities and strongly quasi-injective modules*, Ann. di Mat. pura e appl. **129** (1981), 187–230.

New Mexico State University, Las Cruces, New Mexico 88003

E-mail address: ray@nmsu.edu

University of Connecticut, Storrs, Connecticut 06269

E-mail address: vinson@uconnvm.bitnet

University of Connecticut, Storrs, Connecticut 06269

E-mail address: wjwick@uconnvm.bitnet

Contemporary Mathematics
Volume **171**, 1994

Extending a Splitting Criterion on Mixed Modules

OTTO MUTZBAUER AND ELIAS TOUBASSI

1. Introduction

In this paper we extend the application of a splitting criterion on mixed modules obtained in [**3**] to its utmost limit. We remove two limitations on the class of modules satisfying the main theorem in [**3**]. First, we completely remove the condition that the torsion submodule is a direct sum of cyclics. In this paper the torsion submodule is arbitrary. Second, we relax the condition on the quotient p-rank from 0 to any finite rank. More significantly we show that the generalization in this direction reaches its limit with finite quotient p-rank; when the quotient p-rank is infinite the splitting question becomes undecidable using this criterion. In one theorem we provide examples of the pathologies that can occur when the quotient p-rank is infinite. We note that the results of this paper hold for local mixed abelian groups.

2. Preliminaries

Let R denote a *discrete valuation domain*, i. e. a local principal ideal domain with prime p. All modules are always understood to be R-modules. A module G is said to *split* if its torsion submodule $\mathbf{t}\,G$ is a direct summand. If L is any free pure submodule of the module G, then by [**2**, Section 32, Exercise 7] there is a free pure submodule F containing L such that $G/(\mathbf{t}\,G \oplus F)$ is torsion-free divisible. Such a pure free module F with torsion-free divisible quotient $G/(\mathbf{t}\,G \oplus F)$ is called *relatively maximal pure free in G*. A pure free submodule is relatively maximal if and only if $(F + \mathbf{t}\,G)/\mathbf{t}\,G$ is basic in $G/\mathbf{t}\,G$. For a torsion-free module X define the *p-rank of X* as the dimension of X/pX over the field R/pR. We use this to make two definitions. First we define the *quotient p-rank* of a mixed module G to be the p-rank of the torsion-free quotient $G/\mathbf{t}\,G$. Second

1991 *Mathematics Subject Classification.* 20K21.
This paper is in final form and no version of it will be submitted for publication elsewhere.

we define the *co-p-rank of G relative to F* to be the dimension of the vector space $G/(\mathbf{t}\,G \oplus F)$. By [**2**, Section 32, Exercise 5] the rank of a relatively maximal pure free submodule F in a mixed module equals the quotient p-rank of G and is an invariant of the mixed module. However the co-p-rank of a module is invariant if and only if the quotient p-rank is finite.

If $\mathbf{t}$ is a torsion module then $\{x_i^u \mid i \in \mathbb{N}, u \in I_i\} \subset \mathbf{t}$ is called a *straight basis of $\mathbf{t}$* as in [**1**, 1.4] if $\mathbf{t}[p^i]/\mathbf{t}[p^{i-1}] = \bigoplus_{u \in I_i} R(x_i^u + \mathbf{t}[p^{i-1}])$ for all $i \in \mathbb{N}$, where $\mathbf{t}[p^i] = \{x \in \mathbf{t} \mid \operatorname{ann} x = p^i R\}$ is the p^i-socle of $\mathbf{t}$. If $s_i = \dim_{R/pR} \mathbf{t}[p^i]/\mathbf{t}[p^{i-1}] = |I_i|$ then the sequence $s = (s_i \mid i \in \mathbb{N})$ is an invariant of $\mathbf{t}$. We call the quotient $\mathbf{t}[p^n]/\mathbf{t}[p^{n-1}]$ an *n-section of* $\mathbf{t}$ and s_n the *dimension of an n-section.* The torsion module $\mathbf{t}$ is bounded if and only if there exists a natural number n_0 such that $s_{n_0} = 0$. It follows that $s_n = 0$ for all $n \geq n_0$.

Let T be a free module on $\{t_i^u \mid i \in \mathbb{N}, u \in I_i\}$. Then $\mathbf{t} \cong T/N$ where N is the kernel of the basic epimorphism which maps $t_i^u \mapsto x_i^u$. Such a quotient is called a *representation of* $\mathbf{t}$ relative to a straight basis $\{x_i^u \mid i \in \mathbb{N}, u \in I_i\}$.

Let G be a mixed module with torsion submodule $\mathbf{t}\,G$ with sequence $s = (s_i \mid i \in \mathbb{N})$ of section dimensions and quotient p-rank r. Then there is a relatively maximal pure free submodule F of G of rank r with quotient $G/(\mathbf{t}\,G \oplus F)$ a vector space over K of dimension d. The latter is the co-p-rank of the module G relative to F.

Let I be a set of cardinality d, then the subset

$$B = \{x_i^u, a_{i-1}^k, b_l \mid i \in \mathbb{N}, u \in I_i, k \in I, l \in P\} \subset G$$

is called a *basic generating system of G relative to a straight basis*, if

(1) $\{x_i^u \mid i \in \mathbb{N}, u \in I_i\}$ is a straight basis of $\mathbf{t}\,G$ with $\operatorname{ann} x_i^u = p^i R$ for all $i \in \mathbb{N}, u \in I_i$,

(2) $F = \bigoplus_{l \in P} Rb_l$ is a relatively maximal pure free submodule,

(3) $G/(\mathbf{t}\,G \oplus F) = \bigoplus_{k \in I} K\bar{a}_0^k$,

where $\bar{a}_{i-1}^k = a_{i-1}^k + \mathbf{t}\,G \oplus F$ and $p\bar{a}_i^k = \bar{a}_{i-1}^k$ for all $i \in \mathbb{N}, k \in I$. The definition implies that the elements a_0^k, b_l, $k \in I, l \in P$, are independent and $r = |P|$ and $s_i = |I_i|$ for all $i \in \mathbb{N}$. It is easy to see that

$$G = \langle x_i^u, a_{i-1}^k, b_l \mid i \in \mathbb{N}, u \in I_i, k \in I, l \in P \rangle.$$

A basic generating system intrinsically defines a series of equations with coefficients in R that describe the relations among the generators. Modeling upon these relations we introduce the concept of an abstract array $(\alpha, \eta) = (\alpha_{i-1,j}^{k,u}, \eta_{i-1,l}^k)$ with $i, j \in \mathbb{N}, u \in I_j, k \in I, l \in P$ having entries in R with the condition that the entries of α are in $R \setminus pR$ or 0. Both arrays are assumed to be *row finite in j, u and l*, i. e. for a fixed pair (k, i) there is $\alpha_{i,j}^{k,u} = 0$ for almost all pairs (j, u) and $\eta_{i,l}^k = 0$ for almost all l. We shall refer to α, η and (α, η) as *relation arrays.*

Let $\mathbf{t}$ be a torsion module with sequence $s = (s_j \mid j \in \mathbb{N})$ of section dimensions and let r and d be cardinal numbers and I, I_i and P be sets, such that $|I_i| = s_i$,

$|I| = d$ and $|P| = r$. Then the relation array (α, η) with $i, j \in \mathbb{N}, u \in I_j, k \in I, l \in P$ is said to have *format* (s, r, d). The array α is said to be of *format* (s, d).

We cite results in [**4**] showing that mixed modules can be described by relation arrays and that all relation arrays are realizable.

PROPOSITION 1. ([**4**, 3.1]) *Every mixed module G has a basic generating system*

$$B = \{x_i^u, a_{i-1}^k, b_l \mid i \in \mathbb{N}, u \in I_i, k \in I, l \in P\}$$

and a relation array (α, η) relative to B given by

$$\text{(1)} \qquad pa_i^k = a_{i-1}^k + \sum_{l \in P} \eta_{i-1,l}^k b_l + \sum_{j \in \mathbb{N}} \sum_{u \in I_j} \alpha_{i-1,j}^{k,u} x_j^u \qquad (i \in \mathbb{N}, k \in I).$$

The array η and the torsion elements $\sum_{j \in \mathbb{N}} \sum_{u \in I_j} \alpha_{i-1,j}^{k,u} x_j^u$ are unique relative to B.

Note the following facts. Relation arrays of modules G with divisible quotient $G/\mathbf{t}\,G$ consist only of α. Since we use a straight basis the coefficients $\alpha_{i-1,j}^{k,u}$ of α for the torsion elements $\sum_{j \in \mathbb{N}} \sum_{u \in I_j} \alpha_{i-1,j}^{k,u} x_j^u$ are unique up to pR. Moreover, relation arrays with $\alpha = 0$ belong to splitting modules. This is shown by the following argument. Let $B = \{x_i^u, a_{i-1}^k, b_l \mid i \in \mathbb{N}, u \in I_i, k \in I, l \in P\}$ be a basic generating system of G, then $\mathbf{t}\,G = \langle x_i^u \mid i \in \mathbb{N}, u \in I_i \rangle$ and $G = \mathbf{t}\,G + M$ where $M = \langle a_{i-1}^k, b_l \mid i \in \mathbb{N}, k \in I, l \in P \rangle$. The relations (1) simplify if $\alpha = 0$ and we use this to show $\mathbf{t}\,G \cap M = 0$, i. e. that G splits. Let $z \in \mathbf{t}\,G \cap M$. By (1) we obtain $z = \sum_{k \in I} \lambda_k a_{i_k}^k + \sum_{l \in P} \mu_l b_l$, where $\lambda_k, \mu_l \in R$. Since z is a torsion element there is a natural number m such that $p^m z = 0$ and if there are coefficients $\lambda_k \neq 0$ then we can choose this natural number big enough such that $p^m z = 0$ and $p^m \lambda_k \in p^{i_k} R$ for all of the finitely many $k \in I$ with $\lambda_k \neq 0$. But then

$$0 = p^m z \equiv \sum_{k \in I} p^m \lambda_k a_{i_k}^k \equiv \sum_{k \in I} p^{m - i_k} \lambda_k a_0^k \pmod{\mathbf{t}\,G \oplus F}.$$

Since $\{a_0^k \mid k \in I\}$ is a basis of the vector space $G/(\mathbf{t}\,G \oplus F)$ the coefficients $p^{m-i_k} \lambda_k$ are 0 for all k, i. e. $\lambda_k = 0$ for all $k \in I$. Hence $0 = p^m z = \sum_{l \in P} p^m \mu_l b_l \in F$ and $\mu_l = 0$ for all l, i. e. $z = 0$ as claimed.

THEOREM 2. ([**4**, 3.2]) *Let $\mathbf{t}$ be a torsion module with sequence s of section dimensions. Every relation array (α, η) of format (s, r, d) can be realized by a mixed module with torsion submodule isomorphic to $\mathbf{t}$, quotient p-rank r and co-p-rank d.*

3. Pathologies on Splitting

We begin with some basic facts.

LEMMA 3. *Let H be a torsion-free pure submodule of the mixed module G. Then the torsion submodule of G/H is $\mathbf{t}(G/H) = (H \oplus \mathbf{t}G)/H$. Moreover, G splits if the quotient G/H splits.*

PROOF. Let $g \in G$ and suppose $ng = h+t \in H \oplus \mathbf{t}\,G$, $n \in N$. Then $nmg = mh$ for some natural number m. Thus there is an $h' \in H$ with $nmh' = mh$, since H is pure. Consequently $nh' = h$ and $ng = nh' + t$. This implies $g - h' \in \mathbf{t}\,G$ and hence $g \in H \oplus \mathbf{t}\,G$. Thus $(H \oplus \mathbf{t}\,G)/H$ is the torsion submodule of G/H.

Moreover, if $G/H = \mathbf{t}(G/H) \oplus L/H = (H \oplus \mathbf{t}\,G)/H \oplus L/H$ splits, then $G = \mathbf{t}\,G \oplus L$ splits also. □

The following lemma is easily proved by the modular law of Dedekind.

LEMMA 4. *All submodules of a splitting module that contain the entire torsion submodule also split.*

The relation array $(\alpha, \eta) = (\alpha^{k,u}_{i-1,j}, \eta^{k}_{i-1,l})$ is said to be *small* if the array $\left(\sum_{i=s}^{s+j-1} \alpha^{k,u}_{i,j} p^{i-s}\right)^{k,u}_{s,j}$ is a relation array, i. e. if α is a small relation array in the sense of [**3**]. In general this array is not necessarily a relation array since the row finiteness in j and u is not guaranteed.

Let G be a module with T(G) a direct sum of cyclics and G/T(G) divisible. In [**3**] it was shown that such a G is determined up to isomorphism by its relation array (with respect to any basic system).

Unfortunately this is not the case when the quotient p-rank is infinite. Indeed the next theorem shows that a split mixed module can have any prescribed relation array. This is similar to the type of pathology that is well known in the theory of torsion-free abelian groups.

THEOREM 5. *Let G be a split module with unbounded torsion submodule having a sequence s of section dimensions, and free quotient $G/\,\mathbf{t}G$ of infinite rank f. Let d be any cardinal less than or equal to f and α be any relation array of format (s, d). Then there is basic generating system of G with corresponding relation array (α, η) having format (s, f, d).*

PROOF. Let $G = H \oplus T$, where H is free (and relatively maximal pure) of rank f and $T = \mathbf{t}\,G = \langle x^u_i \mid i \in \mathbb{N}, u \in I_i \rangle$, where I_i are sets of cardinality s_i, the section dimensions of $\mathbf{t}\,G$. We restrict to the case $d = f$. (The other cases are easily treated by suitable indexing of a basis of H.) Let J be a set of cardinality f. We may write H in the form $H = \langle a^k_i \mid i \in \mathbb{N}_0, k \in J \rangle$. We use H and the relation array α of format (s, f) to define

$$F = \langle b_i^k \mid i \in \mathbb{N}_0, k \in J \rangle \quad \text{with} \quad b_{i-1}^k = pa_i^k - a_{i-1}^k - \sum_{j\in\mathbb{N}} \sum_{u\in I_j} \alpha_{i-1,j}^{k,u} x_j^u.$$

It is straightforward to check that F is free. Moreover, F is a relatively maximal pure free submodule since $(F \oplus \mathbf{t}\,G) \cap H = \langle pa_{i+1}^k - a_i^k \mid i \in \mathbb{N}_0, k \in J \rangle$ and

$$\frac{G}{F \oplus \mathbf{t}\,G} \cong \frac{H}{(F \oplus \mathbf{t}\,G) \cap H} \cong K^f,$$

where K is the quotient field of R. Thus $G/F = \langle \bar{x}_i^u, \bar{a}_{i-1}^k \mid i \in \mathbb{N}, u \in I_i, k \in J \rangle$, where $\bar{x}_i^u = x_i^u + F$ and $\bar{a}_i^k = a_i^k + F$, and we have the relations

$$p\bar{a}_i^k = \bar{a}_{i-1}^k + \sum_{j\in\mathbb{N}} \sum_{u\in I_j} \alpha_{i-1,j}^{k,u} \bar{x}_j^u, \quad i \in \mathbb{N}, k \in J.$$

By Lemma 3, $\mathbf{t}(G/F) = (F \oplus \mathbf{t}\,G)/F = \langle \bar{x}_i^u \mid i \in \mathbb{N}, u \in I_i \rangle$ and $\{\bar{x}_i^u, \bar{a}_{i-1}^k \mid i \in \mathbb{N}, u \in I_i, k \in I\}$ is a basic generating system of G/F with corresponding relation array α. Let $\overline{B} = \{x_i^u, a_{i-1}^k, b_{i-1}^k \mid i \in \mathbb{N}, u \in I_i, k \in I\}$. It is straightforward to verify that $\overline{B}$ is a basic generating system with relation array (α, η) of format (s, f, f). $\square$

The above theorem is a decisive set-back to providing an analogue to Theorem 6 in [**3**]. In the latter paper we were able to show that "small" relation arrays were necessary and sufficient for splitting mixed modules in the class $\mathcal{H}$. In view of Theorem 5 above the splitting question becomes undecidable when using relation arrays. All is not lost however as we see in the next section if the quotient p-rank is finite.

4. Splitting Criterion

In this section we give a necessary and sufficient criterion for the splitting of modules of finite quotient p-rank.

THEOREM 6. *For a mixed module G of finite quotient p-rank the following are equivalent:*

(1) *G splits;*
(2) *G has only small relation arrays;*
(3) *G has one small relation array.*

PROOF. (1) $\Rightarrow$ (2): If G splits then one may choose a basic generating system $A = \{x_i^u, a_{i-1}^k, b_l \mid i \in \mathbb{N}, u \in I_i, k \in I, 1 \leq l \leq r\}$ with relation array $\alpha = 0$ and where $F = \langle b_1, \dots, b_r \rangle$ is a relatively maximal pure free submodule of quotient p-rank r. If

$$C = \{y_i^u, c_{i-1}^k, d_l \mid i \in \mathbb{N}, u \in I_i, k \in I, 1 \leq l \leq r\}$$

is another basic generating system of G, where $H = \langle d_1, \dots, d_r\rangle$ is a relatively maximal pure free submodule, we must show that the basic relation array corresponding to C is small. First we form a basic generating system $B' = \{y_i^u, a_{i-1}^k, b_l \mid i \in \mathbb{N}, u \in I_i, k \in I, 1 \le l \le r\}$ which replaces the torsion generators of A. Clearly the relation array corresponding to B' is also 0.

Since $\{a_0^k, b_l \mid k \in I, 1 \le l \le r\}$ is a maximal linearly independent set there is a function $f : I \longrightarrow \mathbb{N}_0$ such that $p^{f(k)}c_0^k = \sum_{h\in I} m_h^k a_0^h + u^k$ with $m_h^k \in R$ and $u^k \in F$. We add the set $\{p^{n_k}c_0^k \mid k \in I, 1 \le n_k \le f(k)\}$ to the basic generating system C and reindex such that $p^{f(k)}c_0^k$ plays the role of "c_0^k". This is again a basic generating system which is small if and only if C is small. In view of our reindexing we may assume that the generators c_0^k in C are linear combinations of the elements a_0^k and of b_l with coefficients in R. Define $e_i^k = \sum_{h\in I} m_h^k a_i^h$. Then the torsion part of the relation array of G relative to the basic generating system $B'' = \{y_i^u, e_{i-1}^k, b_l \mid i \in \mathbb{N}, u \in I_i, k \in I, 1 \le l \le r\}$ is obviously also 0 since $pe_i^k = \sum_{h\in I} m_h^k pa_i^h = \sum_{h\in I} m_h^k a_{i-1}^h + z_{i-1}^k = e_{i-1}^k + z_{i-1}^k$ for all $i \in \mathbb{N}, k \in I$ with suitable $z_{i-1}^k \in F$. In particular, we have $e_0^k \equiv c_0^k$ modulo F. Since the cosets $c_i^k + H + \mathbf{t}\,G$ for $i \in \mathbb{N}$ are uniquely determined by the elements c_0^k we have $e_i^k \equiv c_i^k$ modulo $(F + H + \mathbf{t}\,G)$. So in order to compute the relation array relative to C it remains to replace the e_i^k in the basic generating system B'' within their cosets modulo $F + H + \mathbf{t}\,G$ by $c_i^k = e_i^k + t_i^k + w_i^k$, where $t_i^k \in \mathbf{t}\,G$ and $w_i^k \in F + H$. Write the torsion elements $t_i^k = c_i^k - e_i^k - w_i^k$ in terms of the generators $\{y_j^u \mid j \le 1, u \in I_j\}$ as follows $t_i^k = \sum_{j\ge1}\sum_{u\in I_j} \tau_{i,j}^{k,u} y_j^u$. It is easy to check that the array $(\tau_{i,j}^{k,u})$ satisfies the definition of a relation array. Now

$$\begin{aligned} pc_i^k &= p(e_i^k + t_i^k + w_i^k) = e_{i-1}^k + pt_i^k + pw_i^k + z_{i-1}^k \\ &= c_{i-1}^k - t_{i-1}^k + pt_i^k - w_{i-1}^k + pw_i^k + z_{i-1}^k \\ &= c_{i-1}^k + \sum_{j\ge1}\sum_{u\in I_j} \gamma_{i-1,j}^{k,u} y_j^u - w_{i-1}^k + pw_i^k + z_{i-1}^k, \end{aligned}$$

where $\gamma_{i-1,j}^{k,u} \equiv p\tau_{i,j}^{k,u} - \tau_{i-1,j}^{k,u} \pmod{p^j}$. But $pc_i^k - c_{i-1}^k \in H \oplus \mathbf{t}\,G$, hence $pw_i^k - w_{i-1}^k + z_{i-1}^k \in (F+H) \cap (H \oplus \mathbf{t}\,G) = H \oplus [(F+H) \cap \mathbf{t}\,G]$. The second summand is a torsion submodule of the finitely generated module $F + H$ and so is finitely generated, i. e. bounded. Thus $pw_i^k - w_{i-1}^k + z_{i-1}^k = h_{i-1}^k + s_{i-1}^k$, where $h_{i-1}^k \in H$ and $s_{i-1}^k \in (F+H) \cap \mathbf{t}\,G$ for all $i \in \mathbb{N}, k \in I$.

Since the s_i^k are in a bounded torsion module, it suffices to show that $(\gamma_{i-1,j}^{k,u}) \equiv p(\tau_{i,j}^{k,u}) - (\tau_{i-1,j}^{k,u})$ is small. But

$$\sum_{i=s}^{s+j-1} \gamma_{i,j}^{k,u} p^{i-s} \equiv \sum_{i=s+1}^{s+j} \tau_{i,j}^{k,u} p^{i-s} - \sum_{i=s}^{s+j-1} \tau_{i,j}^{k,u} p^{i-s} \equiv -\tau_{s,j}^{k,u} \pmod{p^j}$$

and the fact that $(\tau_{i,j}^{k,u})$ is a relation array implies that the relation array $(\gamma_{i-1,j}^{k,u})$ is small.

(2) $\Rightarrow$ (3) is obvious.

(3) $\Rightarrow$ (1): Let $\{x_i^u, a_{i-1}^k, b_l \mid i \in \mathbb{N}, u \in I_i, k \in I, l \in P\}$ be a basic generating system of G with small relation array (α, η). Let the pair (k, s) be fixed. By the row finiteness in u and j the elements

$$y_j^{k,s} = \sum_{u \in I_j} \Big(\sum_{i=s}^{s+j-1} \alpha_{i,j}^{k,u} p^{i-s} \Big) x_j^u$$

are well defined. Note that $y_j^{k,s}$ is a sum over the columns of the array α and that for a fixed pair $(k, s) \in I \times \mathbb{N}$ the sum $\sum_{j \geq 1} y_j^{k,s}$ is not well defined in general. However when the relation array α is small, i. e. $\Big(\sum_{i=s}^{s+j-1} \alpha_{i,j}^{k,u} p^{i-s}\Big)_{s,j}^{k,u}$ is a relation array, the sum makes sense. Hence for a fixed pair (k, s) only finitely many elements $y_j^{k,s}$ are not 0 and the sum $\sum_{j \geq 1} y_j^{k,s}$ is defined. The above allows us to introduce new generators b_s^k, $(k, s) \in I \times \mathbb{N}$, where

$$b_s^k = a_s^k + \sum_{j \geq 1} \sum_{u \in I_j} \Big(\sum_{i=s}^{s+j-1} \alpha_{i,j}^{k,u} p^{i-s} \Big) x_j^u. \tag{2}$$

Then we get by Proposition 1

$$\begin{aligned}
pb_s^k &= pa_s^k + p \sum_{j \geq 1} \sum_{u \in I_j} \Big(\sum_{i=s}^{s+j-1} \alpha_{i,j}^{k,u} p^{i-s} \Big) x_j^u \\
&= a_{s-1}^k + \sum_{l \in P} \eta_{s-1,l}^k b_l + \sum_{j \geq 1} \sum_{u \in I_j} \alpha_{s-1,j}^{k,u} x_j^u + \sum_{j \geq 1} \sum_{u \in I_j} \Big(\sum_{i=s}^{s+j-1} \alpha_{i,j}^{k,u} p^{i-s+1} \Big) x_j^u \\
&= b_{s-1}^k + \sum_{l \in P} \eta_{s-1,l}^k b_l - \sum_{j \geq 1} \sum_{u \in I_j} \Big(\sum_{i=s-1}^{s+j-2} \alpha_{i,j}^{k,u} p^{i-s+1} \Big) x_j^u + \sum_{j \geq 1} \sum_{u \in I_j} \alpha_{s-1,j}^{k,u} x_j^u \\
&\qquad + \sum_{j \geq 1} \sum_{u \in I_j} \Big(\sum_{i=s}^{s+j-1} \alpha_{i,j}^{k,u} p^{i-s+1} \Big) x_j^u \\
&= b_{s-1}^k + \sum_{l \in P} \eta_{s-1,l}^k b_l - \sum_{j \geq 1} \sum_{u \in I_j} \Big(\sum_{i=s-1}^{s+j-2} \alpha_{i,j}^{k,u} p^{i-s+1} - \alpha_{s-1,j}^{k,u} - \sum_{i=s}^{s+j-1} \alpha_{i,j}^{k,u} p^{i-s+1} \Big) x_j^u \\
&= b_{s-1}^k + \sum_{l \in P} \eta_{s-1,l}^k b_l + \sum_{j \geq 1} \sum_{u \in I_j} \alpha_{s+j-1,j}^{k,u} p^j x_j^u \\
&= b_{s-1}^k + \sum_{l \in P} \eta_{s-1,l}^k b_l.
\end{aligned}$$

Thus the relation array corresponding to the new basic generating system has $(\alpha_{i-1,j}^{k,u}) \equiv 0$ and G splits. $\square$

Note that (1) and (3) in Theorem 6 are equivalent for arbitrary quotient p-rank. It is well known that if the torsion submodule is bounded then the module splits. This is a trivial consequence of the above theorem because for such modules all relation arrays are small, since there must exist a natural number m such that $\alpha_{i-1,j}^{k,u} = 0$ for all $j \geq m$. The same holds if there is only a bounded part of the torsion submodule involved in the relations.

The following corollary is an easy consequence of Theorem 6 and the above remark.

COROLLARY 7. *For modules with either bounded torsion or finite quotient p-rank splitting is equivalent to all relation arrays being small.*

REFERENCES

1. *K. Benabdallah and K. Honda*, Straight bases of abelian p-groups, Abelian Group Theory, Proceedings, Honolulu (1982/83), Lect. Notes Math., vol. 1006, pp. 556–561.
2. *L. Fuchs*, Infinite Abelian Groups I+II, Academic Press (1970, 1973).
3. *O. Mutzbauer and E. Toubassi*, A splitting criterion for a class of mixed modules, Rocky Mountain J. (1993).
4. *O. Mutzbauer and E. Toubassi*, Hulls of mixed modules with finite quotient p-rank, manuscript (1993).

OTTO MUTZBAUER, MATHEMATISCHES INSTITUT, UNIVERSITÄT WÜRZBURG, AM HUBLAND, 97074 WÜRZBURG, GERMANY
E-mail address: mutzbauer@vax.rz.uni-wuerzburg.d400.de

ELIAS TOUBASSI, DEPARTMENT OF MATHEMATICS, UNIVERSITY OF ARIZONA, TUCSON, ARIZONA 85721, U. S. A.
E-mail address: elias@math.arizona.edu

Contemporary Mathematics
Volume 171, 1994

DIRECT SUMMANDS OF Z^{κ} FOR LARGE κ.

JOHN D. O'NEILL

ABSTRACT. Let Z^{κ} be the direct product of κ copies of the integers. We study direct summands of the abelian group Z^{κ} for $\kappa \geq \mu$ the least measurable cardinal number. In particular we show that any direct summand of Z^{μ} is isomorphic to Z^{α} for some α

Introduction

Let κ be an infinite cardinal number and let Z^{κ} be the direct product of κ copies of the integers Z. Suppose $\phi: Z^{\kappa} \longrightarrow G$ is a homomorphism where G is a slender group. Then ϕ maps almost components of Z^{κ} to 0. If $\kappa < \mu$ (the least measurable cardinal number) and ϕ maps all components of Z^{κ} to 0, then $\phi(Z^{\kappa}) = 0$ (see Theorem 94.4 in [2]). Using this result and the Hom functor Nunke showed in 1962 that for $\kappa < \mu$ any direct summand of Z^{κ} is isomorphic to Z^{α} for some α [4]. More recently homomorphisms like ϕ above for $\kappa \geq \mu$ (if μ exists) have been studied by K. Eda and others and some good results achieved (see Chapter III of [1]; for some generalizations to modules see [5]). In this paper we study

1991 *Mathematics Subject Classification.* 20K25.

direct summands of Z^{κ} for $\kappa \geq \mu$. Among other things we will show (Theorem 12) that any direct summand of Z^{μ} is isomorphic to Z^{a} for some a.

All unexplained terminology concerning set theory and abelian groups may be found in [3] and [2] respectively. In Section I we present some set–theoretic facts which will be used in Section II to study direct products of copies of Z.

I. Set Theory

If J is a set, then $P(J)$ is the family of its subsets.

LEMMA 1. *Let κ be an infinite cardinal number. Let A be a countably complete ideal in the Boolean algebra $P(\kappa)$ containing all singletons. Let B be the Boolean algebra $P(\kappa)/A$. If $X \subsetneq P(\kappa)$, then $[X]$ denotes its image in B. Let D be its set of atoms.*

(*a*) *If A is κ–complete and κ^{+}–saturated, then B is a complete Boolean algebra and A contains an element of size κ.*

(*b*) *If κ is the least measurable cardinal number and $|B|$ is finite, then A is κ–complete.*

(*c*) *If B is complete, then it is atomic if and only if it is isomorphic to $P(D)$.*

(d) *If κ is measurable and A is κ–complete and κ–saturated, then B is atomic.*

(*e*) *If A is κ–complete and $X \subsetneq P(\kappa)$ with $|X| < \kappa$, then $\Sigma\{ [x]\colon x \in X \}$ in B equals $[\Sigma\{ x\colon x \in X \}]$.*

PROOF. (a) B is complete by exercise 35.1 in [3]. By the Corollaire in [7] $P(\kappa)$ contains a subset W of size κ^+ such that $|x| = \kappa$ for each $x \in W$ and $|x \cap y| < \kappa$ for $x \neq y$ in W. Since A contains all subsets of κ of size $< \kappa$, the elements in $Q = \{[x]\colon x \in W - A\}$ are pairwise disjoint in B so $|Q| \leq \kappa$. Hence $W \cap A$ is nonempty.

(b) Since $|B|$ is finite, A is the intersection of a finite number of countably complete non–principal maximal ideals in $P(\kappa)$. Each of these ideals is κ–complete by the dual of Lemma 27.1 in [3].

(c) Statement 25.1 in [8].

(d) B is complete by (a). To show B is atomic we let x be an arbitrary element in $P(\kappa) - A$ and find a subset a of x such that $[a]$ is an atom $\leq [x]$ in B. Let $A_x = P(x) \cap A$. A_x is a κ–complete ideal in $P(x)$ containing all singletons. If a and b are subsets of x whose images are disjoint in $P(x)/A_x$, their images in B are also disjoint since $A_x \subseteq A$. So A_x is κ–saturated in $P(x)$ since A is κ–saturated in P. Now $|x| = \kappa$ which is weakly compact (Lemma 29.7 in [3]). By exercise 35.5 in [3] $P(x)/A_x$ contains an atom, say the image of a for $a \subseteq x$. Hence $[a]$ is an atom $\leq [x]$ in B. Therefore B is atomic.

(e) Exercise 17.16 in [3].

II. Direct Products of Copies of Z

Throughout the rest of the paper we will constantly refer to the situation described in the following premise.

PREMISE 2. For infinite cardinal number κ let $Z^\kappa = \Pi_{i\in\kappa}<e_i> = G \oplus H$. Let ϕ: $Z^\kappa \longrightarrow G$ and ϕ_j: $G \longrightarrow <e_j>$ for $j \in \kappa$ be the natural projections. An element x in Z^κ is written $x = \Sigma_{i\in\kappa} n_i e_i$ with $n_i \in Z$. If $s \subseteq \kappa$, then $e_s = \Sigma_{i\in s} e_i$ and $\Pi(s)$ denotes $\Pi_{i\in s}<e_i>$.

We assume a basic knowledge of slender groups such as is found in [2]. In particular for Z^κ above, if S is a partition of κ and $j \in \kappa$, then $\phi_j(\Pi(s))$ equals 0 for almost all $s \in S$.

DEFINITIONS 3. Assume Premise 2. Let π_j: $\Pi(\kappa) \longrightarrow <e_j>$ be the natural projection. A subset X of $\Pi(\kappa)$ is <u>summable</u> if, for every $j \in \kappa$, $\pi_j(x) = 0$ for almost all $x \in X$. Henceforth "summable" will *always* be in reference to the decomposition $\Pi(\kappa)$ $(= \Pi_\kappa<e_i>)$. Let X be a summable subset of Z^κ. If $x = \Sigma_{i\in\kappa} n_{xi} e_i$ $(n_{xi} \in Z)$ for each $x \in X$, then, for every $j \in \kappa$, $n_{xj} = 0$ for almost all $x \in X$. We write $\Sigma_{x\in X} m_x x$ $(m_x \in Z)$ for the element $\Sigma_{i\in\kappa}(\Sigma_{x\in X} m_x n_{xi}) e_i$ (the sum in parenthesis is finite). Give Z the discrete topology and Z^κ the corresponding product topology. Then $<\Sigma_{x\in X} m_x x\colon m_x \in Z>$ is the <u>closure</u> $\bar{X}$ of X in $\Pi(\kappa)$. X is <u>completely</u> <u>independent</u> if $\Sigma_X m_x x = 0$ always implies $m_x = 0$ for all x. If X is a completely independent summable subset of Z^κ, then we may write $\bar{X} = \Pi_{x\in X}<x>$.

PROPOSITION 4. *Let M be a direct summand of Z^κ. Let X be an infinite summable subset of M. Then M contains a completely independent summable subset Y such that $|Y| = |X|$ and $Z^\kappa =$*

$\Pi_{y\in Y}\langle y\rangle \oplus C$ *where the left product is the closure* $\bar{Y}$ *of* Y *in* $\Pi(\kappa)$ *and* $C \cong Z^a$ *for some* a.

PROOF. Let $M = G$ in Premise 2. By replacing each $x \in X$ by x/n_x for suitable integer n_x we may assume $\langle x\rangle$ is a pure subgroup (hence a direct summand) of G for each $x \in X$. Let $\nu = |X|$. First we find for each ordinal $a < \nu$: a finite subset J_a of κ, $x_a \in X$ and, for $F_a = \oplus_{J_a}\langle e_i\rangle$, a decomposition $F_a = \langle f_a\rangle \oplus G_a$ such that $x_a = f_a + h_a$ for some $h_a \in \Pi(I_{a+1})$ where, for each $\beta < \nu$, $I_\beta = \kappa - \cup_{\gamma<\beta}J_\gamma$. Suppose β is an ordinal $< \nu$ and we have found these things for all $a < \beta$. Since $\cup_{a<\beta}J_a$ has cardinality $< \nu$ and, for each j in this union, $\phi_j(x) = 0$ for almost all x in X, $X \cap \Pi(I_\beta)$ has cardinality ν. Let x_β be any element in this set. By an easy adaptation of the proof of Theorem 19.2 in [2] one can choose a finite subset J_β of I_β and, for $F_\beta = \oplus_{J_\beta}\langle e_i\rangle$, a decomposition $F_\beta = \langle f_\beta\rangle \oplus G_\beta$ so that $x_\beta = f_\beta + h_\beta$ for some h_β in $\Pi(I_{\beta+1})$. By induction we can repeat this process for all ordinals $\beta < \nu$. Let $Y = \{x_a\}_{a<\nu}$. Y is summable and $|Y| = |X|$. Suppose $\Sigma_{a<\nu}n_ax_a = 0$ $(n_a \in Z)$ and $n_\beta \neq 0$ for minimal β. For each $j \in J_\beta$ we have $0 = \phi_j(\Sigma n_ax_a) = \phi_j(n_\beta x_\beta) = \phi_j(n_\beta f_\beta)$ so $n_\beta = 0$, a contradiction. Therefore Y is completely independent and $\bar{Y} = \Pi_{a<\nu}\langle x_a\rangle = \Pi_{y\in Y}\langle y\rangle$. Let $J = \cup_{a<\nu}J_a$. Since the J_a are pairwise disjoint finite sets, $\Pi_{a<\nu}G_a \cong Z^\gamma$ for some γ. We claim $Z^\kappa = \bar{Y} \oplus \Pi_{a<\nu}G_a \oplus \Pi(\kappa - J)$, which will complete the proof. The sum on the right side is clearly direct and contained in Z^κ. Let $z \in Z^\kappa$. We must show z is in the right–hand sum above. For each $\beta < \nu$ note that $\Pi(I_\beta) = <x_\beta> \oplus G_\beta \oplus \Pi(I_{\beta+1})$. By induction for each $\alpha < \nu$ we can find $n_\alpha \in Z$ and $g_\alpha \in G_\alpha$ such that, for each $\beta < \nu$, the element $z - \Sigma_{\alpha<\beta}(n_\alpha x_\alpha + g_\alpha)$ is in $\Pi(I_\beta)$. Let w be the element $\Sigma_{\alpha<\nu}(n_\alpha x_\alpha + g_\alpha)$. For each $\beta < \nu$ then $z - w$ equals $z - \Sigma_{\alpha<\beta}(n_\alpha x_\alpha + g_\alpha) + \Sigma_{\alpha\geq\beta}(n_\alpha x_\alpha + g_\alpha)$ which is in $\Pi(I_\beta)$. Since $\cap_{\beta<\nu} I_\beta = \kappa - J$, the element $z - w$ is in $\Pi(\kappa - J)$. Therefore z is in $\overline{Y} \oplus \Pi_{\alpha<\nu} G_\alpha \oplus \Pi(\kappa - J)$, as desired.

PROPOSITION 5. *Let M be a direct summand of Z^κ for infinite κ. If $|M| = a$, then $Z^\kappa = M \oplus N \oplus Z^\gamma$ where $\gamma \subset \kappa$ and $M \oplus N \cong Z^\beta$ for some $\beta \leq 2^a$.*

PROOF. Let M be G in Premise 2. Let F be the set of all functions: $G \longrightarrow Z$. An element in F has the form $(n_g)_{g\in G}$ where $n_g \in Z$ so $|F| = 2^a$. For $g = \Sigma_{i\in\kappa} n_i e_i \in Z^\kappa$ let $\pi_i(g) = n_i$. Define θ: $\kappa \longrightarrow F$ by $\theta(i) = (\pi_i(g))_{g\in G}$. Let $\sim$ be the equivalence relation on κ defined by $i \sim j$ if and only if $\theta(i) = \theta(j)$. Let Q be the set of equivalence classes in κ. Now Q is a partition of κ and, if $\beta = |Q|$, then $\beta \leq |F| = 2^a$. Let J be a set consisting of exactly one i from each $q \in Q$. Then $Z^\kappa = \Pi_{q\in Q} <e_q> \oplus \Pi(\kappa - J)$ and $\Pi_Q <e_q> \cong Z^\beta$. If $g = \Sigma_{i\in\kappa} n_i e_i \in G$, then $g = \Sigma_{q\in Q} n_q e_q$ where $n_q = n_i$ for each $i \in q$. So $g \in \Pi_Q <e_q>$ and G is contained in this product. Hence $\Pi_Q <e_q> = G \oplus H \cap \Pi_Q <e_q>$.

The following example shows that a subgroup of $Z^{\kappa} = \Pi_{\kappa}<e_i>$ may be isomorphic to Z^{κ} yet not equal the closure of a completely independent summable subset X.

EXAMPLE 6. Let κ be the least measurable cardinal number and write $Z^{\kappa} = \Pi_{\kappa}<e_i>$. Let A be a κ–complete maximal ideal in $P(\kappa)$ containing all singletons. Then (1) $Z^{\kappa} = <e_{\kappa}> \oplus M$ where $M = <\Pi(s)\colon s \in A>$, (2) $M \cong Z^{\kappa}$ yet (3) M is not the closure in $\Pi(\kappa)$ of any completely independent summable subset X.

PROOF. (1) This decomposition of Z^{κ} is well–known and follows easily from the definition of measurable (e.g. see the Remark after Theorem 94.4 in [2] and Example 3 in [6]). (2) If $j \in \kappa$, then $Z^{\kappa} = <e_{\kappa}> \oplus \Pi_{i \neq j}<e_i>$. So $M \cong Z^{\kappa}/<e_{\kappa}> \cong \Pi_{i \neq j}<e_i> \cong Z^{\kappa}$. (3) Suppose M equals the closure $\bar{X}$ in $\Pi(\kappa)$ of a completely independent summable subset X. Write $x = \Sigma_{j \in \kappa} n_{xj} e_j$ $(n_{xj} \in Z)$ for $x \in X$ and $e_i = \Sigma_{x \in X} m_{ix} x$ $(m_{ix} \in Z)$ for $i \in \kappa$. For each $j \in \kappa$ $n_{xj} = 0$ for almost all x. Also, for each $x \in X$, $m_{ix} = 0$ for almost all i (consider the projections: $Z^{\kappa} \longrightarrow M = \bar{X} \longrightarrow <x>$ which is slender). For each i then we have $e_i = \Sigma_X m_{ix} x$ which, by Definitions 3, equals $\Sigma_j (\Sigma_X m_{ix} n_{xj}) e_j$ where the sum in parenthesis is finite and equals δ_{ij} (the Kronecker delta). Now the element $\Sigma_X (\Sigma_{i \in \kappa} m_{ix}) x$ in $\bar{X}$ equals, by definition, $\Sigma_j [\Sigma_X (\Sigma_{i \in \kappa} m_{ix}) n_{xj}] e_j$ which, since the sum in the brackets is finite, equals $\Sigma_j [\Sigma_i (\Sigma_X m_{ix} n_{xj})] e_j = \Sigma_j (\Sigma_i \delta_{ij}) e_j = \Sigma_j e_j = e_{\kappa}$. But e_{κ} is not in M. Therefore $M \neq \bar{X}$.

PROPOSITION 7. *Assume Premise 2.* Suppose *that* $\phi(e_i) = 0$ *for every* $i \in J$, a subset of κ. *Let* $A = \{s \in P(J)\colon \phi(\Pi(s)) = 0\}$. *Let* μ *be the least measurable cardinal number.*

(a) A *is a* μ*-complete ideal in the Boolean algebra* $P(J)$ *containing all singletons.*

(b) *Let* W *be a subset of* $P(J) - A$ *such that* $X \cap Y \in A$ *for all* $X \neq Y$ *in* W. *Then:* (i) $|W| \leq \kappa$ *and* (ii) G *contains a summable subset* X *where* $|X| = |W|$.

(c) A *is a* κ^+*-saturated in* $P(J)$.

(d) *If* $|J| = \kappa$ *and* A *is* κ*-complete, then* A *contains an element of size* κ.

PROOF. (a) Clearly A is an ideal in $P(J)$ containing all singletons. Let S be a subset of A of cardinality $< \mu$. We must show the supremum of S is in A. We may assume the elements in S are pairwise disjoint. Let $x = \Sigma_{s \in S} x_s$ for $x_s \in \Pi(s)$. We must show $\phi(x) = 0$. For each $j \in \kappa$ $\phi_j(x_s) = 0$ for each s so $\phi_j(x) = 0$ by Theorem 94.4 in [2]. Therefore $\phi\,(x) = 0$.

(b) (i) For $j \in \kappa$ let $W_j = \{w \in W\colon \phi_j(\Pi(w)) \neq 0\}$ We will show each W_j is finite. Since W is the union of all W_j, $|W|$ will be $\leq \kappa$. Let $\{w_n\}$, $n \in \omega$, be a countable subset of W. For fixed $j \in \kappa$ it will suffice to show $\phi_j(\Pi(w_n)) = 0$ for almost all n. Since $w_n \cap w_m \in A$ for $n \neq m$, there is a partition Q of $P(J)$ of cardinality $\leq 2^\omega$ such that each $w_n = x_n + y_n$ where $\{x_n\}$ is a set of pairwise disjoint members of Q and each y_n is a sum of $\leq 2^\omega$ elements in $A \cap Q$. Each y_n is in A since A is μ–complete. As a result $\phi_j(\Pi(w_n)) = \phi_j(\Pi(x_n))$ for each n. Since the x_n are pairwise disjoint, $\phi_j(\Pi(w_n)) = 0$ for almost all n. Therefore W_j is finite. (ii) For every $w \in W$ choose $x_w \in \Pi(w)$ such that $\phi(x_w) \neq 0$. Then $X = \{\phi(x_w)\colon w \in W\}$ is summable and $|X| = |W|$.

(c) This follows from (b).

(d) This follows from (c) and from Lemma 1a if we identify J with κ there.

PROPOSITION 8. *Assume Premise 2. If* $H \cong N \oplus Z^{\kappa}$ *for some group* N, *then* $H \cong Z^{\kappa}$.

PROOF. $H \cong N \oplus (Z^{\kappa})^{\kappa} \cong N \oplus (G \oplus N \oplus Z^{\kappa})^{\kappa} \cong (G \oplus N \oplus Z^{\kappa})^{\kappa} \cong (G \oplus H)^{\kappa} = (Z^{\kappa})^{\kappa} \cong Z^{\kappa}$.

PROPOSITION 9. *Let* $Z^{<\kappa} = \{x \in Z^{\kappa}\colon |support(x)| < \kappa\}$ for infinite κ. *The subgroup* $Z^{<\kappa}$ *is not a direct summand of* Z^{κ}.

PROOF. Assume Premise 2 where $H = Z^{<\kappa}$. For $J = \kappa$ in Proposition 7 the ideal A equals $\{s \subseteq \kappa\colon |s| < \kappa\}$ which is κ–complete. By (d) there A contains an element of size κ which is not true. So our assumption is false.

PROPOSITION 10. *Let* M *be a direct summand of* Z^{κ} *where* κ *is the least measurable cardinal number. If* M *contains a summable subset* X *of cardinality* κ, *then* $M \cong Z^{\kappa}$.

PROOF. Let M be H in Premise 2. By Proposition 4 (for $M = H$) Z^{κ} *equals* $\Pi_{y \in Y}\langle y\rangle \oplus C$ where $Y \subseteq H$, $|Y| = \kappa$ and $C \cong Z^{a}$ for some a. By a change of notation we may assume $Z^{\kappa} = \Pi_{i \in \kappa}\langle e_i\rangle$ where, for some $J \subseteq \kappa$, $|J| = \kappa$ and $e_i \in H$ for each $i \in J$. Let A be as in Proposition 7. By (a) and (d) there A

contains an element of size κ. So $H \cong N \oplus Z^{\kappa}$ for some group N. Proposition 8 completes the proof.

PROPOSITION 11. *Assume Premise 2. Suppose* A *is a* κ*-complete* κ*-saturated ideal in* $P(\kappa)$ *containing all singletons and* $B = P(\kappa)/A$ *is atomic. Let* C *be a partition of* κ *such that* $D = \{[c]: c \in C\}$ *is the set of atoms of* B. *Then* $|C| < \kappa$ *and* $Z^{\kappa} = \Pi_{c\in C}\langle e_c\rangle \oplus N$ *where* $N = \langle \Pi_{i\in s}\langle e_i\rangle : s \in A\rangle$.

PROOF. By (a) and (c) of Lemma 1 $B \cong P(D)$. Since A is κ–saturated, $|C| = |D| < \kappa$. Let θ: $P(\kappa) \longrightarrow B$ be the natural map. For fixed $c \in C$ θ maps $P(c)$ to $\{[c], [0]\}$ with kernel $A_c = P(c) \cap A$. Now $|c| = \kappa$ and A_c is a maximal κ–complete ideal in $P(c)$ containing all singletons. It follows as in (1) of Example 6 that $\Pi_{i\in c}\langle e_i\rangle = \langle e_c\rangle \oplus N_c$ where $N_c = \langle \Pi(s): s \in A_c\rangle$. This is true for each $c \in C$ so $Z^{\kappa} = \Pi_{c\in C}\langle e_c\rangle \oplus \Pi_{c\in C}N_c$. We complete the proof by showing that $\Pi_{c\in C}N_c = N$. (a) Let $x = \Sigma x_c \in \Pi_C N_c$ where $x_c \in N_c$. Each $x_c \in \Pi(s_c)$ for some s_c in A_c. Since each $s_c \in A$ and A is κ–complete, $s = \Sigma_C s_c$ is in A. So $x \in \Pi(s) \subseteq N$. (b) Let $s \in A$ and $x \in \Pi(s)$. In B then $[0] = [s] = [\Sigma_C s \cap c] = \Sigma_C[s \cap c]$ (by Lemma 1e). Hence, for each c, $[s \cap c]$ equals 0 and $s \cap c$ is in A_c. Thus $x \in \Pi_{c\in C}(\Pi(s \cap c)) \subseteq \Pi_{c\in C}N_c$.

THEOREM 12. *Let* κ *be the least measurable cardinal number. If* G *is a direct summand of* Z^{κ}, *then* $G \cong Z^{a}$ *for some* a.

PROOF. Assume Premise 2. If G contains a summable subset of cardinality κ, we are done by Proposition 10. Assume then G

contains no such subset. Let $J = \{i \in \kappa\colon \phi(e_i) = 0\}$. Since $X = \{\phi(e_i)\colon i \notin J\}$ is a summable subset of G, $|\kappa - J| = |X| < \kappa$ and $|J| = \kappa$. Let A be as in Proposition 7. By (a) there A is k–complete and by (b) and (c) there and by our assumption on G A is κ–saturated in $P(J)$. Hence $P(J)/A$ is atomic by Lemma 1d for $\kappa = J$. By Proposition 11 $\Pi(J) = M \oplus N$ where $|M| = 2^\gamma$ for some $\gamma < \kappa$ and $\phi(N) = 0$. So $|G| = |\phi(Z^\kappa)| \leq |\Pi(\kappa - J) \oplus M| \leq 2^{|\kappa - J|} + 2^\gamma < \kappa$ (since κ is inaccessible). By Proposition 5 G is a direct summand of a group of the form Z^β where $\beta \leq 2^{|G|} < \kappa$. By Theorem 5 in [4] then the group G is isomorphic to Z^a for some a.

References.

1. P. Eklof and A. Mekler, *Almost Free Modules*, North–Holland, Amsterdam, 1990.
2. L. Fuchs, *Infinite Abelian Groups*, vol.I(1970), vol.II(1973), Academic Press, New York.
3. T. Jech, *Set Theory*, Academic Press, N.Y., 1978.
4. R. J. Nunke, *On direct products of infinite cyclic groups,* Proc. Amer. Math. Soc., *13(1962), 66–71.*
5. J. O'Neill, *Measurable products of modules,* Rend. Sem. Mat. Univ. Padova, 73(1985), 261–269.
6. J. O'Neill, *Direct sums and products of isomorphic abelian groups*, Rocky Mountain J., 17(1987), 573–576.
7. W. Sierpinski, *Sur les suites transfinies finalement disjointes,* Fund. Math. 28(1937), 115–119.
8. R. Sikorski, *Boolean Algebras.* 2nd ed., Springer–Verlag, N.Y., 1964.

Department of Mathematics, University of Detroit
Detroit, Michigan 48219

Contemporary Mathematics
Volume **171**, 1994

Abelian groups as Noetherian modules over their endomorphism rings

AGNES T. PARAS

August 23, 1993

ABSTRACT. Given an abelian group G and its ring of endomorphisms E, equip G with an E-module structure by defining $\alpha.g = \alpha(g)$ for $\alpha \in E$ and $g \in G$. A group G is E-Noetherian if it is Noetherian as an E-module, i.e. every submodule is finitely generated. We describe the structure of torsion-free E-Noetherian groups and show that their endomorphism rings are left Noetherian.

1. Introduction

This paper deals with torsion-free abelian groups of finite rank which are Noetherian when viewed as modules over their endomorphism rings in the natural way. For brevity, we refer to them as E-Noetherian groups. This study was prompted by a remark of Pierce, who pointed out that groups which are finitely generated as modules over their endomorphism rings are not necessarily Noetherian. Consider the group $G = Z \oplus H$, where Z is the group of integers and H is a group such that $\mathrm{Hom}(H, Z) = 0$ and H is not finitely generated as an $\mathrm{End}(H)$-module. Then G is finitely generated as an $\mathrm{End}(G)$-module with generator $(1,0)$ and H is a submodule of G which is not finitely generated. In this paper, the class of all E-Noetherian groups is determined, up to quasi-equality.

Playing a prominent rôle in this characterization are the strongly irreducible groups introduced by Reid in [**7**] and the E-rings introduced by Bowshell and Schultz [**9, 10**]. Descriptions of strongly irreducible groups and E-rings, which are closely related, are included in Section 2 for reference.

Section 3 contains one of our main results, namely, the quasi-decomposition theory of E-Noetherian groups. The proof utilizes some well-known results in

---

1991 *Mathematics Subject Classification.* Primary 20K15; Secondary 16P40.
This paper is in final form and no version of it will be submitted for publication elsewhere.

module theory, e.g., the Wedderburn-Artin Theorems and properties of projective modules, which can be found in [**4**].

There are groups that have left Noetherian endomorphism rings but are not E-Noetherian. In contrast, we show in Section 4 that the endomorphism rings of E-Noetherian groups are necessarily left Noetherian.

We assume familiarity with quasi-equality and quasi-isomorphism. For the quasi-concepts involved, we refer the reader to [**1**], [**2**], [**3**] or [**6**].

All groups are assumed to be torsion-free abelian groups of finite rank.

2. Strongly Irreducible Groups

In this section we describe the groups that will constitute the building blocks in the quasi-decomposition theory of E-Noetherian groups.

DEFINITION 2.1. *A group G is strongly irreducible if G is quasi-equal to each of its non-zero, fully invariant subgroups.*

DEFINITION 2.2. *A group G with $E = End(G)$ is E-finitely generated if there exist finitely many elements $g_1, \dots, g_n$ in G such that $G = \sum_{i=1}^{n} Eg_i$.*

For example, the additive groups of subrings of algebraic number fields are strongly irreducible and it is easy to see that strongly irreducible groups are E-finitely generated. In [**7**], the structure theory of E-finitely generated groups is studied. The following theorem considers a special case of this.

THEOREM 2.1 ([**7**]). *If G is E-finitely generated and E has nil radical zero then G is quasi-equal to a direct sum of strongly irreducible groups G_i such that* $\mathrm{Hom}(G_i, G_j) = 0$ *if $i \neq j$. Conversely, any group with this structure is E-finitely generated and its endomorphism ring has nil radical zero.*

We determine the additive group structure of strongly indecomposable and E-finitely generated groups in terms of E-rings. In [**9**] and [**10**], Schultz and Bowshell provide a complete characterization of E-rings. Of particular interest to us are the torsion-free E-rings.

DEFINITION 2.3. *A ring R with identity is an E-ring if every endomorphism of R^+ is a left multiplication in R, where R^+ is the additive group structure of R.*

It is easy to see that E-rings are necessarily commutative. The next theorem establishes a connection between strongly indecomposable, E-finitely generated groups and torsion-free E-rings. This result is crucial in providing information about the endomorphism rings of E-Noetherian groups.

THEOREM 2.2 ([**7**]). *G is strongly indecomposable and E-finitely generated if and only if G is quasi-isomorphic to the additive group of a strongly indecomposable, torsion-free, finite rank E-ring.*

If G is strongly irreducible then G is quasi-equal to a direct sum of strongly irreducible and strongly indecomposable groups. Notice that if two groups G and H are quasi-equal then $\mathrm{End}(G) \doteq \mathrm{End}(H)$. Since

$$\mathrm{Hom}(\oplus_{i=1}^{n} A_i, \oplus_{j=1}^{m} B_j) \cong \oplus_{i=1}^{n} \oplus_{j=1}^{m} \mathrm{Hom}(A_i, B_j),$$

we can assume without loss of generality that a strongly irreducible E- finitely generated group G is also strongly indecomposable. By Theorem 2.2, we can assume further that G is the additive group of a strongly indecomposable, torsion-free, finite rank E-ring R. If R and S are two such E-rings then $\mathrm{Hom}(R^+, S^+)$ can be viewed as a left S-module. The properties of $\mathrm{Hom}(R^+, S^+)$ are considered in Section 4 using this viewpoint.

3. Quasi-Decomposition of E-Noetherian Groups

Recall that the non-zero, fully invariant subgroups of a strongly irreducible group G are quasi-equal to G. The following lemma shows that strongly irreducible groups are E-Noetherian. A very interesting relation exists between strongly irreducible and E-Noetherian groups as shown in Theorem 3.2 below.

LEMMA 3.1. *Let R be a ring with identity. If A_1 and A_2 are R-modules with A_1 R-finitely generated and $A_1 \doteq A_2$ then A_2 is R-finitely generated.*

PROOF. Let $x_1, \dots, x_k \in A_1$ be a set of R-generators of A_1. Suppose that $nA_1 \subseteq A_2$ and $mA_2 \subseteq A_1$ for some nonzero integers n and m. Then A_2/mn^2A_1 is finite with coset representatives $\{y_1, \dots, y_l\}$. It follows that the set

$$\{mn^2x_1, \dots, mn^2x_k, y_1, \dots, y_l\} \subset A_2$$

generates A_2 as R-module. □

We are now ready to state and prove one of our main results. It describes the structure of E-Noetherian groups up to quasi-isomorphism in terms of the more familiar strongly irreducible groups.

THEOREM 3.2. *A group G is E-Noetherian if and only if G is quasi-equal to a direct sum of strongly irreducible groups.*

PROOF. Suppose G is E-Noetherian. Then every submodule of G is E-finitely generated. Since $Q \otimes E$ is torsion-free of finite rank, write $E \doteq N \oplus S$ as in Theorem 1.4 of [**2**], where $Q \otimes N$ is the Jacobson radical of $Q \otimes E$, $Q \otimes S$ is a semisimple subring of $Q \otimes E$ and the identity element of E is in S. Since N is nilpotent, let k be the least positive integer such that $N^k = 0$. Let $0 \leq i < k-1$. The pure subgroup $< N^iG >_*$ is fully invariant in G since N is an ideal of E. By hypothesis, $<N^iG>_*$ is finitely generated over E.

Let $K_i \overset{def.}{\equiv} < N^iG >_* \ / < N^{i+1}G >_*$. By Theorem 2.1, it suffices to show that $< N^{i+1}G >_*$ is a quasi-summand of $< N^iG >_*$ for $0 \leq i < k-1$ and that $N(\mathrm{End}(K_i))$ is zero for $0 \leq i \leq k-1$.

The group K_i is a finitely generated S-module since if $x_1, \ldots, x_n$ generate $<N^iG>_*$ over E then $<N^iG>_* = \sum_{i=1}^n Ex_i \doteq \sum_{i=1}^n Nx_i + \sum_{i=1}^n Sx_i$. Hence, $K_i \doteq \sum_{i=1}^n S(x_i + <N^{i+1}G>_*)$, where $x_i + <N^{i+1}G>_* \in K_i$. By the preceding lemma, K_i is S-finitely generated. Recall that if $Q \otimes S$ is semisimple then S is Noetherian. By the change of rings theorem (cf. [**8**]),

$$Q \otimes \operatorname{Hom}_S(K_i, S) \cong \operatorname{Hom}_{Q\otimes S}(Q \otimes K_i, Q \otimes S).$$

Since $Q \otimes S$ is a semisimple and Artinian ring, $Q \otimes K_i$ is projective and finitely generated over $Q \otimes S$. Hence, there exist $f_{ij} \in \operatorname{Hom}_{Q\otimes S}(Q\otimes K_i, Q\otimes S)$ and $y_{ij} \in Q\otimes K_i$ ($j = 1, \ldots, n_i$ for some n_i) such that if $y \in Q\otimes S$ then $y = \sum_{j=1}^{n_i} f_{ij}(y).y_{ij}$. Since there are finitely many y_{ij}'s, there is a positive integer l such that $l.y_{ij} \in K_i$ ($j = 1, \ldots, n_i$ and $0 \leq i < k-1$). By the aforementioned isomorphism, there is a positive integer k such that $k.f_{ij} \in \operatorname{Hom}_S(K_i, S)$ ($j = 1, \ldots, n_i$, $0 \leq i < k-1$). Let $m = max\{k, l\}$. Then $m.f_{ij} \in \operatorname{Hom}_S(K_i, S)$, $m.y_{ij} \in K_i$ ($j = 1, \ldots, n_i$, $0 \leq i < k-1$) and $m^2.y = \sum_{j=1}^{n_i} mf_{ij}(y).my_{ij}$ for $y \in K_i$. It is now easy to see that the exact sequence

$$0 \to <N^{i+1}G>_* \to <N^iG>_* \xrightarrow{\nu_i} K_i \to 0$$

almost splits, i.e., $<N^iG>_* \doteq <N^{i+1}G>_* \oplus H_i$, where $K_i \cong H_i \subseteq <N^iG>_*$. So $G \doteq <N^{k-1}G>_* \oplus H_{k-2} \oplus \cdots \oplus H_1 \oplus H_0$. Thus it remains to determine the group structures of $<N^{k-1}G>_*$ and H_i ($0 \leq i < k-1$).

To complete the proof, we show that the endomorphism ring of K_i is quasi-equal to the ring of endomorphisms induced on K_i by E and $N(\operatorname{End}(K_i)) = 0$. Since $<N^iG>_* \doteq <N^{i+1}G>_* \oplus H_i$, we have $\lambda_i : K_i \to H_i$, an isomorphism, such that $\nu_i\lambda_i = m^2.1_{K_i}$. Let $\overline{E_i}$ be the ring of endomorphisms induced on K_i by E. If $\alpha \in E$ and $x \in <N^iG>_*$, define $\overline{\alpha}(\overline{x}) = \overline{\alpha(x)}$, where $\overline{\alpha(x)} \in K_i$. Suppose that $\overline{\beta} \in \operatorname{End}(K_i)$. Then $\beta \overset{def.}{\equiv} \lambda_i\overline{\beta}\nu_i \in \operatorname{Hom}(<N^iG>_*, H_i)$. The elements of $\operatorname{Hom}(<N^iG>_*, H_i)$ may be extended to endomorphisms of

$$G_i \overset{def.}{\equiv} <N^iG>_* \oplus H_{i-1} \oplus \cdots \oplus H_0$$

by defining the image of $H_{i-1} \oplus \cdots \oplus H_0$ to be zero. Note that $G_i \doteq G$ and so $\operatorname{End}(G_i) \doteq E$. Thus we can view $\operatorname{Hom}(<N^iG>_*, H_i)$ as embedded in $\operatorname{End}(G_i)$. Now $m^2\overline{\beta} = \nu_i\lambda_i\overline{\beta}$ and $m^2\overline{\beta}(\overline{x}) = m^2\overline{\beta}\nu_i(x) = \nu_i\lambda_i\overline{\beta}\nu_i(x) = \nu_i\beta(x) = \overline{\beta(x)}$ imply $m^2\overline{\beta}(\overline{x}) = \overline{\beta(x)}$ for $x \in <N^iG>_*$. Hence, $m^2\overline{\beta}$ is induced on K_i by β. Therefore $\operatorname{End}(K_i) \doteq \overline{E_i}$.

Note that the action of E on K_i is equivalent to the action of E/L_i where $N \subseteq L_i \overset{def.}{\equiv} \{\alpha \in E \mid \alpha N^i \subseteq N^{i+1}\}$ and L_i is an ideal of E. Let $N_i^{'}$ be the nil radical of E/L_i and suppose $N_i^{'} = M_i/L_i$ where $L_i \subseteq M_i \subseteq E$. Since Q is a flat Z-module (cf. [**4**, p. 153]), the short exact sequence

$$0 \to L_i/N \to E/N \to E/L_i \to 0$$

tensored with Q yields the short exact sequence

$$0 \to Q \otimes L_i/N \to Q \otimes E/N \to Q \otimes E/L_i \to 0.$$

Since the algebra $Q \otimes E/N$ is semisimple and Artinian,

$$Q \otimes E/N \cong M_{n_1}(\Delta_1) \dot{+} \cdots \dot{+} M_{n_k}(\Delta_k)$$

where Δ_i is a division ring. The only ideals of $Q \otimes E/N$ are the ring direct sums of subcollections of the $M_{n_i}(\Delta_i)$. Hence $Q \otimes E/L_i \cong \dot{+} M_{n_{ij}}(\Delta_{ij})$ and $J(Q \otimes E/L_i) = 0$. It follows that $N(\mathrm{End}(K_i)) = 0$.

Similarly, $N(\mathrm{End}(<N^{k-1}G>_*)) = 0$ since $\mathrm{End}(<N^{k-1}G>_*)$ is quasi-equal to the set of endomorphisms of G restricted to $<N^{k-1}G>_*$. Hence, $K_i \doteqdot H_i$ $(0 \leq i < k-1)$ and $<N^{k-1}G>_*$ are of the form described in Theorem 2.1. Therefore, G is quasi-equal to a direct sum of strongly irreducible groups.

Conversely, suppose that $G \doteq \oplus_{i=1}^n A_i$ where each A_i is strongly irreducible. If $H \neq 0$ is a fully invariant subgroup of G then $H \doteq \oplus_{i=1}^n (H \cap A_i)$. If $(H \cap A_i) \neq 0$ then $\mathrm{End}(A_i)(H \cap A_i) \subseteq (H \cap A_i)$. Since A_i is strongly irreducible, $H \cap A_i$ is quasi-equal to A_i and $\mathrm{End}(A_i)$-finitely generated. Hence, H is $\mathrm{End}(G)$-finitely generated. Since $0 \neq H$ was an arbitrary fully invariant subgroup of G, it follows that G is E-Noetherian. □

REMARK 3.3. *For the proof of necessity, we only need the fact that $<N^iG>_*$ is finitely generated over E for each i. Thus, if G is E-finitely generated then G is E-Noetherian if and only if $<N^iG>_*$ is finitely generated over E for each i.*

4. Endomorphism Rings of E-Noetherian Groups

To complete our characterization of E-Noetherian groups, we consider now their endomorphism rings. If G is E-finitely generated and E is left Noetherian then, certainly, G is E-Noetherian. However, G need not be E-Noetherian if $\mathrm{End}(G)$ is left Noetherian. For example, the ring of integers Z is left Noetherian. If $\mathrm{End}(G) \cong Z$ then G is E-Noetherian if and only if G is a free Z-module. Thus any group G that is not free with $\mathrm{End}(G) \cong Z$ is not E-Noetherian. For example, consider the subgroup $G = <\frac{1}{p} \mid p \text{ is a prime integer}>$ of Q. Here $\mathrm{End}(G) \cong Z$. More generally, it is well known that there are torsion-free groups of essentially any rank with Z as endomorphism ring.

LEMMA 4.1. *Let A and M be groups. Let R be a ring and M be a Noetherian R-module. Then* Hom*(A,M) is Noetherian over R.*

PROOF. Let $\{a_1, \ldots, a_n\}$ be a maximal Z-independent subset in A and let $F = <a_1, \ldots, a_n>$. Note that F is a full, free subgroup of A. Consider the R-homomorphism

$$\beta : \mathrm{Hom}(A, M) \to M^n \quad \text{where } \beta(f) = (f(a_1), \ldots, f(a_n)),$$

for $f \in \text{Hom}(A, M)$. Since F is full in A and M is torsion-free, β is monic. Hence, $\text{Hom}(A, M)$ imbeds in M^n as an R-submodule. Since M is Noetherian, so is M^n. Thus $\text{Hom}(A, M)$ is finitely generated over R, indeed is itself Noetherian over R. $\square$

COROLLARY 4.2. *Let A and M be groups.*

(i) *If M is an E-Noetherian group then* Hom*(A,M) is Noetherian as module over* End*(M).*
(ii) *If M is a strongly indecomposable and strongly irreducible group then* Hom*(A,M) is Noetherian over* End*(M).*

LEMMA 4.3. *Let S be a strongly indecomposable and irreducible group. Let R be a group such that* $\text{Hom}(S, R) \neq 0$ *and* $\cap\{\ker f \mid f \in \text{Hom}(R, S)\} = 0$. *Then S is isomorphic to a subgroup of R.*

PROOF. For $f \in \text{Hom}(R, S)$ define $I(f) = \{f \circ g \mid g \in \text{Hom}(S, R)\}$. Then $I(f)$ is a right ideal of $\text{End}(S)$. Clearly, $I(f) = 0$ if and only if $g(S) \subseteq \ker f$ for all $g \in \text{Hom}(S, R)$. If $I(f) = 0$ for all $f \in \text{Hom}(R, S)$ then

$$g(S) \subseteq \cap\{\ker f \mid f \in \text{Hom}(R, S)\} = 0$$

for all $g \in \text{Hom}(S, R)$. Hence $\text{Hom}(S, R) = 0$. This violates the hypothesis, so $I(f) \neq 0$ for some $f \in \text{Hom}(R, S)$. Since $\text{End}(S)$ is a full subring of a division ring under our hypothesis, we have $I(f) \doteq \text{End}(S)$. Then there is a non-zero integer k such that $k.1_S \in I(f)$, so there exists $g \in \text{Hom}(S, R)$ such that $f \circ g(s) = ks$ for every $s \in S$. Thus g is monic and the result follows. $\square$

COROLLARY 4.4. *Let R and S be strongly indecomposable irreducible groups such that* $\text{Hom}(R, S) \neq 0$ *and* $\text{Hom}(S, R) \neq 0$. *Then* $R \doteq\cong S$.

PROOF. The subgroups

$$\cap\{\ker f \mid f \in \text{Hom}(R, S)\} \text{ and } \cap\{\ker g \mid g \in \text{Hom}(S, R)\}$$

are fully invariant in R and S, respectively, so are full or are 0. They are not full since $\text{Hom}(R, S) \neq 0$ and $\text{Hom}(S, R) \neq 0$, so they are 0. Then Lemma 4.3 applies symmetrically to give monomorphisms $R \hookrightarrow S$ and $S \hookrightarrow R$ which, as is well known, implies $R \doteq\cong S$. $\square$

THEOREM 4.5 ([**1**]). *Suppose that the additive group of a ring R is finite rank, torsion-free. Then R is left Noetherian if and only if $N(R)$ is finitely generated as a left ideal.*

LEMMA 4.6. *Let R and S be torsion-free, finite rank rings with common identity. Suppose R is left Noetherian and $R \doteq S$. Then S is left Noetherian.*

PROOF. First consider the case when $nS \subseteq R \subseteq S$ for some integer n. If I is a left ideal of S then $nI \subseteq R$ is a left R-module. By hypothesis, nI is R-finitely generated. Since $R \subseteq S$ and nI is a left S-module, nI is S-finitely generated. Since I and nI are isomorphic as S-modules, I is S-finitely generated. Thus, if $nS \subseteq R \subseteq S$ then S is left Noetherian.

A second special case is when $nR \subseteq S \subseteq R$ for some integer n. If I is a left ideal of S then nRI is both a left R-module and a left S-module with $nRI \subseteq I \subseteq RI$. Since R is left Noetherian, nRI is R-finitely generated. Let $\{x_1, \dots, x_k\} \subseteq nRI$ such that $\sum_{i=1}^{k} Rx_i = nRI$. Let $J = \sum_{i=1}^{k} Sx_i$. Since J and nRI are both S-modules and J is S-finitely generated, nRI is S-finitely generated. By Lemma 3.1, I is S-finitely generated. So if $nR \subseteq S \subseteq R$ then S is left Noetherian.

Consider the general case $nR \subseteq S$ and $mS \subseteq R$ for some integers n, m. Note that $nmR \subseteq R \cap S \subseteq R$. By the second case, $R \cap S$ is left Noetherian. Now $nmS \subseteq R \cap S \subseteq S$. By the first case S is left Noetherian. □

THEOREM 4.7. *If G is E-Noetherian then* End(G) *is left Noetherian.*

PROOF. Suppose G is E-Noetherian. Then $G \doteq \oplus_i A_i$ such that A_i is strongly indecomposable, strongly irreducible. By Theorem 2.2, we can assume that each A_i is an E-ring. Group the A_i's according to whether they are quasi-isomorphic and call the direct sum of the quasi-isomorphic pieces H_j's. So $G \doteq \oplus_j H_j$. By Corollary 4.4 and the fact that $\mathrm{Hom}(\oplus_i A_i, \oplus_j B_j) \cong \oplus_i \oplus_j \mathrm{Hom}(A_i, B_j)$, if $\mathrm{Hom}(H_i, H_j) \neq 0$, for $i \neq j$, then $\mathrm{Hom}(H_j, H_i) = 0$.

Let $I = \{1, \dots, n\}$ be the index set of the H_j's. Define inductively the following sets: let $I_1 \subseteq I$ be maximal with respect to the property that if $i, j \in I_1$, $l \notin I_1$ then $\mathrm{Hom}(H_i, H_j) = 0$ and $\mathrm{Hom}(H_l, H_i) = 0$. Let $I_k \subseteq I \setminus I_1 \cup \dots \cup I_{k-1}$ be maximal with respect to the property that if $i \neq j \in I_k$ and $l \notin I_1 \cup \dots \cup I_k$ then $\mathrm{Hom}(H_i, H_j) = 0$ and $\mathrm{Hom}(H_l, H_i) = 0$. Let $H_{I_k} = \oplus_{i \in I_k} H_i$ and rearrange $\oplus_j H_j$ as $H_{I_1} \oplus \dots \oplus H_{I_l}$. Let (α_{ij}) be the $l \times l$ matrix with entries $\alpha_{ij} \in \mathrm{Hom}(H_{I_j}, H_{I_i})$. If we view an endomorphism of $\oplus_{j=1}^{l} H_{I_j}$ as a matrix (α_{ij}) acting on an element $(h_{I_j}) \in \oplus_{j=1}^{l} H_{I_j}$ via $(\alpha_{ij})(h_{I_j}) = (\sum_{j=1}^{l} \alpha_{ij} h_{I_j})$, then the prescribed arrangement $\oplus_{j=1}^{l} H_{I_j}$ shows that the matrices (α_{ij}) are lower triangular.

Since $\mathrm{End}(H_{I_j})$ has zero nil radical,

$$N(\mathrm{End}(\oplus_{j=1}^{l} H_{I_j})) = \{(\alpha_{ij}) \in \mathrm{End}(\oplus_{j=1}^{l} H_{I_j}) \mid \alpha_{ij} = 0 \ \text{ for } i \leq j\}.$$

By Corollary 4.2, each $\mathrm{Hom}(A_i, A_j)$ is finitely generated as an A_j-module. Thus, $N(\mathrm{End}(\oplus_{j=1}^{l} H_{I_j}))$ is finitely generated as $\mathrm{End}(\oplus_{j=1}^{l} H_{I_j})$-module. Therefore, by Theorem 4.5, $\mathrm{End}(\oplus_{j=1}^{l} H_{I_j})$ is left Noetherian. It follows from Lemma 4.6 that End(G) is left Noetherian. □

References

1. D.M. Arnold, *Finite Rank Torsion Free Abelian Groups and Rings*, Springer-Verlag, Lect. Notes in Math. vol. 931, NY, 1982.
2. R.A. Beaumont and R.S. Pierce, *Torsion-Free Rings*, Illinois J. Math., 5 (1961), 61-98.
3. L. Fuchs, *Infinite Abelian Groups*, Vol. II, Academic Press, NY, 1973.
4. N. Jacobson, *Basic Algebra II*, W.H. Freeman and Co., NY, 1980.
5. R.S. Pierce, *Abelian groups as modules over their endomorphism rings*, Proceedings Univ. of Conn. Workshop on Abelian Groups, 1989, 45-59.
6. J.D. Reid, *On the ring of quasi-endomorphisms of a torsion-free group*, Topics in Abelian Groups, Scott, Foresman, Chicago 1963, 51-68.
7. J.D. Reid, *Abelian groups finitely generated over their endomorphism rings*, Lect. Notes in Math., vol. 874, Springer-Verlag, NY, 1981, 41-52.
8. I. Reiner, *Maximal Orders*, Academic Press, London, 1975.
9. P. Schultz, *The endomorphism ring of the additive group of a ring*, J. Austral. Math. Soc., 15 (1973) 60-69.
10. P. Schultz and R. Bowshell, *Unital rings whose additive endomorphisms commute*, Math. Ann., 228 (1977), 197-214.

Department of Mathematics, Wesleyan University, Middletown, CT 06459

Current address: Department of Mathematics, University of the Philippines, Diliman, Quezon City, Philippines

E-mail address: aparas@eagle.wesleyan.edu

Contemporary Mathematics
Volume **171**, 1994

Isomorphism of Butler groups at a prime

FRED RICHMAN

ABSTRACT. For T a finite lattice of types, and R a subring of $\mathbf{Q}$, we show that the category of Butler T-groups, with homomorphisms $R \otimes \mathrm{Hom}(A, B)$, is isomorphic to a category of antirepresentations of T by the functor taking A to $R \otimes A$ with $\tau \in T$ represented by the pure R-submodule $R \otimes A(\tau)$. This specializes to Butler's theorem on quasi-isomorphism of Butler groups when $R = \mathbf{Q}$, and extends the work of Arnold and Dugas on near isomorphism of Butler groups when $R = \mathbf{Z}_p$.

1. RT-spaces

A Butler group A is classified up to quasi-isomorphism by the vector space $\mathbf{Q} \otimes A$ together with the subspaces $\mathbf{Q} \otimes A(\tau)$, indexed by τ in the type set of A. If we are interested in isomorphism at a prime p, rather than quasi-isomorphism, it is natural to consider the $\mathbf{Z}_p$-module A_p, together with the family of pure submodules $A(\tau)_p$. In order to put these two ideas under one roof, we consider an arbitrary subring R of $\mathbf{Q}$. A torsion-free module over such a ring R is just a torsion-free abelian group that is divisible by the primes that are invertible in R.

If A and B are torsion-free abelian groups, and R is a subring of $\mathbf{Q}$, then we call an element of $R \otimes \mathrm{Hom}(A, B)$ an *R-homomorphism* (not to be confused with an R-module homomorphism, although if A and B are R-modules—an uninteresting case—then the concepts agree). So a $\mathbf{Q}$-homomorphism is a quasi-homomorphism, a $\mathbf{Z}_p$-homomorphism is a homomorphism at p, and a $\mathbf{Z}$-homomorphism is an ordinary homomorphism.

Let T be a finite sublattice of the lattice of types, with operations join and meet denoted by "$\vee$" and "$\wedge$", and R a subring of $\mathbf{Q}$. By a *T-valuation* on a torsion-free R-module V we mean a function v from V to $T \cup \{\infty\}$ so that

(V1) $vx = \infty$ if and only if $x = 0$.
(V2) $v(x + y) \geq vx \wedge vy$.
(V3) $v(rx) = vx$ if $r \neq 0$.
(V4) $\text{type } x = vx \vee \text{type } R$.

1991 *Mathematics Subject Classification.* Primary 20K15.
This paper is in final form and no version of it will be submitted for publication elsewhere

The first three properties coincide with the definition of a T-valuation on a vector space given in [**3**]. If $R = \mathbf{Q}$, then (V4) holds trivially. If $R = \mathbf{Z}$, then (V4) says that the valuation v assigns to each element its type. If $R = \mathbf{Z}_p$, then (V4) says that x has infinite height in V if and only if $(vx)_p = \infty$. It is convenient to have a description in terms of an antirepresentation of T.

THEOREM 1. *If v is a T-valuation on V, and $V(\tau) = \{x \in V : vx \geq \tau\}$, then*

(R1) *$V(\tau)$ is a pure submodule consisting of elements of type at least τ.*
(R2) *$V(\tau_1 \vee \tau_2) = V(\tau_1) \cap V(\tau_2)$. In particular, if $\tau_1 \leq \tau_2$, then $V(\tau_1) \supset V(\tau_2)$.*
(R3) *Each nonzero x in V is in $V(\sigma)$ for some $\sigma \in T$ such that $\sigma \vee \text{type } R = \text{type } x$.*

Conversely if we have a family of submodules of V, indexed by T, that satisfies these three conditions, then we can define a T-valuation v by setting vx equal to $\max\{\tau \in T : x \in V(\tau)\}$ for each nonzero $x \in V$.

PROOF. To verify (R1), note that (V2) implies that $V(\tau)$ is closed under addition, (V3) implies that it is closed under multiplication by elements of R, and is pure, and (V4) implies that its elements have type at least τ. Equation (R2) follows directly from the definition of $V(\tau)$ and of "join." As for (R3), simply choose $\sigma = vx$.

Conversely, suppose (R1), (R2), and (R3) hold. Set $v0 = \infty$ and $vx = \max\{\tau \in T : x \in V(\tau)\}$ for nonzero x. From (R3) this max is over a nonempty subset of T, and (R2) guarantees that it exists. Clearly (V3) follows from (R1). As $x \in V(vx)$ and $y \in V(vy)$, then $x + y \in V(vx \wedge vy)$, by (R1) and the special case of (R2). Hence (V2) holds. From (R3) we see that $\text{type } x \leq vx \vee \text{type } R$, while clearly $\text{type } x \geq \text{type } R$ and (R1) ensures that $\text{type } x \geq vx$. $\square$

From (R2) we see that a T-valuation is determined by the subgroups $V(\tau)$ for those $\tau \in T$ that are not joins of (nonempty) finite sets of smaller elements, the *join-irreducible* elements of T. The least element $\tau_{\min}$ of T is not join-irreducible, being the join of the empty subset of T, but (R3) implies that $V(\tau_{\min}) = V$. Conversely, any antirepresentation of the partially-ordered set of join-irreducible elements of T, that satisfies (R1) and (R3), extends uniquely to a T-valuation. The extension is unique because $V(\tau)$ is the intersection of the subgroups $V(\sigma)$ over those $\sigma \leq \tau$ that are join-irreducible, and this condition defines an extension because T is distributive.

We will refer to a T-valuated Butler R-module as an *RT-space.* A $\mathbf{Q}T$-space is what was called a T-space in [**3**]; it is simply an antirepresentation of the partially ordered set of join-irreducible elements of T as subspaces of a finite-dimensional vector space over $\mathbf{Q}$. A $\mathbf{Z}T$-space is just a Butler T-group. The motivation for this generalization was the study of $\mathbf{Z}_pT$-spaces and their connection with near isomorphism of Butler groups.

The morphisms in the category of RT-spaces are abelian group homomorphisms $f : A \to B$ (which are, of necessity, R-module homomorphisms) such that $f(A(\tau)) \subset B(\tau)$ for each $\tau \in T$. We denote the set of RT-space morphisms from A to B by $\text{Hom}_T(A, B)$.

2. The main theorem

THEOREM 2. *The functor F from the category of Butler T-groups and R-homomorphisms, to the category of RT-spaces, given by setting $F(A) = R \otimes A$, and $F(A)(\tau) = R \otimes A(\tau)$, is an isomorphism.*

To see that $F(A)$ is an RT-space, check (R1), (R2) and (R3), using the fact that $R \otimes A(\tau)$ can be thought of as $RA(\tau)$ inside $R \otimes A$.

The following theorem shows that F is onto. It reduces to [**3,** Theorem 2.4] and [**2,** Theorem 4] when $R = \mathbf{Q}$.

THEOREM 3. *Any RT-space is realizable by a Butler T-group.*

PROOF. Let V be an RT-space. As V is a Butler group, and T is finite, we can find a a finite set $\mathcal{B}$ of pure rank-one subgroups such that each $V(\tau)$ is a sum of elements of $\mathcal{B}$. For B a rank-one subgroup of V, we let vB denote the common value of vx for nonzero x in B. Note that $B \in \mathcal{B}$ is contained in $V(\tau)$ exactly when $vB \geq \tau$, so $V(\tau) = \sum\{B \in \mathcal{B} : vB \geq \tau\}$. Each $B \in \mathcal{B}$ is a pure submodule of $V(vB)$. For each $B \in \mathcal{B}$, choose a subgroup C_B of type vB such that $B = RC_B$. This uses (V4). Let A be generated by the subgroups C_B. Then A is a Butler T-group. We will show that $RA(\tau) = V(\tau)$.

For one containment, note that

$$V(\tau) = \sum\{B \in \mathcal{B} : vB \geq \tau\} = R\sum\{C_B : B \in \mathcal{B} \text{ and } vB \geq \tau\} \subset RA(\tau).$$

For the other, suppose K is a pure rank-one subgroup of $A(\tau)$. Then the pre-image of K in $\bigoplus_{B\in\mathcal{B}} C_B$ is a Butler T-group, hence is the sum of finitely many pure rank-one subgroups $S_1, \ldots, S_n$ with types in T. We may assume that $S_1, \ldots, S_m$ are the subgroups with nonzero images in K. As $S_1 \oplus \cdots \oplus S_m$ maps onto K, which is pure in $A(\tau)$, we have $\operatorname{type} S_1 \vee \cdots \vee \operatorname{type} S_m \geq \tau$. The type of a nonzero coordinate of S_i in $\bigoplus_{B\in\mathcal{B}} C_B$ is at least $\operatorname{type} S_i$, so the image of S_i is contained in $V(\operatorname{type} S_i)$. Thus

$$K \subset \bigcap_{i=1}^{m} V(\operatorname{type} S_i) = V(\operatorname{type} S_1 \vee \cdots \vee \operatorname{type} S_m) \subset V(\tau).$$

Hence $A(\tau) \subset V(\tau)$, whence $RA(\tau) \subset V(\tau)$. □

THEOREM 4. *Let A and B be Butler T-groups. Then the natural map $R \otimes$ Hom$(A, B) \to$ Hom$_T(R \otimes A, R \otimes B)$ is an isomorphism.*

PROOF. We identify A and B with the subgroups $1 \otimes A$ and $1 \otimes B$ of $R \otimes A$ and $R \otimes B$. The natural map of the theorem is clearly an embedding that identifies $\operatorname{Hom}(A, B)$ with those maps in $\operatorname{Hom}_T(R \otimes A, R \otimes B)$ that take A into B. If A is rank-one, then a nonzero homomorphism $f : R \otimes A \to R \otimes B$ is a map of RT-spaces if and only if $\operatorname{type} A \leq \operatorname{type} B$. In that case, there exists a positive integer n, invertible in R, such that $nf(A) \subset B$. Thus $nf \in \operatorname{Hom}(A, B)$, whence $f \in R \otimes \operatorname{Hom}(A, B)$. So the theorem is true when A has rank one, hence when A is a finite direct sum of rank-one groups.

For an arbitrary Butler T-group A, let F be a finite direct sum of rank-one T-groups, mapping onto A with kernel K, and consider the diagram

$$\begin{array}{ccccc} R\otimes \mathrm{Hom}(A,B) & \rightarrow & R\otimes \mathrm{Hom}(F,B) & \rightarrow & R\otimes \mathrm{Hom}(K,B) \\ \downarrow & & \downarrow & & \downarrow \\ \mathrm{Hom}_T(R\otimes A, R\otimes B) & \rightarrow & \mathrm{Hom}_T(R\otimes F, R\otimes B) & \rightarrow & \mathrm{Hom}_T(R\otimes K, R\otimes B) \end{array}$$

The top row is exact, the bottom row is a zero-sequence, and the first map in each row is one-to-one. The vertical maps are one-to-one, and the middle vertical map is an isomorphism as F is a finite direct sum of rank-one groups. An easy diagram chase shows that the first vertical map is onto. □

3. The Arnold-Dugas theory

Arnold and Dugas [**1**] studied Butler groups A with type-set $\{\tau_0, \tau_1, \dots, \tau_n\}$ where $n > 1$ and $\tau_i \wedge \tau_j = \tau_0$ if $i \neq j$. Discarding rank-one summands of type τ_0, they could assume that

$$\frac{A}{A(\tau_1) + \cdots + A(\tau_n)}$$

was finite. As they were interested in near-isomorphism, they studied isomorphism at p. For this purpose they could assume that $p^m A \subset A(\tau_1)+\cdots+A(\tau_n) = A_0$. They also assumed that none of the types τ_i was ∞ at p, so A_p is a free $\mathbf{Z}_p$-module. If A is almost completely decomposable, then A_0 is the *direct* sum of the $A(\tau_i)$.

The free $\mathbf{Z}/p^m\mathbf{Z}$ module $U = A_0/p^m A_0$ inherits some subgroups from A, namely the images U_i of the $A(\tau_i)$ and the image U_{n+1} of $p^m A$. Arnold and Dugas show that you can construct A with specified $U, U_1, \dots, U_n, U_{n+1}$ exactly when $U_1, \dots, U_n$ are pure, disjoint from U_{n+1}, and the sum of the ranks of any two of them does not exceed the rank of U ([**1,** Theorem 2.2(a)]). For A completely decomposable, they show that all the information about A, up to isomorphism at p, is contained in U together with the subgroups $U_1, \dots, U_n, U_{n+1}$. This is all put in the setting of an isomorphism of certain categories ([**1,** Theorem 2.2]).

What does this look like from the perspective of Theorem 2? If A is of the above reduced form, then $V = F(A)$ is a finite-rank free $\mathbf{Z}_p$-module containing the free submodule $V_0 = V(\tau_1) + \cdots + V(\tau_n)$ which contains $p^m V$. A $\mathbf{Z}_pT$-map from V to V' is a $\mathbf{Z}_p$-module homomorphism that takes each $V(\tau_i)$ into $V'(\tau_i)$. Such a map is equivalent to a $\mathbf{Z}_pT$-map from V_0 to V_0' that takes $p^m V$ to $p^m V'$. This is almost the Arnold-Dugas setting, and suffices for the construction of Butler groups from representations. The restriction on $p^m V$, as a subgroup of V_0, is that it contain $p^m V_0$ and that its intersection with each $V(\tau_i)$ be contained in $p^m V_0$. The latter condition follows from the purity of $V(\tau_i)$. The condition on the ranks of the U_i follows from the fact that $V(\tau_1), \dots, V(\tau_n)$ are pairwise disjoint because $V(\tau_i) \cap V(\tau_j) \subset V(\tau_i \vee \tau_j)$.

The constructions of Butler groups in [**1**] can be transferred practically verbatim to the $\mathbf{Z}_pT$-space setting. Instead of working over the ring $R = \mathbf{Z}/p^m\mathbf{Z}$, work over $R = \mathbf{Z}_p$ and throw $p^m U$ into U_{n+1}. Proofs of indecomposability take place over R/pR, so are unchanged.

What happens when we pass to $V_0/p^m V_0 = A_0/p^m A_0$? Now a homomorphism is zero if it takes V_0 to $p^m V_0$, that is, if it is divisible by p^m. Hence Arnold and Dugas shift to the category where those homomorphisms are set to zero. But a $\mathbf{Z}_pT$-map from V_0 to V_0 is an automorphism if and only if it becomes an automorphism in the new category—this follows from the fact $1+pf$ is always a $\mathbf{Z}_pT$-automorphism as its determinant is a unit in $\mathbf{Z}_p$ and the subgroups $V(\tau_i)$ and $p^m V$ are incompressible in V_0 (every automorphism taking these subgroups into themselves maps them onto themselves).

For almost completely decomposable groups there remains the question of whether every map $V_0/p^m V_0 \to V_0'/p^m V_0'$ lifts to a map $V_0 \to V_0'$. We can do that easily and respect the subgroups $V_0(\tau_i)$: simply do it for each $V_0(\tau_i)$. Then $p^m V$ goes to $p^m V'$ because $p^m V + p^m V_0$ goes to $p^m V' + p^m V_0'$.

The simplest relaxation of the Arnold-Dugas conditions is to allow $(\tau_i)_p$ to be infinite. Then V need not be free—it could have a divisible part. Passing to $V_0/p^m V_0$ destroys the divisible part, so we can't operate in that setting without losing information. But we can still work in the category of $\mathbf{Z}_pT$-spaces. However, for almost completely decomposable groups the finiteness of $(\tau_i)_p$ is necessary for indecomposability.

THEOREM 5. *Let A be an almost completely decomposable group with type-set $\{\tau_0, \tau_1, \ldots, \tau_n\}$, where $n > 1$ and $\tau_i \wedge \tau_j = \tau_0$ if $i \neq j$. If $p^m A \subset A(\tau_1) \oplus \cdots \oplus A(\tau_n)$ and $(\tau_i)_p$ is infinite for some $i \geq 1$, then $A(\tau_i)$ is a summand of A.*

PROOF. More generally, let A be an arbitrary torsion-free group with $p^m A \subset B \oplus C \subset A$. It is an easy exercise to show that if B is p-divisible, then $A = B \oplus C_*$ where C_* is the purification of C in A. (This simple argument was pointed out to me by Dave Arnold.) □

REFERENCES

1. Arnold, D. and M. Dugas, *Representations of finite posets and near-isomorphism of finite rank Butler groups*, Rocky Mtn. J. Math. (to appear).
2. Butler, M.C.R., *Torsion-free modules and diagrams of vector spaces*, Proc. London Math. Soc. **18** (1968), 635–652.
3. Richman, F., *Butler groups, valuated vector spaces, and duality*, Rend. Sem. Mat. Univ. Padova **72** (1984), 13–19.

FLORIDA ATLANTIC UNIVERSITY, BOCA RATON FL, 33431

E-mail address: richman@acc.fau.edu

Contemporary Mathematics
Volume **171**, 1994

The braid group action on the set of exceptional sequences of a hereditary artin algebra

CLAUS MICHAEL RINGEL

ABSTRACT. Let A be a hereditary artin algebra with $s = s(A)$ simple modules. The indecomposable A-modules without self-extensions are of great importance, they may be called exceptional modules. Certain sequences $(X_1,...,X_s)$ consisting of exceptional modules will be called complete exceptional sequences. Crawley-Boevey has pointed out that the braid group on $s-1$ generators acts naturally on the set of complete exceptional sequences. In case A is finite-dimensional over an algebraically closed field, he has shown that this action is transitive, using a recent result by Schofield. We are going to present a direct proof which is valid for arbitrary hereditary artin algebras. It follows that the endomorphism rings of exceptional modules are just those rings which occur as endomorphism rings of the simple modules. Also, we will exhibit the relationship between complete exceptional sequences and tilting modules.

1. Exceptional modules

Let A be a hereditary artin algebra with $s = s(A)$ simple modules. We recall that an artin algebra is called *hereditary,* provided that its global dimension is at most 1, thus provided that we have $\mathrm{Ext}^2(X,Y) = 0$ for all A-modules X, Y. Note that the center of a hereditary artin algebra is semisimple. An artin algebra A is said to be *connected* provided that the center of A is a field, say k, and then A is actually a finite dimensional k-algebra. We usually will assume that A is connected. The modules we consider will always be finite length modules.

An A-module M is called *exceptional* provided that M is indecomposable and $\mathrm{Ext}^1(M,M) = 0$. The exceptional modules are of great importance for the representation theory of A. (These modules also have been called indecomposable partial tilting modules, Schur modules, or open bricks; the variety of such names indicates

---

1991 *Mathematical Subject Classification.* Primary 16E60, 16G20, 16G70, 20F36

This paper is in final form, and no version of it will be submitted for publication elsewhere.

that these modules have been considered in various circumstances and that several mathematicians have felt that they deserve a name. Since there does not yet exist a unified terminology for dealing with exceptional modules, we will provide a dictionary at the end of the note. The use of the word 'exceptional' is parallel to the analogous terminology introduced for vector bundles by Rudakov, see [Ru].)

We denote by $S_1, \dots, S_s$ a complete set of simple A-modules (one from each isomorphism class); clearly these modules are exceptional, and our aim is to outline an inductive procedure for obtaining all exceptional A-modules starting from the simple A-modules.

Note that the endomorphism ring of an exceptional A-module is a division ring and one may ask which division rings can arise as endomorphism rings of exceptional modules. We will see that the only ones which arise in this way are the algebras $\operatorname{End}(S_i)$.

Attached to the algebra A is a corresponding Kac-Moody algebra, and thus a root system. Namely, we define a generalized Cartan matrix $\Delta(A) = (\Delta_{ij})_{ij}$ as follows: Given two simple modules S_i, S_j, we have $\operatorname{Ext}^1(S_i, S_j) = 0$ or $\operatorname{Ext}^1(S_j, S_i) = 0$. Let us assume that $i \neq j$ and $\operatorname{Ext}^1(S_j, S_i) = 0$. We may consider $\operatorname{Ext}^1(S_i, S_j)$ as a vector space over the division ring $\operatorname{End}(S_i)$ or over the division ring $\operatorname{End}(S_j)$, and we define

$$\Delta_{ij} = -\dim_{\operatorname{End}(S_i)} \operatorname{Ext}^1(S_i, S_j),$$
$$\Delta_{ji} = -\dim_{\operatorname{End}(S_j)} \operatorname{Ext}^1(S_i, S_j).$$

If we denote the k-dimension of $\operatorname{End}(S_i)$ by d_i, then we have

$$d_i \Delta_{ij} = d_j \Delta_{ji}.$$

This shows that $\Delta(A)$ is a symmetrizable generalized Cartan matrix in the sense of [K3].

Given any $n \times n$-matrix Δ which is a generalized Cartan matrix, we may consider the corresponding root system. Here, we are only interested in the real roots; in the case of a symmetrizable generalized Cartan matrix they may be defined as follows: We consider the n-dimensional real space $\mathbb{R}^n$ with basis $\mathbf{e}_1, \dots \mathbf{e}_n$, and we define a symmetric bilinear form given by $(\mathbf{e}_i, \mathbf{e}_j) = d_i \Delta_{ij}$. For any vector $\mathbf{x} \in \mathbb{R}^n$, with $(\mathbf{x}, \mathbf{x}) \neq 0$, let $r_{\mathbf{x}}$ be the reflection relative to $\mathbf{x}$ with respect to this bilinear form; thus, for $\mathbf{y} \in \mathbb{R}^n$,

$$r_{\mathbf{x}}(\mathbf{y}) = \mathbf{y} - \frac{(\mathbf{y}, \mathbf{x})}{(\mathbf{x}, \mathbf{x})}\mathbf{x}.$$

For $1 \le i \le n$, let $r_i = r_{\mathbf{e}_i}$. Note that any r_i maps the subgroup $\mathbb{Z}^n$ of $\mathbb{R}^n$ into itself. The group generated by these reflections r_i is called the *Weyl group*. By definition, the *real roots* are those elements of $\mathbb{Z}^n$ which belong to the orbits of the base vectors $\mathbf{e}_i$ under the Weyl group. The canonical generators $\mathbf{e}_i$ of $\mathbb{Z}^n$ are called the *simple roots*. For any real root $\mathbf{x}$, the reflection $r_{\mathbf{x}}$ is defined and belongs to W, thus it maps the set of real roots into itself.

We may identify $\mathbb{Z}^n$ with the Grothendieck group $K_0(A)$ of all A-modules modulo exact sequences; the element of $K_0(A)$ attached to the A-module M will be called its *dimension vector* and denoted by $\mathbf{dim}\, M$; we identify $\mathbb{Z}^n$ and $K_0(A)$ so that we have $\mathbf{dim}\, S_i = \mathbf{e}_i$. We will show that the dimension vector $\mathbf{dim}\, X$ of any exceptional module X is a real root for $\Delta(A)$.

The results mentioned above are known (and some are trivial) in the case when k is an algebraically closed field; so our main interest lies in the case of an arbitrary base field k. There are some remarks by Kac [K1,K2] concerning this general case, but no proofs seem to be available. The inductive construction of the exceptional modules has been shown, for k algebraically closed, by Schofield [S]. The operation of the braid group for obtaining all exceptional modules from the simple ones was, again for k algebraically closed, introduced by Crawley-Boevey [CB]. The direct algorithm we present here may be new even in the case of an algebraically closed base field.

The author is endebted to W. Crawley-Boevey, St. König and H. Krause for helpful comments concerning the final presentation of the paper.

2. Preprojective and preinjective modules

We have noted above that the simple A-modules are exceptional modules. Some other exceptional modules are always known. In case A is representation finite (this means: there are only finitely many isomorphism classes of indecomposable modules), then all indecomposable A-modules are exceptional; otherwise, there are countable families of indecomposable modules which are exceptional, namely the preprojective ones and the preinjective ones. Let us recall the corresponding constructions. For $1 \le i \le s$, we denote by P_i the projective cover of S_i, by Q_i the injective envelope of S_i. Then $P_1, \dots, P_s$ is a complete set of indecomposable projective modules, $Q_1, \dots, Q_s$ is a complete set of indecomposable injective modules, and all these modules P_i, Q_i are exceptional.

We denote by $\tau = DTr$ the Auslander-Reiten translation; given any indecomposable non-projective module M, the module τM again is indecomposable, it is the left hand term of an almost split sequence ending in M. Similarly, we denote by $\tau^- = TrD$ its partial inverse: given any indecomposable non-injective module M, the module $\tau^- M$ is indecomposable, it is the right hand term of an almost split sequence starting in M, see [AR]. If M is exceptional, but not projective, then τM is exceptional again; similarly, if M is exceptional, and not injective, then $\tau^- M$ is exceptional. (This follows from the fact that the restriction of τ to the full subcategory of modules without non-zero projektive direct summands is an equivalence onto the full subcategory of modules without non-zero injective direct summands, its inverse being given by τ^-.)

The indecomposable modules of the form $\tau^{-t} P_i$ are said to be *preprojective,* those of the form $\tau^t Q_i$ are said to be *preinjective.* All these modules are exceptional. We note the following: If A is representation finite, then all the indecomposable A-modules are both preprojective as well as preinjective. If A is not representation finite (and connected), then for any $t \in \mathbb{N}_0$ and any $1 \le i \le s$, the modules $\tau^{-t} P_i$

and $\tau^t Q_i$ are indecomposable modules which are pairwise non-isomorphic, and there are additional indecomposable modules, called the *regular* ones.

In case $s(A) = 2$, the exceptional modules are just the preprojective and the preinjective ones; this case will be considered in detail later, since the algorithm for constructing all exceptional modules will be based on dealing with full exact subcategories which are equivalent to the module categories of hereditary artin algebras B with $s(B) = 2$. In case A is representation infinite and $s(A) > 2$ (and A is connected), then there do exist exceptional modules which are neither preprojective nor preinjective, see [R4].

3. The algorithm for obtaining all exceptional modules

For any hereditary artin algebra A, a pair (X, Y) of exceptional A-modules is called an *exceptional pair* provided that we have

$$\mathrm{Hom}(Y, X) = 0, \quad \mathrm{Ext}^1(Y, X) = 0.$$

An exceptional pair (X, Y) will be called *orthogonal,* provided that also

$$\mathrm{Hom}(X, Y) = 0$$

is satisfied. If (X, Y) is an orthogonal exceptional pair, we may consider the category $\mathcal{C}(X, Y)$ of all A-modules M which have a filtration with factors isomorphic to X and Y. This is an exact abelian subcategory with two simple objects, namely X and Y, it is equivalent to the category of all B-modules of a hereditary artin algebra B with $s(B) = 2$ (see, for example [R1], section 1). Since $\mathcal{C}(X, Y)$ is equivalent to the module category for a hereditary artin algebra B, we may consider those objects in $\mathcal{C}(X, Y)$ which correspond under an equivalence to preprojective, or preinjective B-modules, and we will call them preprojective or preinjective objects in $\mathcal{C}(X, Y)$, respectively. Since $\mathcal{C}(X, Y)$ is closed under extensions inside the category of all A-modules, we see that the preprojective and the preinjective objects in $\mathcal{C}(X, Y)$ are exceptional A-modules.

For any natural number n, we define a class $\mathcal{E}_n = \mathcal{E}_n(A)$ of indecomposable modules of length n inductively as follows: Let $\mathcal{E}_1$ be the simple A-modules. Let us asusme now that for some $n > 1$ the classes $\mathcal{E}_1, \ldots, \mathcal{E}_{n-1}$ already have been defined. Let $\mathcal{E}_n$ be the class of indecomposable A-modules M of length n with the following property: there is an orthogonal exceptional pair (X, Y) with both modules X, Y in $\bigcup_{i=1}^{n-1} \mathcal{E}_i$ such that M is preprojective or preinjective in $\mathcal{C}(X, Y)$. Finally, let $\mathcal{E} = \bigcup_{i \geq 1} \mathcal{E}_i$. We obtain in this way a class of exceptional A-modules and we claim that all exceptional A-modules belong to this class:

THEOREM 1. *The class $\mathcal{E}$ is the class of all exceptional A-modules.*

The proof will be given below using the braid group operation on the set of the socalled exceptional sequences.

COROLLARY 1. *If X is an exceptional A-module, then* $\mathrm{End}(X)$ *is isomorphic to* $\mathrm{End}(S)$ *for some simple A-module S.*

Proof: If (X,Y) is an orthogonal pair of exceptional A-modules, and M is preprojective or preinjective in $\mathcal{C} = \mathcal{C}(X,Y)$, then $\mathrm{End}(M)$ is isomorphic to $\mathrm{End}(X)$ or to $\mathrm{End}(Y)$. For example, assume that M is preprojective in $\mathcal{C}$. Observe that Y is simple projective in $\mathcal{C}$. The projective cover Z of X in $\mathcal{C}$ is an indecomposable module with endomorphism ring isomorphic to $\mathrm{End}(X)$. We denote by $\tau_{\mathcal{C}}$ the Auslander-Reiten translation in $\mathcal{C}$. Since M is preprojective in $\mathcal{C}$, it is either of the form $\tau_{\mathcal{C}}^{-t}Y$ for some $t \in \mathbb{N}_0$, and then its endomorphism ring is isomorphic to $\mathrm{End}(Y)$, or else it is of the form $\tau_{\mathcal{C}}^{-t}Z$ for some $t \in \mathbb{N}_0$, and then its endomorphism ring is isomorphic to $\mathrm{End}(Z)$, and therefore isomorphic to $\mathrm{End}(X)$. In a similar way one deals with the case when M is preinjective.

COROLLARY 2. *If X is an exceptional A-module, then* $\mathbf{dim}\, X$ *is a real root for* $\Delta(A)$.

Proof: It is well-known that the bilinear form $\langle -,- \rangle$ on $\mathbb{Z}^n = K_0(A)$ has the following homological interpretation: Given A-modules M, N, let

$$\langle \mathbf{dim}\, M, \mathbf{dim}\, N \rangle = \dim_k \mathrm{Hom}(M,N) - \dim_k \mathrm{Ext}^1(M,N).$$

Then, according to [R1], Lemma 2.2,

$$(\mathbf{dim}\, M, \mathbf{dim}\, N) = \langle \mathbf{dim}\, M, \mathbf{dim}\, N \rangle + \langle \mathbf{dim}\, N, \mathbf{dim}\, M \rangle.$$

Assume now that (X,Y) is an orthogonal pair of exceptional modules, and assume that we know already that $\mathbf{dim}\, X$ and $\mathbf{dim}\, Y$ are real roots. Let W_{XY} be the subgroup of W generated by $r_{\mathbf{dim}\, X}$ and $r_{\mathbf{dim}\, Y}$. Then the dimension vectors of the preprojective and the preinjective objects in $\mathcal{C}$ belong to the orbits of $\mathbf{dim}\, X$ and $\mathbf{dim}\, Y$ under the operation of W_{XY}, see [R1], 3.2. It follows that the dimension vectors of the preprojective and the preinjective objects in $\mathcal{C}$ are real roots for $\Delta(A)$.

4. The exceptional pairs in the case $s = 2$.

Let us now assume that $s = s(A) = 2$, and let P, Q be the simple modules; we may assume that $\mathrm{Ext}^1(P,Q) = 0$.

LEMMA. *If M is an exceptional module, there are exceptional modules M^-, M^+, unique up to isomorphism, such that (M^-, M) and (M, M^+) are exceptional pairs.*

Proof: An exceptional module M is an indecomposable A-module which is either preprojective or preinjective.

In case A is representation finite, let $P_1, \dots, P_m$ be the indecomposable modules, with

$$\mathrm{Hom}(P_i, P_{i+1}) \neq 0, \quad \text{for all} \quad 1 \leq i < m.$$

In particular, we have $P_1 = P, P_m = Q$. Let $P_i^+ = P_{i+1}$ for $1 \leq i < m$, and $P_i^- = P_{i-1}$ for $1 < i \leq m$. Let $P_1^- = P_m$, and $P_m^+ = P_1$.

(Note that in this case $m = 2$, 3, 4 or 6; the corresponding Cartan matrices $\Delta(A)$ are labelled $\mathbb{A}_1 \times \mathbb{A}_1, \mathbb{A}_2, \mathbb{B}_2$ and $\mathbb{G}_2$, respectively – these are just the Cartan matrices arising for the finite-dimensional semisimple Lie algebras of rank 2.)

In case A is representation infinite, there are countable many indecomposable preprojective modules and countably many indecomposable preinjective modules, and we may label them in the following way (see again [R1]): let $\{P_i \mid i \in \mathbb{N}_1\}$ be the indecomposable preprojective modules, and $\{Q_i \mid i \in \mathbb{N}_1\}$ the indecomposable preinjective modules with

$$\operatorname{Hom}(P_i, P_{i+1}) \neq 0, \quad \text{and} \quad \operatorname{Hom}(Q_{i+1}, Q_i) \neq 0 \quad \text{for all} \quad i \geq 1.$$

It follows that $P_1 = P$, $Q_1 = Q$. In this case, $P_i^+ = P_{i+1}$, and $Q_i^- = Q_{i+1}^-$, for all $i \geq 1$; similarly, $P_i^- = P_{i-1}, Q_i^+ = Q_{i-1}$ for all $i \geq 2$, finally, let $P_1^- = Q_1, Q_1^+ = P_1$.

Note that all the exceptional pairs (X, Y) different from (Q, P) satisfy

$$\operatorname{Hom}(X, Y) \neq 0.$$

5. Exceptional sequences and the braid group

A sequence $\mathcal{X} = (X_1, \ldots, X_n)$ is called *exceptional,* provided that any pair (X_i, X_j) with $i < j$ is exceptional. Actually, we only are interested in isomorphism classes of modules, not in the modules themselves: thus, an exceptional sequence will be considered as a sequence of isomorphism classes. An exceptional sequence $\mathcal{X} = (X_1, \ldots, X_n)$ with $n = s(A)$ is said to be *complete.*

Recall that the braid group B_n in $n-1$ generators $\sigma_1, \ldots, \sigma_{n-1}$ is the free group with these generators and the relations $\sigma_i\sigma_{i+1}\sigma_i = \sigma_{i+1}\sigma_i\sigma_{i+1}$ for all $1 \leq i < n-1$, and $\sigma_i\sigma_j = \sigma_j\sigma_i$ for $j \geq i+2$.

Given an exceptional sequence $(X_1, \ldots, X_n)$, we denote by $\mathcal{C}(X_1, \ldots, X_n)$ the closure of the full subcategory with objects $X_1, \ldots, X_n$ under kernels, images, and extensions. Of course, this is an exact abelian subcategory (in case (X, Y) is an orthogonal exceptional pair, this subcategory $\mathcal{C}(X, Y)$ coincides with the full subcategory of all modules with a filtration with factors isomorphic to X and Y).

The main ingredients for the definition of the operation of B_n on the set of complete exceptional sequences are the following three observations due to Crawley-Boevey:

PROPOSITION 1. *Let $(X_1, \ldots, X_n)$ be an exceptional sequence. $\mathcal{C}(X_1, \ldots, X_n)$ is equivalent to the category of all B-modules for some hereditary artin algebra B with $s(B) = n$.*

PROPOSITION 2. *Let $(X_1, \ldots, X_n)$ be an exceptional sequence of A-modules. If (Y, Y') is an exceptional pair in $\mathcal{C}(X_i, X_{i+1})$, then*

$$(X_1, \ldots, X_{i-1}, Y, Y', X_{i+2}, \ldots, X_n)$$

is an exceptional sequence of A-modules.

PROPOSITION 3. *Let $(X_1, \ldots, X_{i-1}, X_{i+1}, \ldots, X_s)$ be an exceptional sequence, where $1 \leq i \leq s$. Then there exists a unique module X_i such that $(X_1, \ldots, X_s)$ is a complete exceptional sequence.*

For a proof of these assertions, we refer to [CB]; his assumption concerning the base field is not used in this part of the paper. We should mention that these results of Crawley-Boevey were motivated by the work mainly of Gorodentsev, Rudakov, and Bondal, but also others, dealing with exceptional sequences of vector bundles (see the papers [GR], [G], [B] and the collection [Ru]); in these papers, a corresponding action of the braid group had been introduced. The natural setting of such braid group actions seems to be in the context of triangulated categories (say the corresponding derived categories). Actually, as Crawley-Boevey has pointed out, one may use the arguments of [G], section 3.3, in order to obtain also the braid group operation on the set of exceptional sequences of a hereditary artin algebra. The proof presented in [CB] is rather straight-forward, it only uses properties of perpendicular categories.

Let (X, Y) be an exceptional pair. Recall that $\mathcal{C}(X, Y)$ is equivalent to the category of all B-modules, where B is a hereditary artin algebra with $s(B) = 2$. Thus, we may consider inside $\mathcal{C}(X, Y)$ the exceptional module Y^+; our notation will be $r(X, Y) = Y^+$, this module is the unique object in $\mathcal{C}(X, Y)$ such that $(Y, r(X, Y))$ is again an exceptional sequence. Similarly, let $l(X, Y) = X^-$ be the unique object in $\mathcal{C}(X, Y)$ such that $(l(X, Y), X)$ is an exceptional pair.

Given an exceptional sequence $\mathcal{X} = (X_1, \ldots, X_n)$ of A-modules and $1 \leq i < n$, we define

$$\sigma_i(\mathcal{X}) = (X_1, \ldots, X_{i-1}, X_{i+1}, r(X_i, X_{i+1}), X_{i+2}, \ldots, X_n),$$

and

$$\sigma_i^{-1}(\mathcal{X}) = (X_1, \ldots, X_{i-1}, l(X_i, X_{i+1}), X_i, X_{i+2}, \ldots, X_n).$$

In this way, we obtain an action of the braid group in $n-1$ generators $\sigma_1, \ldots, \sigma_{n-1}$ on the set of exceptional sequences: If $i + 2 \leq j$, then both $\sigma_i\sigma_j(\mathcal{X})$ and $\sigma_j\sigma_i(\mathcal{X})$ will be equal to

$$(\ldots, X_{i+1}, r(X_i, X_{i+1}), \ldots, X_{j+1}, r(X_j, X_{j+1}), \ldots)$$

(the positions marked are those with index $i, i+1$ and $j, j+1$). On the other hand, let

$$Y = r\big(r(X_i, X_{i+1}), X_{i+2}\big) \quad \text{and} \quad Y' = r\big(r(X_i, X_{i+2}), r(X_{i+1}, X_{i+2})\big).$$

Then

$$\sigma_i\sigma_{i+1}\sigma_i(\mathcal{X}) = (\ldots, X_{i+2}, r(X_{i+1}, X_{i+2}), Y, \ldots)$$

whereas

$$\sigma_{i+1}\sigma_i\sigma_{i+1}(\mathcal{X}) = (\ldots, X_{i+2}, r(X_{i+1}, X_{i+2}), Y', \ldots)$$

(only the positions $i, i+1, i+2$ are labelled).

Since $\big(X_{i+2}, r(X_{i+1}, X_{i+2}), Y\big)$ and $\big(X_{i+2}, r(X_{i+1}, X_{i+2}), Y'\big)$ are complete exceptional sequences in $\mathcal{C}(X_i, X_{i+1}, X_{i+2})$, it follows from Proposition 3 that $Y = Y'$.

6. The reduction theorem

Let $\mathcal{X} = (X_1, \dots, X_n)$ be an exceptional sequence, let $1 \le i < n$. We say that σ_i is a *transposition* for $\mathcal{X}$, provided that for $\mathcal{Y} = \sigma_i \mathcal{X}$, we have $Y_{i+1} = X_i$ (so that $\mathcal{Y}$ is obtained from $\mathcal{X}$ by just transposing X_i and X_{i+1}). The following assertion is easy to verify.

LEMMA. *Let $\mathcal{X} = (X_1, \dots, X_n)$ be an exceptional sequence, let $1 \le i < n$. Then σ_i is a transposition for $\mathcal{X}$ if and only if* $\mathrm{Hom}(X_i, X_{i+1}) = 0$, *and* $\mathrm{Ext}^1(X_i, X_{i+1}) = 0$.

LEMMA. *Let $\mathcal{X} = (X_1, \dots, X_n)$ be an exceptional sequence, let $1 \le i < n$. Then there exists $t \in \mathbb{Z}$ such that $\mathcal{Y} = \sigma_i^t \mathcal{X}$ satisfies* $\mathrm{Hom}(Y_i, Y_{i+1}) = 0$.

PROOF. We apply the considerations of section 4 to $\mathcal{C}(X_i, X_{i+1})$.

We say that σ_i^t is a *proper reduction* for $\mathcal{X}$, provided that $\mathcal{Y} = \sigma_i^t \mathcal{X}$ satisfies $\mathrm{Hom}(Y_i, Y_{i+1}) = 0$, whereas $\mathrm{Hom}(X_i, X_{i+1}) \neq 0$.

An exceptional sequence $\mathcal{X} = (X_1, \dots, X_n)$ is said to be *orthogonal,* provided that we have $\mathrm{Hom}(X_i, X_j) = 0$ for all $i \neq j$.

THEOREM 2. *Any exceptional sequence can be shifted by the braid group action to an orthogonal sequence using only transpositions and proper reductions.*

In order to give the proof, we need some preparations. For any A-module M, let $|M|$ be its length. For a sequence $\mathcal{X} = (X_1, \dots, X_n)$ of A-modules, let $\|\mathcal{X}\| = (|X_{\pi(1)}|, \dots, |X_{\pi(n)}|)$, where π is a permutation of $1, \dots, n$ such that $|X_{\pi(1)}| \ge \cdots \ge |X_{\pi(n)}|$. For sequences $x = (x_1, \dots, x_n)$, $y = (y_1, \dots, y_n)$ in $\mathbb{N}_0$, we write $x \le y$ provided that $x_i \le y_i$, for alle $1 \le i \le n$, and $x < y$, provided that $x \le y$ and $x \neq y$.

LEMMA. *If σ_i is a transposition for $\mathcal{X}$, then $\|\sigma_i \mathcal{X}\| = \|\mathcal{X}\|$. Is σ_i^t is a proper reduction for $\mathcal{X}$, then $\|\sigma_i^t \mathcal{X}\| < \|\mathcal{X}\|$.*

PROOF. Let $\sigma_i \mathcal{X} = (Y_1, \dots, Y_n)$. Then $Y_j = X_j$ for $j \notin \{i, i+1\}$, whereas Y_i, Y_{i+1} are the simple objects in the category $\mathcal{C}(X_i, X_{i+1})$. Since we assume that σ_i^t is a proper reduction, at least one of the modules X_i, X_{i+1} cannot be simple in $\mathcal{C}(X_i, X_{i+1})$. However, if M is an indecomposable non-simple object in $\mathcal{C}(X_i, X_{i+1})$, then M has a filtration with factors of the form X_i and X_{i+1}, and both types of factors appear at least once.

REMARK. If σ_i^t is a proper reduction for $\mathcal{X}$, we have $\|\sigma_i^t \mathcal{X}\| < \|\mathcal{X}\|$, but we cannot demand to have $\|\sigma_i^t \mathcal{X}\| \le \|\sigma_i^{t-1} \mathcal{X}\| \le \cdots \le \|\mathcal{X}\|$. For example, consider a hereditary algebra with two simple modules P, Q, where P is projective, Q injective, such that the injective envelope of P is of length 2, and the projective cover of Q is of length $c+1$, with $c \ge 3$. Let $\mathcal{X} = \sigma_1^2(Q, P)$. Then $\|\mathcal{X}\| = c$, whereas $\|\sigma_1 \mathcal{X}\| = c^2 - c - 1 > c$, and $\|\sigma_1^{-1} \mathcal{X}\| = c + 1 > c$. Of course, σ_1^t with $t = -2$ is a proper reduction for $\mathcal{X}$, and for $c \ge 4$, this is the only possible choice for t.

PROOF OF THEOREM 2. Let $\mathcal{X} = (X_1, \dots, X_n)$ be an exceptional sequence, and assume that $\mathcal{X}$ is not orthogonal. Choose $a < b$ such that $\mathrm{Hom}(X_a, X_b) \neq 0$, but $\mathrm{Hom}(X_i, X_j) = 0$ for the remaining $a \le i < j \le b$. Let φ: $X_a \to X_b$ be a non-zero morphism. According to [HR], we know that φ has to be a monomorphism or an epimorphism.

Consider first the case where φ is a monomorphism. This monomorphism induces epimorphisms

$$\mathrm{Ext}^1(X_b, X_i) \to \mathrm{Ext}^1(X_a, X_i)$$

for all i. The first Ext-group is zero for $b \geq i$, thus the second Ext-group also vanishes for these i. We see that both $\mathrm{Hom}(X_a, X_i) = 0$, and $\mathrm{Ext}^1(X_a, X_i) = 0$ for $a < i < b$, thus

$$\sigma_{b-2} \ldots \sigma_{a+1}\sigma_a \mathcal{X} = (X_1, \ldots, X_{a-1}, X_{a+1}, \ldots, X_{b-1}, X_a, X_b, \ldots, X_n),$$

and always we just use transpositions. Now we apply some power of σ_{b-1}, say σ_{b-1}^t, in order to replace (X_a, X_b) by (Q, P) with $\mathrm{Hom}(Q, P) = 0$, thus we use a proper reduction.

The case when φ is an epimorphism, is treated similarly: The map φ induces epimorphisms

$$\mathrm{Ext}^1(X_i, X_a) \to \mathrm{Ext}^1(X_i, X_b).$$

The first Ext-group is zero for $i \geq a$, thus also the second Ext-group vanishes for these i. We see that both $\mathrm{Hom}(X_i, X_b) = 0$, and $\mathrm{Ext}^1(X_i, X_b) = 0$ for all $a < i < b$, thus

$$\sigma_{a+1} \ldots \sigma_{b-1} \mathcal{X} = (X_1, \ldots, X_{a-1}, X_a, X_b, X_{a+1}, \ldots, X_{b-1}, X_{b+1}, \ldots, X_n),$$

always using just transpositions. We now apply some power σ_a^t of σ_a in order to replace (X_a, X_b) by (Q, P) with $\mathrm{Hom}(Q, P) = 0$; thus we use a proper reduction.

Since a proper reduction always decreases $\|\mathcal{X}\|$, this process has to stop after a finite number of steps, and we obtain some orthogonal exceptional sequence.

7. The orthogonal complete exceptional sequences

THEOREM 3. *The orthogonal complete exceptional sequences are just those exceptional sequences which consist of the simple modules.*

PROOF. Let $\mathcal{X}$ be an exceptional sequence. Let $\mathcal{C}(\mathcal{X})$ be be the smallest subcategory containing $\mathcal{X}$ and being closed under extensions, kernels of epimorphisms and cokernels of monomorphisms. If $\mathcal{X}$ is complete, then $\mathcal{C}(\mathcal{X}) = A$-mod, as Crawley-Boevey [CB] has shown (again without using any assumption on the base field).

Since $X_1, \ldots, X_s$ are orthogonal modules with division rings as endomorphism rings, $\mathcal{C}(\mathcal{X})$ is just the set of A-modules which have a filtration by modules of the form X_i, see [R1] (the process of simplification). This shows that any simple A-module S_j has a filtration by modules of the form X_i, thus for any S_j, there is some $\pi(j)$ with $S_j = X_{\pi(j)}$. Since the length of the sequence $\mathcal{X}$ is $s = s(A)$, it follows that any X_i is simple.

COROLLARY. *Any complete exceptional sequence can be shifted by the braid group action to an exceptional sequence consisting only of simple modules by using only transpositions and proper reductions.*

LEMMA. *The complete exceptional sequences which consist of simple modules can be obtained from each other by the braid group action using only transpositions.*

This can be verified without difficulties.

COROLLARY. *The braid group acts transitively on the set of complete exceptional sequences.*

As mentioned in the introduction, this assertion, for A finite-dimensional over some algebraically closed field, is the main result of the paper [CB] by Crawley-Boevey. His proof relies on investigations by Schofield [S] dealing with semi-invariants of quivers, and this is the only part of the paper which uses the assumption on the base field. Actually, the decisive Lemma 7 in [CB] may also be shown directly, using considerations similar to the ones above.

8. Endomorphism rings of exceptional modules

Given an artinian ring B, let $J(B)$ be its radical, thus $J(B)$ is the maximal nilpotent ideal of B.

THEOREM 4. *Let $\mathcal{X} = (X_1, \dots, X_s)$ be a complete exceptional sequence, let $B = B(\mathcal{X})$ be the endomorphism ring of $\bigoplus_i X_i$. Then $B/J(B)$ is Morita equivalent to $A/J(A)$.*

PROOF. Let $D(\mathcal{X}) = B/J(B)$. Let D_i be the endomorphism ring of X_i, and note that D_i is a division ring, thus $D = \prod_i D_i$. We claim that $D(\mathcal{X})$ and $D(\sigma\mathcal{X})$ are isomorphic, for any braid group element σ. Of course, it is sufficient to consider a generator σ_i, and we may assume that $s = 2$. But in this case the assertion is obvious.

9. Tilting sequences

An exceptional sequence $\mathcal{X} = (X_1, \dots, X_n)$ is said to be *strongly exceptional,* provided that we have $\operatorname{Ext}^1(X_i, X_j) = 0$ for all i, j. A strongly exceptional sequence which is complete may be called a *tilting sequence.*

We say that σ_i is a *p-extension* for $\mathcal{X}$, provided that $\operatorname{Hom}(X_i, X_{i+1}) = 0$ (the letter p shall indicate that p-extensions are the usual procedure to construct indecomposable projective modules starting with simple modules; also in the general case considered here, we use the p-extensions in order to construct relative projective objects inside the subcategory of modules having a filtration with factors of the form X_i.)

THEOREM 5. *Any exceptional sequence can be shifted by the braid group action to a strongly exceptional sequence using only transpositions and p-extensions.*

PROOF. Let $\mathcal{X} = (X_1, \dots, X_n)$ be an exceptional sequence, and suppose it is not strongly exceptional. Choose $a < b$ such that $\operatorname{Ext}^1(X_a, X_b) \neq 0$, and such that $b-a$ is minimal. We choose t with $a \le t \le b$ maximal such that $\operatorname{Hom}(X_a, X_t) \neq 0$. Since $\operatorname{Hom}(X_a, X_b) = 0$, we even have $a \le t < b$. Let φ: $X_a \to X_t$ be a non-zero map.

We claim that φ is an epimorphism. Since $\mathrm{Ext}^1(X_t, X_a) = 0$, φ is a monomorphism or an epimorphism. Let us assume that $a < t$ and that φ is a monomorphism; the map φ induces a surjective map

$$\mathrm{Ext}^1(X_t, X_b) \to \mathrm{Ext}^1(X_a, X_b),$$

but the latter group is non-zero, whereas the first one is zero, by the minimality assumption on $b-a$. Thus, we obtain a contradiction. As a consequence, we see that $\mathrm{Hom}(X_t, X_i) = 0$ for $t < i \leq b$. For, a non-zero map ψ: $X_t \to X_i$ can be composed with φ and will give a non-zero map $X_a \to X_i$, contrary to the maximality of t.

For $t > a$, we have both $\mathrm{Hom}(X_t, X_i) = 0$, and $\mathrm{Ext}^1(X_t, X_i) = 0$, for all $t < i \leq b$; thus the consecutive application of first σ_t, then σ_{t+1}, and so on, finally σ_{b-1} yields just transpositions, and reduces $b - a$ by 1. The assertion follows by induction.

Thus, it remains to consider the case $t = a$. Since $\mathrm{Hom}(X_a, X_i) = 0$, and $\mathrm{Ext}^1(X_a, X_i) = 0$, for all $a < i < b$, the application of first σ_a, then σ_{a+1}, and so on, finally σ_{b-2} yields transpositions, and we obtain in this way an exceptional sequence $\mathcal{Y}$ with $\mathrm{Hom}(Y_{b-1}, Y_b) = 0$, and $\mathrm{Ext}^1(Y_{b-1}, Y_b) \neq 0$ (here, $Y_{b-1} = X_a$). Obviously, σ_{b-1} is a p-extension for $\mathcal{Y}$.

In order to see that the process stops, let us introduce a set $E(\mathcal{X})$ as follows: For any pair (u, v) with $1 \leq u < v \leq n$, let $E(\mathcal{X}; u, v)$ be the factor group of $\mathrm{Ext}^1(X_u, X_v)$ modulo the subgroup generated by the images of the induced maps $\mathrm{Ext}^1(X_u, \zeta)$ where ζ: $X_j \to X_v$ is a map with $j < v$. By definition, a pair (u, v) belongs to $E(\mathcal{X})$ if and only if there exists a sequence $u = u_0 < u_1 < \ldots u_m = v$ with $m \geq 1$ such that $E(\mathcal{X}; u_{i-1}, u_i) \neq 0$ for all $1 \leq i \leq m$. If σ_i is a transposition for $\mathcal{X}$, then the sets $E(\mathcal{X})$ and $E(\sigma_i\mathcal{X})$ clearly will have the same number of elements.

Thus, let σ_i be a p-extension for $\mathcal{X}$; and let $\mathcal{Y} = \sigma_i\mathcal{X}$. We claim that in this case the number of elements of $E(\mathcal{Y})$ decreases by (at least) one. Let $u < i$, and $i+1 < v$. We note the following: We have $E(\mathcal{Y}; u, v) = E(\mathcal{X}; u, v)$. Since $Y_i = X_{i+1}$, it is easy to see that $E(\mathcal{Y}; u, i) = E(\mathcal{X}; u, i+1)$ and $E(\mathcal{Y}; i, v) = E(\mathcal{X}; i+1, v)$. There is an epimorphism $Y_{i+1} \to X_i$ (with kernel a direct sum of copies of Y_i); it induces an isomorphism between $E(\mathcal{Y}; u, i+1)$ and $E(\mathcal{X}; u, i)$. Finally, assume that $E(\mathcal{Y}; i+1, v) \neq 0$. Then we have $E(\mathcal{X}; i, v) \neq 0$ or $E(\mathcal{X}; i+1, v) \neq 0$. In both cases, it follows that the pair (i, v) belongs to $E(\mathcal{X})$. Altogether we see that (u, v) belongs to $E(\mathcal{Y})$ if and only if it belongs to $E(\mathcal{X})$; that (u, i) or (i, v) belongs to $E(\mathcal{Y})$, if and only if $(u, i+1)$, or $(i+1, v)$ belongs to $E(\mathcal{X})$, respectively, and that $(u, i+1)$ or $(i+1, v)$ belongs to $E(\mathcal{Y})$, if and only if (u, i) or (i, v) belongs to $E(\mathcal{X})$, respectively. Of course, $(i, i+1)$ belongs to $E(\mathcal{X})$, but not to $E(\mathcal{Y})$. This completes the proof.

Given an exceptional sequence $\mathcal{X}$, we denote by $\mathcal{F}(\mathcal{X})$ the set of A-modules which have a filtration with factors of the form X_i.

THEOREM 6. *Let $\mathcal{X}$ be an exceptional sequence. Assume that $\mathcal{T}$ is a strongly exceptional sequence such that $\mathcal{X}$ can be shifted by the braid group action to $\mathcal{T}$ using only transpositions and p-extensions. Then the modules T_i are just those indecomposable modules M in $\mathcal{F}(\mathcal{X})$ which satisfy* $\mathrm{Ext}^1(M, X_i) = 0$ *for all i.*

PROOF. Let $\mathcal{X} = (X_1, \ldots, X_n)$. Let $\mathcal{P}(\mathcal{X})$ be the set of A-modules M in $\mathcal{F}(\mathcal{X})$ which satisfy $\mathrm{Ext}^1(M, X_i) = 0$ for all i. If $\mathcal{Y}$ is obtained from $\mathcal{X}$ by a transposition or a p-extension, then on the one hand, all the modules Y_i belong to $\mathcal{F}(\mathcal{X})$, thus

$\mathcal{F}(\mathcal{Y}) \subseteq \mathcal{F}(\mathcal{X})$ (and actually $\mathcal{F}(\mathcal{Y}) \subset \mathcal{F}(\mathcal{X})$ in the case of a p-extension), whereas, on the other hand, we have $\mathcal{P}(\mathcal{X}) \subseteq \mathcal{F}(\mathcal{Y})$, thus $\mathcal{P}(\mathcal{X}) \subseteq \mathcal{P}(\mathcal{Y})$. The latter implies that $\mathcal{P}(\mathcal{X}) \subseteq \mathcal{P}(\mathcal{T})$. Of course, $\mathcal{P}(\mathcal{T}) = \mathcal{F}(\mathcal{T})$ is just the additive subcategory add $\mathcal{T}$ generated by the modules T_i. It is known (see [DR], Theorem 2) that $\mathcal{P}(\mathcal{X})$ has precisely n isomorphism classes of indecomposable objects, thus $\mathcal{P}(\mathcal{X}) = \operatorname{add} \mathcal{T}$.

THEOREM 7. *Let $\mathcal{T} = (T_1, \dots, T_n)$ be a strongly exceptional sequence. The number of exceptional sequences which can be shifted by the braid group action to $\mathcal{T}$ using only transpositions and p-extensions is at most $n!$.*

PROOF. Let $\mathcal{X}$ be an exceptional sequence which can be shifted by the braid group action to $\mathcal{T}$ using only transpositions and p-extensions. Consider T_1. There is a surjective map $T_1 \to X_i$ for some (uniquely defined) i such that the kernel, as well as all the other modules $T_2, \dots T_n$ have filtrations with factors X_j where $j \neq i$. Let $\mathcal{X}'$ be obtained from $\mathcal{X}$ by deleting X_i. Clearly $\mathcal{T}' = (T_2, \dots, T_n)$ is itself strongly exceptional, and $\mathcal{X}'$ can be shifted by the braid group action to $\mathcal{T}'$ using only transpositions and p-extensions. By induction, we know that there are at most $(n-1)!$ possibilities for $\mathcal{X}'$. Since X_i is uniquely determined by $\mathcal{X}'$ and the index i (see [CB], Lemma 2), there are at most n possibilities for X_i, when $\mathcal{X}'$ is fixed. This completes the proof.

10. Perpendicular categories

We have mentioned above that Crawley-Boevey's definition of the braid group operation relies on properties of perpendicular categories. Perpendicular categories have been studied by Geigle-Lenzing [GL] and Schofield [S]; they are defined as follows: Given a collection $\mathcal{C}$ of A-modules, then $\mathcal{C}^{\perp}$ is the full subcategory of all A-modules Y which satisfy both

$$\operatorname{Hom}(C, Y) = 0 = \operatorname{Ext}^1(C, Y) \quad \text{for all} \quad C \in \mathcal{C};$$

similarly, ${}^{\perp}\mathcal{C}$ is the full subcategory of all A-modules X which satisfy both

$$\operatorname{Hom}(X, C) = 0 = \operatorname{Ext}^1(X, C) \quad \text{for all} \quad C \in \mathcal{C}.$$

Since A is hereditary, both subcategories $\mathcal{C}^{\perp}$ and ${}^{\perp}\mathcal{C}$ are exact (this means: they are abelian, and the inclusion functors are exact).

If X is an exceptional A-module, then it is known that $\mathcal{C}^{\perp}$ is equivalent to the category of all B-modules, similarly, ${}^{\perp}\mathcal{C}$ is equivalent to the category of all B'-modules; where B, B' are artin algebras with $s(B) = s(B') = s(A) - 1$, see [S]. Note that the procedures above yield an effective way of computing B and B', as soon as a complete exceptional sequence containing X is given:

Indeed, let $(X_1, \dots, X_s)$ be a complete exceptional sequence with $X = X_i$ for some i. We may shift X to the end of the sequence, thus, without loss of generality, we may assume that $X = X_s$. But in this case, we know from [CB] that $X^{\perp}$ is just $\mathcal{C}(X_1, \dots, X_{s-1})$. Now, using permutations and proper reductions to the sequence $(X_1, \dots, X_{s-1})$, we may transform it to an orthogonal sequence $(Y_1, \dots, Y_{s-1})$. But then the modules $Y_1, \dots, Y_{s-1}$ are just the simple objects of $X^{\perp}$. Finally, using transpositions and p-extensions, we may transform the sequence $(Y_1, \ldots, Y_{s-1})$ to a sequence $(Z_1, \ldots, Z_{s-1})$ such that we have $\mathrm{Ext}^1(Z_i, Y_j) = 0$ for all $1 \leq i, j \leq s-1$. The latter means that the modules Z_i are the indecomposable projective modules in $X^\perp$; and the endomorphism ring of $\bigoplus_{i=1}^{s-1} Z_i$ is the artin algebra B we are looking for.

Of course, in case we deal with algebras over an algebraically closed field k, we may already stop when we have obtained the orthogonal exceptional sequence $(Y_1, \ldots, Y_{s-1})$, since the dimension of the various k-spaces $\mathrm{Ext}^1(Y_i, Y_j)$ yields the quiver of B, and B is just the corresponding path algebra.

Similarly, we also can construct the artin algebra B'.

Dictionary

The terminology *exceptional* and *strongly exceptional* was introduced by Rudakov and his school [Ru] in the analogous situation of vector bundles.

Exceptional modules have also been called *stones* by Kerner [Ke] and *Schur modules* by Unger [U]. Modules with endomorphism ring a division ring have been named *bricks* in [R3].

A sequence $(X_1, \ldots, X_n)$ is exceptional if and only if, first, the set of modules $X_1, \ldots, X_n$ is *standardizable* in the sense of [DR] and, second, the order of the modules refines the intrinsic partial ordering of this standardizable set.

A pair of modules M, N with $\mathrm{Hom}(M, N) = 0$, $\mathrm{Hom}(N, M) = 0$ has been called *orthogonal* in [R1]. A set of modules consisting of pairwise orthogonal modules having divison rings as endomorphism rings has been called *discrete* by Gabriel and de la Peña [GP]. The p-extensions have been used in a decisive way in [R1,R2].

A sequence $(X_1, \ldots, X_n)$ is strongly exceptional if and only if, first, the direct sum of the modules $X_1, \ldots, X_n$ is a multiplicity–free *partial tilting module* as defined by Happel-Unger [HU], and, second, the order of the modules refines the partial ordering given by the existence of non-zero maps. Of course, the tilting sequences correspond in the same way to (multiplicity–free) *tilting modules*.

References

[AR] Auslander,M.; Reiten, I.: Representation theory of artin algebras III. Communications in Algebra 3 (1975), 239-294.

[B] Bondal, A.I.: Representations of associative algebras and coherent sheaves. Izv. Akad. Nauk SSSR Ser Mat. 53 (1989), 25-44.

[CB] Crawley-Boevey, W.: Exceptional sequences of representations of quivers. Proceedings of ICRA VI, Carleton-Ottawa Math. LNS 14 (1992)

[DR] Dlab, V., Ringel, C.M.: The module theoretical approach to quasi-hereditary algebras. In: Representations of Algebras and Related Topics. London Math. Soc. Lecture Note Series 168. Cambridge University Press (1992), 200–224

[GP] Gabriel, P., de la Peña, J.: Quotients of representation-finite algebras. Communications in Algebra 15 (1987), 279-307.

[GL] Geigle, W., Lenzing, H.: Perpendicular categories with applications to representations and sheaves. J.Algebra 144 (1991), 273-343

[G] Gorodentsev, A.L.: Exceptional bundles on surfaces with a moving anticanonical class. Izv. Akad. Nauk SSSR Ser Mat. 52 (1988), 740-757.

[GR] Gorodentsev, A.L., Rudakov, A.N.: Exceptional vector bundles on projective space. Duke Math. J. 54 (1987), 115-130.
[HR] Happel, D., Ringel, C.M.: Tilted algebras. Trans. Amer. Math. Soc. 274 (1982), 399–443.
[HU] Happel, D., Unger, L.: Almost complete tilting modules. Proc. Amer. Math. Soc. 106 (1989), 603-610.
[K1] Kac, V.G.: Infinite root systems, representations of graphs and invariant theory. Invent. Math. 56 (1980), 57-92.
[K2] Kac, V.G.: Infinite root systems, representations of graphs and invariant theory II. J. Algebra 77 (1982), 141-162.
[K3] Kac, V.G.: Infinite dimensional Lie algebras. Second Edition. Cambridge University Press (1985).
[Ke] Kerner, O.: Elementary stones. Communications in Algebra. (To appear)
[R1] Ringel, C.M.: Representations of K–species and bimodules. J. Algebra 41 (1976), 269–302.
[R2] Ringel, C.M.: Reflection functors for hereditary algebras. J. London Math. Soc. 21 (1980), 465-479.
[R3] Ringel, C.M.: Bricks in hereditary length categories. Resultate der Mathematik 6 (1983), 64-70
[R4] Ringel, C.M.: The regular components of the Auslander-Reiten quiver of a tilted algebra. Chinese Ann. Math. B 9 (1988), 1-18.
[Ru] Rudakov, A.N.: Helices and Vector Bundles. London Math. Soc. LNS 148.
[S] Schofield, A.: Semi–invariants of quivers. J. London Math. Soc. 43 (1991), 383-395.
[U] Unger, L.: Schur modules over wild, finite dimensional path algebras with three non isomorphic simple modules. J. Pure Appl. Algebra. 64 (1990), 205-222.

FAKULTÄT FÜR MATHEMATIK, UNIVERSITÄT, D-33613 BIELEFELD
E-mail address: ringel@mathematik.uni-bielefeld.de

Contemporary Mathematics
Volume **171**, 1994

Direct limits of two-dimensional prime spectra

CHRISTEL ROTTHAUS AND SYLVIA WIEGAND

May 27, 1994

Dedicated to the memory of Richard S. Pierce

A major unsolved problem over the past thirty years has been to *characterize* the prime spectra of commutative Noetherian rings–that is, identify properties for a partially ordered set which imply the set is order-isomorphic to the prime spectrum of some Noetherian ring. (For Noetherian rings, the partially ordered set associated with the prime spectrum determines the Zariski topology.) A more specific elusive problem is to characterize as a partially ordered set the prime spectrum of $\mathbb{Q}[x, y]$, the polynomial ring in two variables over the rational numbers. Positive results related to these two problems include Roger Wiegand's characterization [**rW2**] of the prime spectrum of $\mathbb{Z}[x]$, the polynomials in one variable over the integers, and the work of William Heinzer, David Lantz and Sylvia Wiegand, characterizing various spectra related to certain two-dimensional polynomial rings [**HW1**], [**HW2**], [**HLW1**], [**HLW2**]. A characterization of the spectra of two-dimensional countable semilocal Noetherian domains appears in [**sW**]. On the other hand, examples of Heitmann [**H**], McAdam [**Mc**], and Nagata [**N**] show that various desirable properties for the prime spectrum can fail in Noetherian rings.

In this article, we restrict our attention to prime spectra of integral domains of dimension two, usually Noetherian and countable. In order to get partial answers to the two problems above, we examine spectral properties given in [**rW1**] and [**HW1**], and about a dozen other properties of partially ordered sets and mappings between them. We identify some implications between the properties and make observations relevant to the two problems, but many questions remain.

1991 *Mathematics Subject Classification.* 13B02, 13B20, 13E05, 13F05, 13H05.

Both authors extend thanks to the National Science Foundation for support of this research. Sylvia Wiegand thanks Purdue University for its hospitality during academic year 92/93.

This paper is in final form and will not be submitted elsewhere

Our investigation begins with direct limits of prime spectra, since Spec($\mathbb{Q}[x,y]$) is a direct limit of partially ordered sets which were characterized in [**HW1**]. It is intriguing that Spec($\mathbb{Z}[x]$) is also such a direct limit, even though Spec($\mathbb{Z}[x]$) is not order-isomorphic to Spec($\mathbb{Q}[x,y]$) [**rW2**].

For γ_n a set of n primes of $\mathbb{Z}$ and σ_n a set of n primes of $\mathbb{Q}[X]$, the spectra of the polynomial rings over the respective localizations at γ_n and at σ_n are order-isomorphic [**HW1**]; that is, $U_n :=$Spec($\mathbb{Z}_{\gamma_n}[Y]$) $\cong V_n :=$ Spec($\mathbb{Q}[X]_{\sigma_n}[Y]$). In fact the prime spectrum of the polynomial ring in one variable over a countable one-dimensional Noetherian domain with exactly n maximal ideals, is always order-isomorphic to U_n if $n \geq 1$. It follows that the prime spectrum does not behave nicely with respect to limits. In particular, order-isomorphisms between prime spectra are not preserved under inverse limits—because $\mathbb{Z}[Y]$ is an inverse limit of $\mathbb{Z}_{\gamma_n}[Y]$, and $\mathbb{Q}[X,Y]$ is an inverse limit of $\mathbb{Q}[X]_{\sigma_n}[Y]$. (This is discussed in Sections 2 and 3.)

Nevertheless, the natural maps $\tau_n : U_n \to U_{n+1}$ and $\theta_n : V_n \to V_{n+1}$ have the same identifying properties and there are compatible maps from $U_n \to V_n$ at each n. (Section 4.)

In later sections, we look carefully at the properties of the maps τ_n and θ_n to see where the two direct limits diverge. In Section 5 we find sufficient conditions to insure that the direct limit is Spec($\mathbb{Z}[x]$), and in Section 6 we give our first description of partially ordered sets which are direct limits of U_n, τ_n. We examine properties which are related to Noetherian domains in Section 7. The importance of *exceptional* sequences and *orthogonal pairs of sequences*, defined in Section 1, is demonstrated in Section 8; we show that a partially ordered set U is a nested direct limit of U_n, τ_n if and only if U contains an exceptional sequence. The remaining sections contain partial results concerning implications among properties, questions and other miscellaneous items.

This article is divided into ten topics, as follows:

Topics

(1) Background and notation.
(2) $\mathbb{Z}[Y]$ and $\mathbb{Q}[X,Y]$ as inverse limits of $\mathbb{Z}_{\gamma_n}[Y]$ and $\mathbb{Q}[X]_{\sigma_n}[Y]$.
(3) Spec($\mathbb{Z}[Y]$) and Spec($\mathbb{Q}[X,Y]$) as corresponding direct limits.
(4) The canonical mapping $\tau_n : U_n \to U_{n+1}$, where $U_n :=$Spec($\mathbb{Z}_{\gamma_n}[Y]$).
(5) The characterization of Spec($\mathbb{Z}[Y]$) as a limit.
(6) Partially ordered sets which are a direct limit of U_n; special sequences.
(7) Other axioms for countable two-dimensional Noetherian domains.
(8) Exceptional sequences and orthogonal pairs of sequences.
(9) Relationships between axioms.
(10) More observations and questions

1. Background and notation.

First we repeat some notation and definitions from [**HW1**], [**HLW2**] and [**rW1**].

1.1 Notation. For a partially ordered set U of dimension two, the *j-elements* of U are defined recursively as follows: (a) all maximal elements are j-elements and (b) a non-maximal element u is a j-element if u has infinitely many covers which are j-elements. (An element $v > u$ of U is a *cover* of u if there is no element w with $u < w < v$.)

For $u, v \in U$, and T a finite subset of U,

$$\begin{aligned}
\mathcal{M}(U) &= \{\text{maximal elements of } U \text{ of maximal height}\},\\
\mathrm{G}(u) &= \{w \in U \mid w > u\} \qquad (\text{“greater” set}),\\
\mathrm{L}(T) &= \{w \in U \mid w < t, \text{ for all } t \in T\} \qquad (\text{“less” set}),\\
j\text{-set}(u) &= \{j\text{-elements of } U \text{ minimal with respect to containing } u\}, \qquad \text{and}\\
j^{-1}(T) &= \{w \in U \mid \mathrm{G}(w) = T\} = \{w \in U \mid j\text{-set}(U) = T\} \quad (\text{“inverse } j\text{-set”}).
\end{aligned}$$

NOTES. (1) $j^{-1}(T) \subseteq L(T)$.

(2) If u has height one, but is not a j-element, j-set$(u) = \mathrm{G}(u)$.

(3) If R is a commutative ring, $U = \mathrm{Spec}(R)$, and Q is a nonmaximal prime ideal of R which is not an intersection of maximal ideals, then the standard notion of the *j-radical* of Q is the intersection of all maximal ideals $\mathbf{m}$ of R containing Q, or $\cap$ {j-primes P minimal with respect to containing Q}; thus the j-radical of Q is the intersection of the elements of the j-set of Q.

1.2 DEFINITION. *A partially ordered set U is called* **standard** *if the following three conditions are satisfied:*

(unique min): *U has a unique minimal element u_0.*

(countable): *U is countable*

(dim 2): *U has dimension 2.*

In this article we study standard partially ordered sets exclusively. The following two definitions describe those standard partially ordered sets which are most important for our investigations.

1.3 DEFINITION. [**HW1**] *A standard partially ordered set U is called* **countable n-localized integer polynomial** *or* **C$\mathbb{Z}$(n)P** *provided the following properties hold:*

(∞ ht-1 maxes): *There exist infinitely many height-one maximal elements.*

(n-special): *There exist n height-one nonmaximal special j-elements $u_1, u_2, \ldots u_n$ satisfying: (i) $\mathrm{G}(u_1) \cup \cdots \cup \mathrm{G}(u_n) = \mathcal{M}(U)$, (ii) $\mathrm{G}(u_i) \cap \mathrm{G}(u_j) = \emptyset$ for $i \neq j$, (iii) $|G(u_i)| = \infty$ for $i = 1, \ldots, n$*

(finite covers): *For each height-one non-special u, $\mathrm{G}(u)$ is finite.*

(infinite inv.j-sets): *For each nonempty finite subset T of height-two elements of U, $j^{-1}(T)$ is infinite. In other words, $T = \mathrm{G}(u) = j\text{-set}(u)$, for infinitely many height-one elements u.*

In what follows, let $C\mathbb{Z}(n)P$ denote a partially ordered set satisfying the properties in Definition 1.3. As shown in [**HW1**] the spectrum of the polynomial ring (in one variable) over a countable one-dimensional semilocal ring with n maximal ideals is $C\mathbb{Z}(n)P$.

Roger Wiegand [**rW2**] has characterized the partially ordered set $\mathrm{Spec}(\mathbb{Z}[Y])$ as follows:

1.4 DEFINITION. *A standard partially ordered set U is called* **countable integer polynomial** *or* **C$\mathbb{Z}$P** *provided the following properties hold:*

(infinite covers): *For each height-one element u, $\mathrm{G}(u)$ is (countably) infinite.*

(finite mubs): *The minimal upper bound set, $\mathrm{G}(u) \cap \mathrm{G}(v)$, is finite for every pair u and v of distinct height-one elements of U.*

(rad(princ)): *The "radical of a principal" axiom. For each finite subset T of $\mathcal{M}(U)$ and for each finite subset S of height-one elements of U, there exists a height-one element w such that $\mathrm{G}(w) \cap \mathrm{G}(s) \subseteq T \subseteq \mathrm{G}(w)$ for all $s \in S$.*

NOTES. (1) Suppose U is the prime spectrum of a two-dimensional ring R, I is a finite intersection of maximal ideals of R, and (S) is the sum of a finite (nonzero) number of height-one prime ideals of R contained in I. Then the "radical of a principal" axiom above, *(rad(princ))*, implies that there exists a height-one prime ideal w of R so that I is contained in the radical of $w + (S)$. In the case of a unique factorization domain, this is equivalent to saying that in the quotient ring $\bar{R} = R/(S)$, the image $\bar{I}$ of I is the radical of a principal ideal.

(2) In $\mathrm{Spec}(\mathbb{Q}[x, y])$, *(rad(princ))* is *not* satisfied, nor is *(rad(princ) for singles)*, below in Definition 1.5. For example, if $t = (x - 1, y - 1)$ and $S = \{(x^3 - y^2)\}$, there is no w [**rW2**].

We now list the other properties and conditions which we study in this article. They will be explained further in the later sections, but we have put their definitions in this section for easy reference. Throughout we suppose that U and W are standard partially ordered sets.

1.5 DEFINITION. *Axioms related to the radical of a principal axiom of Definition 1.4.*

(rad(∞-princ)): *The "radical of infinitely many principals" axiom. For each finite subset T of height-two elements of U and for each finite subset S of height-one elements of U, there exist infinitely many height-one elements w such that $\mathrm{G}(w) \cap \mathrm{G}(s) \subseteq T \subseteq \mathrm{G}(w)$ for all $s \in S$.*

(rad(princ) for singles): *"The radical of a principal axiom holds when the set of height two elements is a singleton". For every finite set S of height-one elements in U and for every height-two maximal $t \in U$, there exists a height-one* *element* $w \in U$ *such that*

$$\mathrm{G}(w) \cap \mathrm{G}(s) \subseteq \{t\}, \ \forall s \in S, \ \text{and } t \in \mathrm{G}(w).$$

(smaller comaximals): *"Every height-two element comaximal to a set S contains a height-one element comaximal to S". For every finite set S of height-one elements in U and for every height-two maximal $t \in U$ with $t \notin \bigcup_{s \in S} \mathrm{G}(s)$, there exists a height-one element $w \in U$ such that*

$$\mathrm{G}(w) \cap \mathrm{G}(s) = \emptyset, \ \forall s \in S, \ \text{and } t \in \mathrm{G}(w).$$

(∞ smaller comaximals): *"Every height-two element comaximal to a set S contains infinitely many height-one elements comaximal to S". For every finite set S of height-one elements in U and for every height-two maximal $t \in U$ with $t \notin \mathrm{G}(s)$, $\forall s \in S$, the set*

$$\{w \in U \mid \text{height } (w) = 1, \mathrm{G}(w) \cap \mathrm{G}(s) = \emptyset, \ \forall s \in S, \quad \text{and} \quad t \in \mathrm{G}(w)\}$$

is infinite.

Obviously *(rad(princ))* implies *(rad(princ) for singles)* implies *(smaller comaximals)*.

1.6 PROPOSITION. *Let U be a standard partially ordered set with the (infinite covers) property of Definition 1.4. Then:*

(a) *(rad(princ))* $\Longleftrightarrow$ *(rad(∞-princ))*
(b) *(rad(princ))* $\Longrightarrow$ *(∞ smaller comaximals)*

PROOF. To show the nontrivial implication of (a), assume the contrary. Let T be a finite subset of $\mathcal{M}(U)$ and S be a set of height-one elements of U for which there are just finitely many w with $\mathrm{G}(w) \cap \mathrm{G}(s) \subseteq T \subseteq \mathrm{G}(w)$, for all $s \in S$. Enumerate them: $w_1, w_2, \ldots, w_t$. Now let $S' = S \cup \{w_1, w_2, \ldots, w_t\}$. By hypothesis, there exists a w' such that $\mathrm{G}(w') \cap \mathrm{G}(s) \subseteq T \subseteq \mathrm{G}(w')$, for all $s \in S'$. Now $w \neq w_i$, for all i, because otherwise $\mathrm{G}(w_i) = \mathrm{G}(w_i) \cap \mathrm{G}(w_i) \subseteq T$, a contradiction to *(infinite covers)*. But w' satisfies $\mathrm{G}(w') \cap \mathrm{G}(s) \subseteq T \subseteq \mathrm{G}(w')$, a contradiction to $w_1, w_2, \ldots, w_t$ the only such elements.

For (b) let $S = \{u_1, \ldots, u_n\}$ and $T = \{t\}$, where t is a height-two element comaximal to all elements of S. By *(rad(princ))* there is a height-one element $w \in U$ with $\mathrm{G}(u_i) \cap \mathrm{G}(w) \subseteq T = \{t\} \subseteq \mathrm{G}(w)$. But for all $i \leq n$, $t \notin \mathrm{G}(u_i)$; hence $\mathrm{G}(u_i) \cap \mathrm{G}(w) = \emptyset$ and $T = \{t\} \subseteq \mathrm{G}(w)$. Using *(rad(princ))* again a similar argument to (a) shows that there are infinitely many height-one comaximal elements.

The axioms in Definition 1.7 arise naturally when considering, for each n, the natural mapping from the prime spectrum of polynomials over the integers localized outside the first n primes to the prime spectrum of polynomials over the integers localized outside the first $n+1$ primes. This is explained further in Section 3.

1.7 DEFINITION. *Axioms for mappings. Let W_n, respectively W_m, be standard partially ordered sets which satisfy the conditions in Definition 1.3 for $C\mathbb{Z}(n)P$, respectively $C\mathbb{Z}(m)P$, with $n < m$. Let $u_1, \ldots, u_n$ denote the special elements of W_n from axiom (n-special). For convenience, $u_1, \ldots, u_n, \ldots, u_m$ will denote the special elements of W_m. An order-preserving mapping $\tau : W_n \longrightarrow W_m$ is called* **CP-stable** *if the following axioms are satisfied:*

(stay-special): *"The special elements stay special"; that is, $\tau(u_i) = u_i$, for $i = 1, \ldots, n$.*

(order-embed): *"τ is an order-embedding"; more precisely, τ is an order-isomorphism $\tau : W_n \to W_m - (\{u_{n+1}, \ldots u_m\} \cup \bigcup_{i=n+1}^{m} G(u_i))$.*

NOTES. If τ is an injective order-preserving map, *(stay-special)* is always satisfied. This follows from the fact that in a $C\mathbb{Z}(n)P$-set the special elements u_i are characterized by $|G(u_i)| = \infty$. For *(order-embed)* to hold in addition, each $G(u_i)$ must map *onto* $G(u_i)$.

The next definition arises when considering a partially ordered set W as a direct limit of partially ordered sets which are, for each n, the prime spectrum of polynomials over the integers localized outside the first n primes. The axioms in Definition 1.8 are studied in Section 5 and again in Section 7.

1.8 DEFINITION. *Axioms for limits of CP-stable maps. We consider a sequence $\{W_n\}_{n=1}^{\infty}$ of partially ordered sets which satisfy the axioms in Definition 1.3 for $C\mathbb{Z}(n)P$ and a direct system of CP-stable maps*

$$(*) \qquad \tau_{n,k} : W_n \to W_k, \; n, k \in \mathbb{N}, \; k > n,$$

For $n \leq k$, identify W_n with a subset of W_k, via $\tau_{n,k}$.

(L-no max ht.1s): *For every n, and every height-one $w \in W_n$, there exists a $k \in \mathbb{N}$, $k \geq n$, such that $\tau_{n,k}(w)$ is not maximal in W_k; that is, every height-one element is eventually non-maximal.*

(L-mub-constant): *For every $n \in \mathbb{N}$, and every pair of distinct height-one elements $u, w \in W_n$, there exists an $N \in \mathbb{N}$, $N \geq n$, such that, for every $k \geq N$,*

$$\{t \in W_k | \, t \geq \tau_{n,k}(u), \, t \geq \tau_{n,k}(w)\} = \{t \in W_N | \, t \geq \tau_{n,N}(u), \, t \geq \tau_{n,N}(w)\};$$

that is, the set of minimal upper bounds for a given pair of height-one elements is eventually constant.

In addition, we have a condition on the greater sets:

(L-more covers): *For every $n \in \mathbb{N}$, and every height-one element $u \in W_n$, $u \neq u_i$, $1 \leq i \leq n$, there exists an $N \in \mathbb{N}$, such that for every $k \geq N$, $\tau_{n,k}(\{t \in W_n | \, t > u\}) \subsetneq \{t \in W_k | \, t > \tau_{n,k}(u)\}$; that is, for each height-one element u that is not a special element of W_n, there exists an N so that the set of covers of u is strictly bigger in W_k whenever $k > N$.*

The following two conditions are relevant for the sets W_n to approach a limit that satisfies the radical of a principal axiom (rad(princ)) of Definition 1.4. For *finite sets* S, $T \subseteq W = \varinjlim W_n$, *let* $n_{(S,T)} \in \mathbb{N}$ *denote the smallest positive integer* n *such that* $S, T \subseteq W_n$.

(L-smaller comaximals): *"A height-two element which is comaximal to a set* S *contains a height-one which is comaximal to* S *in the limit." For every finite set* S *of height-one elements in* $W = \varinjlim W_n$ *and for every height-two maximal* $t \in W$, $t \notin \bigcup_{s \in S} \mathrm{G}(s)$, *this set is nonempty:*

$$\bigcap_{k \geq n_{(S,\{t\})}} \{w \in W_k \mid ht(w) = 1, \mathrm{G}(w) \cap \mathrm{G}(s) = \emptyset \text{ (in } W_k), \forall s \in S, \text{ and } t \in \mathrm{G}(w)\}.$$

(L-rad(princ)): *"In the limit every finite intersection of maximals is the radical of a principal in a quotient." For every finite set* S *of height-one primes in* $W = \varinjlim W_n$ *and for every nonempty finite set* T *of height-two maximal elements of* W, *this set is nonempty:*

$$\bigcap_{k \geq n_{(S,T)}} \{w \in W_k \mid ht(w) = 1, \mathrm{G}(w) \cap \mathrm{G}(s) \subseteq T \subseteq \mathrm{G}(w) \text{ (in } W_k), \forall s \in S\}.$$

NOTES. The *(L-rad(princ))* axiom implies *(L-smaller comaximals)*, and *(L-more covers)* implies *(L-no max ht.1s)*. Axiom *(L-more covers)* states that every height-one element $u \in W_n$, except the u_i's, for $i = 1, \ldots, n$, is in the limit comaximal to at most finitely many of the height-one elements $\{u_i\}_{i \in \mathbb{N}}$, the union of all the special elements of the $C\mathbb{Z}(k)P$-sets W_k. This particular situation occurs when we investigate $\operatorname{Spec} \mathbb{Z}[Y]$, respectively $\operatorname{Spec} \mathbb{Q}[X, Y]$, as a limit of $C\mathbb{Z}(n)P$-sets. Although *(L-more covers)* is rather strong and weaker forms of it can possibly be imposed on the direct limit, we do not know any example of a (Noetherian) ring whose spectrum can be written as a limit of $C\mathbb{Z}(n)P$-sets such that axiom *(L-more covers)* is not satisfied.

The next three definitions describe properties of standard partially ordered sets which are a direct limit of $C\mathbb{Z}(n)P$-sets under CP-stable maps. Certainly such sets must have infinitely many "special" elements.

1.9 DEFINITION. **(∞-special)** *Let* U *be a two-dimensional partially ordered set. An infinite sequence* $\{u_n | n \in \mathbb{N}\}$ *of height-one elements of* U *is called a* **special** *sequence provided:*

(i) $\mathrm{G}(u_i) \cap \mathrm{G}(u_j) = \emptyset$, *for* $i, j \in \mathbb{N}$, $i \neq j$.

(ii) $\mathcal{M}(U) = \bigcup_{n \in \mathbb{N}} \mathrm{G}(u_n)$.

(iii) $\mathrm{G}(u_i)$ *is infinite, for each* $i \in \mathbb{N}$.

In case a special sequence exists, the partially ordered set U *is said to satisfy* **(∞-special)**.

Special sequences are studied in Sections 6, 7 and 10. In Section 8, we discuss a stronger condition linking the concept of special sequences to the radical of a principal axiom:

1.10 DEFINITION. *The special sequence* $\{u_n | n \in \mathbb{N}\}$ *is called* **exceptional** *if* (*) *below holds:*

$$(*) \quad \forall \text{ finite set } S \subseteq \{u_n | n \in \mathbb{N}\} \text{and } \forall \text{ finite set } T \subseteq \mathcal{M}(U),$$
$$\Gamma(S,T) = \{w \in U \mid ht(w) = 1, \mathrm{G}(w) \cap \mathrm{G}(s) \subseteq T \subseteq \mathrm{G}(w), \forall s \in S\} \text{ is infinite.}$$

1.11 DEFINITION. *We say that* U *is a* **nested direct limit of CZ(n)P's** *if there exists, for each* $n \in \mathbb{N}$, *a subset* W_n *of* U *such that:*

(i) $W_n \cong C\mathbb{Z}(n)P$.

(ii) For every n, $W_n \subseteq W_{n+1}$.

(iii) The inclusion maps $W_n \xrightarrow{\epsilon_{n,n+1}} W_{n+1}$ *are CP-stable (Definition 1.7).*

(iv) $U = \varinjlim_{n\in\mathbb{N}} W_n = \bigcup_{n\in\mathbb{N}} W_n$.

More precise terminology for this situation is that U *is a* **nested direct limit of CZ(n)P's with respect to** $\{\mathbf{W_n}\}_{\mathbf{n}\in\mathbb{N}}$.

We consider two additional conditions on standard partially ordered sets. The first one, Definition 1.12, is satisfied in Spec $\mathbb{Z}[Y]$, respectively Spec $\mathbb{Q}[X,Y]$, since every maximal ideal in the corresponding rings is two-generated.

1.12 DEFINITION. **The pair of elements condition**. *Let* U *be a partially ordered set of dimension two. The pair of elements condition is that for each height-two element* m *there exists some pair of height-one elements,* w_1, w_2 *so that* m *is the unique element bigger than both* w_1 *and* w_2.

The next definition deals with a property that clearly holds true in Spec $\mathbb{Z}[Y]$ and Spec $\bar{\mathbb{Q}}[X,Y]$, where $\bar{\mathbb{Q}}$ is the algebraic closure of $\mathbb{Q}$ (see section 8). However, we do not know if the condition of Definition 1.13 is satisfied for Spec $\mathbb{Q}[X,Y]$.

1.13 DEFINITION. **The orthogonal pair of special sequences condition**. *Let* U *be a partially ordered set of dimension two. We say that* U *satisfies the orthogonal pair of special sequences condition or contains an orthogonal pair of special sequences provided*

(i) U *contains two special sequences* $\{u_i | i \in \mathbb{N}\}$ *and* $\{v_j | j \in \mathbb{N}\}$ *(as in Definition 1.9), and*

(ii) For every pair $i, j \in \mathbb{N}$, $\mathrm{G}(u_i) \cap \mathrm{G}(v_j)$ *contains exactly one maximal element of* U.

Finally, we present a few axioms which always hold for the prime spectra of Noetherian rings of dimension ≥ 2.

1.14 DEFINITION. *Other axioms for a Noetherian ring. Let* U *be a standard partially ordered set.*

(N-∞ ht.1s): *"There are infinitely many elements of height one"; that is, the set* $\{u \in U | \ height(u) = 1\}$ *is infinite.*

(N-infinite less sets): *"The "less" set of a height-two element is infinite." For every $t \in U$ of height two, the set $\mathrm{L}(t) = \{u \in U |\ height(u) = 1 \text{ and } u < t\}$ is infinite.*

Obviously, *(N-infinite less sets)* $\Longrightarrow$ *(N-∞ ht.1s)*

2. $\mathbb{Z}[Y]$ and $\mathbb{Q}[X,Y]$ as inverse limits.

Let $\{p_i\}_{i\in\mathbb{N}}$, respectively $\{q_i\}_{i\in\mathbb{N}}$, be an enumeration of the maximal ideals in $\mathbb{Z}$, respectively $\mathbb{Q}[X]$. Put $\gamma_n = \{p_1, \ldots, p_n\}$ and $\sigma_n = \{q_1, \ldots, q_n\}$. Set

$\mathbb{Z}_{\gamma_n} = S_{\gamma_n}^{-1}\mathbb{Z}$, with $S_{\gamma_n} = \mathbb{Z} - \cup\gamma_n$, and $\mathbb{Q}[X]_{\sigma_n} = T_{\sigma_n}^{-1}\mathbb{Q}[X]$, $T_{\sigma_n} = \mathbb{Q}[X] - \cup\sigma_n$.

Note that, for $m \geq n \in \mathbb{N}$, $\gamma_n \subset \gamma_m$ and $\sigma_n \subset \sigma_m$. Thus we have canonical morphisms:

$$\varphi_{m,n} : \mathbb{Z}_{\gamma_m}[Y] \to \mathbb{Z}_{\gamma_n}[Y] \text{ and } \psi_{m,n} : \mathbb{Q}[X]_{\sigma_m}[Y] \to \mathbb{Q}[X]_{\sigma_n}[Y].$$

When $m = n + 1$, set $\varphi_n = \varphi_{m,n}$, and define ψ_n similarly.

2.1 PROPOSITION. *$\mathbb{Z}[Y] = \varprojlim(\mathbb{Z}_{\gamma_n}[Y])$ and $\mathbb{Q}[X,Y] = \varprojlim(\mathbb{Q}[X]_{\sigma_n}[Y])$.*

PROOF. By definition,

$$\varprojlim \mathbb{Z}_{\gamma_n}[Y] = \{(w_n)_{n\in\mathbb{N}} | \varphi_{m,n}(w_m) = w_n \text{ for } m \geq n\} \subset \prod_{n\in\mathbb{N}} \mathbb{Z}_{\gamma_n}[Y],$$
$$\varprojlim \mathbb{Q}[X]_{\sigma_n}[Y] = \{(v_n)_{n\in\mathbb{N}} | \psi_{m,n}(v_m) = v_n \text{ for } m \geq n\} \subset \prod_{n\in\mathbb{N}} \mathbb{Q}[X]_{\sigma_n}[Y].$$

Obviously, for every $n \in \mathbb{N}$, we have canonical morphisms δ_n, ϵ_n (localization) and i_n, j_n, which induce morphisms δ, ϵ as shown:

$$\begin{array}{ccc} \mathbb{Z}[Y] & \xrightarrow{\delta_n} & \mathbb{Z}_{\gamma_n}[Y] \\ {=}\downarrow & & \uparrow{i_n} \\ \mathbb{Z}[Y] & \xrightarrow{\delta} & \varprojlim \mathbb{Z}_{\gamma_n}[Y] \end{array} \qquad \begin{array}{ccc} \mathbb{Q}[X,Y] & \xrightarrow{\epsilon_n} & \mathbb{Q}[X]_{\sigma_n}[Y] \\ {=}\downarrow & & \uparrow{j_n} \\ \mathbb{Q}[X,Y] & \xrightarrow{\epsilon} & \varprojlim \mathbb{Q}[X]_{\sigma_n}[Y] \end{array}.$$

Now δ and ϵ are clearly injective.

CLAIM. δ and ϵ are surjective.

PROOF OF CLAIM. Let $(w_n)_{n\in\mathbb{N}} \in \varprojlim \mathbb{Z}_{\gamma_n}[Y]$. Then, for all $n \in \mathbb{N}$:

$$w_n = \frac{a_n}{b_n};\ a_n \in \mathbb{Z}[Y];\ b_n \in \mathbb{Z} - \cup\{p_1, \ldots, p_n\}.$$

We can assume each pair (a_n, b_n) is relatively prime. If some b_n is not invertible in $\mathbb{Z}$, then there is an $m \geq n$, and a $j \leq m$ such that $b_n \in p_j \in \gamma_m$. On the other hand, $\varphi_{m,n}(w_m) = w_n$; that is $\frac{a_m}{b_m} = \frac{a_n}{b_n}$, and $b_m \notin p_j, b_n \in p_j$. But this implies $a_n \in p_j$, a contradiction to a_n, b_n relatively prime; hence b_n is invertible. Thus there exists $w \in \mathbb{Z}[Y]$ with $\delta(w) = (w_n)_{n\in\mathbb{N}} = (w)_{n\in\mathbb{N}}$. A similar argument shows that ϵ is surjective.

3. Spec($\mathbb{Z}[Y]$) and Spec($\mathbb{Q}[X,Y]$) as corresponding direct limits.

In [**HW1**], it was shown that if R is a countable one-dimensional Noetherian domain with exactly n maximal ideals, and n is greater than 1, then $\mathrm{Spec}(R[Y])$ is order-isomorphic to $C\mathbb{Z}(n)P$.

3.1 PROPOSITION. *Spec($\mathbb{Z}[Y]$) and Spec($\mathbb{Q}[X,Y]$) are direct limits (unions) of $C\mathbb{Z}(n)P$.*

PROOF. Both $\mathrm{Spec}(\mathbb{Z}_{\gamma_n}[Y])$ and $\mathrm{Spec}(\mathbb{Q}[X]_{\sigma_n}[Y])$ are order-isomorphic to $C\mathbb{Z}(n)P$, for every $n \in \mathbb{N}$. The canonical morphisms $\varphi_n := \varphi_{n+1,n} : \mathbb{Z}_{\gamma_{n+1}}[Y] \to \mathbb{Z}_{\gamma_n}[Y]$ and $\psi_n := \psi_{n+1,n} : \mathbb{Q}[X]_{\sigma_{n+1}}[Y] \to \mathbb{Q}[X]_{\sigma_n}[Y]$ imply continuous maps on the prime spectra:

$$\varphi_n^* : \mathrm{Spec}\,(\mathbb{Z}_{\gamma_n}[Y]) \to \mathrm{Spec}\,(\mathbb{Z}_{\gamma_{n+1}}[Y]), \text{ and}$$
$$\psi_n^* : \mathrm{Spec}\,(\mathbb{Q}[X]_{\sigma_n}[Y]) \to \mathrm{Spec}\,(\mathbb{Q}[X]_{\sigma_{n+1}}[Y]).$$

Moreover from the canonical maps:

$$\delta_n : \mathbb{Z}[Y] \to \mathbb{Z}_{\gamma_n}[Y] \text{ and } \epsilon_n : \mathbb{Q}[X,Y] \to \mathbb{Q}[X]_{\sigma_n}[Y],$$

we obtain

$$\delta_n^* : \mathrm{Spec}\,(\mathbb{Z}_{\gamma_n}[Y]) \to \mathrm{Spec}\,(\mathbb{Z}[Y]) \text{ and } \epsilon_n^* : \mathrm{Spec}\,(\mathbb{Q}[X]_{\sigma_n}[Y]) \to \mathrm{Spec}\,(\mathbb{Q}[X,Y]).$$

Furthermore δ_n^* and ϵ_n^* are compatible with φ_n^* and ψ_n^*; that is,

$$\delta_n^* = \delta_{n+1}^* \cdot \varphi_n^* \text{ and } \epsilon_n^* = \epsilon_{n+1}^* \cdot \psi_n^*.$$

Therefore the direct limit (union) $U = \varinjlim \mathrm{Spec}\,(\mathbb{Z}_{\gamma_n}[Y])$, maps (preserving order) into $\mathrm{Spec}\,(\mathbb{Z}[Y])$ via δ^*. Each φ_n^* and δ_n^* is injective, and thus δ^* is injective.

Similarly $\epsilon^* : V = \varinjlim \mathrm{Spec}\,(\mathbb{Q}[X]_{\sigma_n}[Y]) \to \mathrm{Spec}(\mathbb{Q}[X,Y])$ is injective.

CLAIM. *δ^* and ϵ^* are surjective.*

PROOF OF CLAIM. Suppose $p, w \in \mathrm{Spec}\,(\mathbb{Z}[Y])$, with $\mathrm{ht}(p) \geq 1$, $\mathrm{ht}(w) = 2$, and $p \leq w$. Then $w \cap \mathbb{Z} = p_n$, for some $n \in \mathbb{N}$, that is, $p_n[Y] \in \mathrm{Spec}\,(\mathbb{Z}[Y])$ is the only height-one extended prime beneath w. Hence p and w are the images under δ_n^* of the prime ideals corresponding to p and w in $\mathrm{Spec}((\mathbb{Z}_{\gamma_n}[Y]))$. Thus δ^* is surjective. Similarly ϵ^* is surjective.

QUESTION. *What distinguishes* $\mathrm{Spec}\,(\mathbb{Z}[Y])$ *and* $\mathrm{Spec}\,(\mathbb{Q}[X,Y])$ *in the limit?*

4. The canonical mapping.

Our goal is to describe $\tau_n : U_n \to U_{n+1}$, where U_n is $\mathrm{Spec}(\mathbb{Z}_{\gamma_n}[Y])$, and τ_n is the mapping induced by localization. For every n, U_n is $C\mathbb{Z}(n)P$; let $u_1, \ldots, u_n$

denote the n special elements of U_n from axiom *(n-special)* of (1.3). For convenience, $u_1, \ldots, u_n$ will also denote the n special elements of U_{n+1} corresponding to the same primes of $\mathbb{Z}[Y]$. For every n and every subset $S \subseteq \{1, \ldots, n\}$, let

$\Upsilon_n(S) = \{u \mid \text{height}(u) = 1, u \neq u_i \text{ for all } i \text{ and } \mathrm{G}(u) \cap \mathrm{G}(u_j) \neq \emptyset \iff j \in S\}$.

That is, Υ_n is a sort of characteristic function, lumping together all height-one elements except for the special elements of the partially ordered set, which are in different categories relative to being less than elements of the various $\mathrm{G}(u_j)$. Note that $\Upsilon_n(\emptyset) = \{$ maximal height-one elements of U_n $\}$.

4.1 PROPOSITION. *For every pair of partially ordered sets W_n and W_{n+1} which are $C\mathbb{Z}(n)P$ and $C\mathbb{Z}(n+1)P$ respectively, containing special elements $\{u_i\}$, and every CP-stable map $\alpha_n \colon W_n \to W_{n+1}$ the following conditions hold:*

(1) *Let Ω_i be the infinite set of height-one maximal elements of $W_i, i = n, n+1$. Then*
$\Omega_n = \Omega_n^1 \cup \Omega_n^2$, *where* $|\Omega_n^1| = \infty = |\Omega_n^2|$, $\alpha_n(\Omega_n^1) = \Omega_{n+1}$,
$\alpha_n(\Omega_n^2) \subseteq \{$ *non-maximal height-one primes of* $W_{n+1}\}$.
$x \in \Omega_n^2 \Rightarrow \mathrm{G}(\alpha_n(x)) \subseteq \mathrm{G}(u_{n+1})$ *in* W_{n+1}.

(2) *For each subset $S \subseteq \{1, \ldots, n\}$, it is possible to write $\Upsilon_n(S) = \Upsilon_{n,1}(S) \cup \Upsilon_{n,2}(S)$ (disjoint sets), so that α_n is a bijection : $\Upsilon_{n,1}(S) \to \Upsilon_{n+1}(S)$ and a bijection : $\Upsilon_{n,2}(S) \to \Upsilon_{n+1}(S \cup \{n+1\})$.*

For every nonempty finite set T of height-two elements of W_n such that $T \cap \mathrm{G}(u_i) \neq \emptyset \iff i \in S$, we have that $j^{-1}(T) \cap \Upsilon_{n,1}(S)$ is infinite and $j^{-1}(T) \cap \Upsilon_{n,2}(S)$ is infinite.

PROOF. (1) Let $t \in \mathrm{G}(u_{n+1})$. Then by *(infinite inv. j-sets)* of Definition 1.3, $|j^{-1}(t)| = \infty$, and $j^{-1}(t) - \{u_{n+1}\}$ is in the image under α_n by *(order-embed)* of Definition 1.7. But $\alpha_n(U_n - \Omega_n) \cap j^{-1}(t) = \emptyset$, since α_n is an order-isomorphism. Hence $|\alpha_n(\Omega_n) \cap j^{-1}(t)| = \infty$. Therefore, $|{\Omega_n}^2| = \infty$. Moreover, since

$$C\mathbb{Z}(n)P \xrightarrow{\cong} C\mathbb{Z}(n+1)P - \{\{u_{n+1}\} \cup \mathrm{G}(u_{(n+1)})\}$$

and since $C\mathbb{Z}(n+1)P$ contains infinitely many maximal height-one elements, $|\Omega_n^1| = \infty$.

Note that $x \in \Omega_n^2$ implies that $\alpha_n(x)$ is a non-maximal height-one element. Again because of *(order-embed)*, we must have that $\mathrm{G}(\alpha_n(x)) \subseteq \mathrm{G}(u_{n+1})$.

For (2), let $S \subseteq \{1, \ldots, n\}$. Now define

$\Upsilon_{n,1}(S) = {\alpha_n}^{-1}(\Upsilon_{n+1}(S))$ and $\Upsilon_{n,2}(S) = {\alpha_n}^{-1}(\Upsilon_{n+1}(S \cup \{n+1\}))$.

Since α_n is a bijection, the restriction is one-to-one. Clearly $\alpha_n(\Upsilon_n(S)) \subset \Upsilon_{n+1}(S) \cup \Upsilon_{n+1}(S \cup \{n+1\})$. Also the restriction is onto since all elements of $\Upsilon_{n+1}(S) \cup \Upsilon_{n+1}(S \cup \{n+1\})$ are outside $\mathrm{G}(u_{n+1})$.

The last part follows since the partially ordered sets satisfy *(infinite inv. j-sets)* of Definition 1.3.

4.2 PROPOSITION. *Suppose that $\sigma, \tau : C\mathbb{Z}(n)P \to C\mathbb{Z}(n+1)P$ are CP-stable maps. Then there is an order-preserving bijection: $\varphi : C\mathbb{Z}(n)P \to C\mathbb{Z}(n)P$, such that $\sigma = \tau \cdot \varphi$; that is,*

$$\begin{array}{ccc} U_n = C\mathbb{Z}(n)P & \xrightarrow{\sigma} & C\mathbb{Z}(n+1)P \\ \varphi\big\downarrow & & =\big\downarrow \\ V_n = C\mathbb{Z}(n)P & \xrightarrow{\tau} & C\mathbb{Z}(n+1)P, \end{array}$$

commutes. Moreover, $\varphi(u_i) = u_i$, for $i = 1, \ldots, n$.

PROOF. Define $\varphi(u_i) = u_i$.

(a) For each $1 \leq j \leq n$, let $\{x_i\}_{i\in\mathbb{N}}$ be an enumeration of $\mathrm{G}(u_j)$ in $C\mathbb{Z}(n)P$. Then $\{\sigma(x_i)\}_{i\in\mathbb{N}}$, respectively $\{\tau(x_i)\}_{i\in\mathbb{N}}$, is an enumeration of $\mathrm{G}(u_j)$, in $C\mathbb{Z}(n+1)P$, since σ and τ are bijections on $\mathrm{G}(u_j)$.

Define $\varphi|_{\mathrm{G}(u_j)} : \mathrm{G}(u_j) \to \mathrm{G}(u_j)$ via $x \in \mathrm{G}(u_j)$ goes to $\varphi(x) = y$, where y is chosen so that $\sigma(x) = \tau(y)$. We do this for $j = 1, \ldots, n$.

(b) Denote the height-one maximal elements of U_n and V_n by Ω_n and Γ_n respectively and write $\Omega_n = \Omega_n^1 \cup \Omega_n^2$ and $\Gamma_n = \Gamma_n^1 \cup \Gamma_n^2$, where $\sigma(\Omega_n^1) = \Omega_{n+1}$, $\tau(\Gamma_n^1) = \Omega_{n+1}$, and $\sigma(\Omega_n^2)$, $\tau(\Gamma_n^2) \subseteq$ {non-maximal height-one primes }.

Let $\{x_i\}_{i\in\mathbb{N}}$, respectively $\{y_i\}_{i\in\mathbb{N}}$, be an enumeration of Ω_n^1, respectively Γ_n^1. Then $\{\sigma(x_i)\}_{i\in\mathbb{N}}$, respectively $\{\tau(y_i)\}_{i\in\mathbb{N}}$, is an enumeration of Ω_{n+1}. Define $\varphi|_{\Omega_n^1} : \Omega_n^1 \to \Gamma_n^1$ via $\varphi(x_i) = y_j$, where $\sigma(x_i) = \tau(y_j)$.

To define φ on Ω_n^2, let

$\Delta_{n+1} = \{y \in C\mathbb{Z}(n+1)P |\ \mathrm{height}(y) = 1 \text{ and } \mathrm{G}(y) \subseteq \mathrm{G}(u_{n+1})\}$.

Let $\{x_i\}_{i\in\mathbb{N}}$, respectively $\{y_i\}_{i\in\mathbb{N}}$, be an enumeration of Ω_n^2, respectively Γ_n^2. Then $\{\sigma(x_i)\}_{i\in\mathbb{N}}$, respectively $\{\tau(y_i)\}_{i\in\mathbb{N}}$, is an enumeration of Δ_{n+1}, by (1) and (3) of Proposition 4.1.

Define $\varphi|_{\Omega_n^2} : \Omega_n^2 \to \Gamma_n^2$ via $\varphi(x_i) = y_j$, where $\sigma(x_i) = \tau(y_j)$.

(c) Let $\{T_i\}_{i=0}^{\infty}$ be all finite nonempty subsets of the set of height-two elements of $C\mathbb{Z}(n+1)P$, which are not contained in $G(u_{n+1})$.

For $T \subseteq C\mathbb{Z}(k)P$, define $j_k^{-1}(T) = \{v \in C\mathbb{Z}(k)P$ such that $v < t \iff t \in T\}$. Then $|j_k^{-1}(T_i)| = \infty$, for all $1 \leq i < \infty$.

Since σ and τ are bijections to $C\mathbb{Z}(n+1)P - \mathrm{G}(u_{n+1}) - \{u_{n+1}\}$, the sets $\sigma^{-1}(j_{n+1}^{-1}(T_i))$ are countably infinite and disjoint. The same is true for the sets $\tau^{-1}(j_{n+1}^{-1}(T_i))$. On each set $\sigma^{-1}(j_{n+1}^{-1}(T_i))$ in U_n, define φ to be a matching

$$\varphi : \sigma^{-1}(j_{n+1}^{-1}(T_i)) \to \tau^{-1}(j_{n+1}^{-1}(T_i)).$$

Since $C\mathbb{Z}(n)P$ is a disjoint union of

$$\{0\} \cup \{u_1, \ldots, u_n\} \cup (\bigcup_{i=1}^{n} \mathrm{G}(u_i)) \cup (\bigcup \{j_n^{-1}(T) | T \text{ finite } \subseteq \mathcal{M}(C\mathbb{Z}(n)P)\},$$

it is straightforward to check that φ is well-defined on all of U_n and an order-preserving bijection.

4.3 PROPOSITION. *The maps*

$$\varphi_n^* : \mathrm{Spec}(\mathbb{Z}_{\gamma_n}[Y]) \to \mathrm{Spec}(\mathbb{Z}_{\gamma_{n+1}}[Y]), \text{ and}$$
$$\psi_n^* : Spec(\mathbb{Q}[X]_{\sigma_n}[Y]) \to Spec(\mathbb{Q}[X]_{\sigma_{n+1}}[Y])$$

are CP-stable (Definition 1.7).

PROOF. The *(stay-special)* axiom is clear.
For *(order-embed)*, note:

$$\mathbb{Z}_{\gamma_n}[Y] = (S_{\gamma_n}^{-1}\mathbb{Z})[Y] \cong S_{\gamma_n}^{-1}(\mathbb{Z}[Y]),\ \mathbb{Q}[X]_{\sigma_n}[Y] = (T_{\sigma_n}^{-1}\mathbb{Q}[X])[Y] \cong T_{\sigma_n}^{-1}\mathbb{Q}[X,Y].$$

Here u_{n+1} and $\mathrm{G}(u_{n+1})$ correspond in $\mathbb{Z}_{\gamma_{n+1}}[Y]$ to:

$$p_{n+1} \text{ (prime in } \mathbb{Z}) \text{ and } \{t \subseteq \mathbb{Z}[Y] \,|\, p_{n+1} \subseteq t, \text{ height}(t) = 2\},$$

and similarly to q_{n+1} and $\mathrm{G}(q_{n+1})$ in $\mathbb{Q}_{\sigma_{n+1}}[Y]$. The canonical maps:

$$\varphi_n^* : \mathrm{Spec}(\mathbb{Z}_{\gamma_n}[Y]) \to \mathrm{Spec}(\mathbb{Z}_{\gamma_{n+1}}[Y]) \text{ and}$$
$$\psi_n^* : \mathrm{Spec}(\mathbb{Q}[X]_{\sigma_n}[Y]) \to \mathrm{Spec}(\mathbb{Q}[X]_{\sigma_{n+1}}[Y])$$

clearly are order-preserving isomorphisms onto the primes of $\mathrm{Spec}(\mathbb{Z}_{\gamma_{n+1}}[Y])$, respectively $\mathrm{Spec}(\mathbb{Q}[X]_{\sigma_{n+1}}[Y])$, which do not contain p_{n+1}, respectively q_{n+1}.

4.4 COROLLARY. *Let $n \in \mathbb{N}$. Then there is a sequence of order-preserving isomorphisms:*

$$\delta_i : Spec(\mathbb{Z}_{\gamma_i}[Y]) \xrightarrow{\cong} Spec(\mathbb{Q}[X]_{\sigma_i}[Y])$$

for $i \leq n$, such that the following diagram commutes;

$$\begin{array}{ccccc}
Spec(\mathbb{Z}_{\gamma_1}[Y]) & \xrightarrow{\varphi_1^*} \cdots & Spec(\mathbb{Z}_{\gamma_{n-1}}[Y]) & \xrightarrow{\varphi_{n-1}^*} & Spec(\mathbb{Z}_{\gamma_n}[Y]) \\
\cong\downarrow \delta_1 & & \cong\downarrow \delta_{n-1} & & \cong\downarrow \delta_n \\
Spec(\mathbb{Q}[X]_{\sigma_1}[Y]) & \xrightarrow{\psi_1^*} \cdots & Spec(\mathbb{Q}[X]_{\sigma_{n-1}}[Y]) & \xrightarrow{\psi_{n-1}^*} & Spec(\mathbb{Q}[X]_{\sigma_n}[Y]).
\end{array}$$

(Here δ_i is taken to be a map identifying $Spec(\mathbb{Z}_{\gamma_i}[Y])$ and $Spec(\mathbb{Q}[X]_{\sigma_i}[Y])$, which are both $C\mathbb{Z}(i)P$.)

PROOF. Use decreasing induction:
By [**HW1**], δ_n exists. Suppose δ_i is constructed. To construct δ_{i-1}, consider:

$$(*)\quad \begin{array}{ccccc}
C\mathbb{Z}(i-1)P \cong & \mathrm{Spec}(\mathbb{Z}_{\gamma_{i-1}}[Y]) & \xrightarrow{\varphi_{i-1}^*} & \mathrm{Spec}(\mathbb{Z}_{\gamma_i}[Y]) & \cong C\mathbb{Z}(i)P \\
 & & & \downarrow \delta_i & \\
C\mathbb{Z}(i-1)P \cong & \mathrm{Spec}(\mathbb{Q}[X]_{\sigma_{i-1}}[Y]) & \xrightarrow{\psi_{i-1}^*} & \mathrm{Spec}(\mathbb{Q}[X]_{\sigma_i}[Y]) & \cong C\mathbb{Z}(i)P.
\end{array}$$

Consider ψ^*_{i-1} and $\delta_i \circ \varphi^*_{i-1}$ as order-preserving maps from $C\mathbb{Z}(i-1)P$ into $C\mathbb{Z}(i)P$. By Proposition 4.3, ψ^*_{i-1} is CP-stable.

The following lemma shows that $\delta_i \circ \varphi^*_{i-1}$ are CP-stable maps.

4.5 LEMMA. *Let*

$$\varphi : C\mathbb{Z}(n)P \rightarrow C\mathbb{Z}(n+1)P$$

$$\delta : C\mathbb{Z}(n+1)P \rightarrow C\mathbb{Z}(n+1)P$$

be order-preserving maps such that

(a) φ is CP-stable,

(b) δ is an order-isomorphism with $\delta(u_i) = u_i$ for all $i = 1, \ldots, n+1$.

Then $\delta \circ \varphi$ is CP-stable.

PROOF OF LEMMA 4.5. The axiom *(stay-special)* is clear.

For *(order-embed)*, note that if $\delta : C\mathbb{Z}(n+1)P \rightarrow C\mathbb{Z}(n+1)P$ is an order-preserving isomorphism with $\delta(u_i) = u_i$, for all $i = 1, \ldots, n+1$, then $\delta|_{C\mathbb{Z}(n+1)P-(\{u_{n+1}\}\cup \mathrm{G}(u_{n+1}))}$ is an order-preserving isomorphism from $C\mathbb{Z}(n+1)P - (\{u_{n+1}\} \cup \mathrm{G}(u_{n+1}))$ onto $C\mathbb{Z}(n+1)P - (\{u_{n+1}\} \cup \mathrm{G}(u_{n+1}))$. Now

$$C\mathbb{Z}(n)P \xrightarrow{\varphi} C\mathbb{Z}(n+1)P - (\{u_{n+1}\} \cup \mathrm{G}(u_{n+1})) \xrightarrow{\delta|} C\mathbb{Z}(n+1) - \{u_{n+1}\} \cup \mathrm{G}(u_{n+1})).$$

To complete the proof of Corollary 4.4 we use Proposition 4.2 to find an order-preserving isomorphism δ_{i-1} with $\delta_{i-1}(u_j) = u_j, \forall j = 1, \ldots, i-1$ such that the diagram:

$$\begin{array}{ccc} \mathrm{Spec}(\mathbb{Z}_{\gamma_{i-1}}[Y]) & \xrightarrow{\varphi^*_{i-1,i}} & \mathrm{Spec}(\mathbb{Z}_{\gamma_i}[Y]) \\ \delta_{i-1} \downarrow\cong & & \cong\downarrow \delta_i \\ \mathrm{Spec}(\mathbb{Q}[X]_{\sigma_{i-1}}[Y] & \xrightarrow{\psi^*_{i-1,i}} & \mathrm{Spec}(\mathbb{Q}[X]_{\sigma_i}[Y]) \end{array}$$

commutes.

5. Characterization of Spec($\mathbb{Z}[Y]$) as a limit.

First we consider the directed system $\{\mathrm{Spec}(\mathbb{Z}_{\gamma_n}[Y])\}_{n\in\mathbb{N}}$ with direct limit $\mathrm{Spec}(\mathbb{Z}[Y])$.

For $n, k \in \mathbb{N}, n \leq k$, identify $\mathrm{Spec}(\mathbb{Z}_{\gamma_n}[Y])$ with its image in $\mathrm{Spec}(\mathbb{Z}_{\gamma_k}[Y])$ via $\varphi^*_{n,k}$ (induced by $\varphi_{n,k}$ from Section 2).

For clarity, let $\mathrm{G}_k(z)$ denote $\{u \in \mathrm{Spec}(\mathbb{Z}_{\gamma_k}[Y]) \mid u > z\}$, for $z \in \mathrm{Spec}(\mathbb{Z}_{\gamma_k}[Y])$, and let $\mathrm{G}_\infty(z)$ denote $\{u \in \mathrm{Spec}(\mathbb{Z}[Y]) \mid u > z\}$.

Now suppose $S \subseteq \mathrm{Spec}(\mathbb{Z}_{\gamma_n}[Y])$ is a finite set of non-maximal height-one prime ideals and $T \subseteq \mathrm{Spec}(\mathbb{Z}_{\gamma_n}[Y])$ is a finite set of maximal height-two prime ideals.

Define for $k \geq n$:

$$\Gamma_k(S,T) = \{w \in \mathrm{Spec}(\mathbb{Z}_{\gamma_k}[Y]) \mid \mathrm{G}_k(w) \cap \mathrm{G}_k(s) \subseteq T \subseteq \mathrm{G}_k(w), \forall s \in S\}.$$

(Here we identify S, T with $\varphi^*_{n,k}(S)$, $\varphi^*_{n,k}(T)$.)

It follows from the axiom *(rad(princ))* for Spec($\mathbb{Z}[Y]$) in Definition 1.4, that there exists a w in Spec($\mathbb{Z}[Y]$) such that $\mathrm{G}_\infty(w) \cap \mathrm{G}_\infty(s) \subseteq T \subseteq \mathrm{G}_\infty(w)$, for all $s \in S$. Therefore, if $N \geq n$ is large enough that $w \in \mathrm{Spec}(\mathbb{Z}_{\gamma_N}[Y])$ and $k \geq N$, then

$$\mathrm{G}_k(w) \cap \mathrm{G}_k(s) \subseteq \mathrm{G}_\infty(w) \cap \mathrm{G}_\infty(s) \subseteq T \subseteq \mathrm{G}_k(w), \text{ for all } s \in S.$$

That is,

$$\bigcap_{k \geq N} \Gamma_k(S, T) \neq \emptyset \qquad \text{for some } N \geq n.$$

5.1 REMARKS AND NOTATION. *More generally, consider a direct system of order-preserving maps:*

$$(*) \qquad \tau_{n,n+1} : C\mathbb{Z}(n)P \to C\mathbb{Z}(n+1)P, \; n \in \mathbb{N}$$

such that $\tau_{n,n+1}$ are CP-stable.

(1) Let $\tau_{n,k} = \tau_{k-1,k} \cdot \ldots \cdot \tau_{n,n+1}, \forall k \geq n$.

(2) For every $u \in C\mathbb{Z}(n)P$ of height one, and every $k \geq n$ let

$$\mathrm{G}_k(u) = \{w \in C\mathbb{Z}(k)P |\, w \geq \tau_{n,k}(u)\}$$

. *(3) In general, for $n \leq k$, we identify $C\mathbb{Z}(n)P$ with a subset of $C\mathbb{Z}(k)P$, via $\tau_{n,k}$, and $C\mathbb{Z}(n)P$ with a subset of $\varinjlim C\mathbb{Z}(n)P = U$ via τ_n.*

5.2 PROPOSITION. *Let*

$(*)$ $\tau_{n,n+1} : C\mathbb{Z}(n)P \to C\mathbb{Z}(n+1)P; n \in \mathbb{N}$

be a direct system of CP-stable maps $\tau_{n,n+1}$ for all $n \in \mathbb{N}$. Suppose furthermore that $(*)$ satisfies (L-mub-constant), and (L-more covers) of Definition 1.8. Then $U = \varinjlim C\mathbb{Z}(n)P$ is a standard partially ordered set which satisfies the axioms (infinite covers) and (finite mubs) of Definition 1.4.

PROOF. The axioms *(unique min)*, *(countable)* and *(dim 2)* from Definition 1.2 are immediate.

For *(infinite covers)*: Let $u \in U$ be a height-one element of U, such that $u \neq u_i$, for all $i \in \mathbb{N}$. Suppose $\mathrm{G}(u)$ (in U) were finite. Then there would be an $n \in \mathbb{N}$ such that $u \in C\mathbb{Z}(n)P, \mathrm{G}(u) \subseteq C\mathbb{Z}(n)P$, and $\mathrm{G}_n(u) = \mathrm{G}(u)$. But by *(L-more covers)*, there is a $k \in \mathbb{N}$ such that $\tau_{n,k}(\mathrm{G}_n(u)) \subsetneq \mathrm{G}_k(\tau_{n,k}(u)) = \mathrm{G}_k(u)$. This contradicts $\mathrm{G}_k(u) \subseteq \mathrm{G}(u) \subseteq \mathrm{G}_n(u)$. Thus *(infinite covers)* holds.

For *(finite mubs)*, let $u, v \in U$ with $ht(u) = ht(v) = 1$. Then there is an $n \in \mathbb{N}$ with $u, v \in C\mathbb{Z}(n)P$ and by *(L-mub-constant)* there is an $N \geq n$ such that $\tau_{N,k}(\mathrm{G}_N(u) \cap \mathrm{G}_N(v)) = \mathrm{G}_k(u) \cap \mathrm{G}_k(v)$. If $u \neq u_i$ or $v \neq u_i$ in $C\mathbb{Z}(n)P$, then $\mathrm{G}_N(u)$ or $\mathrm{G}_N(v)$ is finite, and so $\mathrm{G}(u) \cap \mathrm{G}(v) = \mathrm{G}_k(u) \cap \mathrm{G}_k(v) = \mathrm{G}_N(u) \cap \mathrm{G}_N(v)$ is finite, for every $k \geq N$.

If $u = u_i$ and $v = v_j$, for some i, j, suppose $N > i, j$. Then, in $C\mathbb{Z}(N)P$, $\mathrm{G}_N(u_i) \cap \mathrm{G}_N(u_j) = \emptyset$, and hence $\mathrm{G}(u) \cap \mathrm{G}(v) = \emptyset$ in U, by *(L-mub-constant)*.

5.3 LEMMA. *Let $\tau_{n,n+1} : C\mathbb{Z}(n)P \longrightarrow C\mathbb{Z}(n+1)P$ be a CP-stable map and let $u \in C\mathbb{Z}(n)P$ be a height-one element. Then $\tau_{n,n+1}(\mathrm{G}_n(u)) = \mathrm{G}_{n+1}(u) \cap (\bigcup_{i=1}^{n} \mathrm{G}_{n+1}(u_i))$.*

PROOF. This follows because $\tau_{n,n+1} : C\mathbb{Z}(n)P \xrightarrow{\cong} C\mathbb{Z}(n+1)P - (\{u_{n+1}\} \cup \mathrm{G}(u_{n+1}))$.

5.4 NOTATION. *Let $\tau_{n,n+1} : C\mathbb{Z}(n)P \longrightarrow C\mathbb{Z}(n+1)P$ be a direct system of order-preserving maps such that:*

(a) *Each $\tau_{n,n+1}$ is CP-stable.*

(b) *(L-mub-constant) and (L-more covers) of Definition 1.8 are satisfied.*

Consider $U = \varinjlim C\mathbb{Z}(n)P$ as a union of $C\mathbb{Z}(n)P$ via $\tau_{n,n+1}$. (Note that $\tau_{n,n+1}$ is injective by CP-stable.)

Let $S \subseteq U$ be a finite set of height-one primes, and $T \subseteq U$ a finite set of (height-two) maximal elements. Define

$$\Gamma(S,T) = \{\, w \in U |\, ht(w) = 1, \mathrm{G}(s) \cap \mathrm{G}(w) \subseteq T \subseteq \mathrm{G}(w), \forall s \in S\,\},$$

and, for all $n \in \mathbb{N}$,

$$\Gamma_n(S,T) = \{\, w \in C\mathbb{Z}(n)P |\ \mathrm{G}_n(s) \cap \mathrm{G}_n(w) \subseteq T \cap C\mathbb{Z}(n)P \subseteq \mathrm{G}_n(w), \forall s \in S\,\}.$$

(Recall $\mathrm{G}_n(w) = \{\, v \in C\mathbb{Z}(n)P \,|\, v > w\,\}$.)

5.5 PROPOSITION. *In the notation of 5.4, if $T \neq \emptyset$, then for every $k > n_{(S,T)} \in \mathbb{N}$, we have $\Gamma_k(S,T) \supseteq \Gamma_{k+1}(S,T)$, where $\Gamma_k(S,T)$ and $\Gamma_{k+1}(S,T)$ are considered as subsets of U. If $T = \emptyset$, then $\Gamma_k(S,T) \cup \{u_{k+1}\} \supseteq \Gamma_{k+1}(S,T)$.*

PROOF. First we prove:

CLAIM. $\Gamma_{k+1}(S,T) \cap C\mathbb{Z}(k)P \subseteq \Gamma_k(S,T)$.

PROOF OF CLAIM. Let $w \in \Gamma_{k+1}(S,T) \cap C\mathbb{Z}(k)P$. Then by Lemma 5.3, $\mathrm{G}_k(w) = \mathrm{G}_{k+1}(w) \cap (\bigcup_{i=1}^{k} \mathrm{G}_{k+1}(u_i))$; $\mathrm{G}_k(s) \cap \mathrm{G}_k(w) \subseteq \mathrm{G}_{k+1}(s) \cap \mathrm{G}_{k+1}(w) \subseteq T, \forall s \in S$.

Now $T \subseteq \bigcup_{i=1}^{k} \mathrm{G}_{k+1}(u_i)$; thus $T \subseteq \mathrm{G}_{k+1}(w) \cap (\bigcup_{i=1}^{k} \mathrm{G}_{k+1}(u_i)) = \mathrm{G}_k(w)$, proving the claim.

Continuing the proof of Proposition 5.5, we show that $\Gamma_{k+1}(S,T) \subseteq C\mathbb{Z}(k)P$, if $T \neq \emptyset$. If $w \in \Gamma_{k+1}(S,T) - C\mathbb{Z}(k)P$ then $ht(w) = 1$ and so $w = u_{k+1}$. (Recall that u_{k+1} is the only height-one prime ideal which is in $C\mathbb{Z}(k+1)P$ but not in $C\mathbb{Z}(k)P$.) However, $T \subseteq \bigcup_{i=1}^{k} \mathrm{G}_{k+1}(u_i)$, so $\mathrm{G}_{k+1}(u_{k+1}) = \mathrm{G}_{k+1}(w) \not\supseteq T$, since $T \neq \emptyset$, so we have a contradiction.

In case $T = \emptyset$, then $\Gamma_{k+1}(S,T) \subseteq C\mathbb{Z}(k)P \cup \{u_{k+1}\}$, so the result holds by the claim.

NOTES. If $T = \emptyset$, $S = \{u_1\}$, $k \neq 1$, then $u_k \in \Gamma_k(S,T) - \Gamma_{k-1}(S,T)$. In general, when $T \neq \emptyset$, $\Gamma_{k-1}(S,T)$ is larger than $\Gamma_k(S,T)$. For example, if $\gamma_1 =$

$\{(2)\}$, $\gamma_2 = \{(2), (3)\}$, $S = \{(Y+2)\}$, and $T = \{(2,Y)\}$ in $\mathrm{Spec}(\mathbb{Z}_{\gamma_1}[Y]) = \mathrm{Spec}(\mathbb{Z}_{(2)}[Y])$, then $w = (Y+8)$ is in $\Gamma_1(S,T) - \Gamma_2(S,T)$, since $\mathrm{G}_1((Y+8)) = \{(2,Y)\}$, but $\mathrm{G}_2((Y+8)) \cap \mathrm{G}_2((Y+2)) = \{(3, Y+2), (2,Y)\} \not\subseteq \{(2,Y)\}$.

5.6 PROPOSITION. *Let S,T be as above, $T \neq \emptyset$. Then*

$$\Gamma(S,T) = \bigcap_{k \geq n_{(S,T)}} \Gamma_k\ (S,T),\ \Gamma(S,\emptyset) = (\bigcap_{k \geq n_{(S,\emptyset)}} \Gamma_k\ (S,\emptyset)) \cup (\Gamma(S,\emptyset) \cap \{u_i | i > n_{(S,\emptyset)}\}).$$

(Note that $\Gamma(S,T)$ might be empty.)

PROOF. First we show the first "$\subseteq$": Let $w \in \Gamma(S,T)$ and $k \geq n_{(S,T)}$. Then $S, T \subseteq C\mathbb{Z}(k)P$. Now

$$T \subseteq \bigcup_{i=1}^{k} \mathrm{G}_k(u_i) \subseteq \bigcup_{i=1}^{k} \mathrm{G}(u_i), \text{ and } \mathrm{G}(u_n) \cap \mathrm{G}(u_k) = \emptyset, \text{ for all } n > k.$$

Thus $T \cap \mathrm{G}(u_n) = \emptyset$, for all $n > k$, and $\emptyset \neq T \subseteq \mathrm{G}(w)$ implies $w \neq u_n$, for all $n > k$; hence $w \in C\mathbb{Z}(k)P$. Now $T \subseteq \mathrm{G}(w) \cap C\mathbb{Z}(k)P = \mathrm{G}_k(w)$. Also for every $s \in S$, $\mathrm{G}_k(w) \cap \mathrm{G}_k(s) \subseteq \mathrm{G}(w) \cap \mathrm{G}(s) \subseteq T$. Therefore $w \in \Gamma_k(S,T)$.

Next we show "$\supseteq$": Let $w \in \bigcap_{k \geq n_{(S,T)}} \Gamma_k(S,T)$. Let $t \in \mathrm{G}(s) \cap \mathrm{G}(w)$, for some $s \in S$. There exists a $k \geq n_{(S,T)}$ such that $t \in C\mathbb{Z}(k)P$. Then $t \in \mathrm{G}_k(s) \cap \mathrm{G}_k(w) \subseteq T$ in $C\mathbb{Z}(k)P$. Hence $\mathrm{G}(s) \cap \mathrm{G}(w) \subseteq T$ for all $s \in S$. For $k \geq n_{(S,T)}$, we have $T \subseteq \mathrm{G}_k(w) \subseteq \mathrm{G}(w)$. We have shown $w \in \Gamma(S,T)$.

Now let $w \in \Gamma(S,\emptyset)$; then $\mathrm{G}(w) \cap G(s) = \emptyset, \forall s \in S$. Choose $k \geq n_{(S,\emptyset)}$. If $w \in C\mathbb{Z}(k)P$, then $\mathrm{G}_k(s) \cap \mathrm{G}_k(w) \subseteq \mathrm{G}(w) \cap G(s) = \emptyset, \forall s \in S$, so $w \in \Gamma_k(S,\emptyset)$. If $w \notin C\mathbb{Z}(k)P$, then $w = u_i$, for some $i > k \geq n_{(S,\emptyset)}$. Therefore $w \in \Gamma_k(S,\emptyset) \cup (\Gamma(S,\emptyset) \cap \{u_i | i > n_{(S,\emptyset)}\})$..

Let $w \in \Gamma_k(S,\emptyset)$, and suppose $t \in \mathrm{G}(s) \cap \mathrm{G}(w)$, for all $s \in S$. Choose $k \geq n_{(S,\emptyset)}$ with $t \in C\mathbb{Z}(k)P$. Then $t \in \mathrm{G}_k(s) \cap \mathrm{G}_k(w) = \emptyset$, a contradiction. Thus $\mathrm{G}(s) \cap \mathrm{G}(w) = \emptyset$, so $w \in \Gamma(S,\emptyset)$.

5.7 THEOREM. *Let $\tau_{n,n+1} : C\mathbb{Z}(n)P \to C\mathbb{Z}(n+1)P$ be a direct system of order-preserving maps with direct limit $U = \varinjlim C\mathbb{Z}(n)P$. Then $U \cong \mathrm{Spec}\,\mathbb{Z}[Y]$ if each $\tau_{n,n+1}$ is a CP-stable map, and axioms (L-mub-constant),(L-more covers) and (L- rad(princ)) of Definition 1.8 are satisfied.*

PROOF. By Proposition 5.2, the axioms given imply that U is standard and satisfies *(infinite covers)* and *(finite mubs)* of Definition 1.4. For the axiom *(rad(princ))* of Definition 1.4, we need that $\Gamma(S,T) \neq \emptyset$, for every pair S and T, finite sets of height-one and height-two elements, respectively. But by Proposition 5.6 and *(L- rad(princ))*,

$$\Gamma(S,T) \supseteq \bigcap_{k \geq n_{(S,T)}} \Gamma_k(S,T) \neq \emptyset.$$

6. Partially ordered sets which are a direct limit of $C\mathbb{Z}(n)P$; special sequences.

6.1 PROPOSITION. *Let U be a standard partially ordered set and suppose that $\{u_i\}_{i\in\mathbb{N}}$ are infinitely many distinct height-one elements of U. Put*

$$W_n = U - \bigcup_{i=n+1}^{\infty} (\mathrm{G}(u_i) \cup \{u_i\}).$$

Suppose that for every $n \in \mathbb{N}$ there is an (order-preserving) isomorphism: $\lambda_n : W_n \xrightarrow{\cong} C\mathbb{Z}(n)P$ with $\lambda_n(u_i) = u_i \in C\mathbb{Z}(n)P$ n-special, i.e. W_n is a $C\mathbb{Z}(n)P$-set with special elements $u_1, \ldots, u_n$. Then for $n, m \in \mathbb{N}$ with $n \leq m$ the canonical embedding $W_n \hookrightarrow W_m$ is CP-stable (Definition 1.7).

PROOF. The *(stay-special)* axiom is trivially satisfied.
For *(order-embed)*:

$$\begin{aligned} W_n &= U - \bigcup_{i\geq n+1} (G(u_i) \cup \{u_i\}) \\ &= U - \bigcup_{i=n+1}^{m} (G(u_i) \cup \{u_i\}) - \bigcup_{i\geq m+1} (G(u_i) \cup \{u_i\}) \\ &= (U - \bigcup_{i\geq m+1} (G(u_i) \cup \{u_i\})) - \bigcup_{i=n+1}^{m} (G(u_i) \cup \{u_i\}) \\ &= W_m - (\{u_{n+1}, \ldots, u_m\} \cup \bigcup_{i=n+1}^{m} G(u_i)) \end{aligned}$$

6.2 REMARK. *With the hypotheses of Proposition 6.1, it is not necessarily true that*
$U \cong \varinjlim_{n\in\mathbb{N}} W_n$.

EXAMPLE. If

$$t \in \bigcap_{i=1}^{\infty} G(u_i) \quad \text{then} \quad t \notin \bigcup_{i=1}^{\infty} W_i \cong \varinjlim_{n\in\mathbb{N}} W_n.$$

6.3 THEOREM. *Suppose that U and $W_n, n \in \mathbb{N}$ are as given in 6.1. Then the following are equivalent:*

(a) $U = \bigcup_{n\in\mathbb{N}} W_n \cong \varinjlim_{n\in\mathbb{N}} W_n$.
(b) *U is a nested direct limit of $C\mathbb{Z}(n)P$'s with respect to $\{W_n\}_{n\in\mathbb{N}}$.*
(c) *For every infinite subset $\Gamma \subseteq \mathbb{N}$,*

$$\bigcap_{i\in\Gamma} G(u_i) = \emptyset$$

(d) *$G(u_i) \cap G(u_j) = \emptyset$ for all $i, j \in \mathbb{N}$ with $i \neq j$*
(e) *$\{u_i\}_{i\in\mathbb{N}}$ is a special sequence in U.*

PROOF. $(a) \iff (b)$: Proposition 6.1 implies conditions $(i), (ii), (iii)$ of Definition 1.11. The remaining condition, (iv) of Definition 1.11, is exactly (a).
$(a) \implies (c)$: Let $\Gamma \subseteq \mathbb{N}$ be an infinite set with

$$\bigcap_{i\in\Gamma} G(u_i) \neq \emptyset$$

and let $t \in \bigcap_{i\in\Gamma} G(u_i)$. Then $t \notin W_n$ for all $n \in \mathbb{N}$ and hence $U \neq \bigcup_{n\in\mathbb{N}} W_n$.
$(c) \implies (a)$: Obviously every height-one element $t \in U$ is contained in $\bigcup W_n$. Let $t \in U$ be an element of height two. By (c) t is contained in at most finitely many $G(u_i)$, hence there is $N \in \mathbb{N}$ with $t \notin \mathrm{G}(u_k)$ for all $k \geq N$. This implies $t \in W_k$ for all $k \geq N$.
$(c) \iff (d)$: If $t \in G(u_i) \cap G(u_j)$ for some $i, j \in \mathbb{N}$ with $i \neq j$ then by (c) t is contained in at most finitely many $G(u_k)$. Let $N \in \mathbb{N}$ with $t \notin G(u_k)$ for all $k \geq N$. But then $t \in W_N \cap G(u_i) \cap G(u_j)$ contradicting $W_N \cong C\mathbb{Z}(N)P$.
$(e) \implies (d)$: follows immediately from the definition of a special sequence.
$(d) \implies (e)$: Since $W_n \cong C\mathbb{Z}(n)P$ it is obvious that $G(u_n)$ is infinite for all $n \in \mathbb{N}$. It remains to show that

$$\mathcal{M}(U) = \bigcup_{i\in\mathbb{N}} G(u_i).$$

Let t be a height two maximal element of U. Since t is contained in at most one $G(u_i)$, there is an $N_1 \in \mathbb{N}$ with $t \notin G(u_k)$ for all $k \geq N_1$. Also $t > w$ for some height-one element $w \in U$ and there is an $N_2 \in \mathbb{N}$ with $w \in W_k$ for all $k \geq N_2$. This implies for $k \geq \max(N_1, N_2)$ that t is a height-two element of W_k. Again since $W_k \cong C\mathbb{Z}(k)P$ we obtain that $t \in G(u_i) \cap W_k$ for some $i \leq k$.

6.4 COROLLARY. *Suppose that U is a standard partially ordered set and that $\{u_i\}_{i\in\mathbb{N}}$ is a special sequence in U. For $n \in \mathbb{N}$ put*

$$W_n = U - \bigcup_{i=n+1}^{\infty} (G(u_i) \cup \{u_i\})$$

and suppose that there is an order-isomorphism $\lambda_n : W_n \xrightarrow{\cong} C\mathbb{Z}(n)P$ for all $n \in \mathbb{N}$ with $\lambda_n(u_i)$ n-special. Then U is a nested direct limit of $C\mathbb{Z}(n)P$'s with respect to $\{W_n\}_{n\in\mathbb{N}}$.

6.5 PROPOSITION. *Let U be a standard partially ordered set, $W_n \subseteq U, n \in \mathbb{N}$ subsets in U such that U is a nested direct limit of $C\mathbb{Z}(n)P$'s with respect to $\{W_n\}_{n\in\mathbb{N}}$. Then U admits a special sequence $\{u_i\}_{i\in\mathbb{N}}$ and*

$$W_n = U - \bigcup_{i=n+1}^{\infty} (G(u_i) \cup \{u_i\})$$

PROOF. The special elements u in a $C\mathbb{Z}(n)P$-set are uniquely determined by $|G(u)| = \infty$. This implies that the canonical embedding $W_n \hookrightarrow W_m$ for $m \geq n$ maps the set of special elements $\{u_1, \dots, u_n\}$ of W_n into the set of special

elements of W_m. Let $\{u_i\}_{i\in\mathbb{N}}$ be the infinite sequence of elements in U with $\{u_1,\ldots,u_n\}$ special in W_n. Then:
(a) $|G(u_i)| = \infty$ in U since $|G(u_i)\cap W_n| = \infty$ for $i \le n$.
(b) $G(u_i)\cap G(u_j) = \emptyset$ for $i,j\in\mathbb{N}$ and $i\ne j$ since $G(u_i)\cap G(u_j)\cap W_n = \emptyset$ and $\bigcup W_n = U$
(c) Let $t\in\mathcal{M}(U)$ be a height-two maximal element of U. Then $t\in W_n$ for some $n\in\mathbb{N}$ and for n sufficiently large t is of height two in W_n. Therefore $t\in G(u_i)\cap W_n$ for some $i\le n$.
This shows that $\{u_i\}_{i\in\mathbb{N}}$ is a special sequence in U.
Finally, if $t\in W_n$ then $t\notin\bigcup_{i\ge n+1}(G(u_i)\cup\{u_i\})$ since the embedding $W_n\hookrightarrow W_m$ is CP-stable. Conversely, if $t\in U-\bigcup_{i\ge n+1}(G(u_i)\cup\{u_i\})$ then $t\in W_m$ for some $m\in\mathbb{N}$. We may assume that $m\ge n$. Since $t\not> u_i$ for all $i\ge n+1$ we also obtain that $t\notin W_m\cap\bigcup_{i=n+1}^{m}(G(u_i)\cup\{u_i\})$ implying that $t\in W_n$.

6.6 Proposition. *Let $\{W_n,\phi_{n,m}\}_{n,m\in\mathbb{N}}$ be a direct system of standard partially ordered sets with (i) $W_n\cong C\mathbb{Z}(n)P$ and (ii) $\phi_{n,m}: W_n\longrightarrow W_m$ CP-stable for $n,m\in\mathbb{N}, n\le m$. Let*

$$U\cong \varinjlim_{n\in\mathbb{N}}\{W_n,\phi_{n,m}\}$$

Then:

(a) *U satisfies axiom (N-infinite less sets) from Definition 1.14.*
(b) *U satisfies (infinite covers) $\Longleftrightarrow$ $\{W_n,\phi_{n,m}\}$ satisfies (L-more covers).*
(c) *U satisfies (finite mubs) $\Longleftrightarrow$ $\{W_n,\phi_{n,m}\}$ satisfies (L-mub-constant).*
(d) *U satisfies (smaller maximals) $\Longleftrightarrow$ $\{W_n,\phi_{n,m}\}$ satisfies (L-smaller co-maximals)*
(e) *U satisfies (rad(princ)) $\Longleftrightarrow$ $\{W_n,\phi_{n,m}\}$ satisfies (L-rad(princ).*

Proof. straightforward

Note. A standard partially ordered set U is a direct limit of $C\mathbb{Z}(n)P$ if and only if there exists a countably infinite special sequence $\{u_n | n\in\mathbb{N}\}$ of height-one elements in U and $W_n = U - \bigcup_{i\ge n+1}^{\infty}(\mathrm{G}(u_i)\cup\{u_i\})\cong C\mathbb{Z}(n)P$.

Questions. 1. For what rings R, is Spec(R) a direct limit of $C\mathbb{Z}(n)P$'s?

2. If U is a direct limit of $C\mathbb{Z}(n)P$'s, *not* every maximal element necessarily has height two. It could be that $j^{-1}(T)$ is infinite or finite or empty, for various finite sets T. There could be two height-one elements in infinitely many maximal, height-two elements, and so U would not be the prime spectrum of a Noetherian ring. If we required that U be a Noetherian spectrum, so that *(finite mubs)* of Definition 1.4 holds— $\mathrm{G}(u)\cap\mathrm{G}(v)$ is finite for each pair u,v, as well as that U be a direct limit of $C\mathbb{Z}(n)P$'s, what rings R would have Spec(R) a direct limit of $C\mathbb{Z}(n)P$'s?

3. If U is standard and satisfies *(infinite covers)*, and *(finite mubs)* and U is a direct limit of $C\mathbb{Z}(n)P$'s, then what can be said about U?

7. Other axioms for countable two-dimensional Noetherian domains.

7.1 PROPOSITION: THE EXISTENCE OF A SPECIAL SEQUENCE. *Let U be a standard partially ordered set which satisfies (infinite covers), (finite mubs), (smaller comaximals) and (N-∞ ht.1s). Let $u_1 \in U$ be an element of height one. Then U admits a special sequence $\{u_n | n \in \mathbb{N}\}$ which starts with u_1.*

PROOF. Let $\{t_n | n \in \mathbb{N}\} = \mathcal{M}(U)$ be an enumeration of the height-two elements of U.

CLAIM 1. $|\mathcal{M}(U) - \mathrm{G}(u_1)| = \infty$.

PROOF OF CLAIM 1. By *(N-∞ ht.1s)*, there exists a height-one element $u \in U$, $u \neq u_1$. By *(infinite covers)*, $|\mathrm{G}(u)| = \infty$. By *(finite mubs)*, $|\mathrm{G}(u) \cap \mathrm{G}(u_1)| < \infty$. Hence $|\mathrm{G}(u) - \mathrm{G}(u_1)| = \infty$, proving the claim.

Now let $i_1 \in \mathbb{N}$ be minimal with $t_{i_1} \notin \mathrm{G}(u_1)$. By *(smaller comaximals)*, there exists a height-one element $u_2 \in U$ such that $t_{i_1} \in \mathrm{G}(u_2)$ and $\mathrm{G}(u_1) \cap \mathrm{G}(u_2) = \emptyset$.

CLAIM 2. Suppose now that $\{u_1, \dots, u_n\}$ have been constructed with $\mathrm{G}(u_i) \cap \mathrm{G}(u_j) = \emptyset$, for $i \neq j$. Then $|\mathcal{M}(U) - (\mathrm{G}(u_1) \cup \cdots \cup \mathrm{G}(u_n))| = \infty$.

PROOF OF CLAIM 2. By *(N-∞ ht.1s)*, there exists a height-one element $u \in U$, $u \neq u_i$ for $i = 1, \dots, n$. By *(infinite covers)*, $|\mathrm{G}(u)| = \infty$.
By *(finite mubs)*, $|\mathrm{G}(u) \cap (\bigcup_{i=1}^{n} \mathrm{G}(u_i))| < \infty$.
Hence $|\mathrm{G}(u) - (\bigcup_{i=1}^{n} \mathrm{G}(u_i))| = \infty$, proving Claim 2.

Now let $i_n \in \mathbb{N}$ be minimal with respect to $t_{i_n} \notin \bigcup_{i=1}^{n} \mathrm{G}(u_i)$. Then by *(smaller comaximals)*, there exists a height-one element $u_{n+1} \in U$ such that $t_{i_n} \in \mathrm{G}(u_{n+1})$ and $\mathrm{G}(u_i) \cap \mathrm{G}(u_{n+1}) = \emptyset$, for $1 \leq i \leq n$.
By this means, we have constructed an infinite *special* sequence of height-one elements of U.

7.2 PROPOSITION. *The following axioms hold when $U \cong$ Spec(R), for R a two-dimensional countable Noetherian domain: (unique min), (countable), (dim 2) of Definition 1.2 (i.e. U is standard); (finite mubs) of Definition 1.4; (N-infinite less sets) and hence (N-∞ ht.1s) of Definition 1.14; and (smaller comaximals) of Definition 1.5. If U satisfies (infinite covers) then U admits an infinite special sequence.*

PROOF. All the axioms in the first statement are obvious except for *(smaller comaximals)*. For this, let $u_1, \dots, u_n \in U$ with height$(u_i) = 1$. Suppose $t \in U$ has height two and that $t \notin \cup \mathrm{G}(u_i)$, over all $i = 1, \dots, n$; that is, u_i and t are comaximal in Spec(R). Now considering t and the u_i as prime ideals in R, $u_1 \cdot \ldots \cdot u_n$ and t are comaximal and so there exist $a \in u_1 \cdot \ldots \cdot u_n$ and $b \in t$ such that $1 = a + b$. Let w be a height-one prime ideal containing b and contained in t. Then $t \in \mathrm{G}(w)$ and $\mathrm{G}(w) \cap \mathrm{G}(u_i) = \emptyset$, for every $1 \leq i \leq n$. The last statement follows from Proposition 7.1.

7.3 QUESTIONS. *(1) Do the axioms in the first statement of 7.2 characterize the prime spectrum of a countable Noetherian domain of dimension two?*
(2) Let U be a standard partially ordered set which satisfies (infinite covers), finite mubs), and (smaller comaximals).

(a) When is U a limit of $C\mathbb{Z}(n)P$'s?

(b) Suppose $U = \varinjlim C\mathbb{Z}(n)P$ and $\{v_1, \ldots, v_n\}$ is a finite set of height-one elements of U with $\mathrm{G}(v_i) \cap \mathrm{G}(v_j) = \emptyset$ for $i \neq j$. By the proof of Proposition 7.1 $\{v_1, \ldots, v_n\}$ can be extended to a special sequence of U. When can $\{v_1, \ldots, v_n\}$ be extended to a set of exceptional height-one elements of U?

7.4 REMARKS. *Let U be a standard partially ordered set which satisfies (infinite covers), (finite mubs), (smaller comaximals) (or (∞ smaller comaximals)), and (N-∞ ht.1s). Let $\{v_1, \ldots, v_n\}$ be a finite set of height-one elements of U such that $\mathrm{G}(v_i) \cap \mathrm{G}(v_j) = \emptyset$, for $i \neq j$.*

(1) Then, as shown in Proposition 7.1, $\{v_1, \ldots, v_n\}$ can be extended to an infinite special sequence of height-one elements of U.

(2) However U might not be a limit of $C\mathbb{Z}(n)P$'s, as in the example below.

EXAMPLE. The partially ordered set U pictured below satisfies all the axioms mentioned but is not a limit of $C\mathbb{Z}(n)P$'s:

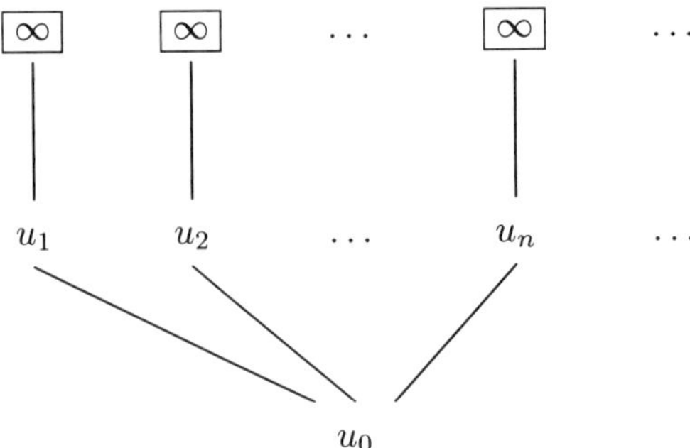

(3) Also (infinite covers), (finite mubs), (smaller comaximals) and (N-infinite less sets) seem not to be enough to write U as a limit of $C\mathbb{Z}(n)P$'s with respect to any infinite special sequence of height-one elements, as shown by the following example:

EXAMPLE. Consider a partially ordered set U like this picture:

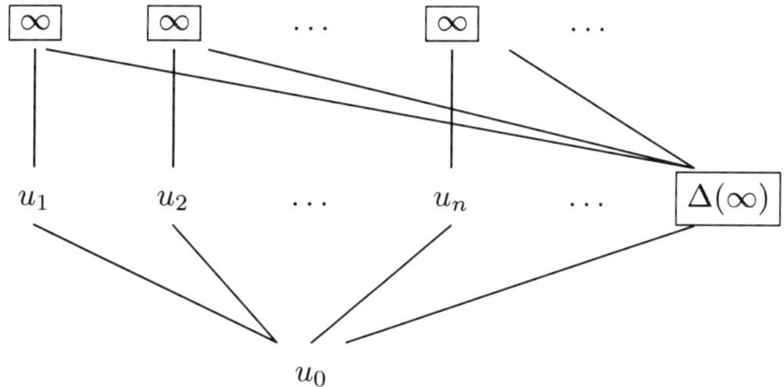

Here the infinite set Δ is to satisfy: For every $v \in \Delta$ and for every $n \in \mathbb{N}$, the set $\mathrm{G}(v) \cap \mathrm{G}(u_n)$ is nonempty and finite.

7.5 REMARK. *Let U be a standard partially ordered set satisfying (infinite covers), (finite mubs), (∞ smaller comaximals) and (N-infinite less sets) from Definitions 1.4, 1.5 and 1.14. Suppose that $\{u_n | n \in \mathbb{N}\}$ is an infinite special sequence of height-one elements of U as in Definition 1.9 and that v is a height-one element of U, $v \neq u_n$, for all $n \in \mathbb{N}$. Then by (infinite covers), $|\mathrm{G}(v)| = \infty$, and by (finite mubs), $|\mathrm{G}(v) \cap \mathrm{G}(u_n)| < \infty$. Hence for infinitely many $n \in \mathbb{N}, \mathrm{G}(v) \cap \mathrm{G}(u_n) \neq \emptyset$.*

7.6 PROPOSITION. *Let U be a standard partially ordered set satisfying (infinite covers), (finite mubs) (Definition 1.4), (N-infinite less sets) (Definition 1.14) and (smaller comaximals) (Definition 1.5). Suppose that $\{u_n | n \in \mathbb{N}\}$ is an infinite special sequence of height-one elements of U and put*

$$W_n = U - (\bigcup_{i \geq n+1} \mathrm{G}(u_i) \cup \{u_i\}) \text{ for all } n \in \mathbb{N}.$$

Then W_n is a two-dimensional partially ordered set which satisfies all but the last axiom of Definition 1.3; in particular, the set of height-one maximals and the set of height-one non-maximal elements are both infinite. Thus W_n looks like the diagram below.

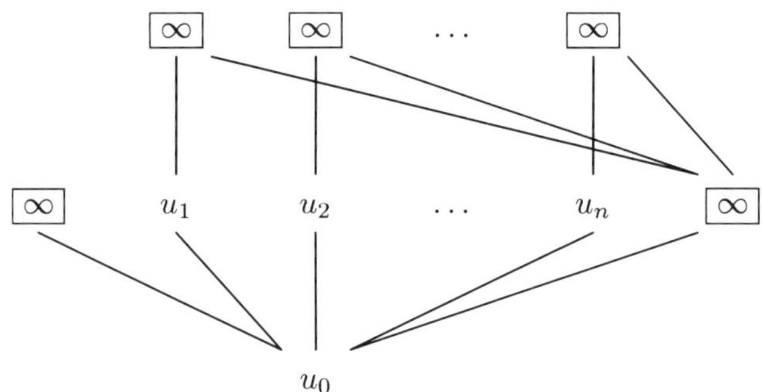

(The relationships of the lower right boxed section are unknown and too complicated to display. Each point in that box is less than a finite non-zero number of height-two elements.)

NOTE. It is not clear and is even unlikely that *(infinite inv. j-sets)* of $C\mathbb{Z}(n)P$ is satisfied.

PROOF OF PROPOSITION. Clearly W_n is standard.

To see that there are infinitely many height-one maximals, *(∞ ht-1 maxes)*, let $t \in \mathrm{G}(u_{n+1})$; then by *($\infty$ smaller comaximals)* there are infinitely many height-one elements v_i, $i \in \mathbb{N}$ such that

(i) $t \in \mathrm{G}(v_i)$,

(ii) $\mathrm{G}(v_i) \cap \mathrm{G}(u_j) = \emptyset$, for $i \in \mathbb{N}$, $1 \leq j \leq n$.

Hence the set of $\{v_i,\ i \in \mathbb{N},\ v_i \neq u_{n+1}\}$ is contained in W_n and the v_i's are maximal in W_n.

For *(finite covers)*, let u be a non-maximal height-one element of W_n with $u \neq u_i$, for all $1 \leq i \leq n$. Then

$$\mathrm{G}_n(u) = \mathrm{G}(u) \cap W_n = \bigcup_{i=1}^{n} (\mathrm{G}(u) \cap \mathrm{G}(u_i)).$$

Hence $\mathrm{G}_n(u)$ is finite.

That there are infinitely many non-maximal height-one elements follows from *(N-infinite less sets)*, namely, for any $t \in G(u_n)$ there are infinitely many height-one elements $v \in U$ with $v < t$.

REMARK. Axiom *(infinite inv. j-sets)* for $C\mathbb{Z}(n)P$ seems to require some *(rad(princ))* condition on the special sequence $\{u_n | n \in \mathbb{N}\}$.

8. Exceptional sequences and orthogonal pairs of sequences.

In this section we investigate connections between the existence of exceptional sequences (Definition 1.10) in a standard partially ordered set U and representations of U as a *nested* direct limit of $C\mathbb{Z}(n)P$'s.

8.1 THEOREM. *Let U be a standard ordered set satisfying (infinite covers), and (finite mubs) of Definition 1.4. Then U is a nested direct limit of $C\mathbb{Z}(n)P$'s as in Definition 1.11 if and only if U admits an exceptional sequence (Definition 1.10).*

PROOF. Suppose $U = \varinjlim_{n\in\mathbb{N}} W_n = \bigcup_{n\in\mathbb{N}} W_n$, where $W_n \cong C\mathbb{Z}(n)P$ and the canonical embedding $W_n \subseteq W_{n+1}$ is CP-stable.

By Proposition 6.5 there is a special sequence $\{u_n | n \in \mathbb{N}\} \subseteq U$ such that $u_1, \ldots, u_n$ are the n special elements of W_n for each n and

$$W_n = U - \bigcup_{i=n+1}^{\infty} (G(u_i) \cup \{u_i\})$$

Suppose that S is a finite subset of $\{u_n | n \in \mathbb{N}\}$, that T is a finite set of height-two maximal elements of U, and further that n is chosen so that $S, T \subseteq W_n$. Now $W_n \cong C\mathbb{Z}(n)P$ and by axiom *(infinite inv. j-sets)* for $C\mathbb{Z}(n)P$, the set $j_n^{-1}(T) = \{w \in W_n | \mathrm{G}_n(w) = T\}$ is infinite.

But $j_n^{-1}(T) \subseteq \Gamma(S,T) = \{w \in U \mid \text{height } w = 1, \mathrm{G}(w) \cap \mathrm{G}(s) \subseteq T \subseteq \mathrm{G}(w), \forall s \in S\}$. Therefore $\{u_n | n \in \mathbb{N}\}$ is exceptional.

Conversely, suppose U is standard and satisfies *(infinite covers)* and *(finite mubs)*, and that $\{u_n | n \in \mathbb{N}\}$ is an exceptional sequence for U.

Put $W_n = U - \bigcup_{i\geq n=1} \{\mathrm{G}(u_i) \cup \{u_i\}\}$.

Obviously $W_n \subseteq W_{n+1}$.

To show that $W_n \cong C\mathbb{Z}(n)P$, we check that the axioms of Definition 1.3 hold. Obviously W_n is standard. For *(∞ ht-1 maxes)*, to show that there are infinitely many height-one maximal elements, let $t \in \mathrm{G}(u_{n+1})$. Set $T = \{t\}$; $S = \{u_1, \ldots, u_n\}$. By the assumption that $\{u_n | n \in \mathbb{N}\}$ is an exceptional sequence for U,

$$\Gamma(S, \{t\}) = \{w \in U \mid ht(w) = 1, \mathrm{G}(w) \cap \mathrm{G}(u_i) \subseteq \{t\} \subseteq \mathrm{G}(w), \forall i\} \text{ is infinite.}$$

However, $t \in \mathrm{G}(u_{n+1})$ implies that $t \notin \mathrm{G}(u_i), \forall i, 1 \leq i \leq n$, which implies that $\mathrm{G}(w) \cap \mathrm{G}(u_i) = \emptyset$, for $w \in \Gamma(S, \{t\})$. Thus there are infinitely many $w \neq u_{n+1}$, $w \in W_n$ such that $\mathrm{G}(w) \cap \mathrm{G}(u_i) = \emptyset$, $\forall i, 1 \leq i \leq n$. Therefore each of these w is a maximal height-one element of W_n.

The axiom *(n-special)* holds because the exceptional sequence $\{u_n | n \in \mathbb{N}\}$ for U is also a special sequence.

For *(finite covers)*, let $u \in W_n$ be a non-maximal height-one element of W_n which is not a special element (that is, $u \neq u_i$, for all $1 \leq i \leq n$). Then

$$\mathrm{G}_n(u) = \{t \in W_n | u < t\} = \bigcup_{i=1}^{n} (\mathrm{G}(u) \cap \mathrm{G}(u_i)), \text{ where } \mathrm{G}(u) = \{t \in U | u < t\}.$$

Hence $\mathrm{G}_n(u) = \mathrm{G}(u) \cap W_n$ is finite.

For *(infinite inv. j-sets)*, note that, for every finite subset T of height-two maximal elements of W_n, $j^{-1}(T) = \Gamma(\{u_1, \dots, u_n\}, T)$.

8.2 THEOREM. *Let U be a standard partially ordered set satisfying (infinite covers), (finite mubs), and (rad(princ)) of Definition 1.4. Then*

(i) U admits an exceptional sequence of height-one elements.

(ii) Every special sequence of height-one elements in U is exceptional.

PROOF. By [**rW2**] we know that $U \cong \mathrm{Spec}(\mathbb{Z}[Y])$. Hence U is a *nested* direct limit of $C\mathbb{Z}(n)P$'s and by Proposition 8.1 U admits an exceptional sequence. Now let $\{u_n | n \in \mathbb{N}\}$ be a special sequence for U, let T be a finite set of height-two elements of U and let S be a finite set of height-one elements $\{u_n | n \in \mathbb{N}\}$.

By Proposition 1.6 U satisfies *(rad(∞ princ))*: the set $\Gamma(S,T) = \{w \in U |$ height$(w) = 1$, and $\mathrm{G}(w) \cap \mathrm{G}(s) \subseteq T \subseteq \mathrm{G}(w), \forall s \in S\}$ is infinite. Thus $\{u_n | n \in \mathbb{N}\}$ is an exceptional sequence.

8.3 PROPOSITION. *If K is an algebraically closed countable field, there exists an orthogonal pair of special sequences in $U = \mathrm{Spec}(K[x,y])$.*

PROOF. Enumerate $K = \{\alpha_i\}_{i=1}^{\infty}$. For each i, let $u_i = x - \alpha_i$, $v_i = y - \alpha_i$. Obviously $\{u_i\}_{i \in \mathbb{N}}$ and $\{v_i\}_{i \in \mathbb{N}}$ are orthogonal special sequences in $\mathrm{Spec}(K[X,Y])$.

8.4 THEOREM. *Let U be standard partially ordered set satisfying (infinite covers) and (rad(princ)) of Definition 1.4. Then U satisfies the orthogonal pair of special sequences condition.*

NOTE. We already know that the orthogonal pair of special sequences condition holds whenever all the axioms of Definition 1.4 hold, since it holds for $\mathrm{Spec}(k[x,y])$, and the axioms of Definition 1.4 characterize $\mathrm{Spec}(k[x,y])$, where k is the algebraic closure of a finite field. We show the orthogonal pair of sequences property is implied by *(infinite covers)* and *(rad(princ))* directly.

PROOF. First we enumerate the maximal elements: $\mathcal{M}(U) = \{t_1, t_2, \dots\}$. Each of these has height two by *(infinite covers)*. The proof will be complete once we have proved that there are sequences $\{u_i\}_{i \in \mathbb{N}}$ and $\{v_i\}_{i \in \mathbb{N}}$ of height-one elements of U satisfying Statement $(*_n)$ for each $n \geq 1$:

STATEMENT $(*_n)$. *For every n, there exist $u_1, u_2, \dots, u_n$ and $v_1, v_2, \dots, v_n$ such that*

(i) For every pair i, j, $1 \leq i, j \leq n$, $G(u_i) \cap G(v_j)$ is a singleton $\{t_{i,j}\}$, for some $t_{i,j}$.

(ii) For every $i \neq j$, $G(u_i) \cap G(u_j) = \emptyset$; $G(v_i) \cap G(v_j) = \emptyset$.

(iii) $\{t_1, \dots t_n\} \subseteq \{t_{i,j} | 1 \leq i, j \leq n\}$.

First we show $(*_n)$ for $n = 1$: Since t_1 has height two, there exists a u_1 of height one, with $u_1 < t$. Choose $v_1 = w$ from the radical of a principal axiom corresponding to $S = \{u_1\}$, $T = \{t_1\}$. Then $\mathrm{G}(u_1) \cap \mathrm{G}(v_1) = \{t_1\}$.

Now suppose $n \geq 1$ and Statement $(*_n)$ holds; that is, we have $u_1, u_2, ..., u_n$ and $v_1, v_2, ..., v_n$ satisfying (i), (ii) and (iii). We produce u_{n+1} and v_{n+1} so that the two extended sequences satisfy (i), (ii) and (iii), yielding Statement $(*_{n+1})$.

Let t be the first element of $\mathcal{M}(U) - \{t_{i,j} | 1 \leq i, j \leq n\}$.

Case 1: There exists an i so that $u_i < t$. Then $t \neq t_{i,j}, \forall i, j$, so $v_j \not< t, \forall j$. Choose, $\forall r \neq i$, $1 \leq r \leq n$, an element $x_r \in \mathcal{M}(U) - \{t_{i,j}\}$ such that $u_r < x_r$, using the *(infinite covers)* property. Take $x_i = t$. Again, $v_j \not< x_r, \forall j$. Let $S = \{u_1, v_1, \ldots, u_n, v_n\}$, $T = \{x_1, \ldots, x_n\}$ $(x_i = t)$. Choose v_{n+1} using the radical of a principal axiom, so that $\mathrm{G}(v_{n+1}) \cap \mathrm{G}(s) \subseteq T \subseteq \mathrm{G}(v_{n+1})$, $\forall s \in S$.

Now choose $\forall m$, $1 \leq m \leq n+1$, $y_m \in \mathcal{M}(U) - \{t_{i,j}\}$ such that $v_m < y_m$, using the (infinite covers) property. Let $S = \{u_1, v_1, \ldots, u_n, v_n, v_{n+1}\}$, $T = \{y_1, \ldots, y_n, y_{n+1}\}$. Choose u_{n+1} using the radical of a principal axiom, so that $\mathrm{G}(u_{n+1}) \cap \mathrm{G}(s) \subseteq T \subseteq \mathrm{G}(u_{n+1}) \, \forall s \in S$. Then the extended sequences work.

Case 2: If $v_i < t$, for some i, use the symmetric argument.

Case 3: For every i, $u_i \not< t$ and $v_i \not< t$. Choose, $\forall m$, $1 \leq m \leq n+1$, $y_m \in \mathcal{M}(U) - \{t_{i,j}\}$ such that $v_m < y_m$, using the *(infinite covers)* property. Let $S = \{u_1, v_1, \ldots, u_n, v_n\}$, $T = \{y_1, \ldots, y_n, t\}$. Choose u_{n+1} using the radical of a principal axiom, so that $\mathrm{G}(u_{n+1}) \cap \mathrm{G}(s) \subseteq T \subseteq \mathrm{G}(u_{n+1}), \forall s \in S$. Find v_{n+1} similarly. Then $t \in \mathrm{G}(u_{n+1})$, $t \notin \mathrm{G}(u_i)$, $\forall i \leq n$.

8.5 REMARK. *By Proposition 7.2 and Remark 7.5 we know that if K is a countable field (possibly finite) and $f(X, Y)$ is an irreducible element of $K[X, Y]$, then there exists a special sequence in $U = \mathrm{Spec}(K[X, Y])$ beginning with $f(X, Y)K[X, Y]$. If K is the algebraic closure of a finite field, then* $\mathrm{Spec}(K[X, Y])$ *is isomorphic to* $\mathrm{Spec}(\mathbb{Z}[Y])$ *by* [**rW2**] *and hence, following Proposition 8.2, there is an exceptional sequence starting at $f(X, Y)K[X, Y]$. However, if K is $\mathbb{Q}$ or $\bar{\mathbb{Q}}$, the algebraic closure of $\mathbb{Q}$, the only exceptional sequences we know are given by the prime ideals of $K[X]$ and their images under transformations of the variables.*

8.6 PROPOSITION. *Suppose $U \cong$* $\mathrm{Spec}(R[X])$*, where R is a one-dimensional Noetherian domain with countably infinitely many maximal ideals. Then there exists a countably infinite exceptional sequence in U.*

PROOF. Let $\{\mathbf{m}_i\}_{i \in \mathbb{N}}$ be the set of maximal ideals of R. Take $u_i = \mathbf{m}_i[X]$. We show the sequence $\{u_i\}_{i \in \mathbb{N}}$ is exceptional in $\mathrm{Spec}(R[X])$. Clearly it is a special sequence.

Suppose that $S \subseteq \{\mathbf{m}_1[X], \ldots, \mathbf{m}_n[X]\}$ in $\mathrm{Spec}(R[X])$. Let $T = \{t_1, \ldots, t_r\}$ be a finite set of height-two maximal elements; say that $t_j = (\mathbf{m}_{j(i)}, g_j(X))$, where $g_j(X)$ is a polynomial in $R[X]$ which is irreducible mod $\mathbf{m}_{j(i)}$, since $(R/\mathbf{m}_{j(i)})[X]$ is a principal ideal domain. For convenience we suppose that the $\mathbf{m}_{j(i)}$ are among the $\mathbf{m}_1, \ldots, \mathbf{m}_n$. Let R' be R localized outside the union of the $\mathbf{m}_1, \ldots, \mathbf{m}_n$. Then R' is a semilocal domain with exactly n maximal ideals and so $\mathrm{Spec}(R'[X])$ satisfies the axioms in Definition 1.3 [**HW1**], in particular, *(infinite inv. j-sets)*. Thus in $\mathrm{Spec}(R'[X])$, there are infinitely many height-one prime ideals Q contained in only the maximal ideals $(\mathbf{m}_{j(i)}, g_j(X))R'[X]$, where $1 \leq j \leq r$. These Q correspond to height-one prime ideals P of $R[X]$ which are contained in the t_j, $1 \leq j \leq r$; that is $T \subseteq \mathrm{G}(P)$, for each P. Now let $s \in S$, that is, $s = \mathbf{m}_i[X]$, for some $1 \leq i \leq n$. Then $t \in \mathrm{G}(s) \cap \mathrm{G}(P)$ implies that $t' = tR'[X]$ is a maximal ideal of R' such that $Q = PR'$ is contained in t'. Since Q is only contained in maximal ideals of R' of form $(\mathbf{m}_{j(i)}, g_j(X))R'[X]$, then $t \in T$. Thus $\mathrm{G}(s) \cap \mathrm{G}(P) \subseteq T$, for all $s \in S$.

8.7 Remark. *If R is a countable one-dimensional domain, then* $\mathrm{Spec}(R[x])$ *is a direct limit of $C\mathbb{Z}(n)P$'s, unless R is a Henselian local domain.*

Proof. If R has only finitely many maximal ideals, but is not Henselian, then by [**HW1**], $\mathrm{Spec}(R[x])$ is $C\mathbb{Z}(n)P$, so it is a direct limit. If R is Henselian, then by [**HW1**], $\mathrm{Spec}(R[x])$ is not $C\mathbb{Z}(n)P$; the last axiom *(infinite inv. j-sets)* of Definition 1.3 is not satisfied. If R has infinitely many maximal ideals, then $R[x]$ has a countably infinite exceptional sequence by 8.6 and the obvious localizations are $C\mathbb{Z}(n)P$, so again it is a direct limit.

9. Relationships between axioms.

9.1 Proposition. *Let U be a standard partially ordered set satisfying axiom (infinite covers) from Definition 1.4. If U satisfies (rad(princ)) then U also satisfies axiom (N-infinite less sets) of Definition 1.14.*

Proof. Let t be a height-two element of U. If $L(\{t\}) = \{u \in U | u < t\}$ is infinite, we are done. If not, then *(rad(princ))* implies that the set $\Gamma(L(\{t\}), \{t\}) \neq \emptyset$. Now $w \in \Gamma(L(\{t\}), \{t\})$ yields that, for every $s \in L(\{t\}), G(w) \cap G(s) \subseteq \{t\}$. However $w \notin L(\{t\})$ since $G(w) \cap G(w)$ is infinite by *(infinite covers)*.

In what follows, let U be a standard partially ordered set satisfying *(infinite covers)*, and *(finite mubs)* of Definition 1.4. Consider the condition $(*)$ below as well as the other conditions defined earlier.

$(*)$ There exists a sequence of exceptional elements *and* every special sequence is exceptional.

By 8.2, *(rad(princ))* of Definition 1.4 $\Longrightarrow$ $(*)$.

Certainly $(*)$ $\Longrightarrow$ There exists a sequence of exceptional elements (Definition 1.10).

There exists a sequence of exceptional elements $\Longleftrightarrow$ U is a *nested* limit of $C\mathbb{Z}(n)P$s from Definition 1.11, by 8.1.

U is a nested limit of $C\mathbb{Z}(n)P$s $\Longrightarrow$ There exists a special sequence (Definitions 1.11 and 1.9).

(rad(princ)) $\Longrightarrow$ *(N-infinite less sets)* of Definition 1.14, by Proposition 9.1.

(N-infinite less sets) $\Longrightarrow$ *(N-∞ ht.1s)* of Definition 1.14.

(rad(princ)) $\Longrightarrow$ *(∞ smaller comaximals)* $\Longrightarrow$ *(smaller comaximals)* of Definition 1.5 by Proposition 1.6.

By Proposition 7.1, *(N-∞ ht.1s)* and *(smaller comaximals)* together $\Longrightarrow$ *(∞-special)* of Definition 1.9.

Pictorially, the implications look like this:

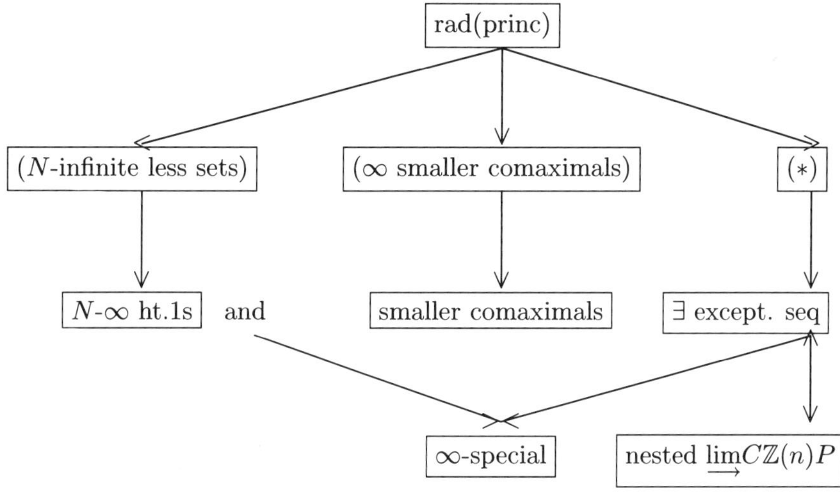

QUESTIONS.

(a) If R is a two-dimensional Noetherian domain such that $U = \mathrm{Spec}(R)$ is standard and satisfies *(infinite covers)* and *(finite mubs)* of Definition =1.4, then does U satisfy *(∞ smaller comaximals)* of Definition 1.5?

(b) Are there examples showing the converses in the diagram are not true?

(c) Note that $\mathrm{Spec}(\mathbb{Q}[X,Y])$ and $\mathrm{Spec}(\bar{\mathbb{Q}}[X,Y])$ do not satisfy condition $(*)$; that is, not every special sequence in $\mathrm{Spec}(\mathbb{Q}[X,Y])$ or $\mathrm{Spec}(\bar{\mathbb{Q}}[X,Y])$ is exceptional. By Remark 8.5, $\mathrm{Spec}(\mathbb{Q}[X,Y])$ has a special sequence beginning with $X^3 - Y^2$. Such a sequence could not be exceptional, because when $S = \{(X^3 - Y^2)\}$ and $T = \{(X-1, Y-1)\}$, then $\Gamma(S,T) = \emptyset$ [**rW2**]. What conditions would insure that a special sequence in $\mathrm{Spec}(\mathbb{Q}[X,Y])$ could be expanded to an exceptional sequence?

9.2 PROPOSITION. *Let U be a standard partially ordered set satisfying (infinite covers), (finite mubs), (N-∞ ht.1s) of Definition 1.14, (smaller comaximals) and suppose that every special sequence is exceptional. Then for every height-one element $u \in U$ and every finite set T of height-two elements, this set is infinite:*

$$\Gamma(\{u\}, T) = \{w \in U \mid \ \textit{height}(w) = 1, \ \textit{and} \ \mathrm{G}(w) \cap \mathrm{G}(u) \subseteq T \subseteq \mathrm{G}(w)\}.$$

(That is, (rad(princ)) of Definition 1.4 holds for $S = \{u\}$.)

PROOF. Since U satisfies *(N-∞ ht.1s)* and *(smaller comaximals)* by Proposition 7.1, the set $\{u\}$ can be extended to a special sequence $\{u_n \mid n \in \mathbb{N}\}$ with $u_1 = u$. By assumption the sequence $\{u_n \mid n \in \mathbb{N}\}$ is exceptional and the assertion follows.

9.3 PROPOSITION. *Let R be a two-dimensional Noetherian domain. Then $U = \mathrm{Spec}(R)$ satisfies (∞ smaller comaximals) of Definition 1.5.*

PROOF. Let $u_1, \ldots, u_m$ be a set of height-one elements of U, and let t be a height-two element outside $\cup \mathrm{G}(u_i)$. Choose $\mu \in u_1 \cap \cdots \cap u_m$ such that the element $\rho = 1 - \mu \in t$. We construct $\rho_i = \rho + \tau_i \mu \in t$, for all $i \in \mathbb{N}$ such that the ρ_i have no common prime divisor.

Let $\rho_1 = \rho$. Pick $\tau_2 \in t$ and outside every minimal prime divisor of ρ_1. Set $\rho_2 = \rho_1 + \tau_2 \mu$.

Now suppose that $\rho_1, \ldots, \rho_n$ have been constructed such that $\rho_i = \rho + \tau_i \mu \in t$, *and* the minimal prime divisors of the ρ_i are mutually distinct.

Let $\Gamma = \{g \mid g \text{ is a minimal prime divisor of } \rho_i, \text{ for some } 1 \leq i \leq n\}$.

Then Γ is a finite set. Pick $\sigma_{n+1} \in t - \bigcup_{g \in \Gamma} g$.

NOTE. $\mu \notin g$, for every $g \in \Gamma$, since otherwise $1 \in g$.

CASE 1. $\tau_i - \sigma_{n+1} \notin \bigcup_{g \in \Gamma} g$.

In this case, set $\tau_{n+1} = \sigma_{n+1}$.

CASE 2. $\tau_i - \sigma_{n+1} \in \bigcup_{g \in \Gamma} g$.

In this case, write $\Gamma = \Delta_1 \cup \Delta_2$, where

$\Delta_1 = \{g \in \Gamma \mid \tau_i - \sigma_{n+1} \in g\}; \Delta_2 = \{g \in \Gamma \mid \tau_i - \sigma_{n+1} \notin g\} = \Gamma - \Delta_1$.

If $\Delta_2 \neq \emptyset$, pick $\beta \in t \cap \bigcap_{g \in \Delta_2} g - \bigcap_{g \in \Delta_1} g$.

If $\Delta_2 = \emptyset$, pick $\beta \in t - \bigcap_{g \in \Delta_1} g$.

Put $\tau_{n+1} = \sigma_{n+1} + \beta$. Then $\tau_i - \tau_{n+1} \notin \bigcup_{g \in \Gamma} g$, for all $1 \leq i \leq n$.

Now define $\rho_{n+1} = \rho + \tau_{n+1} \mu$.

CLAIM. The minimal prime divisors of $\rho_1, \ldots, \rho_{n+1}$ are mutually distinct.

PROOF OF CLAIM. Suppose for some i with $1 \leq i \leq n$, ρ_i and ρ_{n+1} have a common prime divisor g. Then $g \in \Gamma$, which implies that $(\tau_i - \tau_{n+1})\mu = \rho_i - \rho_{n+1} \in g$, a contradiction.

Now every ρ_i is an element of t. Suppose $v_i \subseteq t$ is a height-one prime with $\rho_i \in v_i$. Then $t \in \mathrm{G}(v_i)$ and $\mathrm{G}(u_j) \cap \mathrm{G}(v_i) = \emptyset$, for every $1 \leq j \leq n$.

Moreover $v_i \neq v_j$ for $i \neq j$. This shows that $U = \mathrm{Spec}(R)$ satisfies ((∞ smaller comaximals).

10. More observations and questions.

Let U be a standard partially ordered set.

QUESTION 1. *When do exceptional sequences exist (Definition 1.10)?*

Certainly *(infinite covers), (finite mubs), (rad(princ))* imply the existence of an exceptional sequence, because they imply the partially ordered set is order-isomorphic to $\mathrm{Spec}(\mathbb{Z}[Y])$, which has such sequences. Does a weaker axiom together with *(infinite covers)*, and *(finite mubs)* imply it–say *(smaller comaximals)*, or possibly *(smaller comaximals)* with the orthogonal pair of special sequences condition or some other condition, or possibly *(rad(princ) for singles)* with some other condition?

For arbitrary partially ordered sets, what conditions would insure that any element could be expanded to an exceptional sequence?

QUESTION 2. *Does the orthogonal pair of special sequences condition (Definition 1.13) hold for* Spec($\mathbb{Q}[X,Y]$)*? Does some generalization of the (rad(princ)) axiom hold in* Spec($\mathbb{Q}[X,Y]$)*?*

Here is an idea (incomplete) for showing that the orthogonal pair of special sequences condition does not hold: If it did, then the generators for the elements of $\{u_i | i \in \mathbb{N}\}$ or $\{v_j | j \in \mathbb{N}\}$ would have arbitrarily high degree in y. Use an extension of the observation that (Y^2+1, X^2+1) is contained in two maximal ideals rather than just one.

QUESTION 3. *If U satisfies (infinite covers)* and *(finite mubs)* of Definition 1.4, and *(smaller comaximals)* of Definition 1.5, then does *((∞ smaller comaximals)* of Definition 1.5 hold?

That is, if there exists one element v, then do there exist infinitely many?

QUESTION 4. *If $U = \varinjlim C\mathbb{Z}(n)P$ and (L-mub-constant) of Definition 1.8 holds, then is $U \cong$* Spec(R)*, for R a Noetherian domain?*

What about adding *(L-more covers)*?

QUESTION 5. *If U satisfies (infinite covers) and (finite mubs) of Definition 1.4 and (infinite less sets for singletons), then is U a Noetherian spectrum?*

Note that Krull's Principal Ideal Theorem implies that *(infinite less set for singletons)* must hold for the prime spectrum of a Noetherian ring.

QUESTION 6. *If $U = \varinjlim C\mathbb{Z}(n)P$, $U \cong$* Spec(R)*, and (L-more covers) holds in U, then is every prime ideal of R an intersection of maximal ideals?*

QUESTION 7. *For which two-dimensional Noetherian rings, does* Spec(R) *satisfy the pair of elements condition (Definition 1.12) ?*

For example, when R is a unique factorization domain, this would say that each maximal ideal is two-generated up to radical.

QUESTION 8. *When the pair of elements condition is satisfied, can the two elements u and/or v be part of a prescribed set?*

QUESTION 9. *Is* Spec($\mathbb{Z}[Y]$) *embedded in U, for every partially ordered set U which is a $\varinjlim C\mathbb{Z}(n)P$ and which satisfies (L-more covers) of Definition 1.8?*

QUESTION 10. *Describe the different homeomorphic spaces which can occur as a limit of U_n via τ_n. Which of these spaces are the prime spectra of rings, respectively Noetherian rings? What conditions would insure that the representation of a partially ordered set U as a limit of U_n's is unique?*

REFERENCES

[HW1] W. Heinzer and S. Wiegand, *Prime ideals in two-dimensional polynomial rings*, Proc. Amer. Math. Soc. **107(3)** (1989), 577-586.

[HW2] W. Heinzer and S. Wiegand, *Prime ideals in polynomial rings over one-dimensional domains*, preprint.

[HLW1] W. Heinzer, D. Lantz, and S. Wiegand, *Projective lines over one-dimensional semilocal domains and spectra of birational extensions*, Algebraic Geometry and Applications (C. Bajaj, ed.), Collection of papers from Abhyankar's 60th Birthday conference, Springer Verlag, New York (to appear).

[HLW2] W. Heinzer, D. Lantz, and S. Wiegand, *Prime ideals in birational extensions of polynomial rings*, Contemporary Mathematics (to appear).

[H] R. Heitmann, *Prime ideal posets in Noetherian rings,*, Rocky Mountain J. Math. **7(4)** (1977), 667.

[M] H. Matsumura, *Commutative Ring Theory*, Cambridge Studies in Advanced Mathematics **8**, Cambridge University Press, 1989.

[Mc] S. McAdam, *A noetherian example*, Communications in Algebra **4** (1976), 245-247.

[N] M. Nagata, *On the chain problem of prime ideals*, Nagoya Math. J. **10** (1956), 51-64.

[rW1] R. Wiegand, *Homeomorphisms of affine surfaces over a finite field*, J. London Math. Soc.(2) **18** (1978), 28-32.

[rW2] R. Wiegand, *The prime spectrum of a two-dimensional affine domain*, J. Pure Appl. Alg. **40** (1986), 209–214.

[sW] S. Wiegand, *Intersections of prime ideals in Noetherian rings*, Communications in Algebra **11(16)** (1983), 1853-1876.

MICHIGAN STATE UNIVERSITY, EAST LANSING, MI 48824

E-mail address: rotthaus@ mth.msu.edu

UNIVERSITY OF NEBRASKA LINCOLN, LINCOLN, NE 68588-0323

E-mail address: swiegand@ unlinfo.unl.edu

Contemporary Mathematics
Volume **171**, 1994

When is an Abelian p–group determined by the Jacobson radical of its endomorphism ring?

PHILLIP SCHULTZ

Dedicated to the memory of Dick Pierce

ABSTRACT. Let G be an abelian p–group. Then G is determined by the Jacobson radical $\mathbf{J}$ of its endomorphism ring unless G is the direct sum of a divisible group of infinite rank and an elementary group of much smaller rank. This is proved by establishing some Galois connections among the lattices of the fully invariant subgroups of G and the annihilator ideals of $\mathbf{J}$.

§1. Introduction

In [**HPS**] it was shown that if G and G' are abelian p–groups which have unbounded basic subgroups and whose endomorphism rings have Jacobson radicals $\mathbf{J}$ and $\mathbf{J}'$ with torsion ideals $\mathbf{t}(\mathbf{J})$ and $\mathbf{t}(\mathbf{J}')$ respectively, then $\mathbf{t}(\mathbf{J}) \cong \mathbf{t}(\mathbf{J}')$ if and only if $G \cong G'$. Examples were given to show that if $G = B \oplus D$ where B is elementary and D divisible, then G may not be determined by $\mathbf{t}(\mathbf{J})$ or even by $\mathbf{J}$. The purpose of this paper is to show just when G is determined by $\mathbf{J}$. It turns out that the counter–examples given in [**HPS**] are essentially the only ones. That is, G is determined by $\mathbf{J}$ unless $G = B \oplus D$ where B is elementary of rank $r > 0$ and D is divisible of infinite rank $d > 2^r$. In fact, if G is not of this form but $G = B \oplus D$ with B bounded and D divisible, a stronger property holds, namely the structure of G can be recovered from that of $\mathbf{J}$. On the other hand, the results of [**HPS**] show that if G has an unbounded basic subgroup, then any isomorphism from $\mathbf{J}$ to $\mathbf{J}'$ is induced by an isomorphism from G to G'.

1991 *Mathematics Subject Classification.* 20K30.
This paper is in final form and no version of it will be submitted for publication elsewhere

Hausen and Johnson [**HJ**] have shown that the latter property fails for non–zero bounded groups with the exception of a few explicit finite 2–groups.

The main tool in the proof is an ideal theoretic representation in $\mathbf{J}$ of the fully invariant subgroup lattice of G. In order to avoid set–theoretic complications, I assume the Generalised Continuum Hypothesis (GCH).

I wish to acknowledge helpful remarks from Professors Jutta Hausen and Cheryl E Praeger.

§2. Preliminary reductions

The following results are all well–known, and can be found for example in [**F1**] or [**F2**]. Let G be an abelian p–group with a bounded basic subgroup, so G may be decomposed as $G = B \oplus D$, where:

(1) D is a divisible group of rank r_d;
(2) If $B \neq 0$, then $B = \oplus_{j\in[1,k]} B_j$, where k is a positive integer and for all $j \in [1, k]$, B_j is homocyclic of exponent m_j and rank $r_j > 0$, and $1 \leq m_1 < m_2 < \cdots < m_k$.

In order to recover G from $\mathbf{t}(\mathbf{J})$ or $\mathbf{J}$, it suffices to recover the parameters k, r_d, m_j and r_j for all $j \in [1, k]$. In this section, I show that k, r_d and m_j for all $j \in [1, k]$ are determined by $\mathbf{J}$. I also dispose of the cases in which G is divisible or homocyclic or $G = B \oplus D$ with B elementary.

LEMMA 2.1. *Let G be a non–zero abelian p–group. Then $\mathbf{J} = 0$ if and only if G is elementary.*

PROOF. Let I be the ideal of $\mathcal{E}(G)$ generated by multiplication by p. Since multiplication by p is central in $\mathcal{E}(G)$, $I = p\mathcal{E}(G)$. For any $p\eta \in I$, let $(p\eta)' = \sum_{i=1}^{\infty}(-1)^i p^i \eta^i \in I$. This expression makes sense, for if $a \in G$ has order p^j, then $a(p\eta)' = \sum_{i=1}^{j-1}(-1)^i p^i a\eta^i$. Furthermore, $(p\eta)'$ lies in $p\mathcal{E}(G) = I$. Now it is routine to check that $(p\eta)'$ is the quasi–inverse of $p\eta$, so $I \subseteq \mathbf{J}$. Hence $\mathbf{J} = 0$ if and only if $p\mathcal{E}(G) = 0$ which in turn is true if and only if $pG = 0$.

LEMMA 2.2. *Let G be an abelian p–group. Then $\mathbf{J} \neq 0$ but $\mathbf{t}(\mathbf{J}) = 0$ if and only if G is non–zero divisible. In the latter case, the rank of G is determined by $\mathbf{J}$.*

PROOF. Suppose G is divisible of rank $r > 0$. By Lemma 2.1, $\mathbf{J} \neq 0$ and by Chapter 43 of [**F1**], $\mathcal{E}(G)$ is a torsion–free p–adic module, so $\mathbf{t}(\mathbf{J}) = 0$.

Conversely, let G be an abelian p–group such that $\mathbf{t}(\mathbf{J}) = 0$ but $\mathbf{J} \neq 0$. Then $0 \neq G = R \oplus D$, where R is reduced and D is divisible. Suppose $R \neq 0$. If $D = 0$, then either R is elementary, in which case $\mathbf{J} = 0$, or R contains a cyclic summand of order $\geq p^2$, in which case multiplication by p restricted to this summand is a non–zero element of finite order in $\mathbf{J}$. In either case we have a contradiction, so $D \neq 0$. But then the injection of any cyclic summand of R into D is a non–zero element of $\mathbf{J}$ of finite order, another contradiction. Hence $R = 0$.

It remains to show that if G is divisible, then $\mathbf{J}$ determines the rank r of G. Write $\mathcal{E}$ for $\mathcal{E}(G)$, and note that $\mathbf{J} = p\mathcal{E} = \mathrm{Ann}(G[p])$, the annihilator in $\mathcal{E}$ of $G[p]$. Since J is torsion-free, $\mathbf{J}/p\mathbf{J} = p\mathcal{E}/p^2\mathcal{E} \cong \mathcal{E}/p\mathcal{E} \cong \mathcal{E}(G[p])$. Hence $\mathbf{J}/p\mathbf{J}$ is finite if and only if $\mathbf{J}/p\mathbf{J}$ is the zero ring on $\oplus_{r^2}\mathbf{Z}(p)$ and $\mathbf{J}/p\mathbf{J}$ is infinite if and only if it is the zero ring on $\oplus_{2^r}\mathbf{Z}(p)$.

By the GCH, r is determined by $\mathbf{J}$.

LEMMA 2.3. *Let G be a homocyclic p-group of exponent $m > 1$ and rank r. Then $\mathbf{t}(\mathbf{J}) = \mathbf{J}$ and $\mathbf{J}$ determines m and r.*

PROOF. By [**P, Theorem 13.6**], $\mathbf{J} = \mathbf{t}(\mathbf{J}) = p\mathcal{E}(G)$. Hence as additive groups, $\mathbf{J} \cong p\mathrm{Hom}(\oplus_r\mathbf{Z}(p^m), \oplus_r\mathbf{Z}(p^m)) \cong \prod_r \oplus_r\mathbf{Z}(p^{m-1})$. If r is finite, then $\mathbf{J} \cong \oplus_{r^2}\mathbf{Z}(p^{m-1})$ while if r is infinite, then $\mathbf{J} \cong \oplus_{2^r}\mathbf{Z}(p^{m-1})$. In either case, m and r are determined by $\mathbf{J}$.

LEMMA 2.4. *Let G be an abelian p-group with non-zero divisible subgroup D. Let $\mathbf{K}$ be the Jacobson radical of $\mathcal{E}(D)$. Then $G = B \oplus D$ where B is a non-zero bounded p-group if and only if $\mathbf{J} = \mathbf{t}(\mathbf{J}) \oplus \mathbf{K}$, where $\mathbf{t}(\mathbf{J})$ is a bounded ideal and $\mathbf{K}$ is a maximal torsion-free additive direct summand of $\mathbf{J}$.*

PROOF. Suppose $G = B \oplus D$ with $0 \neq B$ bounded. Since D is a fully invariant subgroup of G, $\mathcal{E}(G) \cong \mathrm{Hom}(B, B \oplus D) \oplus \mathcal{E}(D)$, where the first term is bounded and the second a torsion-free ideal which is maximal as a torsion-free additive subgroup of $\mathcal{E}(G)$. Let π be the corresponding projection of $\mathcal{E}(G)$ onto $\mathcal{E}(D)$ restricted to $\mathbf{J}$. Then π maps $\mathbf{J}$ onto $\mathbf{K} \subseteq \mathcal{E}(D)$ and has $\mathbf{t}(\mathbf{J})$ in its kernel. On the other hand, if η is in the kernel of π, then $\eta|D = 0$, so η has finite order and hence lies in $\mathbf{t}(\mathbf{J})$. Thus the kernel is precisely $\mathbf{t}(\mathbf{J})$. The mapping π splits, since $\mathbf{t}(\mathbf{J})$ is a bounded pure subgroup of $\mathbf{J}$, and for the same reason, $\mathbf{K}$ is maximal as a torsion-free subgroup of $\mathbf{J}$.

Conversely, if G has an unbounded reduced summand B, then G contains cyclic summands of arbitrarily large order. Let $G = \langle a \rangle \oplus A$ where $\langle a \rangle$ is a summand of order $\geq p^n$ for some $n \in \mathbf{N}$. Then the endomorphism which is multiplication by p on $\langle a \rangle$ and 0 on A is an element of $\mathbf{t}(\mathbf{J})$ of order $\geq p^{n-1}$. Thus $\mathbf{t}(\mathbf{J})$ is unbounded.

THEOREM 2.5. *Let $G = B \oplus D$ where B is elementary of rank $r_1 > 0$ and D is divisible of rank $r_d > 0$. Then r_d is determined by $\mathbf{J}$ and r_1 is determined by $\mathbf{t}(\mathbf{J})$ together with r_d if and only if r_d is finite or $2^{r_1} > r_d$.*

PROOF. By Lemma 2.4, $\mathbf{J}$ has a maximal torsion-free direct summand $\mathbf{K}$ which is $\mathbf{J}(\mathcal{E}(D))$ and hence determines r_d by Lemma 2.2. Now every non-zero element η of $\mathbf{t}(\mathbf{J})$ has order p. By [**P, Theorem 13.6**] η raises heights and so maps B into D and annihilates D. Hence $\mathbf{t}(\mathbf{J})$ is the zero ring on the additive elementary group $\mathrm{Hom}(B, D)$ This group has rank $r_1 r_d$ if r_1 is finite and $2^{r_1} r_d$ otherwise. Hence if r_d is known, so is r_1 unless r_d is infinite and no less than 2^{r_1}.

In view of Lemmas 2.1 to 2.4 and Theorem 2.5, I shall assume in the rest of this paper that $G = \oplus_{j\in[1,k]} B_j \oplus D$ where D is divisible and each B_j is homocyclic of exponent m_j and rank r_j, $1 \leq m_1 < m_2 < \cdots < m_k$ and $k > 1$ or $m_1 > 1$. I show next that the parameters k and m_j are determined by $\mathbf{t}(\mathbf{J})$, using the digraph $D(G)$ of G constructed in [**PS**]. This construction was for bounded groups only, so I first deal with that case and then describe the modifications necessary to deal with the non–bounded case.

The vertices of $D(G)$ are the sections of G of the form $B_r[p^s]/B_r[p^{s-1}]$. There is a path in $D(G)$ from $B_r[p^s]/B_r[p^{s-1}]$ to $B_{r'}[p^{s'}]/B_{r'}[p^{s'-1}]$ if and only if there is an endomorphism in $\mathbf{t}(\mathbf{J})$ inducing a mapping of $B_r[p^s]/B_r[p^{s-1}]$ into $B_{r'}[p^{s'}]/B_{r'}[p^{s'-1}]$. An endomorphism $\eta \in \mathbf{t}(\mathbf{J})$ is called elementary if for some cyclic summands $\langle a\rangle$ and $\langle b\rangle$, $G = \langle a\rangle \oplus A = \langle b\rangle \oplus B$, $\eta|_A = 0$ and $G\eta \subseteq \langle b\rangle$. An elementary endomorphism is called irreducible if it is not the product of two elementary endomorphisms and l–elementary if it is the product of l but not more than l irreducible elementaries. It was shown in [**PS, Theorem 2.9**] that there is a path of length l in the digraph $D(G)$ between vertices $B_r[p^s]/B_r[p^{s-1}]$ and $B_{r'}[p^{s'}]/B_{r'}[p^{s'-1}]$ if and only if there exists an l–elementary endomorphism which induces a mapping of $B_r[p^s]/B_r[p^{s-1}]$ into $B_{r'}[p^{s'}]/B_{r'}[p^{s'-1}]$. Furthermore, the proof of [**PS, Corollary 2.10**] shows that there exists $\eta \in \mathbf{t}(\mathbf{J})^l \setminus \mathbf{t}(\mathbf{J})^{l+1}$ inducing a mapping of an element of $B_r[p^s]/B_r[p^{s-1}]$ to an element of $B_{r'}[p^{s'}]/B_{r'}[p^{s'-1}]$ if and only if there is an l–elementary endomorphism with the same property.

The construction of $D(G)$ can be extended to groups G for which $D \neq 0$ by adjoining for each $s \in \mathbf{N}$ new vertices $D[p^{s+1}]/D[p^s]$, vertical arrows from $D[p^{s+1}]/D[p^s]$ to $D[p^s]/D[p^{s-1}]$, and for all $i \in [1, k-1]$, horizontal arrows from $B_k[p^{i+1}]/B_k[p^i]$ to $D[p^{i+1}]/D[p^i]$. Since elements of $\mathbf{t}(\mathbf{J})$ annihilate D and increase socle heights, the correspondence between paths in $D(G)$ and elements of $\mathbf{t}(\mathbf{J})$ described above continues to hold for the modified construction.

The next theorem shows how the parameters k and m_j, $j \in [1, k]$, can be determined from $\mathbf{t}(\mathbf{J})$.

THEOREM 2.6. *Let $G = \oplus_{j\in[1,k]} B_j \oplus D$ where B_j is homocyclic of exponent m_j and D is divisible.*

(1) $k = \begin{cases} \max\{n : \mathbf{t}(\mathbf{J})[p].\mathbf{t}(\mathbf{J})^{n-1} \neq 0\} & \text{if } D=0, \\ \max\{n : \mathbf{t}(\mathbf{J})[p].\mathbf{t}(\mathbf{J})^{n} \neq 0\} & \text{otherwise.} \end{cases}$

(2) *For all $j \in [1, k-1]$,*

$$m_j = \begin{cases} \max\{\exp(\eta) : \eta = \eta_1\eta_2\ldots\eta_{k-j} \in \mathbf{t}(\mathbf{J})^{k-j} \text{ with} \\ \quad \exp(\eta) = \exp(\eta_1) = \exp(\eta_2) = \cdots = \exp(\eta_{k-j}\} & \text{if } D=0, \\ \max\{\exp(\eta) : \eta = \eta_1\eta_2\ldots\eta_{k-j-1} \in \mathbf{t}(\mathbf{J})^{k-j+1} \text{ with} \\ \quad \exp(\eta) = \exp(\eta_1) = \exp(\eta_2) = \cdots = \exp(\eta_{k-j+1}\} & \text{otherwise.} \end{cases}$$

(3) $m_k = \begin{cases} \exp(\mathbf{t}(\mathbf{J})) + 1 & \text{if } D=0, \\ \exp(\mathbf{t}(\mathbf{J})) & \text{otherwise.} \end{cases}$

PROOF. (1) $k-1$ is the length of the unique path in $D(G)$ from $B_1[p]$ to $B_k[p]$ if $D=0$, and k is the length of the unique path from $B_1[p]$ to $D[p]$ otherwise. By the remarks above, these lengths are realised by the endomorphisms described.

(2) If $D=0$, the expression on the right hand side defines the largest exponent of a product of $(k-j)$ elements of $\mathbf{t}(\mathbf{J})$, all of the same exponent. By the results of [**PS**], this is the maximum s for which there is a horizontal $(k-j-1)$–path in $D(G)$ passing through some $B_r[p^s]/B_r[p^{s-1}]$. There is a unique such path, going from $B_j[p^{m_j}]/B_j[p^{m_j-1}]$ to $B_k[p^{m_j}]/B_k[p^{m_j-1}]$, so this maximum exponent is m_j. The argument for $D \neq 0$ is similar.

(3) If $D=0$, the largest exponent in $\mathbf{t}(\mathbf{J})$ is achieved by an elementary endomorphism corresponding to the path connecting $B_k[p^{m_k}]/B_k[p^{m_k-1}]$ to $B_k[p^{m_k-1}]/B_k[p^{m_k-2}]$, so this exponent is m_k-1.

If $D \neq 0$, the largest exponent in $\mathbf{t}(\mathbf{J})$ is achieved by the elementary irreducible corresponding to the horizontal arrow connecting $B_k[p^{m_k}]/B_k[p^{m_k-1}]$ to $D[p^{m_k}]/D[p^{m_k-1}]$, so this exponent is m_k.

§3. The lattice of fully invariant subgroups

In order to find the ranks r_j it is not enough to consider only the digraph $D(G)$. In addition, we must use the lattice of fully invariant subgroups of G. Throughout this section, let $G=B\oplus D$ as described in §2. For any $j\in[1,k-1]$ the **j–th gap** $m_{j+1}-m_j$ is denoted by g_j, and, if $D\neq 0$, the **k–th gap** g_k is defined to be ∞. Let $\mathcal{F}=\mathcal{F}(G)$ denote the lattice of all fully invariant subgroups of G under the inclusion relation. Fuchs [**F2, Theorem 67.1**] describes a combinatorial characterization of $\mathcal{F}$ due to Kaplansky, but for this paper it is more convenient to use a modification of another description due to Moore and Hewett [**MH**]. Their characterization is for groups with no elements of infinite height, but it can be extended to non–reduced groups whose reduced part is bounded.

A **layer sequence** for G is a function $\ell:[1,k+1]\longrightarrow \mathbf{N}\cup\{\infty\}$ satisfying

(1) $0\le \ell(1)\le m_1$;

(2) for $j\in[2,k+1]$, $\ell(j-1)\le \ell(j)\le \ell(j-1)+g_{j-1}$,

where the last term $\ell(k+1)$ is simply omitted if $D=0$. As usual, we let $n+\infty=\infty$ for all $n\in\mathbf{N}$.

It is straightforward to check that the set $\mathcal{L}=\mathcal{L}(G)$ of layer sequences for G forms a complete distributive lattice under the pointwise operations. For $\ell\in\mathcal{L}$, let $G[\ell]=\oplus_{j\in[1,k]}B_j[p^{\ell(j)}]\oplus D[p^{\ell(k+1)}]$, where $D[p^\infty]$ means D itself.

THEOREM 3.1. *The mapping $\ell\longleftrightarrow G[\ell]$ is a lattice isomorphism of $\mathcal{L}$ onto $\mathcal{F}$.*

PROOF. I show first that for all $\ell\in\mathcal{L}$, $G[\ell]\in\mathcal{F}$. It is clear that the summand $D[p^{\ell(k+1)}]$ is fully invariant in G. Let $\langle a\rangle$ be a cyclic summand of maximal order in G, and let $\eta\in\mathcal{E}(G)$. Then $b=p^{m_k-\ell(k)}a\in G[\ell]$ and $b\eta$ has exponent $\le \ell(k)$ and height $\ge m_k-\ell(k)$. But every such element is in $G[\ell]$. Now let

$c = e + d \in G[\ell]$, where $e \in \oplus_{j\in[1,k]} B_j[p^{\ell(j)}]$ and $d \in D[p^{\ell(k+1)}]$. Then e has exponent $\leq \ell(k)$ and height $\geq m_k - \ell(k)$, so $e = b\lambda$ for some $\lambda \in \mathcal{E}(G)$. Hence $c\eta = b\lambda\eta + d\eta$. By the remarks above, both $b\lambda\eta$ and $d\eta$ are in $G[\ell]$.

Next, it is routine to verify that $\ell \leq \ell'$ in $\mathcal{L}$ implies $G[\ell] \leq G[\ell']$ in $\mathcal{F}$.

It remains to show that each $H \in \mathcal{F}$ has the form $H = G[\ell]$ for some $\ell \in \mathcal{L}$. Now $H = E \oplus F$, where $E = H \cap B$ and $F = H \cap D$. Since F is fully invariant in G, $F = D[p^{\ell(k+1)}]$ for some $\ell(k+1) \in \mathbf{N} \cup \{\infty\}$, so if $E = 0$ we are done.

Assume that $E \neq 0$, and for each $j \in [1,k]$, let $\ell(j)$ be the maximum exponent of an element of $B_j \cap E$. Since H is fully invariant in G, every element of B_j of exponent $\leq \ell(j)$ is in E, so for all $j \in [1,k]$, $B_j[p^{\ell(j)}] \subseteq E$ and hence $\oplus_{j\in[1,k]} B_j[p^{\ell(j)}] \subseteq E$. Conversely, let $0 \neq x \in E$, say $x = \sum_{j\in[1,k]} b_j$, with $b_j \in B_j$. Since H is fully invariant, each $b_j \in E$, so $b_j \in B_j[p^{\ell(j)}]$ and hence $x \in \oplus_{j\in[1,k]} B_j[p^{\ell(j)}]$.

Certainly $0 \leq \ell(1) \leq m_1$. Suppose it has been shown that $\ell(j-1) \leq \ell(j) \leq \ell(j-1) + g_{j-1}$ for some $j \in [2, k-1]$. Let $\langle a \rangle$ be a cyclic summand of G of exponent $\ell(j+1)$. Then $p^{m_{j+1}-\ell(j+1)}a \in E$, so every element of B of exponent $\leq \ell(j+1)$ and height $\geq m_{j+1} - \ell(j+1)$ is in E. Hence $\ell(j) \leq \ell(j+1) \leq \ell(j) + g_j$. Furthermore, every element of D of exponent $\leq \ell(k) \in H$, so $\ell(k) \leq \ell(k+1)$. We have constructed a layer sequence ℓ such that $H = G[\ell]$.

§4. The Galois connections

Throughout this section, let $G = B \oplus D$ as described in §2. I now establish two Galois connections between the lattices $\mathcal{F} = \mathcal{F}(G)$ of fully invariant subgroups of G and $\mathcal{I} = \mathcal{I}(G)$ of ideals of $\mathbf{t}(\mathbf{J})$.

To simplify the notation, we let for all $H \in \mathcal{F}$ and, for all $I \in \mathcal{I}$,

(1) $H' = \{\eta \in \mathbf{t}(\mathbf{J}) : H\eta = 0\}$,
(2) $H^* = \{\eta \in \mathbf{t}(\mathbf{J}) : G\eta \subseteq H\}$,
(3) $I' = \{g \in G : gI = 0\}$ and
(4) $I^* = GI$.

Note that H' and H^* are in $\mathcal{I}$ and I' and I^* are in $\mathcal{F}$.

In particular, it is trivial to verify that $G' = \mathbf{0}$, the zero ideal, $G^* = \mathbf{t}(\mathbf{J})$, $\mathbf{t}(J)^* = G[p^{m_k-1}]$ and

$$\mathbf{t}(\mathbf{J})' = \begin{cases} p^{m_k-1}B & \text{if } G \text{ is reduced} \\ D & \text{otherwise.} \end{cases}$$

I define H'' to mean $(H')'$, I'^* to mean $(I')^*$, and so on.

The word *annihilator* is frequently used in this context of rings acting on groups, but I shall reserve it for another use. Namely, for all $I \in \mathcal{I}$, the **left annihilator** of I, denoted $\mathcal{L}I$, means $\{\lambda \in \mathbf{t}(\mathbf{J}) : \lambda I = 0\}$, and the **right annihilator** of I, denoted $\mathcal{R}I$, means $\{\rho \in \mathbf{t}(\mathbf{J}) : I\rho = 0\}$.

The first Galois connection. The proofs of the following facts are completely routine, and follow the usual pattern of constructing a Galois correspondence (see for example [**B, Chapter V, Theorem 20**]):

(1) If $H \in \mathcal{F}$ then $H' \in \mathcal{I}$.
(2) If $I \in \mathcal{I}$ then $I' \in \mathcal{F}$.
(3) If $H \subseteq K$ in $\mathcal{F}$ then $K' \subseteq H'$ in $\mathcal{I}$.
(4) If $I \subseteq J$ in $\mathcal{I}$ then $J' \subseteq I'$ in $\mathcal{F}$.
(5) For all $H \in \mathcal{F}$, $H \subseteq H''$.
(6) For all $I \in \mathcal{I}$, $I \subseteq I''$.
(7) For all $H \in \mathcal{F}$, $H' = H'''$.
(8) For all $I \in \mathcal{I}$, $I' = I'''$.

An element H of $\mathcal{F}$ is called **closed** if $H = H''$; an element I of $\mathcal{I}$ is called **closed** if $I = I''$. It follows that $H \in \mathcal{F}$ is closed if and only if $H = I'$ for some $I \in \mathcal{I}$ and $I \in \mathcal{I}$ is closed if and only if $I = H'$ for some $H \in \mathcal{F}$. From the facts above, the set of closed elements of the respective lattices themselves inherit a lattice structure, and there is a lattice anti–isomorphism between the lattice of closed elements of $\mathcal{F}$ and the lattice of closed elements of $\mathcal{I}$. I now identify the first of these lattices:

THEOREM 4.1. *$H \in \mathcal{F}$ is closed if and only if $D \subseteq H$.*

PROOF. Suppose $D \subseteq H$. By (5) above, $H \subseteq H''$, so it suffices to show that if $a \notin H$, then $a \notin H''$, or equivalently, that if $a \notin H$, then there exists some $\eta \in H'$ such that $a\eta \neq 0$. This is certainly true if $H = G$, so suppose H is a proper subgroup and let $a \notin H$, say $a = \sum b_j + d$ with each $b_j \in B_j$ and $d \in D$. By 3.1 above, $H = G[\ell]$ for some layer sequence ℓ, so at least one $b_i \notin B[p^{\ell(i)}]$. Say for example that b_i has exponent $> \ell(i)$. Take $\eta \in \mathbf{t}(\mathbf{J})$ such that $B_j\eta = 0$ for all $j \neq i$ and η acts as multiplication by $p^{\ell(i)}$ on B_i. Then $\eta \in H'$, but $a\eta \neq 0$. Thus $H = H''$ is closed.

Now suppose that $D \not\subseteq H$ and let $\ell(k+1) = n < \infty$, so $D[p^n]$ is the largest subgroup of D contained in H. Take $a \in D$ of order $n+1$. Then every element of $\mathbf{t}(\mathbf{J})$ maps a to zero, so $a \in H''$ but clearly $a \notin H$.

Our next object is to identify the closed ideals of $\mathbf{t}(\mathbf{J})$. Clearly, they are annihilator ideals, in the sense that they annihilate some fully invariant subgroup of G, but they are also annihilators in the ring theoretic sense. For reasons which will become clear later, the zero ideal $\mathbf{0}$ and the whole ring $\mathbf{t}(\mathbf{J})$ must be dealt with separately.

LEMMA 4.2. *$\mathbf{0}$ and $\mathbf{t}(\mathbf{J})$ are closed ideals.*

PROOF. $\mathbf{0}' = G$ and $G' = \mathbf{0}$;

$$\mathbf{t}(\mathbf{J})' = \begin{cases} p^{m_k - 1}G & \text{if } G \text{ is reduced} \\ D & \text{otherwise.} \end{cases}$$

If G is reduced, then $p^{m_k-1}G' = \mathbf{t}(\mathbf{J})$. If not, then $D' = \mathbf{t}(\mathbf{J})$.

In the next two lemmas, I characterise the closed ideals ring theoretically.

LEMMA 4.3. *Let $I \in \mathcal{I}$. The following are equivalent:*

(1) *For some $H = G[\ell] \in \mathcal{F}$ with $\ell(k) < m_k$, $I = H'$.*
(2) *For some $H = G[\ell] \in \mathcal{F}$ with $\ell(k) < m_k$, $I = \mathcal{R}H^*$.*
(3) *For some $K \in \mathcal{I}$, $I = \mathcal{R}K$.*

PROOF. (1) $\Longrightarrow$ (2) Suppose $I = H'$ where $H = G[\ell]$ for some layer sequence ℓ such that $\ell(k) < m_k$. Let $\eta \in H'$. Then for all $\mu \in H^*$, $G\mu\eta \subseteq H\eta = 0$, so $\eta \in \mathcal{R}H^*$. Conversely, let $\eta \in \mathcal{R}H^*$. Let $\langle a \rangle$ be a cyclic summand of G of maximal order, and let $G = \langle a \rangle \oplus A$. Since $a \notin H$, by the definition of layer sequence, no cyclic summand of G of maximal order is contained in H, so by [**PS, Theorem 2.9**], for all $h \in B \cap H$, there exists $\mu \in \mathbf{J}$ with $a\mu = h$ and $A\mu = 0$. Thus every element of $B \cap H$ is in $G\mu$ for some $\mu \in H^*$, while $(D \cap H)\eta = 0$ since $\eta \in \mathbf{t}(\mathbf{J})$. Hence $H\eta \subseteq G\mathbf{H}^*\eta = 0$, so $\eta \in H' = I$.
(2) $\Longrightarrow$(3) is obvious.
(3) $\Longrightarrow$ (1) Suppose $I = \mathcal{R}K$, and let $H = K^*$. Then $HI = GKI = 0$, so $I \subseteq H'$. Conversely, let $\eta \in H'$. Then for all $\mu \in K$, $G\mu\eta \subseteq H\eta = 0$, so $\mu\eta = 0$ and hence $\eta \in \mathcal{R}K = I$. Since $H = GK$, H contains no element of exponent m_k, so $H = G[\ell]$ with $\ell(k) < m_k$.

THEOREM 4.4. *The closed ideals in $\mathcal{I}$ are the right annihilator ideals in $\mathcal{I}$ together with $\mathbf{0}$ and $\mathbf{t}(\mathbf{J})$.*

PROOF. By Lemma 4.2, $\mathbf{0}$ and $\mathbf{t}(\mathbf{J})$ are closed. By Lemma 4.3, all right annihilator ideals are closed and all proper closed ideals are right annihilators. Note that neither $\mathbf{0}$ nor $\mathbf{t}(\mathbf{J})$ are right annihilators.

Denote by $\mathcal{J}$ the lattice of right annihilator ideals in $\mathbf{t}(\mathbf{J})$. For any $I \in \mathcal{J}$, $I = H'$ for a unique $H = G[\ell] \in \mathcal{F}$ satisfying $\ell(k) < m_k$. Denote this I by $I(\ell)$. Let $\mathcal{L}^*$ denote the lattice of layer sequences for G satisfying $\ell(k) < m_k$ and $\mathcal{F}^*$ the lattice of fully invariant subgroups of G of the form $G[\ell]$ for $\ell \in \mathcal{L}^*$.

COROLLARY 4.5. *The mapping $\ell \leftrightarrow I(\ell)$ is an anti–isomorphism between the lattices $\mathcal{L}^*$ and $\mathcal{J}$.*

REMARK. Since the lattice $\mathcal{L}^*$ is completely determined by the parameters k and m_j, which in turn are determined by $\mathbf{t}(\mathbf{J})$, we may regard ideals of $\mathbf{t}(\mathbf{J})$ of the form $G[\ell]'$ as known.

4.6 The second Galois connection. I use Lemma 4.3 to establish a Galois connection from $\mathcal{J}$ to itself. The mappings $I \mapsto \mathcal{L}I$ and $I \mapsto \mathcal{R}I$ are clearly lattice anti–isomorphisms satisfying conditions (1) to (8) of §4.1, with $'$ replaced alternately by $\mathcal{R}$ and $\mathcal{L}$. Thus the closed ideals for the Galois connection determined by these maps are precisely those I for which $\mathcal{LR}I = I = \mathcal{RL}I$. I show now that every I in $\mathcal{J}$ is closed with respect to these operations. (Note however that $\mathcal{L}\mathbf{0}= \mathcal{R}\mathbf{0}= \mathbf{t}(\mathbf{J})$ while $\mathcal{R}\mathbf{t}(\mathbf{J}) = \mathcal{L}\mathbf{t}(\mathbf{J}) = \mathbf{t}(\mathbf{J})^{n-1}$, where $\mathbf{t}(\mathbf{J})$ has Loewy length n, so $\mathbf{0}$ is not closed. This is no contradiction, since $\mathbf{0}\notin \mathcal{J}$). As in the previous section, we need a separate Lemma to deal with $I = \mathbf{t}(\mathbf{J})$.

LEMMA 4.7. *$\mathbf{t}(\mathbf{J})$ is closed.*

PROOF. $\mathcal{L}\mathbf{t}(\mathbf{J}) = \mathcal{R}\mathbf{t}(\mathbf{J}) = \mathbf{t}(\mathbf{J})^{n-1}$, where n is the Loewy length of $\mathbf{t}(\mathbf{J})$, and $\mathcal{R}\mathbf{t}(\mathbf{J})^{n-1} = \mathcal{L}\mathbf{t}(\mathbf{J})^{n-1} = \mathbf{t}(\mathbf{J})$.

LEMMA 4.8. *Let $H \in \mathcal{F}^*$, and let $I = H'$. Then $\mathcal{L}I = H^*$.*

PROOF. By Lemma 4.3, $I = \mathcal{R}H^*$. Let $\mu \in H^*$. Then for all $\eta \in I$, $\mu\eta = 0$, so $\mu \in \mathcal{I}$. Conversely, suppose $\mu \notin H^*$. Then there exists $a \in G$ such that $a\mu \notin H = H'' = \{g \in G : g\eta = 0 \ \forall \eta \in I\}$. Thus there is some $\eta \in I$ for which $a\mu\eta \neq 0$, so $\mu \notin \mathcal{L}I$.

THEOREM 4.9. *For all $I \in \mathcal{I}$, $I \in \mathcal{J}$ if and only if $\mathcal{L}\mathcal{R}I = I = \mathcal{R}\mathcal{L}I$.*

PROOF. Let $I \in \mathcal{J}$, so $I = H'$ for some $G \neq H \in \mathcal{I}$. By Lemma 4.8, $\mathcal{L}I = H^*$ and by Lemma 4.3, $\mathcal{R}H^* = I$, so $\mathcal{R}\mathcal{L}I = I$. Now let $K = GI \in \mathcal{F}^*$, so $I = K^*$. Then by Lemma 3.4, $\mathcal{R}I = K' \in \mathcal{I}^*$, so by Lemma 4.8, $\mathcal{L}\mathcal{R}I = K^* = I$.

Conversely, if $I = \mathcal{R}\mathcal{L}I$, then $I \in \mathcal{J}$ by definition.

REMARK. As in the Remark following Corollary 4.5, we may assume ideals of $\mathbf{t}(\mathbf{J})$ of the form $G[\ell]^*$ are known.

§5. J determines G if $m_1 \neq 1$

In §2 I showed that $\mathbf{t}(\mathbf{J})$ determines the parameters k and m_j of G. In this section, I show that if B_1 is not elementary, the left and right annihilators of certain ideals $I(\ell)$ of $\mathcal{J}$ determine the parameters r_j. The idea is to single out, for each $j \in [1, k]$ a section of $\mathbf{t}(\mathbf{J})$ isomorphic as additive group to $\mathrm{Hom}(B_j, B_j[p])$. The latter group is elementary of rank determined by r_j.

LEMMA 5.1. *For $j \in [2, k]$, and, if $m_1 > 1$ for $j = 1$, let*

(1) $V_j = p^{m_j-1}G[p]^* \cap G[p^{m_j-1}]'$ and

(2) $U_j = p^{m_j}G[p]^* \cap G[p^{m_j}]'$.

Define

$$\theta_j : V_j/U_j \longrightarrow \mathrm{Hom}(B_j, B_j[p])$$

by

$$\theta_j : \eta + U_j \mapsto \eta|_{B_j}.$$

Then θ_j is an additive isomorphism.

PROOF. I start by defining a potential inverse for θ_j: for any $f \in \mathrm{Hom}(B_j, B_j[p])$, let $f\phi_j$ be $f + U_j$ on B_j and zero on the complement $\oplus_{i\neq j} B_i \oplus D$. It is routine to verify that $f + U_j \in V_j/U_j$. Thus it remains to show that θ_j is a well-defined additive function which is inverse to ϕ_j.

Suppose $\eta + U_j = \mu + U_j \in V_j/U_j$. Then $\eta - \mu \in G[p^{m_j}]$, so $(\eta - \mu)|_{B_j} = 0$. Thus θ_j is well-defined.

Let $\eta = \eta_1 + \eta_2 + \eta_3$, where $\eta_1 = \eta|_{\oplus_{i<j} B_i}$, $\eta_2 = \eta|_{B_j}$ and $\eta_3 = \eta|_{\oplus_{i>j} B_i}$. Then clearly each of η_1, η_2 and η_3 is in V_j. Since $V_j \subseteq G[p^{m_j-1}]'$, $\eta_1 = 0$. Since

$V_j \subseteq p^{m_j-1}G[p]^*$, η_3 raises heights in $\oplus_{i>j}B_i$ by at least m_j, so $\eta_3 \in U_j$. Hence $\eta + U_j = \eta_2 + U_j$.

Now let $\eta_2 = \eta^{(1)} + \eta^{(2)} + \eta^{(3)}$, where $\eta^{(1)}$ is η_2 followed by the projection of G onto $\oplus_{i<j}B_i$, $\eta^{(2)}$ is η_2 followed by the projection of G onto B_2, and $\eta^{(3)}$ is η_2 followed by the projection of G onto $\oplus_{i>j}B_i$. Once again it is routine to verify that each of $\eta^{(1)}$, $\eta^{(2)}$ and $\eta^{(3)}$ is in V_j. Since $\eta^{(1)} \in p^{m_j-1}G^*$, $\eta^{(1)} = 0$. Since $\eta^{(3)} \in p^{m_j}G^*$, $\eta^{(3)} \in U_j$. Hence $\eta + U_j = \eta^{(2)} + U_j$. But $\eta^{(2)}$ can be regarded as a map from B_j to $p^{m_j-1}B_j = B_j[p]$, so $\eta\theta_j \in \operatorname{Hom}(B_j, B_j[p])$ and moreover θ_j preserves addition and is a left and right inverse for ϕ_j.

THEOREM 5.2. *For all $j \in [2,k]$, and, if $m_1 > 1$ for $j = 1$, $\mathbf{t}(\mathbf{J})$ determines the parameters r_j.*

PROOF. By Lemma 5.1, it remains to show that the ideals V_j and U_j of $\mathbf{t}(\mathbf{J})$ are determined by $\mathbf{t}(\mathbf{J})$, and not just by the action of $\mathbf{t}(\mathbf{J})$ on G. I define layer sequences as follows:

$$\ell_j(i) = \begin{cases} m_i & \text{for } i \le j \\ m_j & \text{for } i > j \end{cases} \qquad \ell_j^-(i) = \begin{cases} m_i & \text{for } i < j \\ m_j - 1 & \text{for } i \ge j \end{cases}$$

$$\ell_j[p](i) = \begin{cases} 0 & \text{for } i \le j \\ 1 & \text{for } i > j \end{cases} \qquad \ell_j^-[p](i) = \begin{cases} 0 & \text{for } i < j \\ 1 & \text{for } i \ge j. \end{cases}$$

It is routine to verify that $G[\ell_j] = G[p^{m_j}]$, $G[\ell_j^-] = G[p^{m_j-1}]$, $G[\ell_j[p]] = p^{m_j}G[p]$ and $G[\ell_j^-[p]] = p^{m_j-1}G[p]$. By the remarks following Corollary 4.5 and Theorem 4.9, the ideals V_j and U_j of Lemma 5.1 are determined by $\mathbf{t}(\mathbf{J})$. Thus the vector space $\operatorname{Hom}(B_j, B_j[p])$ is determined by $\mathbf{t}(\mathbf{J})$ and its dimension is r_j^2 if finite and 2^{r_j} if infinite. Hence for each $j \in [1,k]$, r_j is determined by $\mathbf{t}(\mathbf{J})$.

By Lemmas 2.1 and 2.4, r_d is determined by $\mathbf{J}$ and by Theorem 2.6, each m_i is determined by $\mathbf{J}$. Hence Theorem 5.2 completes the proof that $\mathbf{J}$ determines G except for the case that B_1 is elementary.

§6. J determines G if $k > 1$

In this final section, we may assume that $m_1 = 1$ but $k > 1$. To simplify the notation, I denote D by B_{k+1} and r_d by r_{k+1}.

LEMMA 6.1. (a) *For all $j \in [2,k]$,* $\operatorname{Hom}(B_1, B_j) \cong \dfrac{p^{m_j-1}G[p]^*}{p^{m_j}G[p]^* \cap G[p]'}$.

(b) $\operatorname{Hom}(B_1, B_{k+1}) \cong \dfrac{p^{m_k}G[p]^*}{G[p]'}$.

PROOF. First note that if $\eta \in \operatorname{Hom}(B_1, B_j)$, then η raises heights in the socle of G and has order p. Thus each $\operatorname{Hom}(B_1, B_j)$ is an additive elementary subgroup of $\mathbf{t}(\mathbf{J})$.

For $j \in [2, k+1]$, define $\phi_j : p^{m_j-1}G[p]^* \longrightarrow \operatorname{Hom}(B_1, B_j)$ by $\phi_j : \eta \mapsto \iota\eta\pi_j$, where ι is the injection of B_1 into G and π_j is the canonical projection of G onto B_j. Since every element of $\operatorname{Hom}(B_1, B_j)$ the restriction to B_1 of a homomorphism of G into $p^{m_j-1}G[p]$, ϕ_j is an additive epimorphism. For $j \in [2,k]$, the kernel of $\phi_j = \{\eta \in p^{m_j-1}G[p]^* : G\eta \in p^{m_j}G[p] \text{ and } B_1\eta = 0\}$. Hence $\mathrm{Hom}(B_1, B_j) \cong \dfrac{p^{m_j-1}G[p]^*}{p^{m_j}G[p]^* \cap G[p]'}$.

The proof that $\mathrm{Hom}(B_1, B_{k+1}) \cong \dfrac{p^{m_k}G[p]^*}{G[p]'}$ is similar but simpler.

I call r_1 **large** if for some $j \in [2, k+1]$, $\dim \mathrm{Hom}(B_1, B_j) > r_j$. Otherwise r_1 is called **small**. Lemma 6.1 shows that this property of r_1 is determined by the lattice of closed ideals of $\mathbf{t}(\mathbf{J})$.

LEMMA 6.2. *If r_1 is large or if for some $j \in [1, k+1]$, r_j is finite, then r_1 is determined by $\mathbf{t}(\mathbf{J})$.*

PROOF. By the remarks following Corollary 4.5 and Theorem 4.9, the ideals of $\mathbf{t}(\mathbf{J})$ involved in the representation of $\mathrm{Hom}(B_1, B_j)$ in Lemma 6.1 are determined by $\mathbf{t}(\mathbf{J})$. Choose j for which $\dim \mathrm{Hom}(B_1, B_j) > r_j$ or r_j is finite. Since $\dim \mathrm{Hom}(B_1, B_j) = \begin{cases} r_1.r_j & \text{if } r_1 \text{ is finite} \\ 2^{r_1}.r_j & \text{otherwise} \end{cases}$, we have $r_1 = \dim \mathrm{Hom}(B_1, B_j)/r_j$ if $\mathrm{Hom}(B_1, B_j)$ is finite, and $2^{r_1} = \dim \mathrm{Hom}(B_1, B_j)$ if $\mathrm{Hom}(B_1, B_j)$ is infinite. By the GCH, r_1 is determined by $\mathbf{t}(\mathbf{J})$.

LEMMA 6.3. *If r_1 is small, then r_1 is determined by $\mathbf{t}(\mathbf{J})$.*

PROOF. Let $k' = k$ if $D = 0$ and $k+1$ otherwise. Now $G[p]^* = \mathbf{t}(\mathbf{J}) \cap \mathrm{Hom}(G, G[p])$, so $(G[p]^*)^{k'}$ is the ideal of $\mathbf{t}(\mathbf{J})$ generated by all products of k' homomorphisms in $\mathbf{t}(\mathbf{J})$ which map G into $G[p]$. Any such homomorphism annihilates D, so can be regarded as having domain the bounded part of G. It follows from [**PS, Theorem 2.9**] that every such product factors through B_1. Thus $(G[p]^*)^{k'}$ has dimension at most r_1 if r_1 is finite, and at most 2^{r_1} if r_1 is infinite. On the other hand, since r_1 is small, there are elements of $G[p]^*$ having image the whole of B_1, so in both cases the upper bounds on the dimension of $(G[p]^*)^{k'}$ are attained. Thus

$$\dim(G[p]^*)^{k'} = \begin{cases} r_1 & \text{if finite} \\ 2^{r_1} & \text{otherwise,} \end{cases}$$

so r_1 is determined by $\mathbf{t}(\mathbf{J})$.

The results can now be summarised in:

THEOREM 6.4. *Let G be an abelian p-group. Then G is determined by the Jacobson radical of its endomorphism ring if and only if pG divisible implies that*

(1) *$pG \neq 0$, and*
(2) *rank(pG) is finite or rank$(pG) < 2^{\mathrm{rank}(G/pG)}$.*

PROOF. By Theorem 1.1 of [**HPS**] we may assume that G is bounded modulo its divisible subgroup. By Theorem 2.6, the parameters k and m_j for $j \in [1, k+1]$ are determined by $\mathbf{J}$. By Lemmas 2.2 and 2.4, r_d is determined by $\mathbf{J}$. By Theorem

5.2, r_j is determined by $\mathbf{J}$ if $j > 1$ or if $m_1 > 1$ and by Lemmas 6.2 and 6.3, r_1 is determined by $\mathbf{J}$ if $k > 1$.

The only remaining possibility is that $k = 1$. In that case, Lemma 2.1 and Theorem 2.5 ensure that G is determined by $\mathbf{J}$ if and only if (1) and (2) are satisfied.

References

[B] G. Birkhoff, *Lattice Theory*, Amer. Math. Soc. Colloquium Publications, Vol. 25, 1967.

[F1] L. Fuchs, *Infinite Abelian Groups*, Vol. 1, Academic Press, New York, 1970.

[F2] L. Fuchs, *Infinite Abelian Groups*, Vol. 2, Academic Press, New York, 1973.

[HJ] J. Hausen and J. A. Johnson, *Abelian p–groups strongly determined by the Jacobson radical of their endomorphism rings*, Journal of Algebra (to appear).

[HPS] J. Hausen, C. E. Praeger and P. Schultz, *Most Abelian p–groups are determined by the Jacobson radical of their endomorphism rings*, Math. Z (to appear).

[MH] J. Douglas Moore and Edwin J. Hewett, *On fully invariant subgroups of abelian p–groups*, Comment. Math. Univ. St. Pauli **20 (2)** (1971), 97–106.

[P] R. S. Pierce, *Homomorphisms of primary abelian groups*, Topics in Abelian Groups, Scott, Foresman, 1963, pp. 215–310.

[PS] C. E. Praeger and P. Schultz, *The Loewy length of the Jacobson radical of a bounded endomorphism ring*, Abelian groups and non–commutative rings, Amer. Math. Soc., Contemporary Methematics 130, 1992, pp. 349–360.

The University of Western Australia, Nedlands, Western Australia, 6009

E-mail address: schultz@maths.uwa.oz.au

Contemporary Mathematics
Volume **171**, 1994

SIMILARITIES AND DIFFERENCES BETWEEN ABELIAN GROUPS AND MODULES OVER NON-PERFECT RINGS

JAN TRLIFAJ

ABSTRACT. The notions of a Whitehead test module, an almost free module, and a slender module are investigated over arbitrary non-perfect rings. Methods that have originally been developed for abelian groups are used to show similarities, but also important differences, appearing in the general case. In particular, a class of countable hereditary non-perfect rings is described over which all semisimple modules are slender.

In the past several years, methods originally developed for the investigation of abelian groups have been used to study the more general case of modules over non-left perfect rings (see e.g. [**EM**]). Recall that a ring R is *non-left perfect* provided there is a strictly decreasing countably infinite chain of principal right ideals of R. When generalizing results concerning particular properties of abelian groups, we often meet the following possibilities:

(1) the results generalize to modules over arbitrary non-left perfect rings, while the corresponding results for modules over left perfect rings do not hold;

(2) the results generalize only to modules over certain non-left perfect rings (e.g. the countable hereditary ones);

(3) there is no generalization to R-modules even if R is a countable hereditary non-left perfect ring.

The main objective of the present paper is to demonstrate these possibilities by three important notions of contemporary module theory: that of a Whitehead

---

1991 *Mathematics Subject Classification.* Primary 20K40, 20K35; Secondary 16P70, 16E60.

Key words and phrases. non-perfect ring, slender module, almost free module, Whitehead test module.

This research has partially been supported by grant FDR-0579 of the Czech Ministry of Education.

This paper is in final form and no version of it will be submitted for publication elsewhere.

test module, of an almost free module, and of a slender module. We show that these notions provide examples of possibilities (1), (2), and (3), respectively.

Each of the notions is treated in a separate section, and the necessary definitions appear there. Before proceeding to the sections, we fix our basic notation. First, $\mathbb{Z}$ denotes the ring of all integers, and $\mathbb{Q}$ the $\mathbb{Z}$-module of all rational numbers. Let R be an associative ring with unit. Then R is said to be *(von Neumann) regular* provided each $r \in R$ has a pseudo-inverse $s \in R$ (i.e. $r = rsr$). It is easy to see that left- (or right-) perfect and regular rings coincide with the *semisimple* rings (= finite ring direct products of full matrix rings over skew-fields). The non-perfect regular rings include e.g. all full endomorphism rings of infinite dimensional linear spaces. Since factors, products and direct limits of regular rings are regular, many other examples of non-perfect regular rings can be constructed starting from the above mentioned ones (see e.g. [**G**]).

The category of all (unitary left R-) modules is denoted by R-Mod. Homomorphisms in module categories are written as acting on the opposite side from scalars. A module M is *simple* provided 0 and M are the only submodules of M. A module M is *semisimple* provided M is a sum of its simple submodules. The injective hull of a module M is denoted by $I(M)$. A system of modules $(M_\alpha; \alpha < \lambda)$ is said to be a *continuous chain* provided $M_0 = 0$, $M_\alpha \subseteq M_{\alpha+1}$ for all $\alpha < \lambda$, and $M_\alpha = \cup_{\beta<\alpha} M_\beta$ for all limit ordinals $\alpha < \lambda$. A continuous chain with $M = \cup_{\alpha<\lambda} M_\alpha$ is a *κ-filtration* of M provided $gen(M_\alpha) < \kappa$ for all $\alpha < \lambda$.

A closed and unbounded subset of a regular uncountable cardinal λ is said to be a *cub*. A subset $E \subseteq \lambda$ is *stationary* in λ provided $E \cap C \neq \emptyset$ for each cub C.

For basic properties of these notions, we refer to [**AF**], [**EM**], and [**G**].

§1 Whitehead Test Modules

We start with the notion of a (Whitehead) test module, defined in [**E2, §11**]:

1.1 Definition. Let R be a ring and N be a module. Then N is said to be a *(Whitehead) test module* provided for all $M \in Mod$-R, M is projective iff $Ext_R(M, N) = 0$.

First, we consider the left perfect ring case:

1.2 Proposition. *Let R be a left perfect left hereditary ring. Then each free module is a test module.*

Proof. Clearly, it suffices to prove that R is a test module. Proving indirectly, assume $Ext_R(M, R) = 0$ and M is non-projective. Let

$$0 \to K \xrightarrow{\nu} P \to M \to 0$$

be a projective cover of M, i.e. a short exact sequence with P projective and K a superfluous submodule of P. By the premise, $K \neq 0$ and K has a maximal submodule, L. Then K/L is a simple module, hence a quotient of R. Since R is left hereditary, $Ext_R(M, K/L) = 0$. Let $\pi \in Hom_R(K, K/L)$ be the projection. Since the sequence

$$\ldots \to Hom_R(P, K/L) \xrightarrow{Hom_R(\nu, K/L)} Hom_R(K, K/L) \to Ext_R(M, K/L) = 0$$

is exact, there is a $\phi \in Hom_R(P, K/L)$ such that $\phi \restriction K = \pi$. Then $Ker(\phi)$ is a maximal submodule of P, and $K \subseteq Rad(P) \subseteq Ker(\phi) \subset P$. Thus, $\pi = \phi \restriction K = 0$, a contradiction. □

Of course, 1.2 does not extend to arbitrary left perfect rings. For example, consider a QF-ring R which is not semisimple (e.g. $R = \mathbb{Z}/\mathbb{Z}p^n$, p a prime and $n > 1$). Then each projective module is injective, hence not a test module. Nevertheless, we have

1.3 THEOREM. *Let R be a left perfect ring. Then there is a proper class of test modules.*

PROOF. Note that any module containing a summand which is a test module is likewise a test module. Thus, it suffices to construct a single test module in *R-Mod*. Denote by D the direct sum of a representative set of all simple modules. Let M be a module such that $Ext_R(M, D) = 0$. If M is not projective and $0 \to K \to P \to M \to 0$ is a projective cover of M, then $K \neq 0$ and K has a maximal submodule, L. The same argument as in 1.2 shows that $Ext_R(M, K/L) \neq 0$, whence $Ext_R(M, D) \neq 0$. Therefore, M is projective, and D is a test module. □

Now, we proceed to non-left perfect rings. In order to see the similarities with the abelian group case, we recall the set theoretic principles that play important role here:

1.4 DEFINITION. (i) Let κ be an infinite cardinal and E be a subset of κ such that $E \subseteq \{\alpha < \kappa; cf(\alpha) = \aleph_0\}$. Let $(n_\nu; \nu \in E)$ be a *ladder system* , i.e. for each $\nu \in E$, let $(n_\nu(i); i < \aleph_0)$ be a strictly increasing sequence of non-limit ordinals less that ν such that $\sup_{i<\aleph_0} n_\nu(i) = \nu$.

(ii) Let κ be a cardinal such that $cf(\kappa) = \aleph_0$. Consider the following assertion UP_κ: "there exist a stationary subset E of κ^+ satisfying $E \subseteq \{\alpha < \kappa^+; cf(\alpha) = \aleph_0\}$ and a ladder system $(n_\nu; \nu \in E)$ such that for each cardinal $\lambda < \kappa$ and each sequence $(h_\nu; \nu \in E)$ of mappings from $\aleph_0$ to λ there is a mapping $f : \kappa^+ \to \lambda$ such that $\forall\, \nu \in E\ \exists\, j < \aleph_0\ \forall\, j < i < \aleph_0 : f(n_\nu(i)) = h_\nu(i)$".
Denote by UP the assertion "UP_κ holds for every uncountable cardinal κ such that $cf(\kappa) = \aleph_0$".

(iii) Let κ be a regular uncountable cardinal and E be a stationary subset of κ. Denote by $\diamondsuit_\kappa(E)$ the *Jensen's diamond*, i.e. the assertion

" Let A be any set of cardinality κ and $(A_\alpha; \alpha < \kappa)$ a κ-filtration of A. Then there is a system $\{S_\alpha; \alpha < \kappa\}$ such that $S_\alpha \subseteq A_\alpha$ for all $\alpha < \kappa$, and the set $\{\alpha \in E; X \cap A_\alpha = S_\alpha\}$ is stationary in κ, for every $X \subseteq A$."
Denote by JD the assertion "$\diamondsuit_\kappa(E)$ holds for all regular uncountable cardinals κ and all stationary subsets E of κ".

UP is inconsistent with JD, but both UP and JD are well-known to be consistent with ZFC + GCH (see [**ES, §2**] and [**EM, VI**]).

1.5 THEOREM. *Assume UP. Then there is no test module over any non-left perfect ring.*

PROOF. By [**ES, Corollary 2.2**] or [**T1, Corollary 1.6**]. $\square$

Assuming JD, there is a similarity between $\mathbb{Z}$-Mod and R-Mod provided R is left hereditary:

1.6 THEOREM. *Assume JD. Let R be a left hereditary ring and put $\kappa = card(R) \times \aleph_0$. Then each free module of rank $\geq \kappa$ is a test module.*

PROOF. Assume M is a $\leq \kappa$ generated module such that $Ext_R(M, R^{(\kappa)}) = 0$. Let K be a submodule of $R^{(\kappa)}$ such that $M \simeq R^{(\kappa)}/K$. By the premise, $gen(K) \leq \kappa$. Since R is left hereditary, we infer that $Ext_R(M, K) = 0$. Denote by ν the embedding of K into $R^{(\kappa)}$. We have the exact sequence

$$\ldots \to Hom_R(R^{(\kappa)}, K) \xrightarrow{Hom_R(\nu,K)} Hom_R(K, K) \to Ext_R(M, K) = 0.$$

In particular, $id_K = \nu\pi$ for some $\pi \in Hom_R(R^{(\kappa)}, K)$. Hence, $Ker(\pi)$ is a summand of $R^{(\kappa)}$, and $Ker(\pi) \simeq R^{(\kappa)}/Im(\nu) \simeq M$ is projective.

By induction on λ, we prove the projectivity of each module M such that $gen(M) = \lambda$ and $Ext_R(M, R^{(\kappa)}) = 0$. We have already proved the result for $\lambda \leq \kappa$. Assume λ is regular and $\lambda > \kappa$. Then the induction step follows from the induction premise and [**E1, Theorem 1.5**]. If λ is singular, $\lambda > \kappa$, the general singular compactness theorem [**H, §4**] applies (with "free" = "a direct sum of countably generated projectives" = projective, by a classical result of Kaplansky). $\square$

In fact, a weaker version of JD (called a generalized weak diamond) is enough for the proof of 1.6. Moreover, for certain classes of left hereditary non-left perfect rings, the claim of 1.6 can be strengthened to "each free module is a test module" (this is of course true for $R = \mathbb{Z}$, but also if R is a simple non-semisimple regular ring such that R has countable dimension over its center). More details on this will appear in [**CT**]. Nevertheless, the following problem remains open:

1.7 PROBLEM. *For what classes of non-left perfect non-left hereditary rings is the existence of test modules consistent? (Or, vice versa, for what classes of non-left perfect non-left hereditary rings is it true in ZFC that there are no test modules?)*

§2 Almost Free Modules

2.1 DEFINITION. Let κ be a regular uncountable cardinal and M be a module.

(i) M is *almost κ-free* if there is a set, $\mathcal{C}$, of free submodules of M such that:

(1) $gen(X) < \kappa$ for all $X \in \mathcal{C}$, and

(2) for each subset $A \subseteq M$ with $card(A) < \kappa$ there is some $X \in \mathcal{C}$ with $A \subseteq X$.

(ii) M is *κ-free* provided it is almost κ-free and among the sets $\mathcal{C}$ of free submodules satisfying (1) and (2) there is at least one satisfying

(3) $\mathcal{C}$ is closed under unions of well-ordered chains of length $< \kappa$.

(iii) M is *strongly κ-free* provided it is almost κ-free and among the sets $\mathcal{C}$ of free submodules satisfying (1) and (2) there is at least one such that $0 \in \mathcal{C}$ and (2+) for any subset $A \subseteq M$ with $card(A) < \kappa$ and any $X \in \mathcal{C}$ there is some $Y \in \mathcal{C}$ such that $X \cup A \subseteq Y$ and Y/X is free.

Clearly, any free module is both κ-free and strongly κ-free. If R is a free ideal ring (e.g. $R = \mathbb{Z}$), then any strongly κ-free module is κ-free. A similar result holds true for left hereditary rings:

2.2 THEOREM. *Let κ be a regular uncountable cardinal. Let R be a left hereditary ring such that $(card(R))^+ < \kappa$ and R is not left perfect. Then each $\leq \kappa$-generated strongly κ-free module is κ-free.*

PROOF. Let M be a strongly κ-free module and $\mathcal{C}$ be as in 2.1(iii). In view of (2), we can assume that $gen(M) = \kappa$. By [**EM, IV, Proposition 1.11**], M has a strictly increasing κ-filtration $(M_\alpha; \alpha < \kappa)$ such that for all $\alpha < \beta < \kappa$, $card(M_{\alpha+1}/M_\alpha) \geq card(M_\alpha)$, $M_{\alpha+1} \in \mathcal{C}$, $M_{\beta+1}/M_{\alpha+1}$ is free, and $card(M_\alpha) \geq (card(R))^+$ provided $\alpha > 0$. By 2.1(ii), it suffices to prove that M_α is free for all $\alpha < \kappa$. This is clear in the case when either $\alpha = 0$, or α is non-limit, or $cf(\alpha) = \aleph_0$. Since R is left hereditary, it is also clear that each M_α is projective. Assume $cf(\alpha) > \aleph_0$ and put $\mu = card(M_\alpha)$. Of course, $\aleph_0 < \mu = gen(M_\alpha) < \kappa$. First, suppose there is some $\beta < \alpha$ such that $card(M_\beta) = \mu$. Then $card(M_{\beta+1}) = \mu$ and $M_{\beta+1}$ is a free summand of M_α. We have $M_{\beta+1} \simeq R^{(\mu)}$ and $M_\alpha = M_{\beta+1} \oplus X_\alpha$ for a projective module X_α. Since $gen(X_\alpha) \leq \mu$ and $\mu \geq \aleph_0$, Eilenberg's trick ([**EM, I, Lemma 2.3**]) implies $M_\alpha \simeq X_\alpha \oplus R^{(\mu)} \simeq R^{(\mu)}$, and M_α is free.

Hence, we may assume that $card(M_\beta) < \mu$ for all $\beta < \alpha$, and that $\alpha \leq \mu$. If μ is a singular cardinal, we may also assume that $card(M_\beta) < card(M_{\beta+1})$ for all $\beta < \alpha$.

Since M_α is projective, we have $M_\alpha = \oplus \sum_{\gamma < \mu} P_\gamma$, where P_γ is projective and countably generated for all $\gamma < \mu$. For each $\beta < \alpha$, let A_β be the smallest subset of M such that $M_\beta \subseteq \oplus \sum_{\gamma \in A_\beta} P_\gamma$. Put $c_\beta = card(A_\beta)$. Clearly, $c_\beta = gen(M_\beta) < \mu$. Moreover, $(A_\beta; \beta < \alpha)$ is a continuous chain of subsets of μ such that $\mu = \cup_{\beta < \alpha} A_\beta$. Fix $\beta < \alpha$. Put $Q_\beta = \oplus \sum_{\gamma \in A_\beta} P_\gamma$.

If μ is regular, then $\alpha = \mu$ and there is a sequence, $\{\beta_i; i < \aleph_0\}$, of non-limit ordinals such that $\beta < \beta_0 < \cdots < \beta_i < \beta_{i+1} < \cdots < \alpha$ and $M_\beta \subseteq Q_\beta \subseteq M_{\beta_0} \subseteq Q_{\beta_0} \subseteq \cdots \subseteq M_{\beta_i} \subseteq Q_{\beta_i} \subseteq M_{\beta_{i+1}} \subseteq Q_{\beta_{i+1}} \subseteq \ldots$. Put $\gamma = \gamma_\beta = sup_{i < \aleph_0} \beta_i$. Then $\gamma < \alpha$. Moreover, $Q_\gamma = M_\gamma$ is free, and $Q_\gamma = F_\gamma \oplus G_\gamma$, where $F_\gamma = M_{\beta_0}$ and G_γ is a free module of rank $\geq c_{\beta_0}$.

If μ is singular, we put $\beta_0 = \beta$ and $\gamma = \gamma_\beta = \beta + 1$. Then M_γ is a free summand in Q_γ of rank c_γ, i.e. $Q_\gamma = M_\gamma \oplus X_\gamma$, where X_γ is projective and $gen(X_\gamma) \leq c_\gamma$.

By Eilenberg's trick, Q_γ is free. Moreover, there are free submodules $F_\gamma \simeq R^{(c_\beta)}$ and $G_\gamma \simeq R^{(c_\gamma)}$ of Q_γ such that $Q_\beta \subseteq F_\gamma \subset F_\gamma \oplus G_\gamma = Q_\gamma$.

Anyway, put $Y_\gamma = \oplus\sum_{\delta \in A_\gamma \setminus A_\beta} P_\delta$. Then $Q_\gamma = Q_\beta \oplus Y_\gamma$, and $F_\gamma = Q_\beta \oplus (F_\gamma \cap Y_\gamma)$. Since $F_\gamma \cap Y_\gamma$ is projective and $gen(F_\gamma \cap Y_\gamma) \leq c_{\beta_0}$, Eilenberg's trick implies $(F_\gamma \cap Y_\gamma) \oplus G_\gamma \simeq G_\gamma$. Thus, $Y_\gamma \simeq Q_\gamma/Q_\beta \simeq (F_\gamma \cap Y_\gamma) \oplus G_\gamma$ is free.

Finally, define a strictly increasing continuous function $f : \lambda \to \alpha$ by $f(0) = 0$, $f(\beta + 1) = \gamma_{f(\beta)}$ for all $\beta < \lambda$, and by $f(\beta) = sup_{\delta<\beta} f(\delta)$ for all limit $\beta < \lambda$. Then $(Q_{f(\beta)}; \beta < \lambda)$ is a μ-filtration of M_α such that $Q_{f(\beta+1)}/Q_{f(\beta)}$ is free for all $\beta < \lambda$. Hence, $M_\alpha = \cup_{\beta<\lambda} Q_{f(\beta)}$ is free. □

2.3 COROLLARY. *Let R be a countable left hereditary non-left perfect ring. Let κ be a regular uncountable cardinal. Then each $\leq \kappa$-generated strongly κ-free module is κ-free.*

PROOF. If $\kappa > \aleph_1$, we apply Theorem 2.2. If $\kappa = \aleph_1$ and M is a strongly $\aleph_1$-free module such that $gen(M) \leq \aleph_1$, then M has an $\aleph_1$-filtration $(M_\alpha; \alpha < \aleph_1)$ as in [**EM, IV, Proposition 1.11**]. Since all $0 < \alpha < \aleph_1$ are either non-limit or of cofinality $\aleph_0$, all M_α, $\alpha < \aleph_1$, are free, and M is $\aleph_1$-free. □

2.4 DEFINITION. Let κ be an infinite cardinal and K be a skew-field. Denote by L the right linear K-space of dimension κ and let $S = End(L_K)$, i. e. S is the ring of all linear transformations of L. Put $T = \{f \in S; rank(f) < \kappa\}$. It is well-known that T is the unique maximal two-sided ideal of S. Put $R_\kappa = S/T$.

It is easy to see that projective = free, over R_κ. Nevertheless, R_κ is not left hereditary (i.e. not a free ideal ring). In contrast with 2.2, we have

2.5 THEOREM. *Let $\kappa > \aleph_1$ be a regular cardinal. Then R_κ is a regular non-left (right) perfect ring. There exists a non-projective strongly κ-free R_κ-module M_κ such that $gen(M_\kappa) = \kappa$ and M_κ is not κ-free.*

PROOF. By [**T2, Theorem**]. □

§3 Slender Modules

3.1 DEFINITION. Let R be a ring. A module M is said to be *slender* provided for every R-homomorphism $f : R^{\aleph_0} \to M$, $1_n f = 0$ for all but finitely many $n < \aleph_0$. Here, $\{1_n; n < \aleph_0\}$ denotes the canonical R-basis of the free submodule $R^{(\aleph_0)}$ of $R^{\aleph_0}$. Clearly, any submodule of a slender module is slender. No non-zero injective module is slender (see e.g. [**EM,III,Proposition 1.4**]).

A module M is *cocyclic* provided there is $0 \neq m \in M$ such that any R-homomorphism $f : M \to N$ with $m \notin Ker(f)$ is a monomorphism. The notion of a cocyclic module is clearly dual to that of a cyclic module. The cocyclic modules are exactly the subdirectly irreducible elements of R-Mod. Moreover, they coincide with the modules M such that $S \subseteq M \subseteq I(S)$ for a simple module S ([**W, 14.8**]).

First, we recall the Nunke's structure theorem for slender groups which characterizes them by forbidden subgroups:

3.2 THEOREM. *A $\mathbb{Z}$-module M is slender iff M does not contain a copy of $\mathbb{Q}$, $\prod_{n<\aleph_0}\mathbb{Z}$, the cyclic group of order p, or the group of all p-adic integers, for any prime p.*

PROOF. Well-known (see e.g. [**EM, IX, Corollary 2.4**]). □

In particular, all slender groups are torsion free. Moreover, for arbitrary commutative rings (countable or uncountable), there are no slender modules with a non-zero socle. In particular, there are no non-zero semisimple slender modules, and slender modules of finite length. This follows from

3.3 PROPOSITION. *Let R be an arbitrary ring. Let I be a two-sided ideal such that I is a maximal left ideal in R. Then no module containing a copy of the simple module R/I is slender.*

PROOF. Since R/I is a skew-field, the left R/I-module $M = R/I$ is injective. In particular, there is an R-homomorphism $f : R^{\aleph_0} \to M$ such that $Ker(f) \supseteq I^{\aleph_0}$ and $1_n f \neq 0$ for all $n < \aleph_0$. Hence, the module M is not slender, and neither is any module containing a copy of M. □

We are going to deal with essential differences that appear in the structure theory of slender modules over general countable hereditary non-left perfect rings. Our examples will be (von Neumann) regular rings. We recall their basic properties:

3.4 LEMMA. *Let R be a regular ring. Then each finitely generated left (right) ideal of R is generated by an idempotent. Moreover, each countably infinitely generated left (right) ideal is generated by a countable set of orthogonal idempotents. Each projective module is a direct sum of cyclic modules.*

PROOF. Well-known (see e.g. [**G, §1 and §2**]). □

3.5 DEFINITION. Let κ be an infinite cardinal. A ring R is said to be left (right) *κ-hereditary* provided each left (right) ideal I with $gen(I) \leq \kappa$ is projective. Hence, each regular ring is left and right $\aleph_0$-hereditary.

3.6 LEMMA. *Let R be a regular ring such that each left ideal is countably generated. Then R is a left hereditary ring. Moreover, the module $M^{\aleph_0}/M^{(\aleph_0)}$ is injective for all $M \in R\text{-}Mod$.*

PROOF. The first assertion is clear by 3.5. W.l.o.g. we can assume that R is not semisimple. By Baer's criterion, it is enough to extend any $\phi \in Hom_R(I, M^{\aleph_0}/M^{(\aleph_0)})$ into some $\varphi \in Hom_R(R, M^{\aleph_0}/M^{(\aleph_0)})$, for each left ideal I of R. This is possible by 3.4 provided $gen(I) < \aleph_0$. If $gen(I) = \aleph_0$, then 3.4 implies there is an infinite set, $\{e_i; i < \aleph_0\}$, of orthogonal idempotents in R such that $\oplus\sum_{i<\aleph_0} Re_i = I$. Let $\phi \in Hom_R(I, M^{\aleph_0}/M^{(\aleph_0)})$. Then $e_i\phi = (e_i m_n^i)_{n<\aleph_0} + M^{(\aleph_0)}$, where $m_n^i \in M$ for all $i < \aleph_0$ and $n < \aleph_0$. Define $\varphi \in Hom_R(R, M^{\aleph_0}/M^{(\aleph_0)})$ by $1\varphi = (p_n)_{n<\aleph_0} + M^{(\aleph_0)}$, where $p_n = \sum_{i\leq n} e_i m_n^i$ for all $n < \aleph_0$. Then $\varphi \restriction I = \phi$, q.e.d. □

Both the cocyclic and the slender modules are in some sense "small". If R is as in 3.6, the two classes are closely related:

3.7 LEMMA. *Let R be a regular ring such that each left ideal is countably generated. Let M be a cocyclic module. Then M is slender iff M is not injective.*

PROOF. Since no non-zero injective module is slender, the direct implication holds true. On the other hand, let M be a non-injective cocyclic module. Let S be a simple module such that $S \subseteq M \subseteq I(S)$ and $M \neq I(S)$.
Proving indirectly, assume M is not slender. Then there is a $\varphi \in Hom_R(R^{\aleph_0}, M)$ such that the set $C = \{n < \aleph_0; 1_n\varphi \neq 0\}$ is infinite, $\{1_n; n < \aleph_0\}$ being the canonical free R-basis of $R^{(\aleph_0)}$. Enumerating the elements of C, we have $C = \{c_i; i < \aleph_0\}$. Since S is essential in M and S is simple, there are some $0 \neq s \in S$ and $x_i \in R$, $i < \aleph_0$, such that $x_i(1_{c_i}\varphi) = s$ for each $i < \aleph_0$. Define $\psi \in Hom_R(R^{\aleph_0}, R^{\aleph_0})$ by $((r_n)_{n<\aleph_0})\psi = (r'_n)_{n<\aleph_0}$, where $r'_n = r_n$ provided $n \in \aleph_0 \backslash C$ and $r'_{c_i} = r_{c_i}x_i$ for all $i < \aleph_0$. Then $\psi\varphi \in Hom_R(R^{\aleph_0}, M)$, $(1_n)\psi\varphi = 0$ provided $n \in \aleph_0 \setminus C$ and

$$(*) \qquad 1_{c_i}\psi\varphi = s \qquad \text{for all } i < \aleph_0.$$

Define $\xi \in Hom_R(R^{\aleph_0}, R^{\aleph_0})$ by $((r_n)_{n<\aleph_0})\xi = (r'_n)_{n<\aleph_0}$, where $r'_n = r_n$ provided $n \in (\aleph_0 \backslash C) \cup \{c_0\}$ and $r'_{c_{i+1}} = r_{c_{i+1}} - r_{c_i}$ for all $i < \aleph_0$. Then $\xi\xi' = \xi'\xi = id_{R^{\aleph_0}}$, where $\xi' \in Hom_R(R^{\aleph_0}, R^{\aleph_0})$ is defined by $((r_n)_{n<\aleph_0})\xi' = (r'_n)_{n<\aleph_0}$; $r'_n = r_n$ provided $n \in (\aleph_0 \setminus C) \cup \{c_0\}$ and $r'_{c_{i+1}} = \sum_{j=0}^{i+1} r_{c_j}$ for all $i < \aleph_0$. Put $\theta = \xi\psi\varphi \in Hom_R(R^{\aleph_0}, M)$. Then $(1_n)\theta = 0$ for all $n < \aleph_0$. Hence, θ induces a homomorphism $\bar{\theta} \in Hom_R(R^{\aleph_0}/R^{(\aleph_0)}, M)$ via $((r_n)_{n<\aleph_0} + R^{\aleph_0})\bar{\theta} = ((r_n)_{n<\aleph_0})\theta$. By 3.6, R is left hereditary and $R^{\aleph_0}/R^{(\aleph_0)}$ is injective. Thus, also $Im(\bar{\theta})$ is injective, $Im(\bar{\theta}) = 0$ and $\theta = 0$. Then $\psi\varphi = \xi'\theta = 0$, a contradiction with $(*)$. □

Of course, any countable regular ring satisfies the premises of 3.6 and 3.7. Countable regular rings are of special interest, since any regular ring is a direct limit of its finitely generated (and hence countable) regular subrings. For the simple ones, we get a result contrasting with 3.2 and 3.3:

3.8 THEOREM. *Let R be a simple countable regular ring such that R is not semisimple. Then all semisimple modules, and all modules of finite length, are slender.*

PROOF. First, we prove that no non-zero countable module N is injective: Since R is not semisimple, there is an infinite set, $\{e_n; n < \aleph_0\}$, of orthogonal idempotents in R. Note that $e_nN \neq 0$ for all $n < \aleph_0$ (Otherwise, $0 = Re_nN = Re_nRN = RN = N$, since R is a simple ring). Hence, we have

$$card(Hom_R(\oplus \sum_{n<\aleph_0} Re_n, N)) \geq 2^{\aleph_0} \text{ and } card(Hom_R(R, N)) = card(N) \leq \aleph_0.$$

Denote by ν the embedding of $\oplus \sum_{n<\aleph_0} Re_n$ into R. Then the homomorphism $\nu' = Hom_R(\nu, N)$ is not onto. Since the sequence

$$Hom_R(R, N) \xrightarrow{\nu'} Hom_R(\oplus \sum_{n<\aleph_0} Re_n, N) \to Ext_R(R/ \oplus \sum_{n<\aleph_0} Re_n, N) \to 0$$

is exact, we have $Ext_R(R/\oplus\sum_{n<\aleph_0} Re_n, N) \neq 0$, and N is not injective.

In particular, no simple module is injective. By 3.7, each simple module is slender. The first assertion is now a consequence of [**L, Theorem 3(2)**]. The second assertion follows from the obvious fact ([**EM, p.78**]) that the class of all slender modules is closed under extensions. □

So far, we have been dealing with left hereditary regular rings such that $gen(I) \leq \aleph_0$ for all left ideals I. In the remaining part of this section, we shall deal with the more general case of hereditary, and κ-hereditary, regular rings. First, there are no bounds for the size of $gen(I)$ for general hereditary regular rings:

3.9 EXAMPLE. For each infinite cardinal κ there are a left and right hereditary regular ring R_κ and a two-sided ideal I_κ such that I_κ is a maximal left (right) ideal and $gen(I_\kappa) \geq \kappa$.

PROOF. Consider a skew-field K, a right linear K-space L of dimension κ, and the ring of all linear trasformations $S = End(L_K)$. Put $F = \{f \in S; rank(f) < \aleph_0\}$ and $T = \{t \in S; \exists k \in K \forall l \in L : f(l) = k.l\}$. Let $R = R_\kappa$ be the subring of S generated by F and T. Then R is well-known to be a regular ring such that $F = I_\kappa$ is the left (right) socle of R (see e.g. [**F, Example 19.47**]). Moreover, L is a simple module and $F \cong L^{(\lambda)}$ for some $\lambda \geq \kappa$. In particular, F is a semisimple projective module and $K \cong R/F$ (as rings). Note that each $r \in R$ can be written as $r = f + t$ for some $f \in F$ and $t \in T$.

Let I be a left ideal of R. If $I \subseteq F$, then I is a summand of F, and I is projective. Assume there is some $e \in I \setminus F$. W.l.o.g. we can assume that $e^2 = e$. Then $e = f + t$ for some $f \in F$ and $t \in T$, whence $t^2 - t \in F \cap T$, and $t^2 = t = 1$. Then $I = (R(1-e) \cap I) \oplus Re = (Rf \cap I) \oplus Re$. But Rf is a finitely generated semisimple module, and so is $Rf \cap I$. Thus, I is finitely generated and projective. Similarly, each right ideal of R is projective. □

Note that by 3.3, the simple module $M = R_\kappa/I_\kappa$ from 3.9 is neither a left nor a right slender module.

3.10 LEMMA. *Let κ be a regular uncountable cardinal. Let R be a regular left κ-hereditary ring. Let C be a left ideal such that $gen(C) = \aleph_0$. Then there is a left ideal D such that $C \subseteq D$, $gen(D) = \aleph_0$ and D is a summand in any left ideal $I \supseteq D$ such that $\aleph_0 \leq gen(I) \leq \kappa$.*

PROOF. Proving indirectly, assume there is a left ideal C with $gen(C) = \aleph_0$ such that there is no D with the required properties. By induction on $\alpha < \aleph_1$, we construct a strictly increasing continuous chain, $(C_\alpha; \alpha < \aleph_1)$, of left ideals such that $gen(C_\alpha) = \aleph_0$ and $C \subseteq C_\alpha$ for all $\alpha < \aleph_1$.

First, $C_0 = C$. Assume $C \subseteq C_\alpha$ is defined. By the assumption on C and C_α, there is a left ideal J_α containing C_α such that $\aleph_0 \leq gen(J_\alpha) \leq \kappa$ and C_α is not a summand in J_α. By 3.4, J_α is a direct sum of principal left ideals, $J_\alpha = \oplus\sum_{\beta<\lambda} Rx_\beta$ for some $\aleph_0 \leq \lambda \leq \kappa$. Then there is a countable subset $A \subseteq \lambda$ such that $C_\alpha \subseteq \oplus\sum_{\beta\in A} Rx_\beta = C_{\alpha+1}$. Since $C_{\alpha+1}$ is a summand of J_α,

$C_{\alpha+1}/C_\alpha$ is not projective. If $\alpha < \aleph_1$ is a limit ordinal, we put $C_\alpha = \cup_{\gamma<\alpha} C_\gamma$. Now, $I = \cup_{\alpha<\aleph_1} C_\alpha$ is a left ideal of R, $gen(I) = \aleph_1 \leq \kappa$, and the set $\{\alpha < \aleph_1; C_{\alpha+1}/C_\alpha$ is not projective$\}$ is stationary in $\aleph_1$. By **[EM, VI, Proposition 1.7]**, I is not projective, a contradiction. □

We finish by showing that 3.10 may fail for $\kappa = \aleph_0$, i.e. for arbitrary regular rings. For instance, a self-injective regular ring is $\aleph_1$-hereditary iff it is semisimple:

3.11 THEOREM. *Let R be a regular left self-injective ring and D be a left ideal such that $gen(D) = \aleph_0$. Then there is a left ideal I such that $D \subseteq I$, $gen(I) = \aleph_0$ and D is not a summand in I. In particular, R is left $\aleph_0$-hereditary, but not left $\aleph_1$-hereditary.*

PROOF. Let $\{f_{mn}; m < \aleph_0, n < \aleph_0\}$ be a set of orthogonal idempotents such that $D = \oplus \sum_{m<\aleph_0, n<\aleph_0} Rf_{mn}$. For each $n < \aleph_0$, let $D_n = \oplus \sum_{m<\aleph_0} Rf_{mn}$. By **[G, Lemma 9.7]**, there exist orthogonal idempotents $\{e_n; n < \aleph_0\}$ such that D_n is an essential submodule of Re_n for all $n < \aleph_0$. Then $I = \oplus \sum_{n<\aleph_0} Re_n$ satisfies $gen(I) = \aleph_0$, $D \neq I$, D is essential in I, whence D is not a summand in I. By 3.10, R is not $\aleph_1$-hereditary. □

REFERENCES

[AF] F.W.Anderson and K.R.Fuller, *Rings and Categories of Modules*, 2nd edition, Springer, New York, 1991.

[CT] R.Colpi and J.Trlifaj, *Test Modules and Representable Equivalences* (in preparation).

[E1] P.C.Eklof, *Homological algebra and set theory*, Trans.Amer.Math.Soc. **227** (1977), 207-225.

[E2] P.C.Eklof, *Set-theoretic methods: the uses of gamma invariants*, Abelian Groups (Proc. Conf. Curaçao 1991), M.Dekker, New York, 1993, pp. 43-53.

[EM] P.C.Eklof and A.H.Mekler, *Almost Free Modules*, North-Holland, New York, 1990.

[ES] P.C.Eklof and S.Shelah, *On Whitehead modules*, J. of Algebra **142** (1991), 492-510.

[F] C.Faith, *Algebra II - Ring Theory*, Springer, New York, 1976.

[G] K.R.Goodearl, *Von Neumann Regular Rings*, 2nd edition, Krieger, Melbourne, 1991.

[H] W.Hodges, *In singular cardinality, locally free algebras are free*, Alg. Universalis **12** (1981), 205-220.

[L] E.L.Lady, *Slender rings and modules*, Pac.J.Math. **49** (1973), 397-406.

[T1] J.Trlifaj, *Non-perfect rings and a theorem of Eklof and Shelah*, Comment.Math.Univ.Carolinae **32** (1991), 27-32.

[T2] J.Trlifaj, *Strong incompactness for some non-perfect rings*, Proc.Amer.Math.Soc. (1994) (to appear).

[W] R.Wisbauer, *Foundations of Module and Ring Theory*, Gordon and Breach, Philadelphia, 1991.

DEPARTMENT OF ALGEBRA, FACULTY OF MATHEMATICS AND PHYSICS,
CHARLES UNIVERSITY,
SOKOLOVSKA 83, 186 00 PRAGUE 8,
THE CZECH REPUBLIC

E-mail address: TRLIFAJ@CSPGUK11.BITNET

Contemporary Mathematics
Volume 171, 1994

A Functor from Mixed Groups to Torsion-Free Groups

W.J. Wickless

to Dick, a fine mathematician and friend

The Class $\mathcal{G}$

Throughout the paper all groups will be abelian. If we say that G is a mixed group we always will mean that G is an "honest" mixed group, that is, if $T = T(G)$ is the torsion subgroup of G, then $0 \neq T \neq G$. If G is mixed then "rank G" means "torsion-free rank G". We consider mixed groups in a class $\mathcal{G}$ defined as follows.

Definition 1 The class $\mathcal{G}$ is the class of all reduced mixed groups G such that:

1) G has finite rank and

2) G/T is divisible and

3) G is self-small ($\mathrm{Hom}(G,\oplus G_i) = \oplus_i \mathrm{Hom}(G,G_i)$ for any family of groups $\{G_i\}$ such that each $G_i \cong G$).

For a mixed group G and a prime p, let $T_p = T_p(G)$, be the p-torsion subgroup of G. The following theorem provides a useful alternate characterization of the class $\mathcal{G}$.

Subject Classification: Primary 20K21, Secondary 20K15
The paper is in final form and no version of it will be submitted elsewhere.

Theorem 1 For a reduced mixed group G the following are equivalent:

I) $G \in \mathcal{G}$

II) a) The group G/T is a nonzero finite dimensional Q-vector space and

b) each T_p is finite and

c) the inclusion map $\oplus T_p \to \Pi T_p$ can be extended to a pure embedding of G into ΠT_p and

d) if π_p is the projection of ΠT_p onto any fixed T_p and F is some maximal rank free subgroup of G, then $\pi_p(F) = T_p$ for almost all p. (In the statement of d) and henceforth, we suppress the pure embedding referred to in c) and regard G as a pure subgroup of ΠT_p.)

Proof See [AGW], Section 2.

The groups in $\mathcal{G}$ were first studied in [GW] in connection with the problem of finding mixed groups G such that the endomorphism ring E = E(G) is a von Neumann regular ring or a right principal projective ring. The paper [GW] was motivated by earlier papers of Rangaswamy [R1], [R2], [R3] and of Fuchs and Rangaswamy [FR] which investigated Baer, right principal projective and von Neumann regular endomorphism rings. In [GW] and [AGW] the following rational algebra played a key role.

Definition 2 Let $G \in \mathcal{G}$ and let V = G/T. Define

$A(G) = \{\bar{\alpha} \in \mathrm{End}_Q(V) \mid \bar{\alpha} \text{ is induced by } \alpha \in \mathrm{End}_Z(G)\}$.

It is easy to check that, for each $G \in \mathcal{G}$, A(G) will be a Q-subalgebra of End(V).

In [GW] it was shown that if A(G) is semisimple then E(G) is right principal projective. In [AGW] the flat dimension of groups in $\mathcal{G}$ as E-modules was computed. Specifically, the flat dimension of the module ${}_{E(G)}G$ was proved to be equal to the flat dimension of the module ${}_{A(G)}V$.

Let V be an arbitrary finite dimensional Q-vector space. The problem of characterizing those Q-subalgebras $A \subset \mathrm{End}_Q(V)$ which can be realized as A = A(G) for some $G \in \mathcal{G}$ was posed in [AGW]. More precisely, the question was considered: For which A does there exist a $G \in \mathcal{G}$ and an isomorphism $G/T \cong V$ such that, under identification via the isomorphism, A = A(G). In [VW-1] it was proved that A = A(G) in this sense if and only if there exists a full rank locally free subgroup $H \subset V$ such that A coincides with the quasi-endomorphism ring of H. Here

QEnd(H) is regarded as a subalgebra of $\mathrm{End}_Q(V)$ in the natural way. Thus the mixed group realization problem of [AGW] was shown to be equivalent to a torsion-free realization problem. This later problem of realizing an abstract subalgebra $A \subset \mathrm{End}(V)$ as $A = \mathrm{QEnd}(H)$ for $H \subset V$ was first studied in [PV].

The category $Q\mathcal{G}$

Our present paper grew out of a question of Manfred Dugas as to whether the correspondence $G \longmapsto H$ between mixed groups and torsion-free groups, used in [VW-1] to prove the equivalence of the mixed and torsion-free realization problems, could be made into some sort of functor between categories of mixed groups and torsion-free groups. I would like to thank him here for posing this very helpful question.

For $G' \in \mathcal{G}$ denote $T' = T(G')$, $T_p' = T_p(G')$ and $V' = G'/T'$.

Definition 3 The category $Q\mathcal{G}$ is the category with objects groups in $\mathcal{G}$ and morphisms $\mathrm{Hom}_{Q\mathcal{G}}(G,G') = \{\overline{\alpha} \in \mathrm{Hom}_Q(V,V') \mid \overline{\alpha} \text{ is induced by } \alpha \in \mathrm{Hom}_Z(G,G')\}$.

It is routine to check that $Q\mathcal{G}$ is an additive category. In this context the algebra A(G) is just $\mathrm{Hom}_{Q\mathcal{G}}(G,G)$. With notation as above, the kernel of the natural map $\alpha \to \overline{\alpha}$ from $\mathrm{Hom}_Z(G,G')$ to $\mathrm{Hom}_Q(V,V')$ is $\mathrm{Hom}_Z(G,T')$. Thus, for all G,G', the abelian groups $\mathrm{Hom}_{Q\mathcal{G}}(G,G')$ and $\mathrm{Hom}_Z(G,G')/\mathrm{Hom}_Z(G,T')$ are isomorphic. Hence $Q\mathcal{G}$ can be regarded as a full subcategory of the category WALK ([W-1]).

List the primes in their natural order. For $G \in \mathcal{G}$ and $k \geq 1$ denote $G_k = \oplus_{1\leq i\leq k} T_{p_i}$ and $G_k^* = G \cap \Pi_{p>p_k} T_p$ (recall that $G \subset \Pi T_p$). Since each T_p is finite, we have, for each k, $G = G_k \oplus G_k^*$.

Our next theorem gives some alternate characterizations of isomorphism in $Q\mathcal{G}$.

Theorem 2 Let $G,G' \in \mathcal{G}$. Then the following are equivalent:

a) $G \cong G'$ in $Q\mathcal{G}$

b) There exists k such that $G_k^* \cong G'^*_k$ as abelian groups.

c) There exists subgroups $H \subset G$, $H' \subset G'$, each of bounded index such that $H \cong H'$ as abelian groups.

Proof a) $\to$ b) Let $G, G' \in \mathcal{G}$ and $f\colon G \to T' = T(G')$. Since G is pure in ΠT_p and $\Pi T_p'$ is algebraically compact, we can extend f to an element $\Pi f_p \in \mathrm{Hom}(\Pi T_p, \Pi T_p')$. Let $F = \{x_1, ..., x_n\}$ be a maximal rank free subgroup of G.

Since $f(G) \subset T'$ and F is finitely generated, then $f(F) \subset \oplus_{i=1}^{s} T_{p_i}'$ for some $s > 0$. That is, denoting $x_{ip} = \pi_p(x_i)$, $f_p(x_{ip}) = 0$ for $1 \leq i \leq n$ and $p > p_s$. Using condition d) of Theorem 1, we have $f_p = 0$ for all sufficiently large $p \geq p_s$. We have shown that $\mathrm{Hom}_Z(G,T')$ is a torsion group.

Hence, tensoring Q into the exact sequence

$$0 \to \mathrm{Hom}(G,T') \to \mathrm{Hom}(G,G') \to \mathrm{Hom}(G,G')/\mathrm{Hom}(G,T') \to 0$$

yields the isomorphism $Q \otimes [\mathrm{Hom}(G,G')/\mathrm{Hom}(G,T')] \cong Q \otimes \mathrm{Hom}(G,G')$. It follows easily that if $G \cong G'$ in WALK then $G \cong G'$ in the category QAb, the category with objects abelian groups and morphisms $\mathrm{Hom}_{QAb}(G,G') = Q \otimes \mathrm{Hom}_Z(G,G')$.

Suppose $G \cong G'$ in $Q\mathcal{G}$. Then, by the remark following the definition of the category $Q\mathcal{G}$, $G \cong G'$ in WALK. From the previous paragraph, $G \cong G'$ in QAb. By the standard argument, there exists $f \in \mathrm{Hom}_Z(G,G')$, $g \in \mathrm{Hom}_Z(G',G)$ and a positive integer n, such that $gf = n\, 1_G$ and $fg = n\, 1_H$. If p_k is the largest prime occurring in the prime factorization of n, then the restrictions of the maps f, g to G_k^*, $G_k'^*$ will be inverse group isomorphisms.

The implication b) $\to$ c) is trivial.

For c) $\to$ a) suppose there exist subgroups $H \subset G$, $H' \subset G'$ and $t > 0$ such that $H \cong H'$ as abelian groups and with $t\, G \subset H$, $t\, G' \subset H'$. If p_k is the largest prime occurring in the prime factorization of t, then $H_k^* = G_k^*$ and $H_k'^* = G_k'^*$. Moreover, the isomorphism θ from H to H' will carry $H_k^* = G_k^*$ isomorphically onto $H_k'^* = G_k'^*$. Let $\overline{\theta} : V \to V'$ be the map induced by $0 \oplus \theta : G_k \oplus G_k^* \to G_k' \oplus G_k'^*$. It is easy to check that $\overline{\theta}$ is an isomorphism in $Q\mathcal{G}$. This completes the proof of the proposition.

Definition 4 A type τ is called a **nontrivial locally free type** if $\tau = [(k_p)]$ with $k_p < \infty$ for all primes p and $k_p > 0$ for infinitely many p. For short we say such a τ is an **nlf type.**

Definition 5 Let $\tau = [(k_p)]$ be an nlf type. A group $G \in \mathcal{G}$ is said to have outer type less than or equal to τ $(\mathrm{OT}(G) \leq \tau)$ if $p^{k_p} T_p(G) = 0$ for almost all primes p.

Definition 6 For an nlf type τ let $Q\mathcal{G}_\tau$ be the full subcategory of $Q\mathcal{G}$ with objects mixed groups $G \in \mathcal{G}$ such that $\mathrm{OT}(G) \leq \tau$. Let $Q\mathfrak{T}_\tau$ be the category with objects locally free torsion-free groups H of outer type less than or equal to τ

$(OT(H) \leq \tau)$ with no rank one summand of type τ, and morphisms $\mathrm{Hom}_{Q\mathcal{H}_\tau}(H,H') = \mathrm{QHom}(H,H')$.

It is easy to check that, for each nlf type τ, the categories $Q\mathcal{G}_\tau$ and $Q\mathcal{H}_\tau$ defined above are additive categories. We now can prove our main theorem.

Theorem 3 For each nlf type τ there is a rank-preserving category equivalence between the categories $Q\mathcal{G}_\tau$ and $Q\mathcal{H}_\tau$.

Proof Let $\tau = [(k_p)]$ be an nlf type. We will construct rank-preserving additive functors $f = f_\tau : Q\mathcal{G}_\tau \to Q\mathcal{H}_\tau$ and $g = g_\tau : Q\mathcal{H}_\tau \to Q\mathcal{G}_\tau$ such that $gf \approx 1_{Q\mathcal{G}_\tau}$ and $fg \approx 1_{Q\mathcal{H}_\tau}$.

Let $G \in Q\mathcal{G}_\tau$. Choose $F \subset G$ a maximal free subgroup. By Theorem 1 (d), the natural projection map $\pi_p : \Pi T_p \to T_p$ will map F onto T_p for almost all p. For these p there is a natural epimorphism $1 \otimes \pi_p : F_p = Z_p \otimes F \to T_p$. Let N_p be the kernel of $1 \otimes \pi_p$. Because $OT(G) \leq \tau$, we have $p^{k_p} F_p \subset N_p$ for almost all p. We define a locally free group $H = f(G)$ with $F \subset H \subset QF$ by defining each localization H_p. For primes p such that N_p has been defined and such that $p^{k_p} F_p \subset N_p$, set $H_p = p^{-k_p} N_p \subset QF$. For the (finitely many) other primes set $H_p = F_p$. If $H = \cap H_p$ then, plainly, $F \subset H \subset QH$, thus $\operatorname{rank} H = \operatorname{rank} F = \operatorname{rank} G$. Since, for all p, $H_p \subset p^{-k_p} F_p$, it follows that H is locally free with $OT(H) \leq \tau$. Furthermore, if F' is another maximal free subgroup of G, then F' and F are quasi-equal; thus $F'_p = F_p$ for almost all p. Hence, up to quasi-isomorphism, our construction of $H = f(G)$ is independent of the choice of the maximal free subgroup $F \subset G$.

To show that $f(G) \in Q\mathcal{H}_\tau$, we need to prove, additionally, that $f(G)$ can have no rank one summand of type τ. Suppose to the contrary that $f(G) = A \oplus B$, where A is a rank one group of type τ. Choose $0 \neq a \in A \cap F$ such that the p-height of a in $f(G)$ is greater than or equal to k_p for each prime p. It follows that $a \in p^{k_p} [f(G)]_p$ for each p. But, by construction, $p^{k_p} [f(G)]_p = N_p$ for almost all p. Thus, now regarding $a \in F \subset G$, $\pi_p(a) = 0$ for almost all p. Hence, $a \in \oplus T_p$. But $a \in F$ and $F \cap \oplus T_p = 0$, a contradiction.

Let $\varphi \in \mathrm{Hom}_{Q\mathcal{G}_\tau}(G,G')$ for some $G, G' \in Q\mathcal{G}_\tau$. Choose a map $\alpha \in \mathrm{Hom}_Z(G,G')$ such that α induces φ. Let $F \subset G$ and $F' \subset G'$ be maximal free subgroups. Since F is finitely generated and $F' \subset G'$ is a maximal free subgroup, it follows that $t\,\alpha(F) \subset F'$ for some $t > 0$. Thus $\alpha(F_p) \subset F'_p$ for almost all p. Since $\alpha(T_p) \subset T_p'$ for all p, we have $\alpha(N_p) \subset N'_p$ for almost all p. Thus α, regarded in the natural way as an element of $\mathrm{Hom}_Z(QF,QF')$, maps $p^{-k_p} N_p = f(G)_p$ to $p^{-k_p} N'_p = f(G')_p$ for almost all p. Therefore, α induces a quasi-homomorphism, which we call $f(\varphi)$ from $f(G)$ to $f(G')$. It is a simple exercise to see that $f(\varphi)$ is independent of the choice of α such that $\varphi = \overline{\alpha}$. It is also easy to check that, with the above definitions for $f(G)$ and $f(\varphi)$, f is an additive functor from $Q\mathcal{G}_\tau$ to $Q\mathcal{X}_\tau$.

To construct the inverse functor, let $H \in Q\mathcal{X}_\tau$. Since $OT(H) \le \tau$ we can choose a maximal free subgroup $F \subset H$ such that $p^{k_p}(H_p/F_p) = 0$ for all primes p. Moreover, the set $\{F_p/(p^{k_p}H_p) \mid p \text{ a prime}\}$ is an infinite set of finite p-groups; otherwise $H_p = p^{-k_p} F_p$ for almost all p, implying that H is homogeneous completely decomposable of type τ, contrary to the definition of $Q\mathcal{X}_\tau$. Denote $T_p = F_p/(p^{k_p}H_p)$ and define g(H) to be the pure subgroup of ΠT_p generated by $T = \oplus T_p$ and the canonical image F_1 of F in ΠT_p. (This pure subgroup is just the inverse image of $Q[(F_1 + T)/T]$ under the natural map $\Pi T_p \to \Pi T_p/T$.)

Note that $F \cong F_1$ since, if $x \in F$ with $x \in p^{k_p} H_p$ for almost all p, then $\mathrm{type}(x) \ge \tau$. Since $OT(H) \le \tau$, the pure subgroup generated by x would be a rank one summand of type τ ([VW-2], Proposition 1.7), contrary to hypothesis. Hence rank g(H) = rank H and g(H) is an honest mixed group. It is easy to see that $g(H) \in Q\mathcal{G}_\tau$ and that, up to isomorphism in $Q\mathcal{G}_\tau$, g(H) is independent of the choice of the maximal free subgroup $F \subset H$.

Let a be a quasi-homorphism from H into H', for some H, H' in $Q\mathcal{X}_\tau$. Let $F \subset H$, $F' \subset H'$ be maximal free subgroups. Then, $t\,a(F) \subset F'$ for $t > 0$ and, arguing as before, a induces a map a_p from $T_p = F_p/(p^{k_p} H_p)$ to $T_p' = F'_p/(p^{k_p} H'_p)$ for almost all p. Set $a_p = 0$ for the finitely many exceptional primes p, and denote $\alpha = \Pi a_p \in \mathrm{Hom}(\Pi T_p, \Pi T_p')$. If F_1, F_1' are the canonical images of F, F' in ΠT_p, $\Pi T_p'$, then $s\,\alpha(F_1) \subset F_1'$ for $s > 0$. Since g(H') is pure in $\Pi T_p'$ it follows that $\alpha(F_1) \subset g(H')$. Plainly $\alpha(\oplus T_p) \subset \oplus T_p' \subset g(H')$. Now a simple computation shows that $\alpha[g(H)] \subset g(H')$. (Or see [GW], Lemma 3.0.) Thus we can regard $\alpha \in \mathrm{Hom}_Z[g(H),g(H')]$. Define $g(a) = \overline{\alpha}$, where $\overline{\alpha}$ is the map from $V = g(H)/T[g(H)]$ to $V' = g(H')/T[g(H')]$ induced by α. It is not hard to see that the map $\overline{\alpha}$, as constructed, is independent of the choices of F, F'. It is easy to check that, with these definitions, $g : \mathcal{X}_\tau \to \mathcal{G}_\tau$ is an additive functor.

Finally, a routine verification shows that $fg \approx 1_{Q\mathcal{X}_\tau}$ and $gf \approx 1_{Q\mathcal{G}_\tau}$.

Applications

The category QTF whose objects are torsion-free abelian groups of finite rank and whose morphisms are quasi-homomorphisms is well known to be a Krull-Schmidt category; that is every object in QTF has, up to isomorphism in QTF and permutation of summands, a unique decomposition into a direct sum of indecomposables in QTF. Since, for each nlf type τ, the category $Q\mathcal{X}_\tau$ is a full subcategory of QTF and since the objects in $Q\mathcal{X}_\tau$ are closed under abelian group direct summands, each category $Q\mathcal{X}_\tau$ is also Krull-Schmidt. We can use our category equivalences to make the same claim for our mixed group category $Q\mathcal{G}$.

Theorem 4 The category $Q\mathcal{G}$ is a Krull-Schmidt category. The category $Q\mathcal{G}$ contains indecomposables of arbitrary finite rank.

Proof Since the groups in $\mathcal{G}$ have finite nonzero rank, each object of $Q\mathcal{G}$ certainly can be decomposed as a finite direct sum of indecomposables in $Q\mathcal{G}$. Suppose $G = \oplus_{1\le i\le n} K_i$ and $G = \oplus_{1\le j\le m} L_j$ are direct decompositions of $G \in Q\mathcal{G}$ into indecomposables. Let τ be an nlf type large enough so that G, the K_i's and the L_j's are all objects in $Q\mathcal{G}_\tau$. The two direct sum decompositions, regarded in the category $Q\mathcal{G}_\tau$, remain direct sum decompositions of G into indecomposables. Since $Q\mathcal{G}_\tau \approx Q\mathcal{X}_\tau$ and $Q\mathcal{X}_\tau$ is a Krull-Schmidt category, then so is $Q\mathcal{G}_\tau$. Thus the two decompositions are equivalent in $Q\mathcal{G}_\tau$, hence in $Q\mathcal{G}$, and the proof of the first claim is complete. Since the functors f, g preserve ranks the second claim is obvious.

Remark We remark that it is easy to prove, directly from the definitions, that if K is a group direct summand of $G \in \mathcal{G}$, then either $K \in \mathcal{G}$ or K is finite. Thus $G \in Q\mathcal{G}$ is indecomposable if and only if, for all group direct sum decompositions $G = G_1 \oplus G_2$, either G_1 or G_2 is finite.

For any nlf type τ let QTF_τ be the full subcategory of QTF whose objects are tffr groups H with $OT(H) \le \tau$. Let A be a rank one group of type τ. A classical result of Warfield ([W-2]) is that the functor $w_A(H) = \mathrm{Hom}_Z(\_\_,A)$ is a duality on QTF_τ, that is a contravariant functor on QTF_τ such that $w_A{}^2 \approx 1_{QTF_\tau}$.

Let $Q\mathcal{X}_{\tau}^{o}$ be the full subcategory of QTF_{τ} with objects tffr groups of outer type $\leq \tau$, having no summands isomorphic to either A or Z. Since the functor w_A interchanges copies of A and Z, the restriction of w_A to $Q\mathcal{X}_{\tau}^{o}$ provides a duality on $Q\mathcal{X}_{\tau}^{o}$. We can use this fact, together with our category equivalences, to define a "Warfield Duality" for certain categories of mixed groups.

Definition 7 For a nlf type $\tau = [(k_p)]$ let $G_0(\tau)$ be the pure subgroup of $\Pi\, Z(p^{k_p})$ generated by $\oplus\, Z(p^{k_p})$ and the element (1, 1, 1, ...,). Let $Q\mathcal{G}_{\tau}^{o}$ be the full subcategory of $Q\mathcal{G}_{\tau}$ with objects groups in $Q\mathcal{G}_{\tau}$ having no summand quasi-isomorphic to $G_0(\tau)$.

Theorem 5 For each nlf type τ the category $Q\mathcal{G}_{\tau}^{o}$ admits a duality.

Proof Note that $G_0(\tau)$ is the image of Z under our functor $g : Q\mathcal{X}_{\tau} \to Q\mathcal{G}_{\tau}$. Thus, $Q\mathcal{X}_{\tau}^{o} \approx Q\mathcal{G}_{\tau}^{o}$ via the restriction of the functors g, f. Let $D : Q\mathcal{G}_{\tau}^{o} \to Q\mathcal{G}_{\tau}^{o}$ be defined as a composition of functors $D = g\, w_A\, f$. It is not difficult to check that D is a contravariant functor on $Q\mathcal{G}_{\tau}^{o}$ such that $D^2 \approx 1_{Q\mathcal{G}_{\tau}^{o}}$.

It follows directly from Theorem 1 of [M] that rank one groups in $\mathcal{G}$ are determined up to isomorphism by their torsion subgroups. For higher rank groups in $\mathcal{G}$ with isomorphic torsion subgroups, we obtain a result which is no surprise. First we need a definition.

Definition 8 Call a group $G \in Q\mathcal{G}_{\tau}$ **completely decomposable** if it is a direct sum of rank one groups in QG_{τ} (equivalently, if $f(G) \in Q\mathcal{X}_{\tau}$ is (quasi) isomorphic to a completely decomposable torsion-free group).

Theorem 6 For each nlf type $\tau = (k_p)$ there are uncountably many non-isomorphic rank two completely decomposable groups $G \in QG_{\tau}$, each with torsion subgroup isomorphic to $T = \oplus_p Z(p^{k_p})$.

Proof Employing an uncountable family $\{P_{\gamma}\}$ of almost disjoint infinite subsets of the infinite set of primes $P_+ = \{p \mid k_p > 0\}$, it is routine to construct rank two completely decomposable torsion-free groups $H_{\gamma} \supset Z \oplus Z$ such that: (1) the H_{γ}'s are pairwise non quasi-isomorphic and (2) each $H_\gamma/(Z \oplus Z) \cong T$. The family $\{g(H_\gamma)\}$ will be as advertised.

In some contrast to Theorem 6, we show that the completely decomposable mixed groups in $\mathcal{G}$ are determined in $\mathcal{G}$ by their Z-endomorphism rings.

Theorem 7 Let $G, G' \in Q\mathcal{G}$ with G completely decomposable. Suppose the rings $End_Z(G)$ and $End_Z(G')$ are isomorphic and that rank G = rank G'. Then the groups G and G' are isomorphic.

Proof Since $G \in \mathcal{G}$ we can regard $E = End_Z(G)$ as a pure subring of ΠE_p where $E_p = End_Z(T_p)$. Moreover, in this identification, each E_p will be the ideal of p-torsion elements of E. (See [GW].) Thus, employing the obvious notation, if $E \cong E'$, then each $E_p \cong E_p'$. By the Baer-Kaplansky Theorem, $T_p \cong T_p'$ for all p. That is, G and G' will have isomorphic torsion subgroups. Thus, in view of Theorem 2, to prove our result it will suffice to prove that $G \cong G'$ in $Q\mathcal{G}$.

Suppose rank $G = n$. Since G is completely decomposable we can, again employing Theorem 2, choose a subset of orthogonal idempotents $\{e_1,...,e_n\}$ of E such that $G = \oplus_{1\le i\le n} e_iG \oplus L$, with each $e_iG \in \mathcal{G}$ of rank one and L finite. Since $E \cong E'$ there is a corresponding set of orthogonal idempotents $\{e_1',...,e_n'\}$ in E'. Thus we can obtain an analogous group direct sum decomposition $G' = \oplus_{1\le i\le n} e_i'G' \oplus L'$ with $L' = (1 - \Sigma_{1\le i\le n} e_i')\, G'$ also finite.

To prove that $G \cong G'$ in $Q\mathcal{G}$ we will prove that, as groups, $e_iG \cong e_i'G'$ for all i. First, for each i, $End_Z(e_iG) \cong e_iEe_i \cong e_i'E'e_i' \cong End_Z(e_i'G')$. Since rank $e_iG = 1$, $End_Z(e_i'G') \cong End_Z(e_iG)$ is infinite. Thus, we cannot have $e_i'G'$ finite. By the remark following Theorem 4, each $e_i'G' \in Q\mathcal{G}$. In particular, rank $e_i'G' \ge 1$ for all i. Since rank G' = rank $G = n$, it follows that rank $e_i'G' = 1$ for all i. Looking back to the argument in the first paragraph of this proof, since e_iG and $e_i'G'$ are groups in $Q\mathcal{G}$ with isomorphic endomorphism rings, then e_iG and $e_i'G'$ have isomorphic torsion subgroups. Because e_iG and $e_i'G'$ are of rank one, in view of the remarks preceding Theorem 6, we can conclude that $e_iG \cong e_i'G'$ for all $1 \le i \le n$. Hence the proof of Theorem 7 is complete.

Examples Let $T = \oplus_p Z(p)\, u_p$ (the symbol u_p is just a placeholder). Partition the set of primes P into disjoint infinite subsets $P = P_1 \cup P_2 \cup P_3 = P_1 \cup P_4 \cup P_5$. Choose the subsets such that, for $i \in \{2,3\}$ and $j \in \{4,5\}$, both $P_i \backslash P_j$ and $P_j \backslash P_i$ are infinite. We define G_1, G_2, G_3 in $\mathcal{G}$ as follows. Let G_1 be the pure subgroup of $\Pi = \Pi_p Z(p)\, u_p$ generated by T and the torsion-free element (u_p). Let

G_2 be the pure subgroup of Π generated by T and (x_p), (y_p) where, for $p \in P_1$, $x_p = y_p = u_p$, for $p \in P_2$, $x_p = u_p$ and $y_p = 0$, and, for $p \in P_3$, $x_p = 0$ and $y_p = u_p$. Let G_3 be the pure subgroup of Π generated by T and (z_p), (w_p) where, for $p \in P_1$, $z_p = w_p = u_p$, for $p \in P_4$, $z_p = u_p$ and $w_p = 0$, and, for $p \in P_5$, $z_p = 0$ and $w_p = u_p$. For $1 \le i \le 3$, looking at $\mathrm{End}_{Q\mathcal{G}}(G_i)$, it is not hard to show that each $\mathrm{End}_Z(G_i)$ is isomorphic to the subring R of the ring $\Pi_p Z/pZ$ described as follows : $R = \{(t_p) \mid t_p \equiv q \pmod p$ for some fixed $q \in Q$ and almost all $p\}$. Considering G_1 and G_2, we see that groups in $\mathcal{G}$ of different ranks can have isomorphic endomorphism rings. Thus, the hypothesis rank G = rank G' of Theorem 7 was necessary. The groups G_2 and G_3 are examples of (non completely decomposable) non isomorphic groups with isomorphic endomorphism rings and equal ranks. Thus, the assumption that G be completely decomposable was also necessary for Theorem 7.

REFERENCES

[AGW] U. Albrecht, H. P. Goeters and W. Wickless, The flat dimension of mixed abelian groups as E-modules, to appear in Rocky Mt. J. Math.

[FR] L. Fuchs and K. Rangaswamy, On generalized regular rings, Math Z., 107 (1968), 71-81.

[GW] S. Glaz and W. Wickless, Regular and principal projective endomorphism rings of mixed abelian groups, to appear in Comm. in Algebra.

[M] C. Megibben, On mixed groups of torsion-free rank one, Ill. J. Math., 11 (1967), 134-44.

[PV] R. Pierce and C. Vinsonhaler, Quasi-realizing modules, Lecture Notes in Pure and Applied Math., 146 (1993), 219-29.

[R-1] K. Rangaswamy, Representing Baer rings as endomorphism rings, Math Ann., 190 (1970), 167-76.

[R-2] _______________, Regular and Baer rings, Proc. A.M.S., 42 (1974), 354-58.

[R-3] _______________, Abelian groups with endomorphic images of special types, J. Alg., 6 (1967), 271-80.

[VW-1] C. Vinsonhaler and W. Wickless, Realizations of finite dimensional algebras over the rationals, to appear in Rocky Mt. J. Math.

[VW-2] ______________, Projective and injective classes of completely decomposable groups, Springer-Verlag Lecture Notes in Math., 1006 (1983), 144-63.

[W-1] R. Warfield, The structure theory of mixed abelian groups, Springer-Verlag Lecture Notes in Math., 616 (1977), 1-38.

[W-2] ______________, Homomorphisms and duality for torsion-free groups, Math Z., 107 (1968), 189-200.

University of Connecticut
Storrs, Ct., 06269
USA

Contemporary Mathematics
Volume 171, 1994

A CHARACTERIZATION OF A CLASS OF BUTLER GROUPS II

PETER D. YOM

We will consider a class of Butler groups which appear as pure subgroups of completely decomposable torsion-free groups of finite rank. Let $A_1,\ldots,A_n$ $(n \geq 2)$ be subgroups of the additive group of rationals $\mathbf{Q}$ which contain the integers $\mathbf{Z}$. Then $\mathcal{G}(A_1,\ldots,A_n)$ is the kernel of the summation map $A_1 \oplus \cdots \oplus A_n \to \Sigma A_i \subseteq \mathbf{Q}$. The class of Butler groups of the form $\mathcal{G}(A_1,\ldots,A_n)$ was first investigated by F. Richman [**5**]. D. Arnold and C. Vinsonhaler introduced quasi-representing graphs [**1**] for groups of the form $\mathcal{G}(A_1,\ldots,A_n)$ and used them to produce complete sets of numerical quasi-isomorphism invariants for the subclass of CT-groups [**2**, Theorem 2.3] and later for strongly indecomposable groups [**3**, Theorem 6]. In [**6**] the author developed an algorithm called the *two-vertex exchange* [**6**, Theorem 6] for the class of strongly indecomposable groups $\mathcal{G}(A_1,\ldots,A_n)$. Two-vertex exchanges produce n-tuples $(B_1,\ldots,B_n)$ from $(A_1,\ldots,A_n)$ so that $\mathcal{G}(B_1,\ldots,B_n)$ is quasi-isomorphic to $\mathcal{G}(A_1,\ldots,A_n)$. In this paper, using quasi-representing graphs, we will show that the same two-vertex exchange also characterizes groups

Mathematical Subject Classification Primary 20k15
This paper is in final form, and no version of it will be submitted for publication elsewhere.

$\mathcal{G}(A_1,\ldots,A_n)$ without the strongly indecomposability condition. We give a characterization theorem for the class of CT-groups and some applications.

1. TWO-VERTEX EXCHANGE. Throughout the paper, the term *group* will mean torsion-free abelian group of finite rank. Standard notation and terminology are as in [**1**]. For a positive integer n, let $\Omega_n = \{1,...,n\}$ and S_n be the symmetric group of order n. An n-tuple $(A_1,\ldots,A_n)$ of subgroups of $\mathbf{Q}$ is called *trimmed* if $A_i \subseteq \Sigma\{A_j | \ j \neq i\}$ for each $i \in \Omega_n$. Whenever we employ the notation $\mathcal{G}(A_1,\ldots,A_n)$ it is tacitly assummed that the n-tuple $(A_1,\ldots,A_n)$ is trimmed. If $B_i \cong A_{\rho(i)}$ for some $\rho \in S_n$ then $\mathcal{G}(A_1,\ldots,A_n)$ is quasi-isomorphic to $\mathcal{G}(B_1,\ldots,B_n)$. Hence, the quasi-isomorphism class of $G = \mathcal{G}(A_1,\ldots,A_n)$ is determined by the types τ_i = type A_i. Hereafter we will refer to the group G as an n-tuple $(\tau_1,\ldots,\tau_n)$, keeping in mind that it is defined up to quasi-isomorphism. In this notation we say an n-tuple $(\sigma_1,\ldots,\sigma_n)$ is quasi-isomorphic to G if there is a group $\mathcal{G}(B_1,\ldots,B_n)$ quasi-isomorphic to G and σ_i = type B_i for each $i \in \Omega_n$.

Let $G = (\tau_1,\ldots,\tau_n)$ and define C_G be the complete graph with vertices $\tau_1,\ldots,\tau_n$ and edges $i{:}j$ labeled by $\tau_i \wedge \tau_j$ for $i \neq j$. A *representing graph* for G is any subgraph of C_G that is obtained by iteration of the algorithm: if a graph contains a circuit S with edge labels $\geq \tau$ and at least one edge labeled by τ, then remove an edge of S labeled by τ. A labeled graph is a *quasi-representing graph* for a Butler group H if it is a representing graph for some $(\sigma_1,\ldots,\sigma_n)$ which is quasi-isomorphic to H. Suppose T is a quasi-representing graph for a Butler group H, with n vertices and with edges $i{:}j$ labeled by τ_{ij}. If we let $\tau_i = \sup\{\tau_{ij} \mid i \neq j,\ i{:}j \in \mathrm{T}\}$ for each $i \in \Omega_n$, then T is a representing graph for $(\tau_1,\ldots,\tau_n)$, and $(\tau_1,\ldots,\tau_n)$ is quasi-isomorphic to H [**1**, Theorem 1.6].

Let $\mathcal{Q}(G)$ be the set of quasi-representing graphs for G. Given T $\in \mathcal{Q}(G)$ and U $\in \mathcal{Q}(H)$, define $\mathcal{L}$(T,U) be the set of label-preserving bijections φ from the edges of T to the edges of U such that S is a circuit in T iff $\varphi(S)$ is a circuit in U. For a nonempty subset X of Ω_n, we write $\tau^X = \sup\{\tau_i \mid i \in X\}$.

LEMMA 1. [**1**, Corollary 1.7] *Let* T $\in \mathcal{Q}(G)$, U $\in \mathcal{Q}(H)$. *If* $\mathcal{L}$(T,U) *is nonempty then* G *is quasi-isomorphic to* H.

LEMMA 2. *Let* $G = (\tau_1,\ldots,\tau_n)$. *If* $\tau_k \wedge \tau_m \leq \tau_1$ *for some* k,m *with* $2 \leq k \neq m \leq n$, *then there is a quasi-representing graph for* G *in* $C_G \setminus \{k{:}m\}$.

Proof. Let $S = \tau_k,\tau_m,\tau_1$ be a circuit in C_G with edges $k{:}m, m{:}1, 1{:}k$. Then $\tau_k \wedge \tau_m \leq \tau_1 \wedge \tau_k$ and $\tau_k \wedge \tau_m \leq \tau_1 \wedge \tau_m$ by the hypothesis. By the algorithm for quasi-representing graphs, the edge $k{:}m$ may be deleted from S. Thus, there is a quasi-representing graph for G in $C_G \setminus \{k{:}m\}$. ■

COROLLARY 3. *Let* $G = (\tau_1,\ldots,\tau_n)$. *Suppose* $\{X,Y\}$ *is a partition of* $\{3,\ldots,n\}$ *such that* $\tau^X \wedge \tau^Y \leq \tau_1 \wedge \tau_2$. *Then there exists* T $\in \mathcal{Q}(G)$ *such that* (i) $k{:}m \notin$ T *for all* $k \in X$, $m \in Y$ (ii) *any path in* T *connecting* τ_k *and* τ_m *contains* τ_1 *or* τ_2.

Proof. By recursive applications of Lemma 2, there is a quasi-representing graph T for G in $C_G \setminus \{k{:}m \mid k \in X, m \in Y\}$. Since T is a connected graph and $\{X,Y\}$ is a partition of $\{3,\ldots,n\}$, any path in T connecting τ_k and τ_m must contain τ_1 or τ_2. ■

Let $G = (\tau_1,\ldots,\tau_n)$ and $T \in \mathcal{Q}(G)$. For each type τ, define the subgraph $T(\tau)$ be the set of all edges labeled $\geq \tau$. It is easy to see that $T(\tau)$ is connected. For each type τ, define an equivalence relation on vertices of T as follows: τ_i and τ_j are τ-equivalent in G provided $i = j$ or τ_i and τ_j are connected by a path in $T\backslash T(\tau)$.

THEOREM 4. (Two-vertex exchange) *Let* $G = (\tau_1,\tau_2,\tau_3,\ldots,\tau_n)$ *and* $H = (\sigma_1,\sigma_2,\tau_3,\ldots,\tau_n)$ *with* $\tau_i \neq \sigma_j$ *for* $i, j \in \{1,2\}$. *Then the following are equivalent:*

(a) *G and H are quasi-isomorphic;*

(b) rank $G/G[\tau]$ = rank $H/H[\tau]$ *and* rank $G(\tau)$ = rank $H(\tau)$ *for each type* τ;

(c) *There is a partition* $\{X,Y\}$ *of* $\{3,\ldots,n\}$ *such that*

(i) $\tau^X \wedge \tau^Y \leq \tau_1 \wedge \tau_2$.

(ii) $\sigma_1 = (\tau_1 \wedge \tau_2) \vee (\tau_2 \wedge \tau^X) \vee (\tau_1 \wedge \tau^Y)$.

(iii) $\sigma_2 = (\tau_1 \wedge \tau_2) \vee (\tau_1 \wedge \tau^X) \vee (\tau_2 \wedge \tau^Y)$.

(d) $\mathcal{L}(T,U)$ *is nonempty for some* $T \in \mathcal{Q}(G)$ *and* $U \in \mathcal{Q}(H)$.

Proof. (a)$\Rightarrow$(b) is clear. (b)$\Rightarrow$(c) By Corollary 2.1(b) in [**1**], rank$G(\tau)$+1 = the number of τ_i's with $\tau_i \geq \tau$. Let $\tau = \tau_1 \wedge \tau_2$. Since rank$G(\tau)$ = rank$H(\tau)$, it follows that $\sigma_1, \sigma_2 \geq \tau$. Thus, $\tau_1 \wedge \tau_2 \leq \sigma_1 \wedge \sigma_2$ and, by a symmetrical argument, we have $\tau_1 \wedge \tau_2 = \sigma_1 \wedge \sigma_2$. A similar argument gives rise to functions $e, f\colon \{3,\ldots,n\} \to \{1,2\}$ such that

$$\left.\begin{array}{r}\tau_1 \wedge \tau_i = \sigma_{e(i)} \wedge \tau_i \text{ for } 3 \leq i \leq n \\ \tau_2 \wedge \tau_i = \sigma_{f(i)} \wedge \tau_i \text{ for } 3 \leq i \leq n \text{ and } \{e(i), f(i)\} = \{1,2\}\end{array}\right\} \tag{1.1}$$

We will show that e can be chosen to satisfy

$$\text{if } 3 \leq i, j \leq n \text{ and } \tau_i \wedge \tau_j \nleq \tau_1 \wedge \tau_2, \text{ then } e(i) = e(j) \tag{1.2}$$

In this case, we let $X = \{i \mid e(i) = 1\}$ and $Y = \{i \mid e(i) = 2\}$. Then $\{X,Y\}$ is a partition of $\{3,\dots,n\}$ and (1.2) implies that (i) is satisfied. Since H is trimmed and $\tau_1 \wedge \tau_2 = \sigma_1 \wedge \sigma_2$, (ii) and (iii) follow as consequences of (1.1).

We now use the hypothesis that rank $G/G[\tau]$ = rank $H/H[\tau]$ to verify that e can be chosen to satisfy (1.2). By Corollary 2.1(e) in [1], rank $G/G[\tau] = k(\tau)-1$, where $k(\tau)$ is the number of τ-equivalence classes. Note that $\{\tau_i\}$ is a τ_i-equivalence class for each $i \in \Omega_n$. Since rank $G/G[\tau]$ = rank $H/H[\tau]$, it follows that σ_1, σ_2 are in distinct τ-equivalence classes in H for $\tau = \tau_1$ or τ_2. Similarly τ_1, τ_2 are in distinct τ-equivalence classes in G for $\tau = \sigma_1$ or σ_2.

Fix i and j such that $\tau_i \wedge \tau_j \nleq \tau_1 \wedge \tau_2$ and we prove that $e(i)$ can be chosen to be equal to $e(j)$. In view of (1.1), we have $\tau_1 \wedge \tau_i = \sigma_k \wedge \tau_i$, $\tau_2 \wedge \tau_i = \sigma_m \wedge \tau_i$ and $\tau_1 \wedge \tau_j = \sigma_m \wedge \tau_j$, $\tau_2 \wedge \tau_j = \sigma_k \wedge \tau_j$, where $\{k,m\} = \{1,2\}$. We claim that $\tau_i \wedge \tau_j \nleq \tau_1$. Otherwise, $\tau_i \wedge \tau_j = \tau_1 \wedge \tau_i \wedge \tau_j = \sigma_k \wedge \tau_i \wedge \sigma_m \wedge \tau_j \leq \sigma_k \wedge \sigma_m = \tau_1 \wedge \tau_2$. This is a contradiction to the assumption. A similar argument shows that $\tau_i \wedge \tau_j \nleq \tau_2, \sigma_1$, or σ_2.

Next, we show that either $\tau_1 \wedge \tau_i = \tau_2 \wedge \tau_i$ or $\tau_1 \wedge \tau_j = \tau_2 \wedge \tau_j$. In which case $e(i)$, respectively $e(j)$, can be chosen to be either 1 or 2, hence chosen so that $e(i) = e(j)$, as desired. Assume $\tau_1 \wedge \tau_i \nleq \tau_2 \wedge \tau_i$ and we will show that $\tau_1 \wedge \tau_j = \tau_2 \wedge \tau_j$. First, observe that $\tau_1 \wedge \tau_i \nleq \tau_2$. If $\tau_1 \wedge \tau_j \nleq \tau_2 \wedge \tau_j$, then $\tau_1 \wedge \tau_j \nleq \tau_2$. Hence, we now have a sequence of vertices $\sigma_k,\tau_i,\tau_j,\sigma_m$ connecting σ_1, σ_2 and the edges $k{:}i$, $i{:}j$, $j{:}m$ are labeled $\nleq \tau_2$ (because $\tau_1 \wedge \tau_i = \sigma_k \wedge \tau_i$ and $\tau_1 \wedge \tau_j = \sigma_m \wedge \tau_j$ from the above paragraph). This is a contradiction to the fact that σ_1, σ_2 are in distinct τ_2-equivalence classes in H. Therefore, we must have that $\tau_1 \wedge \tau_j \leq \tau_2 \wedge \tau_j$. We can use a similar argument and the fact that τ_1, τ_2 are in distinct σ_m- equivalence classes in G to show that $\tau_2 \wedge \tau_j \leq \tau_1 \wedge \tau_j$ (In this case, the sequence of vertices $\tau_1,\tau_i,\tau_j,\tau_2$ has edges $1{:}i$, $i{:}j$, $j{:}2$ with labels $\nleq \sigma_m$). Hence, if $\tau_1 \wedge \tau_i \nleq \tau_2 \wedge \tau_i$, we have that $\tau_1 \wedge \tau_j = \tau_2 \wedge \tau_j$ as desired. The other case is similar.

(c)⇒(d) By Corollary 3, assume T is a quasi-representing graph for G in $C_G \setminus \{k{:}m \mid k \in X,\ m \in Y\}$. Since the lattice of types is distributive, we can show that

$$\left.\begin{array}{c} \sigma_1 \wedge \sigma_2 = \tau_1 \wedge \tau_2 \\ \sigma_1 \wedge \tau_k = \tau_1 \wedge \tau_k \text{ and } \sigma_2 \wedge \tau_k = \tau_2 \wedge \tau_k \text{ if } k \in X \\ \sigma_1 \wedge \tau_m = \tau_2 \wedge \tau_m \text{ and } \sigma_2 \wedge \tau_m = \tau_1 \wedge \tau_m \text{ if } m \in Y \end{array}\right\} \tag{1.3}$$

Since $\sigma_1 \wedge \sigma_2 = \tau_1 \wedge \tau_2$, it follows that with the same partition $\{X,Y\}$ of $\{3,\ldots,n\}$ we have $\tau^X \wedge \tau^Y \leq \sigma_1 \wedge \sigma_2$. Let $W = C_H \setminus \{k{:}m \mid k \in X,\ m \in Y\}$. We will show that $\mathcal{L}(T,U)$ is nonempty for some quasi-representing graph U for H with $U \subseteq W$.

In the sequel, we will write edges $i{:}j$ with label equal to $\tau_i \wedge \tau_j$ by $\tau_i\tau_j$. By (1.3), we define a label-preserving bijection φ between the edges of C_G and C_H as follows:

$$\left.\begin{array}{c} \varphi(\tau_i\tau_j) = \tau_i\tau_j \text{ for } 3 \leq i \neq j \leq n \text{ and } \varphi(\tau_1\tau_2) = \sigma_1\sigma_2 \\ \varphi(\tau_1\tau_k) = \sigma_1\tau_k \text{ and } \varphi(\tau_2\tau_k) = \sigma_2\tau_k \text{ if } k \in X \\ \varphi(\tau_2\tau_m) = \sigma_1\tau_m \text{ and } \varphi(\tau_1\tau_m) = \sigma_2\tau_m \text{ if } m \in Y \end{array}\right\} \tag{1.4}$$

We claim that $U = \varphi(T) \in \mathcal{Q}(H)$ and $\mathcal{L}(T,U)$ is nonempty. We will first show $U \in \mathcal{Q}(H)$. By Theorem 1.6 in [1] it suffices to show that there is no circuit in $U(\tau)$ with an edge labeled by τ and if e is an edge not in U and labeled by τ, then there is a circuit in $U(\tau) \cup \{e\}$.

Let τ be an arbitrary type and suppose S is a circuit in $U(\tau)$ with an edge labeled by τ. Then $\varphi^{-1}(S)$ is a circuit in $T(\tau)$ with an edge labeled by τ. Since T is a quasi-representing graph for G, this is impossible. Thus, there is no circuit in $U(\tau)$ with an edge labeled by τ. Let e be an edge in $W\setminus U$ labeled by τ and e' be the corresponding edge in T labeled by τ as defined in (1.4) above. For example, if $e = \sigma_1\tau_i$ for $i \in X$ (or Y) then $e' = \tau_1\tau_i$ (or $\tau_2\tau_i$). Note that e' is not an edge in T because if e' were an edge in T, then $e = \varphi(e') \in \varphi(\mathrm{T})$ is an edge in U. Since $\mathrm{T} \in \mathcal{Q}(G)$, there is a circuit S in $\mathrm{T}(\tau) \cup \{e'\}$. By (1.4), it is clear that there is a circuit in $\varphi(S) \cup \{e\} \subset \mathrm{U}(\tau) \cup \{e\}$. Thus, $\mathrm{U} \in \mathcal{Q}(H)$.

Note that φ is a label-preserving bijection between the edges of T and U. We will show that S is a circuit in T iff $\varphi(S)$ is a circuit in U. Let S be a simple circuit in T. If $S \subseteq \{\tau_3,\ldots,\tau_n\}$, then $\varphi(S) = S$. Suppose $S \cap \{\tau_1,\tau_2\} = \tau_1$. Since $k{:}m \notin S$ for all $k \in X$ and $m \in Y$, it follows that $S = \tau_1,\tau_{x_1},\ldots,\tau_{x_t}$ for $x_i \in X$ (respectively Y). That is, S has edges $1{:}x_1,x_1{:}x_2,\ldots,x_t{:}1$ with $t \geq 3$. In this case, $\varphi(S) = \sigma_1,\tau_{x_1},\ldots,\tau_{x_t}$ (respectively $\sigma_2,\tau_{x_1},\ldots,\tau_{x_t}$) is a circuit. Similarly, we can treat the case $S \cap \{\tau_1,\tau_2\} = \tau_2$. If $\tau_1\tau_2$ is an edge of S and $S = \tau_1,\tau_2,\tau_{x_1},\ldots,\tau_{x_t}$ with $x_i \in X$ (respectively Y), then $\varphi(S) = \sigma_1,\sigma_2,\tau_{x_1},\ldots,\tau_{x_t}$ (respectively $\sigma_2,\sigma_1,\tau_{x_1},\ldots,\tau_{x_t}$) is a circuit. Suppose $\tau_1\tau_2$ is not an edge of S. Then $S = \tau_1,\tau_{x_1},\ldots,\tau_{x_r},\tau_2,\tau_{y_1},\ldots,\tau_{y_t}$ with $x_i \in X$ and $y_i \in X$ (or Y). Hence, $\varphi(S) = \sigma_1,\tau_{x_1},\ldots,\tau_{x_r},\sigma_2,\tau_{y_1},\ldots,\tau_{y_t}$ (or $\sigma_1,\tau_{x_1},\ldots,\tau_{x_r},\sigma_2,\tau_{y_t},\ldots,\tau_{y_1}$) is a circuit. And similarly, we can treat the case $x_i \in Y$ and $y_i \in X$ (or Y). Therefore, in all cases, $\varphi(S)$ is a circuit in $\varphi(\mathrm{T})$. With symmetric arguments, we can show that if $\varphi(S)$ is a circuit in $\varphi(\mathrm{T})$ then S is a circuit in T. Hence, we showed S is a circuit in T iff $\varphi(S)$ is a circuit in $\varphi(\mathrm{T})$. Therefore, $\mathcal{L}(\mathrm{T},\mathrm{U})$ is nonempty.

Finally, (d)$\Rightarrow$(a) Lemma 1. ■

Remark. Let G and H be as in Theorem 4, then we say H is obtained from G by a *two-vertex exchange*. Let φ be defined as (1.4), then $\mathrm{U} = \varphi(\mathrm{T})$ is a quasi-representing graph for H and $\mathcal{L}(\mathrm{T},\mathrm{U})$ is nonempty iff $\mathrm{T} \in C_G \setminus \{k{:}m \mid k \in X, m \in Y\}$.

Throughout the rest of the paper, $G = (\tau_1,\ldots,\tau_n)$, $n > 2$, and $\mathrm{T} \in \mathcal{Q}(G)$. We say T is *reduced* if any two edges of T belong to a circuit and if e, f are edges with label $e \leq$ label f then e belongs to a circuit not containing f. If G is strongly indecomposable, then each quasi-representing graph for G is reduced and conversely [**3**, Theorem 3]. We call a subgraph B of a reduced T a *block* if B is the intersection of all circuits in T containing B.

LEMMA 5. *Let* T *be a reduced quasi-representing graph. Then*

(a) *The set of blocks forms a partition of the set of edges of* T.

(b) *Any two edges in a block have incomparable labels.*

Proof. (a) Define a relation on the set of edges in T as follows: $e \sim f$ iff e and f are in the same block. Reflexivity and symmetry are clear. Suppose $e \sim f$ and $f \sim g$. Then by the definition, $e \sim f$ implies e and f are in the same block, hence, any circuit S containing e also contains f. Since $f \sim g$, it follows that S also contains g. Hence, e and g are in the same block (i.e., $e \sim g$). Thus, the relation $\sim$ is an equivalence relation.

(b) Let e, f be edges in a block of T with label $e \leq$ label f. Then there exists a circuit $S \subset \mathrm{T}$ which contains e but not f since T is reduced. By the definition of block, S also contains f. This is a contradiction. Therefore, any pair of edges in a block has incomparable labels. ■

LEMMA 6. *If* $\varphi \in \mathcal{L}(\mathrm{T},\mathrm{U})$, *then* B *is a block in* T *iff* $\varphi(B)$ *is a block in* U.

Proof. Let B be a block and $S_1,\ldots,S_m$ be all of the circuits in T containing B. Then by the definition $B = \bigcap_{i \in \Omega_m} S_i$. We claim that $\varphi(S_1),\ldots,\varphi(S_m)$ are all circuits containing

$\varphi(B)$ in U. Suppose S is a circuit in $\mathcal{U}$ which properly contains $\varphi(B)$. Then $\varphi^{-1}(S)$ is a circuit in T containing the block B. Therefore, the $\varphi(S_i)$'s are all of the circuits in U containing $\varphi(B)$. Since φ is bijective, $\varphi(B) = \bigcap_{i \in \Omega_m} \varphi(S_i)$. Thus, $\varphi(B)$ is a block in U. The converse is clear. ■

Note that if $\mathcal{L}$(T,U) is nonempty, then U is obtained by permuting the labels of edges within blocks of T. In particular, a two-vertex exchange performed on G permutes the labels of edges within blocks of T.

2. APPLICATIONS. A group G is called a *CT-group* if its quasi-representing graph has pairwise incomparable labels of edges. It is easy to see that a CT-group has a unique quasi-representing graph.

THEOREM 7. *Let G and H be strongly indecomposable CT-groups, and let* T $\in \mathcal{Q}(G)$, U $\in \mathcal{Q}(H)$. *Then the following are equivalent:*

(a) *G and H are quasi-isomorphic;*

(b) *H is obtained from G via a sequence of two-vertex exchanges.*

(c) $\mathcal{L}$(T,U) *is nonempty.*

Proof. (a)⇒(b) Theorem 10 in [6]. (c)⇒(a) Lemma 1. (b)⇒(c) Suppose there is a sequence of two-vertex exchanges transforming G to H. Then, we have CT-groups $G = G_1, G_2, \ldots, G_{m-1}, G_m = H$ such that G_{i+1} is obtained from G_i by a two-vertex exchange for each $i \in \Omega_{m-1}$. By Theorem 4, $\varphi_i \in \mathcal{L}(T_i, T_{i+1})$, where $T_i \in \mathcal{Q}(G_i)$ and each T_i is uniquely determined. Thus, it follows that $\varphi = \varphi_{m-1} \circ \cdots \circ \varphi_1 \in \mathcal{L}(T,U) = \mathcal{L}(T_1, T_m)$. ■

Remarks. Let $G = (\tau_1,\tau_2,\tau_3,\ldots,\tau_n)$ and $H = (\sigma_1,\sigma_2,\tau_3,\ldots,\tau_n)$ be strongly indecomposable CT-groups, and let $\{X,Y\}$ be a partition of $\{3,\ldots,n\}$

(a) If there is an edge in T between X and Y, then G is not quasi-isomorphic to H by the remark following Theorem 4 and by Theorem 7. In this case, there is no two-vertex exchange for this partition.

(b) Let ζ_i be the type of a subring of $\mathbf{Q}$ generated by $\frac{1}{p_i}$ for prime p_i, and let $\zeta_{ijk} = \sup\{\zeta_i,\zeta_j,\zeta_k\}$. If we let $\tau_1 = \zeta_{125}$, $\tau_2 = \zeta_{34}$, $\tau_3 = \zeta_{145}$, $\tau_4 = \zeta_{235}$ then it is easy to check that $G = (\tau_1,\tau_2,\tau_3,\tau_4)$ is a CT-group and its representing graph T is a circuit with edges labeled by ζ_{15}, ζ_{25}, ζ_3, ζ_4. If we let $X = \{3\}$, $Y = \{4\}$ then there is no edge in T between X and Y but $\zeta_5 = \tau_3 \wedge \tau_4 \nleq \tau_1 \wedge \tau_2 = \text{type}(\mathbf{Z})$. Hence, there is no two-vertex exchange for G for this partition by Theorem 4.

A group G is called an *elementary group* iff for each type γ, either $T(\gamma)$ is empty, a singleton edge, or all of T. Equivalently, G is elementary iff either rank $G(\gamma) = 0$, 1, or n-1 for each type γ.

LEMMA 8. *Let $G = (\tau_1,\ldots,\tau_n)$ be an elementary strongly indecomposable group. Then*

(a) *G is a CT-group.*

(b) *A partition $\{X,Y\}$ of $\Omega_n\backslash\{k,m\}$ has no edges between X and Y in a quasi-representing graph T iff $\tau^X \wedge \tau^Y \leq \tau_k \wedge \tau_m$.*

Proof. (a) Suppose edges e, f of a quasi-representing graph T for G are labeled by σ, τ respectively and $\sigma \leq \tau$. Then, $T(\sigma) = T$ because G is elementary. Since T is reduced, there is a circuit S in T containing both e and f. This is impossible by the definition of quasi-representing graph because we can't have that both $S \subseteq \mathrm{T}(\sigma)$ and S contains e. Therefore, T has pairwise incomparable labels of edges. Consequently, G is a CT-group.

(b) Let $i \in X, j \in Y$ be arbitrary and let $\sigma = \tau_i \wedge \tau_j$. Assume there is no edge in T between X and Y. Since $i{:}j \notin \mathrm{T}$, it follows from the definition of quasi-representing graph that there is a path $P \subseteq \mathrm{T}(\sigma)$ connecting τ_i and τ_j. By the elementary condition, we have $\mathrm{T}(\sigma) = \mathrm{T}$. In particular, $\tau_k, \tau_m \geq \sigma$. Since $i \in X$ and $j \in Y$ are arbitrary, it follows that $\tau^X \wedge \tau^Y \leq \tau_k \wedge \tau_m$. The converse is clear by Corollary 3 and uniqueness of T. ■

A group G is a *circuit group* if its representing graph T is a circuit. Let G be a circuit group and T be labeled by types $\sigma_1,\ldots,\sigma_n$. Then we will write $\tau_n = \sigma_1 \vee \sigma_n$ and $\tau_i = \sigma_i \vee \sigma_{i+1}$ for each $i \in \Omega_{n-1}$. Since any two edges in a block have incomparable labels by Lemma 5(b), a circuit group is a CT-group. Let $\Delta(G) = \{\tau \mid \mathrm{rank} G/G[\tau] = 1$ and $\mathrm{type} G/G[\tau] = \tau\}$. We say n-tuples $(\tau_1,\ldots,\tau_n)$ and $(\sigma_1,\ldots,\sigma_n)$ are equivalent if $\tau_i = \sigma_{\rho(i)}$ for some $\rho \in S_n$.

The next lemma is a generalization of an example due to F. Richman.

LEMMA 9. *Let $G = (\tau_1,\ldots,\tau_n)$ be an elementary circuit group with $\tau_n = \sigma_1 \vee \sigma_n$ and $\tau_i = \sigma_i \vee \sigma_{i+1}$ for each $i \in \Omega_{n-1}$ and $n \geq 3$. Then*

(a) *An elementary circuit group $(\omega_1,\ldots,\omega_n)$ is quasi-isomorphic to G iff for some $\rho \in S_n$, $\omega_n = \sigma_{\rho(1)} \vee \sigma_{\rho(n)}$ and $\omega_i = \sigma_{\rho(i)} \vee \sigma_{\rho(i+1)}$ for each $i \in \Omega_{n-1}$.*

(b) *There are $\dfrac{(n-1)!}{2}$ non-equivalent $(\omega_1,\ldots,\omega_n)$ quasi-isomorphic to G.*

(c) $|\Delta(G)| = \dfrac{n(n-1)}{2}$.

Proof. (a) Let $H = (\omega_1,\ldots,\omega_n)$ and let $U \in Q(H)$ be a circuit with edges labeled by $\gamma_1,\ldots,\gamma_n$ and let $\omega_n = \gamma_1 \vee \gamma_n$ and $\omega_i = \gamma_i \vee \gamma_{i+1}$ for each $i \in \Omega_{n-1}$. Note that T is a circuit with edges labeled by $\sigma_1,\ldots,\sigma_n$. By Theorem 7, G and H are quasi-isomorphic iff $\gamma_i = \sigma_{\rho(i)}$ for some $\rho \in S_n$. Or equivalently, $\omega_n = \sigma_{\rho(1)} \vee \sigma_{\rho(n)}$ and $\omega_i = \sigma_{\rho(i)} \vee \sigma_{\rho(i+1)}$ for each $i \in \Omega_{n-1}$.

(b) For each $\rho \in S_n$, define $H_\rho = (\omega_1,\ldots,\omega_n)$ be an elementary circuit group with $\omega_n = \sigma_{\rho(1)} \vee \sigma_{\rho(n)}$ and $\omega_i = \sigma_{\rho(i)} \vee \sigma_{\rho(i+1)}$ for each $i \in \Omega_{n-1}$. Then H_ρ and G are quasi-isomorphic for each $\rho \in S_n$ by (a). Without loss, identify $T_\rho \in Q(H_\rho)$ as a regular n-gon with edges labeled by $\sigma_{\rho(1)},\ldots,\sigma_{\rho(n)}$. Define an equivalence relation on $\{H_\rho \mid \rho \in S_n\}$ as follows: H_ρ and H_σ are equivalent iff T_ρ is a symmetry of T_σ. Let D_n = the dihedral group of order n. Then, each equivalence class of H_σ has $|D_n|$ elements and each equivalence class gives a new $(\omega_1,\ldots,\omega_n)$ quasi-isomorphic to G. Thus, there are $\frac{|S_n|}{|D_n|} = \frac{(n-1)!}{2}$ non-equivalent $(\omega_1,\ldots,\omega_n)$ quasi-isomorphic to G.

(c) Let $1 \le i \ne j \le n$ be an arbitrary pair of integers, and let $\tau = \sup\{\sigma_i,\sigma_j\}$. Suppose $\sigma_k \ge \tau$ for some $k \ne i,j$ then $\sigma_k \ge \sigma_i$. This contradicts G is a CT-group. Thus, $\sigma_k \not\ge \tau$ for all $k \ne i,j$. Since T is a circuit there are exactly two τ-equivalence classes X, Y in G. That is, rank$G/G[\tau] = 1$. By Theorem 1.10 in [1] type$G/G[\tau] = \sigma^X \wedge \sigma^Y = \tau$, hence $\tau \in \Delta(G)$. A similar argument shows that if $\tau \in \Delta(G)$ then τ is the supreme of labels of two edges from T. Since there are n edges in T, $|\Delta(G)| = \frac{n(n-1)}{2}$. ■

COROLLARY 10. *Suppose $G = (\tau_1,\ldots,\tau_n)$ is a strongly indecomposable CT-group with $n \ge 3$. Then, there are at most $\frac{(n-1)!}{2}$ number of non-equivalent $(\sigma_1,\ldots,\sigma_n)$ quasi-isomorphic to G.*

Proof. (a) We claim that we get the maximum number of $(\sigma_1,\ldots,\sigma_n)$ quasi-isomorphic to G if G is an elementary circuit group. First note that any partition $\{X,Y\}$ satisfying condition (i) of Theorem 4(c) gives a new $(\sigma_1,\ldots,\sigma_n)$ quasi-isomorphic to G. Let $\{X,Y\}$ be an arbitrary partition of n-2 vertices in T. If there is an edge in T between X and Y, then there is no two-vertex exchange for this partition by the remark(a) following Theorem 7. Assume there is no edge in T between X and Y then, by Lemma 8(b), we always have a two-vertex exchange if G is an elementary group. Hence, assume G is elementary. The above argument shows that we obtain the maximum number of two-vertex exchanges if G is circuit. Therefore, we have at most $\frac{(n-1)!}{2}$ non-equivalent $(\sigma_1,\ldots,\sigma_n)$ quasi-isomorphic to G by Lemma 9(b). ■

ACKNOWLEDGMENTS

The author would like to thank Professor C. Vinsonhaler for his many valuable suggestions.

REFERENCES

1. D. Arnold and C. Vinsonhaler, Quasi-representing graphs for a class of torsion-free abelian groups, *Abelian Group Theory*, Gordon and Breach, London (1987), 309-332.

2. D. Arnold and C. Vinsonhaler, Quasi-isomorphism invariants for a class of torsion-free abelian groups, *Houston J. Math.* Vol **15** No **3** (1989), 327-340.

3. D. Arnold and C. Vinsonhaler, Invariants for classes of torsion-free abelian groups, *Proc. Amer. Math. Soc.* **105** (1989), 293-300.

4. L. Fuchs and C. Metelli, On a class of Butler Groups, *Manuscript Math.* **71** (1991), 1-28.

5. F. Richman, An extension of the theory of completely decomposable torsion-free abelian groups, *Trans. Amer. Math. Soc.* **279** (1983), 175-185.

6. P. Yom, A characterization of a class of Butler groups, submitted to *J. Algebra.*

DEPARTMENT OF MATHEMATICS, FORDHAM UNIVERSITY, BRONX, NY 10458

Recent Titles in This Series

(Continued from the front of this publication)

142 **Chung-Chun Yang and Sheng Gong, Editors,** Several complex variables in China, 1993
141 **A. Y. Cheer and C. P. van Dam, Editors,** Fluid dynamics in biology, 1993
140 **Eric L. Grinberg, Editor,** Geometric analysis, 1992
139 **Vinay Deodhar, Editor,** Kazhdan-Lusztig theory and related topics, 1992
138 **Donald St. P. Richards, Editor,** Hypergeometric functions on domains of positivity, Jack polynomials, and applications, 1992
137 **Alexander Nagel and Edgar Lee Stout, Editors,** The Madison symposium on complex analysis, 1992
136 **Ron Donagi, Editor,** Curves, Jacobians, and Abelian varieties, 1992
135 **Peter Walters, Editor,** Symbolic dynamics and its applications, 1992
134 **Murray Gerstenhaber and Jim Stasheff, Editors,** Deformation theory and quantum groups with applications to mathematical physics, 1992
133 **Alan Adolphson, Steven Sperber, and Marvin Tretkoff, Editors,** p-adic methods in number theory and algebraic geometry, 1992
132 **Mark Gotay, Jerrold Marsden, and Vincent Moncrief, Editors,** Mathematical aspects of classical field theory, 1992
131 **L. A. Bokut′, Yu. L. Ershov, and A. I. Kostrikin, Editors,** Proceedings of the International Conference on Algebra Dedicated to the Memory of A. I. Mal′cev, Parts 1, 2, and 3, 1992
130 **L. Fuchs, K. R. Goodearl, J. T. Stafford, and C. Vinsonhaler, Editors,** Abelian groups and noncommutative rings, 1992
129 **John R. Graef and Jack K. Hale, Editors,** Oscillation and dynamics in delay equations, 1992
128 **Ridgley Lange and Shengwang Wang,** New approaches in spectral decomposition, 1992
127 **Vladimir Oliker and Andrejs Treibergs, Editors,** Geometry and nonlinear partial differential equations, 1992
126 **R. Keith Dennis, Claudio Pedrini, and Michael R. Stein, Editors,** Algebraic K-theory, commutative algebra, and algebraic geometry, 1992
125 **F. Thomas Bruss, Thomas S. Ferguson, and Stephen M. Samuels, Editors,** Strategies for sequential search and selection in real time, 1992
124 **Darrell Haile and James Osterburg, Editors,** Azumaya algebras, actions, and modules, 1992
123 **Steven L. Kleiman and Anders Thorup, Editors,** Enumerative algebraic geometry, 1991
122 **D. H. Sattinger, C. A. Tracy, and S. Venakides, Editors,** Inverse scattering and applications, 1991
121 **Alex J. Feingold, Igor B. Frenkel, and John F. X. Ries,** Spinor construction of vertex operator algebras, triality, and $E_8^{(1)}$, 1991
120 **Robert S. Doran, Editor,** Selfadjoint and nonselfadjoint operator algebras and operator theory, 1991
119 **Robert A. Melter, Azriel Rosenfeld, and Prabir Bhattacharya, Editors,** Vision geometry, 1991
118 **Yan Shi-Jian, Wang Jiagang, and Yang Chung-chun, Editors,** Probability theory and its applications in China, 1991
117 **Morton Brown, Editor,** Continuum theory and dynamical systems, 1991
116 **Brian Harbourne and Robert Speiser, Editors,** Algebraic geometry: Sundance 1988, 1991
115 **Nancy Flournoy and Robert K. Tsutakawa, Editors,** Statistical multiple integration, 1991
114 **Jeffrey C. Lagarias and Michael J. Todd, Editors,** Mathematical developments arising from linear programming, 1990

(See the AMS catalog for earlier titles)